Johann R.E. Lutjeharms

The Agulhas Current

J.R.E. Lutjeharms

The Agulhas Current

with 187 figures, 8 in colour

Springer

Professor Johann R.E. Lutjeharms
Department of Oceanography
University of Cape Town
Rondebosch 7700
South Africa

ISBN-10 3-540-42392-3 Springer Berlin Heidelberg New York
ISBN-13 978-3-540-42392-8 Springer Berlin Heidelberg New York

Library of Congress Control Number: 2006926927

Springer is a part of Springer Science+Business Media
springer.com

Printed in Germany

Typesetting: Camera-ready by Marja Wren-Sargent, ADU, University of Cape Town
Figures: Anne Westoby, Cape Town
Satellite images: Tarron Lamont and Christo Whittle, MRSU, University of Cape Town
Technical support: Rene Navarro, ADU, University of Cape Town
Cover design: E. Kirchner, Heidelberg
Production: Almas Schimmel
Printing and binding: Stürtz AG, Würzburg, Germany
Printed on acid-free paper 32/3141/as 5 4 3 2 1 0

In memory of

Günther Dietrich

1911–1972

the first true oceanographer of the Agulhas Current system.

Contents

Preface ix

Acknowledgements xi

Figure credits xiii

1. Introduction 1
   - Early investigations 1
   - Early scientific research 2
   - Modern research interest 3
   - South West Indian Ocean circulation 4
   - Agulhas source currents 5
   - Agulhas Current 6
   - Agulhas Retroflection 7
   - Agulhas Return Current 7
   - South Indian Ocean Current 7
   - Western boundary currents 8
   - Western boundary currents of the southern hemisphere 8
   - Global role of the Agulhas Current 9
   - Influence of the Agulhas Current on local climate and weather 11
   - Biological influence of the Agulhas Current 14
   - "Giant" waves 16
   - Perspective 16

2. Large-scale circulation of the South West Indian Ocean 17
   - The wind regime of the South West Indian Ocean 22
   - Flow patterns of the South West Indian Ocean 23
   - Water masses of the South West Indian Ocean 26
   - Movement of water masses 32
   - Transports in the South West Indian Ocean 41
   - Thermal structure of the South West Indian Ocean 44
   - Modelling large-scale flows of the South West Indian Ocean 46
   - Conclusions 51

3. Sources of the Agulhas Current 53
   - Surface flows 53
   - Flow from east of Madagascar 56
   - Flow through the Mozambique Channel 69
   - Recirculation in a South West Indian Ocean sub-gyre 84
   - Synopsis 89

4. The northern Agulhas Current 91
Kinematics of the northern Agulhas Current 91
Hydrography of the northern Agulhas Current 105
Circulation on the continental shelf 107
Inception of the Natal Pulse 113
Recapitulation 120

5. The southern Agulhas Current 121
Hydrography of the southern Agulhas Current 121
Meanders and shear-edge features 125
Shelf-edge upwelling 132
Hydrography of the Agulhas Bank 138
Atmospheric conditions 146
Summary 150

6. The Agulhas Current retroflection 151
The nature of the Agulhas retroflection 151
Agulhas rings 170
Inter-ocean exchange at the Agulhas retroflection 191
Agulhas Current eddies 198
The dynamics of the Agulhas retroflection 199
Overview 206

7. The Agulhas return flow 209
The Agulhas Return Current 209
The Subtropical Convergence south of Africa 218
Variability and eddy shedding 225
Dynamics of the Agulhas Return Current 228
South Indian Ocean Current 230
Summation 231

8. The greater Agulhas Current: some reflections 233
A global perspective 233
An inventory of things learnt 234
Knowledge gaps remaining 236
A look behind the curtain 237

References 241

Bibliography 269

Name index 295

Subject index 297

Preface

In 1977 Paul Tchernia, the world-renowned French physical oceanographer, wrote a now well-known textbook on global descriptive oceanography[1] based on the lecture series that he had taught for ten years to students of the University of Paris VI. It dealt with all major ocean basins, their general circulation, hydrographic structure, water masses, the origin and formation of such water masses, etc. In its treatment of the anticyclonic circulation of the North Atlantic Ocean, this comprehensive treatise includes seven pages of discussion on the western boundary current of that basin, the Gulf Stream. The analogous current of the South Indian Ocean, the Agulhas Current, as well as its sources, are by contrast dealt with in but two, brief paragraphs.

The reason for this gross disparity is immediately apparent if a comparison is made between what was known about these two – otherwise comparable – currents at the time. Knowledge on the Agulhas Current, as measured by the number of research articles that had been published in the scientific literature, was in fact totally inadequate for more than the rather cursory discussion in the Tchernia text (viz. Figure 1). Even including early papers dealing with the currents of the South Indian Ocean *in toto*, papers on abyssal currents as well as atlases, not more than about 80 publications on the greater Agulhas Current system existed by the late 1970s.

This situation has, for a number of reasons, changed dramatically since then (Figure 1). Since the early 1980s about 13 papers, on average, have been published annually on the physics and chemistry of the Agulhas Current. These contributions have covered a wider range of disciplines

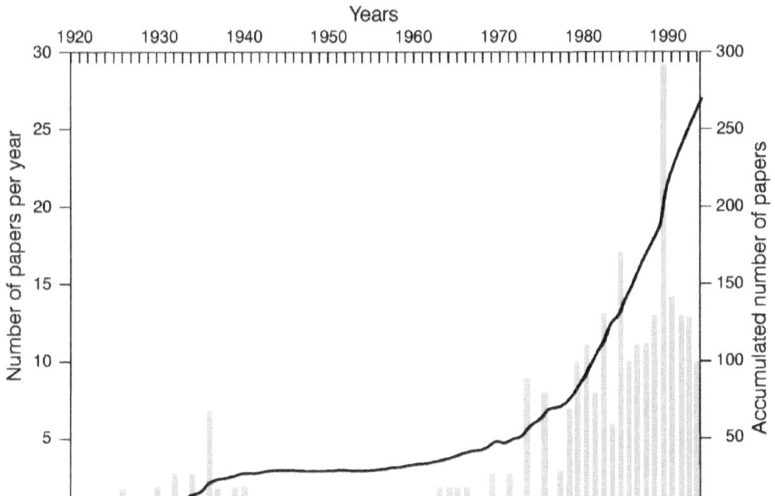

Figure 1. The number of research papers that have appeared per year with, as subject, the Agulhas Current or closely related topics (left ordinate). The continuous curve gives the accumulated total for these papers from 1920 onwards (right ordinate). These numbers are based on a number of bibliographies on the subject[2–5]. Note the peak in the publication rate in the 1930s; a secondary, but smaller peak in the 1960s due to the International Indian Ocean Expedition; and a noteworthy and persistent growth starting in the 1970s.

than ever before and have come from a large number of countries, including South Africa, the United States, Germany, the Netherlands, Britain, France, the Soviet Union, Mozambique and Portugal. Major advances have thus been made in the understanding of the hydrography, the kinematics, the dynamics and the chemistry of the current whilst numerical modelling of the broader current system has developed rapidly and is showing increasing promise.

This enormous increase in knowledge of the Agulhas Current system is quite unparalleled when compared to the contemporaneous growth in knowledge about analogous currents such as the Gulf Stream in the North Atlantic and the Kuroshio in the North Pacific Ocean. However, the annual number of publications concerning the western boundary currents of the northern hemisphere still exceeds by far those on the currents of the southern hemisphere. An analysis, for instance, of all publications that are represented in the *Aquatic Sciences and Fisheries Abstracts* and that include the names of these currents in their abstracts, shows that during the period 1988 to 1993, for example, 1074 papers had appeared on the Gulf Stream, 965 on the Kuroshio, 141 on the Agulhas Current, 71 on the Brazil Current and 69 on the East Australia Current. Those on the Agulhas Current were, however, from the lowest base by far. This implies that the actual growth in understanding of this current over the past two decades has probably exceeded any comparable growth in knowledge on the other currents.

This spurt in new information has led to a number of eminent oceanographers proposing that the time was ripe for a comprehensive, descriptive treatise on the greater Agulhas Current system. Such a monograph, it was suggested, should bring together and synthesise what has been learnt; particularly over these past few decades. More than this, it should be a user-friendly key to unlocking the wealth of information scattered throughout the primary literature. These are the aims of this book.

Acknowledgements

Anyone foolish enough to attempt to write a book soon recognises that no book is the product of the toil of but one individual. Even more so in the case of a wide-ranging review – as this monograph attempts to be – that is by definition the outcome of the efforts of a very large number of people. It is therefore incumbent on me to thank a veritable host of colleagues on whose work this compendium rests. Their names appear as authors to papers in the Reference List and in the Bibliography. A considerable number of them have become personal friends whom I owe an enormous debt of gratitude for many valuable and instructive conversations on the Agulhas Current system that have helped form my thinking. They may in consequence discover many ideas in these pages that originally were theirs and that I may have subconsciously absorbed and made my own. I therefore wish to apologise for any instances where proper acknowledgement may inadvertently be lacking.

Many colleagues have made specific contributions in time, support and help. I thank in particular Professor John Woods of Imperial College, London, for first suggesting and Dr James Baker for emphasising that I should write this book. I am especially grateful to Professor Gerold Siedler of the *Institute für Meereskunde, Christian-Albrechts-Universität*, in Kiel, Germany, and Professor Will de Ruijter of the *Instituut voor Marien en Atmospherisch Onderzoek, Universiteit Utrecht*, in Utrecht, the Netherlands, for hosting me for extended sabbaticals in 1994, in 1999 and in 2005 during which most of this work was done. I am also immensely grateful to a large number of colleagues at both these institutions who made my time there congenial, stimulating and productive. My stays in Kiel were made possible by an *Alexander von Humboldt Preis* from the foundation of that name; my stays in Utrecht by a *Gast-hoogleraarschap* from the IMAU as well as financial support from the University of Cape Town and the South African Foundation for Research Development.

A number of colleagues have helped by bringing unusual and valuable information on particularly historical aspects of the Agulhas Current to my attention. Professor Geoff Brundrit and Dr Pedro Monteiro helped in getting appropriate material on the sailing experiences of Vasco da Gama; Messrs Wilbert Weijer and Gerard van der Schrier volunteered information on Maury and his knowledge of the Agulhas Return Current whilst Professor Gerard Kortum unearthed extraordinary amounts of material on a number of German oceanographers. Miss Tania Harteveld kindly translated some eighteenth century Latin from maps of that period. I particularly appreciate the efforts of a number of colleagues, Mr Wilbert Weijer, Dr Paul van der Vaart, Professors Will de Ruijter and Sandy Harris, as well as Dr Peter Jan van Leeuwen for reading the manuscript and giving me the benefit of their insightful comments and criticism. I am indebted to many who helped in other ways: Mrs Shirley Hutchings for prodigious amounts of typing, Mr Wilhelm Lutjeharms for checking references, Mrs Anne Westoby for so ably drawing the illustrations, Mr Christo Whittle and Miss Tarron Lamont for enthusiastic support with illustrative satellite images and Mrs Marja Wren-Sargent for the meticulous way in which she prepared the manuscript for printing. Mr Adrian Webb prepared a few figures.

No academic goes on extended sabbatical leave, however well planned and prepared, without his closest colleagues having to carry a larger burden. I therefore wish to thank all in the Department of Oceanography at the University of Cape Town who have shouldered some of my tasks in my absence, in particular Professors Geoff Brundrit, Frank Shillington, Chris Reason, Dr John Largier, Mrs Lesley Staegemann, Mrs Helen King, Drs Mathieu Rouault and Isabelle Ansorge and all members of the Ocean Climatology Research Group who made my absences possible and lightened the load.

Figure credits

Raw data for the creation of the satellite images (Figures 3.8, 4.9, 4.24, 5.5, 6.6, 6.9 and 6.35) was acquired by the Satellite Applications Centre (SAC), Pretoria and processed at the MRSU (Marine Remote Sensing Unit, University of Cape Town). SAC is a division of the Council for Scientific and Industrial Research (CSIR).

Permission by the following for the reproduction of figures is gratefully acknowledged:

African Journal of Marine Science
Figures 3.31, 3.32, 5.11, 5.13, 5.18, 5.20, 5.23, 6.14, 6.19

American Geophysical Union
Figures 1.3, 1.4, 2.4, 2.5, 2.10, 2.19, 3.4, 3.5, 3.19, 3.23, 3.30, 3.33, 3.35, 4.1, 4.5, 4.11, 4.24, 4.25, 4.26, 4.27, 5.15, 6.4, 6.8, 6.21, 6.22, 6.25, 6.27, 6.29, 6.30, 6.34, 6.37, 7.8

American Meteorological Society
Figures 2.26, 3.15, 3.20, 3.28, 6.2, 6.5, 6.6, 6.7, 6.9, 6.28, 6.33, 6.40, 7.16, 7.17

Elsevier Ltd
Figures 2.3, 2.8, 2.9, 2.13, 2.14, 2.20, 2.22, 2.28, 2.29, 3.6, 3.7, 3.8, 3.10, 3.11, 3.18, 3.21, 3.22, 3.24, 3.26, 3.34, 4.10, 4.12, 4.15, 4.17, 4.18, 4.20, 4.23, 5.1, 5.2, 5.3, 5.5, 5.6, 5.7, 5.8, 5.9, 5.10, 5.12, 5.14, 5.17, 5.22, 6.12, 6.13, 6.15, 6.17, 6.18, 6.23, 6.24, 6.26, 6.31, 6.32, 6.35, 6.36, 6.38, 6.39, 6.41, 6.42, 6.43, 6.44, 6.45, 7.2, 7.3, 7.4, 7.6, 7.7, 7.9, 7.10, 7.11, 7.12, 7.14, 7.19, 7.20

International Hydrographic Review
Figure 3.16

Journal of Marine Research
Figures 2.15, 4.2, 4.6, 4.8, 4.14, 4.16, 6.1

Lamont-Doherty Earth Observatory of Columbia University
Figures 5.4, 7.5

Leibniz-Institut für Meereswissenschaften
Figures 2.27, 6.20

National Science Foundation
Figures 2.2, 2.6, 2.7, 2.11, 2.12, 2.17, 2.18, 2.21, 3.29

The Oceanographic Society of Japan/ TERRAPUB/Kluwer Academic Publishers
Figure 4.7

Royal Society of South Africa
Figures 5.24, 5.25

South African Geographical Journal
Figures 2.25, 4.3, 4.13

South African Journal of Science
Figures 2.16, 3.9, 3.31, 4.22, 5.16, 5.21, 5.27, 6.10, 6.11, 7.15

Springer-Verlag
Figure 4.19

University of Cape Town Press
Figure 5.19

University of Hawaii Press
Figure 2.24

VnR Pty Ltd
Figure 3.12

1

Introduction

Firmly rooted in the practical needs of shipping and navigation, the study of the ocean as an organised enterprise has a history that precedes the scientific revolution of the seventeenth century. It nevertheless comes as a surprise that one of the very first deep-sea currents described in some detail was the remote Agulhas Current lying off the south-eastern coast of Africa.

In the diary of Admiral Vasco da Gama[6], the first European to have sailed from Europe to India, there is an entry for 16 December 1497 that reads:

> On Saturday we sailed past the last cross and we could see, running on the beach, a couple of people as well as much cattle. As we progressed so the land appeared to improve with more and taller trees. That night we laid up off the Rio do Infante. The following day we set off with following winds until evening when the wind shifted eastwards and we set an offshore tack. We tacked seawards and landwards until Tuesday evening when the westerly wind returned and we laid up offshore for the night to check our position the following day. In the morning we closed on land and at 10 o'clock found ourselves off St Croix Island, 60 leagues back from our last land fix. This was due to a very strong current.
>
> The same day we set off again and with a good following wind which lasted four days, we broke through past the current. We thought this current was going to prevent us achieving our wish [to reach India]. From that day on it pleased God in his mercy that we should make good progress forwards and not backwards and pray that it shall always remain so.

The *last cross* mentioned here was one planted by a predecessor of Vasco da Gama, Bartholomew Dias, at Kwaaihoek, about 100 km north-east of the present city of Port Elizabeth. The *Rio do Infante* probably refers to the present Kei River and *St Croix Island*, still so named, is in Algoa Bay. To the best of our knowledge, this is the first written account of an encounter with the Agulhas Current.

Early investigations

This description by Da Gama of his navigational problems off south-eastern Africa demonstrates quite vividly why a disproportionate interest would be shown in this current for the next few centuries. The major shipping route of that time lay around the southern tip of Africa and once the Cape of Good Hope had been passed, this intense south-westward flow of water remained a major obstacle to be overcome in a passage to India.

One of the very first scientific treatises written about any major current therefore came to be published on the Agulhas Current. Published posthumously, *An Investigation of the Currents of the Atlantic Ocean and Those Which Prevail Between the Indian Ocean and the Atlantic* by Major James Rennell was a milestone in 1832, not only for a fuller understanding of this particular current, but also for ocean science as a whole[7]. The ideas and concepts he put forward were received with acclaim by the best scientific minds of the day[8]. Alexander von Humboldt, the pre-eminent geographer of his day, visited Rennell while on a diplomatic mission in London, and subsequently adopted many of Rennell's concepts on ocean currents in general and on the role of the Agulhas Current in particular. Unfortunately, Von Humboldt's major treatise on ocean currents was never published[9]. It might have altered dramatically the direction in which concepts on the Agulhas Current developed.

Subsequent investigations were still, for a long time, driven by the operational needs of commercial shipping and such studies carried out by members of the *Koninklijk Nederlandsch Meteorologisch Instituut*[10] confirmed and extended the work of Rennell. Nevertheless, innovative research on other aspects of the Agulhas Current was also being carried out by the Dutch at the time. The possible influence of the Agulhas Current on local winds[11], a modern topic that is receiving increasing attention at present[12,13], was one of these. Many of the very early studies have been shown by recent investigations to have been quite prescient in their conclusions, but much was subsequently lost to science and only reclaimed recently[14].

Most of this initial work remained based on collec-

Origin of the name *Agulhas*

As with so many place names that have an ancient derivation, there is uncertainty about the exact origin of the name *Agulhas Current*.

Bartholomew Dias, the first European explorer to round the Cape of Good Hope, gave up his attempt to reach India at the Great Fish River on the south-eastern coast of South Africa and shortly after reached the southern tip of Africa for the first time, on his return journey. This probably occurred on 16 May 1488, the name-day of the Irish monk, St Brendan, since he called this promontory *Ponto de S. Brandão*. He was unaware that this was the southernmost tip of Africa[634–5].

This name for the southernmost cape of the African continent soon fell into disuse although it was employed to denote the present Quoin Point for some time. A map of 1502 by Alberto Cantino, summarising the early Portuguese cartography of the region[634], shows a *Golfo das Agulhas* (Bay of Needles) directly to the east of the cape currently called Cape Agulhas[23], and cartographers subsequently started using variations on the name *Cape Agulhas* with increasing frequency. So, for instance, Hen Hondio in 1631 designated this headland as *C. d' Angulas*, I. Covens and C. Mortier (1660–1730) as *Cap des Aiguilles*, Gerardo and Leonardo Valk (1650–1720) as *C. das Agulhas* and D'Anville in 1763 as *Kaap des Aiguilles*.

Two possible origins for the concept "needles" have been put forward: one, the resemblance of the jagged reefs opposite the cape; two, the observation that the needles of compasses showed no magnetic declination at this point, i.e. magnetic north and true north were identical at this time. The latter origin is more likely, since, for example, authors such as Jan Huijgens van Linschoten in his book *Itinerario* (1595) have called this cape the *Cape of Compasses* or *Needles of the Compass*. Particulary notable in this regard is the inscription on a map of Mathia Hasio (probably redrawn from one by Batiste Homan) of about 1750 describing the cape as *Prom. Acicularum ſc declinatione magnet carentium* (Cape Needle, i.e. without magnetic declination).

The Agulhas Bank, Agulhas Current and the Agulhas Plateau all take their names from the cape. Cartographers I. Covens and C. Mortier (1660–1730) have called the continental shelf south of Africa *Banc du Cap*; by 1746 M. Belin designated it as *Banc des Equilles*. When the name of the cape was first transferred to the current is not entirely clear. Kerhallet (1852)[24] has still indicated a *Courant de Cap*, while Zimmermann (1865)[25] has shown a *Cap Lagullass Strömung*, suggesting the gradual equating of the name of the cape and the current. Rennell in his 1778[27] map has indicated the *Bank of Lagullas*, but has not named the current; but fifty years later (1832[7]) he clearly denoted it *Lagullas Current*, as did Maury[28] in 1855. By 1866 Findlay[29] called it, unequivocally, *Agulhas Current*, as did Krümmel (1882)[30]. Krümmel (1911, p. 672)[31] has in fact stated that south of 30° S latitude the current along the south-eastern coast of Africa is to be called the Agulhas Current, and so it has remained since.

tions of ships' drift obtained from the logbooks of mariners and the occasional measurements of sea surface temperatures by scientifically inclined ship captains. During the subsequent age of global ocean exploration many of the early cruises visited Cape Town such as, for instance, those carried out from the vessels *Challenger*[15] (1873), *Gazelle*[16] (1874), *Waterwich*[17] (1894), *Valdivia*[18] (1898; Figure 1.1), *Gauss*[20] (1903), *Planet*[21] (1906) and *Möwe*[22] (1913).

Many of these carried out a few observational stations near Cape Town to measure hydrographic variables to deep within the water column. During the *Meteor* cruises[32–3] of the 1920s, under the leadership of Defant, and the *Discovery* cruises[34] of the same decade, lines of stations were furthermore carried out from Cape Town. By this combination of efforts an eclectic collection of deep measurements for the region south of Africa was slowly built up.

Early scientific research

The first truly scientific investigation of the southern Agulhas Current, making use of this fortuitous accumulation of deep oceanographic measurements, was carried out by Dietrich in 1935[40]. Not only did he comprehensively describe the Agulhas Current and its manifestations throughout the water column for the first time, he also compared it to what was known about the Gulf Stream[41] and described what he considered to be an Atlantic branch of the current[42]. By this time the science of oceanography had gone through a period of rapid development and expansion, but the results of the newer investigations were largely concentrated in the northern hemisphere[8]. The scientific need for an increased understanding of the Agulhas Current as a geophysical system was not as acute as had better information on the Agulhas Current as an opposing drift at the sea surface previously been for shipping.

Nonetheless, the publications by Dietrich in the 1930s were supported by a considerable number of other papers of German origin. Möller[43] and Michaelis[44] had reported on the circulation of the Indian Ocean in general while Paech[45] had described the seasonal currents around Madagascar. Möller[46] and Thomsen[47] had, furthermore, contributed to an animated debate[48] about the deep circulation of the Indian Ocean. All these and

Major James Rennell

James Rennell is, without any doubt, the pre-eminent pioneer of research on the Agulhas Current.

Born on 3 December 1742 at Upcot in Devonshire, England[35], he lost his father at the age of four. With little formal schooling, he secured an appointment as midshipman in the Royal Navy at the age of fourteen. On his first ship, the *Brilliant*, he was party to the capture of two French privateers. His modest share of the prize money was spent, characteristically, largely on the purchase of books.

Having surveyed the Bay of St Cast, the chain of sand banks that connect Ceylon and the Coromandel coast, as well as various harbour approaches, he was recommended, at the age of 21, to the Governor of Bengal, who was anxious to have his province properly surveyed. During this survey, which took many years under very difficult conditions, Rennell was severely wounded by a tribesman on the Bhuton border in 1766[35]. From this injury he never fully recovered. His fieldwork continued, however, notwithstanding recurrent attacks of malaria and even having single-handedly to kill a leopard with a bayonet after it had mauled six of his men.

At the age of 36, the survey completed, Rennell returned to England. However, his previous employers, the British East India Company, refused to defray the cost of publication of the Bengal Atlas. Publication eventually was made possible by generous public subscription. He spent the rest of his life working on a wide variety of geographical studies, including many on ocean currents[7]. As leading authority on the ocean he was offered the post of First Hydrographer to the British Admiralty, but refused, as he was of the opinion that those duties would have interfered unduly with his research activities[35].

His first large work was the *Chart of the Bank and the Current of Cape Lagullas*[27] which was published in 1778[36], its style and format still largely used in Admiralty Current Charts to this day[37]. In his many analyses and discussions he exhibited an unusual understanding of the variability and complexity of ocean currents, far in advance of thinking at the time[38]. He was the originator of the terms *drift* and *stream*; the former he considered to be wind-driven, the latter due to a head of water set up by the flow of a drift current against a coast.

Major James Rennell, pioneer of the formal study of the Agulhas Current[39]

Rennell died on 29 March 1830 at the age of 87 and was buried in Westminster Abbey, having received many honours during his long career[39]. He had been Fellow of the Royal Societies of both London and Edinburgh, member of the Royal Institute of France, member of the Imperial Academy at St Petersburg as well as member of the Royal Society of Göttingen.

other results on the Indian Ocean were illuminated and integrated in an admirable way by Schott in 1935 in his monumental book *Geographie des Indischen und Stillen Ozeans*[49]. It is one of the extraordinary facts of oceanographic history that most of this German work was subsequently either ignored and forgotten[213], or absorbed and subsumed in works by non-German authors[50].

The Second World War largely terminated the German research on the Agulhas Current (viz. Figure 1, p. ix) and when this current again started attracting substantial research interest in the 1970s, the collective memory of these pre-war German publications seems to have faded almost completely[198].

Modern research interest

The remarkable new growth in scientific interest in the Agulhas Current over the past thirty years has come about due to a number of interrelated factors. First, the very high horizontal temperature gradients of the current, and of its attendant mesoscale eddies, has made it an exceptionally suitable study area for the use of thermal infrared radiance, as observed by orbiting satellite instruments[59]. This has led to a number of significant discoveries[60–4] that has highlighted the central role that this current and its dynamics plays in the global interchange of water masses.

Secondly, the growing concern about global climate change, and the role the ocean may play in ameliorating this anthropogenic disturbance, has stimulated oceanographic research that is not regionally restricted but that has a global viewpoint. Since the Agulhas Current seems to play a key role in inter-ocean exchange[65–8], increasing interest has focused on it. Thirdly, components of the Agulhas Current system have been shown to exhibit one of the largest slopes in sea level[69] of any comparable features in the world ocean. This has led to many studies in which users of the new technology of satellite altimetry[70–4] have concentrated their attention on this ocean region.

As a direct consequence of these and other factors, the growth in publication rate on problems relating to the Agulhas Current has been formidable (Figure 1, p. ix) during the past three decades. What then are the main characteristics of this current that have emerged from this research and how do these compare to those of similar currents in other parts of the world ocean?

South West Indian Ocean circulation

The Agulhas Current forms the western limb of the wind-driven, anti-cyclonic circulation of the south Indian Ocean, carrying warm tropical and subtropical water southward[802]. It is a relatively narrow current, only about 100 km wide. Its water movement is swift; measurements of up to 2.0 to 2.5 m/s having been made on its landward border[75–6], from where the highest

John D.F. Gilchrist and the Agulhas Current

The history of the marine sciences in South Africa is highly unusual in at least one respect. The roots of the subject are not shrouded in the mists of time as in many other countries, but a very specific commencement date can be identified.

In the 1890s the Department of Agriculture of the Cape Colony decided to appoint a trained marine biologist to establish the fisheries potential of the waters around this coast. In 1896 Dr John Gilchrist arrived from Scotland to take up this post[51], the first person in South Africa to take a definite scientific interest in the sea. This is therefore justifiably considered the starting date of the subject in South Africa.

Gilchrist brought a trawler from abroad, renamed the *Pieter Faure*, and proceeded to investigate the local waters. Within a year he discovered the rich fishing grounds for sole on the Agulhas Bank and commercial trawling commenced shortly afterwards[51]. Without doubt those that had appointed him felt their own good judgement fully confirmed. Gilchrist went on to lay the foundation for what was later to become the Sea Fisheries Research Institute, the marine interest of the South African Museum in Cape Town[52] as well as that at the University of Cape Town. Failing health forced him to retire as Professor from this university in 1926 and he died in the same year.

Gilchrist strongly believed in the unity of the ocean sciences and tried to integrate the results of his faunal discoveries with environmental factors. In his presidential address to the South African Association for the Advancement of Science[53], for instance, he emphatically stressed the importance of physical oceanography to an understanding of the marine fauna. On cruises he insisted that temperatures and salinities be measured at each trawling station[54] and he employed innovative methods to map currents by means of drift bottles[55]. The potential therefore existed for Gilchrist to have laid the foundation for physical oceanography and a study of the Agulhas Current system in South Africa. This, regrettably, did not occur[56].

Many of the conclusions he reached on this system where

John Gilchrist, South African marine science pioneer

just plain wrong. Having completed a number of station lines off South Africa's east coast he came to the conclusion that "the warm Mozambique [sic] current off this coast is of no great depth and rather sharply marked off from the underlying strata" [57]. He believed, notwithstanding his own drift results, that the difference of water temperatures to either side of the Cape Peninsula, south of Cape Town, was due to a peeling off of cold water from the "southern drift current"[58]. Warm water from the Agulhas Current was thought to enter False Bay, to the east, but not to round the Cape of Good Hope. Fortunately these erroneous early concepts seem to have had little effect on the future development of physical oceanography in South Africa[56].

Figure 1.1. During the age of heroic exploratory cruises the German research vessel *Valdivia* lies at anchor off Simon's Town in False Bay, near Cape Town, in November 1898 (from Chun, *Aus den Tiefen des Weltmeeres*, 1900[19]). In the accompanying text Chun has incorrectly ascribed the 4 °C to 6 °C difference in water temperature between False Bay and Table Bay purely to the inflow of warm water, from the Agulhas Current, across the Agulhas Bank[19] into False Bay.

velocities have been reported. Its surface temperatures range from 27 °C in summer to 22 °C in winter[77] and usually exhibit an excess in temperature of about 6 °C compared to its direct oceanic environment. At the surface the salinity of the Agulhas Current is about 35.0 to 35.5. Its depth of vertical penetration increases downstream to about the latitude of the Agulhas Bank. Nomenclature on the Agulhas Current has varied, as has the latitudinal extent of current that this name has had to cover. At present it is generally accepted that the current bordering the shelf edge south of 27° S (Figure 1.2), i.e. well south of the mouth of the Mozambique Channel, is the Agulhas Current. It continues to bear this name until it has drastically and permanently altered its flow direction south of Africa. This current forms a natural continuum in all its hydrographic and hydrodynamic characteristics.

In early publications the current along the eastern seaboard of southern Africa was sometimes called the Mozambique Current[24,78]. The continuity, at all times, between the current that reputedly flows southward through the Mozambique Channel and the Agulhas Current has, however, been seriously called into doubt once sufficient hydrographic measurements had been made[79]. Restricting the name Mozambique Current to a southward flow in the Mozambique Channel therefore seemed appropriate at the time. Nevertheless, as so often happens in these cases, the incorrect name Mozambique Current for the Agulhas Current has persisted in that underworld of science: school textbooks and popular scientific publications. This faulty nomenclature complicates an intelligible description of the sources of the Agulhas Current proper.

Agulhas source currents

From the very first description of the surface flow of the south-west Indian Ocean, it has been surmised that the Agulhas Current is fed from two independent sources[26], the stronger being a Mozambique Current, through the Mozambique Channel, with a smaller contribution coming from east of Madagascar, the East Madagascar Current. This portrayal, based largely on records of ships' drifts and sea surface measurements, has had to be considerably modified in the light of more modern observations.

The southward flow along the Mozambican coast has been shown to be intermittent, in the form of eddies, and not to form a continuum with the northern Agulhas Current at all times[79]. The East Madagascar Current is a narrow, intense western boundary current of small dimensions[80] that may retroflect south of Madagascar

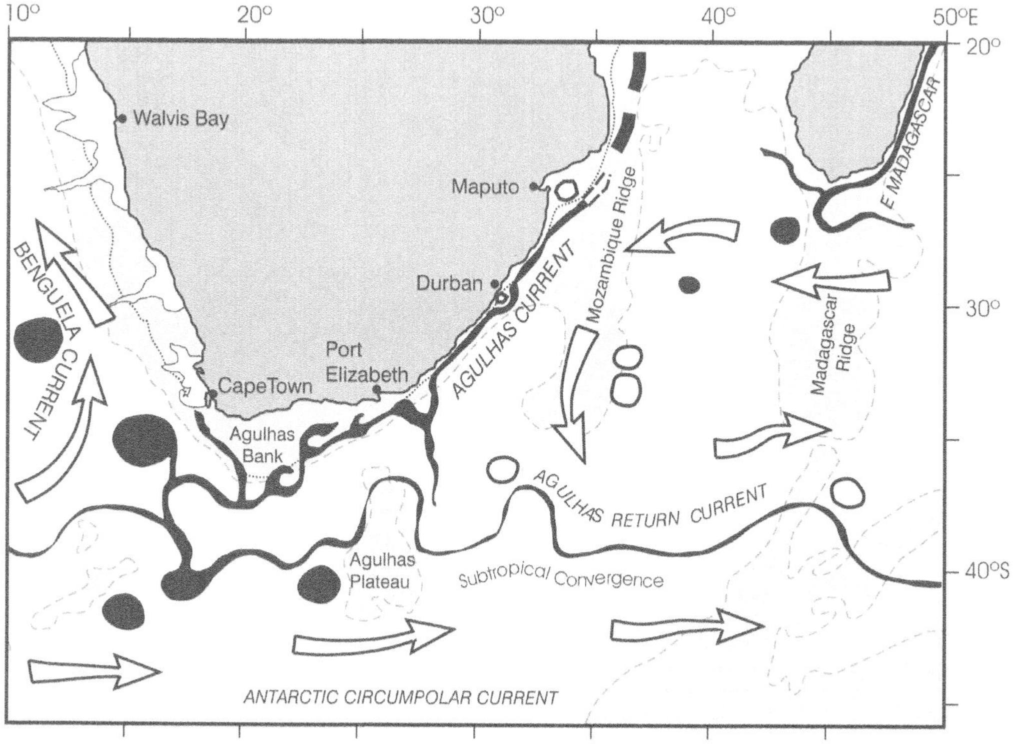

Figure 1.2. A conceptual portrayal of the Agulhas Current system as presently understood. Areas shallower than 3000 m are shaded. The edge of the continental shelf is circumscribed by a dotted line at the 500 m isobath. Intense currents and their component parts are black; the general background circulation is indicated by open arrows. Cyclonic eddies are open; anti-cyclonic rings and eddies black. Note the wind-driven coastal upwelling along the west coast of southern Africa, the tight Agulhas retroflection and the Subtropical Convergence that forms the generic border between the subtropical gyres and the Southern Ocean to the south.

over the shallow Madagascar Ridge (Figure 1.2). Contributions from east of Madagascar may therefore, most likely, be in the form of rings or eddies from the East Madagascar Current as well as the general westward flow of the South Indian subtropical gyre.

A third contributory to the water masses of the Agulhas Current, as it turns out by far the largest, is a sizeable degree of recirculation of water[81–3], particularly at depth, in a South West Indian Ocean sub-gyre. The core of this sub-gyre, judged on portrayals based on all available hydrographic stations[84], lies over the Agulhas Basin, mostly west of the Mozambique Plateau (or Ridge, as in Figure 1.2). The volume flux of the Agulhas Current, insofar as this is known, increases monotonically downstream. In this it is similar to other western boundary currents, such as the Brazil Current[85]. This increase may derive largely from the above-mentioned recirculation in a South West Indian sub-gyre that feeds the Agulhas Current along its full length.

Agulhas Current

Once it has been properly constituted, somewhere north of the city of Durban, the Agulhas Current follows the slope of the continental shelf closely[86]. In the interval between Delagoa Bay (at Maputo) to the north (Figure 1.2) and the city of Port Elizabeth to the south, the continental shelf along the east coast of southern Africa is quite narrow, never exceeding 25 km between the coast and the 200 m isobath. The continental slope is on the whole also very steep and uncomplicated here. This continental slope topography stabilises the Agulhas Current[87] so that it shows no wide meanderings of the type familiar in other boundary currents such as the Gulf Stream and the Kuroshio.

Along this stretch of coastline there is only one location where the shelf is markedly wider; at the coastal offset between Richards Bay and Durban, that is called the Natal Bight. This distinctive feature of coastline morphology is the locus of an infrequent perturbation on the Agulhas Current[62]. This solitary meander, the Natal Pulse, moves downstream with the current and grows rapidly so that at the latitude of Port Elizabeth the current core may be up to 200 km seaward of its usual location. The Natal Pulse may cause a number of major disruptions to the flow regime, including upstream retroflections near Port Elizabeth[64] and early shedding of Agulhas rings farther downstream.

On reaching the wide Agulhas Bank (Figure 1.2), downstream of Port Elizabeth, the behaviour of the landward edge of the Agulhas Current changes noticeably. Extensive meandering across its average path, the growth of substantial shear-edge eddies[88] and the trailing of long warm plumes[89] are characteristic of the current along this part of its south-westward path. The interaction of the rapidly moving water of the current with the comparatively stagnant water of the adjacent expanse of continental shelf is complex and as yet imperfectly understood.

Agulhas Retroflection

Having passed the Agulhas Bank, the Agulhas Current overshoots the African continent and, in a convoluted way, progrades into the South Atlantic Ocean. Here it undergoes a dramatic change in direction[90], retroflecting back into the South Indian Ocean[91] as the Agulhas Return Current[92]. A major and climatologically important[68] leakage of water from the Indian Ocean to the South Atlantic occurs here. First, plumes of surface water become detached from the current proper and drift into the cold Benguela Current[702] (Figure 1.2) as Agulhas filaments[93]. The major leakage, however, comes about through a process of ring shedding at the Agulhas retroflection[65].

The westward progradation of the tight loop of Agulhas water that forms the Agulhas retroflection is a continuous process, with the warm Indian Ocean water extending further and further into the South Atlantic until the Agulhas Current and the flanking Agulhas Return Current coalesce and a ring of the Agulhas Current is occluded. These rings are the largest of their kind observed to date[94] and drift into the South Atlantic Ocean carrying enormous amounts of heat and salt with them. Their lifetimes have been estimated to be more than two years[95].

Agulhas Return Current

That part of the flow of the Agulhas Current that is not lost to the South Atlantic Ocean through the process of ring shedding flows eastwards from the retroflection as the Agulhas Return Current[96]. It usually, but not always, lies in juxtaposition to the Subtropical Convergence[97] (Figure 1.2) in this region. As a result the meridional property gradients of both systems are markedly increased and instabilities in the very strong current shear produce a range of mesoscale eddies[63].

Through this mixing, as well as through an immense loss of heat to the atmosphere[98], the characteristics of the water masses forming the Agulhas Return Current are rapidly changed downstream. In its flow eastwards the Agulhas Return Current crosses a number of major features of bottom topography, including the Agulhas Plateau and the extension of the Mozambique Plateau. These deflect the deep-reaching current towards the equator[99]. The meanders thus caused may, on occasion, persist far to the east, gradually losing meridional amplitude.

South Indian Ocean Current

Once it has accomplished its retroflection and has become part of the deep ocean circulation of the South Indian Ocean flowing eastwards, the remnant of the Agulhas Return Current is often called the South Indian Ocean Current[100] (Figure 1.2), analogous to the South Atlantic Current. There is as yet no clear consensus in the literature on the exact longitude where this change of name would be most appropriate.

Based on the characteristics of the flow along the Subtropical Convergence in the South Indian Ocean, the Kerguelen Plateau, at about 70° E, probably constitutes the most suitable location from where one could start calling the zonal flow the South Indian Ocean Current. At this longitude the hydrographic influence of the Agulhas Current[101] and its waters[96] in the eastward flow has all but been terminated. Most has been lost to recirculation in a South West Indian Ocean sub-gyre. With increasing distance from land, the number of reliable oceanographic measurements also decreases, so that less is known about the Agulhas Return Current as well as the South Indian Ocean Current the further east one focuses one's attention. Nonetheless, wherever meridional lines of oceanographic stations are carried out across the Subtropical Convergence in the South Indian Ocean, at about 40° S, a band of higher zonal velocities is found north of this front[100].

The Agulhas Current system, as briefly described above, has all the main characteristics that define it as a western boundary current. It also has some unusual characteristics that distinguishs its uniqueness within this family of currents.

Western boundary currents

Western boundary currents come in a wide variety of shapes and sizes. The most intensively studied and therefore best known, the Gulf Stream, originates at a well-defined aperture, the Florida Straits, of a regional sea, the Gulf of Mexico. Here the Florida Current and the Antilles Current join, so that at Cape Hatteras, the Gulf Stream is fully constituted, about 120 km wide, and with a surface speed of up to 2.5 m/s. The volume flux of the current increases downstream and is estimated at between 60 and 150×10^6 m^3/s[102]. Downstream of Cape Hatteras it meanders dramatically to both sides of its mean path[103] shedding warm core rings to the landward and cold core rings to its seaward side[105–8]. Downstream of the Grand Banks, off Newfoundland, at about 50° W, the Gulf Stream leaves the shelf edge and moves into the deep North Atlantic Ocean where it is known as the North Atlantic Current. The counterpart of the Gulf Stream in the North Pacific Ocean, the Kuroshio, has many similar current attributes, but also some distinctly different ones.

Kuroshio

The Kuroshio in the North West Pacific starts as a rather shallow drift at about 29° N, between the islands of Taiwan and Luzon, and leaves the continental border at about 35° N[116]. It is about 80 km wide, has a volume transport of between 40 and 72×10^6 m^3/s[117] and a surface speed of between 0.75 and 2.5 m/s[118]. Along the Japanese coast it may exhibit very large meanders. Unlike those of the Gulf Stream, these may persist for many months in more or less the same location and subsequently switch to a new, once more quasi-stable configuration[119].

If this trajectorial bi-modality is unique to the Kuroshio as a western boundary current, so is the seasonality of the Somali Current in the North Indian Ocean. During the North East Monsoon season the flow along the coast of Somalia is in a generally south-westerly direction. During the South West Monsoon season, from May to September, a well-developed anticyclonic gyre has the Somali Current as its western boundary, from about 0° to 7° N. It is only about 300 m deep, but carries 65×10^6 m^3/s with maximum surface speeds of up to 2 m/s[120].

Western boundary currents of the southern hemisphere

By contrast, the East Australian Current, off the Pacific seaboard of Australia, is quite small, both in flux and in lateral dimensions[109–10]. It is thought to extend along the eastern coast of Australia from about 23° to 34° S, where it turns abruptly eastward along the Tasman Front[111]. In this process both cyclonic and anti-cyclonic rings are formed with lateral dimensions of between 200–300 km[112]. Surface speeds of up to 2 m/s have been measured in this current[113].

By far the most diminutive, and least studied, western boundary current, even smaller than the East Australian Current, is the East Madagascar Current. Its inception is thought to occur where the South Equatorial Current of the Indian Ocean impinges on the eastern coast of Madagascar, some of its waters subsequently moving southward. The core of the southern branch of the East Madagascar Current is only 50 km wide, 50 m deep, with an estimated volume flux of 41×10^6 m^3/s in the upper 1500 m[80]. South of Madagascar it is thought at times to retroflect[114], with most of its water drifting back into the central areas of the Indian Ocean.

The only remaining western boundary current of the southern hemisphere is the Brazil Current. The southern Brazil Current has an estimated maximum volume flux of 68×10^6 m^3/s[115]; the flux as well as the depth of the current increasing downstream. As a western boundary current it is unique in meeting a current moving in the opposite direction, the Falkland Current, more or less head on. Both currents then move off in an easterly direction with rings and eddies being shed by both[305].

The northern Brazil Current, on the other hand, crosses the equator and therefore lies partially in the northern hemisphere. It has[687] a mean transport of $16(\pm2) \times 10^6$ m^3/s, but displays considerable variability in this transport. It is a relatively shallow western boundary current, extending only to a depth of about 500 m. In a way very similar to that of the termination of the Agulhas Current, it retroflects. There is evidence[687] that there is a direct relation between the intensity of the North Brazil Current, the latitude of penetration of its retroflection loop and the number of rings shed from this loop. These rings are much weaker than Agulhas rings and few seem to last longer than about four months[684]. Once spawned, they move off at about 13 km/day. On formation they are mostly elliptical[684] with a mean major axis of about 126 km and a minor axis of about 65 km. Direct observations have however shown[688] that some rings may have a diameter of 390 km. Between three to eleven rings are shed per year[686,688] with very strong seasonal variability in the number of rings shed annually. Rings transport an average of 8×10^6 m^3/s. What triggers the shedding of rings in this current? In a high resolution model it has been shown[685] that the North Equatorial Current is barotropically unstable, radiating Rossby waves. The period of these waves in the model depend on the stability of the wind regime. These waves reflect at the Brazilian coast and trigger the creation of about six or seven rings per year.

Characteristics of western boundary currents

This family of currents may therefore be seen to have a number of generic traits, but each current also has some singular and unique characteristics. The Kuroshio exhibits distinct meander modes; the southern Brazil Current has a directly opposing current; the full Somali Current occurs during one season of the year only. It may therefore be expected that the Agulhas Current, forming part of this family of currents, would have certain basic traits similar to the other currents, but also some special features.

Basic resemblances are many. First, but of primary importance, is that western boundary currents are narrow, intense, flow poleward and are driven by the zonally integrated wind-stress curl of the adjacent basins. In all of these the Agulhas Current conforms to the norm, except that the extent of zonal integration across the South Indian Ocean for its driving force is unclear. The subtropical ocean gyre here is not a simple, ocean-wide movement[83] as is to be found in the South or North Atlantic Oceans, but is concentrated to the west and interrupted by the land mass of Madagascar.

Aspects of the Agulhas Current circulation that are unique are legion. Its source flow is interrupted by the presence of the world's largest island, Madagascar, and may possibly be under the influence of a seasonally varying surface flow in that part of the Equatorial and Indian Ocean that is under the influence of the Monsoon regime. The flow path of its upstream component is extremely stabilised by the shelf topography, but its return flow may be destabilised by the unusual location of submarine plateaux and ridges along its route[99]. But by far its most unusual feature is tied to the limited poleward extent of the African continent, and the opportunity this presents for a transfer of warm Agulhas Current water into the South Atlantic Ocean and that this occurs on the eastern side of a subtropical gyre. Not only is this a unique feature of the Agulhas as a western boundary current; it also has global climate implications.

Global role of the Agulhas Current

Carrying vast amounts of heat southward, the Agulhas Current would be expected to play a role in global ocean heat fluxes, and also conceivably in the vertical heat flux, and thus the linkage between ocean and atmosphere. That this latter linkage role is a serious one is immediately apparent from portrayals of the surface energy fluxes over the Atlantic Ocean[121–2], in which the southern termination of the Agulhas Current is especially prominent for its high heat fluxes to the atmosphere.

In any study of the global ocean–atmosphere circulation the role of the Agulhas Current would therefore probably play a definite part. Such investigations and modelling have become of great importance in order to study, in particular, variations in the global circulation, particularly with the expectation that a substantial variation in this ocean–atmosphere circulation may be imminent and may be man-induced through the so-called "green-house" effect[123–5]. One of the global oceanic circulation processes that has received considerable attention, as part of this new area of interest, has been the concept of a global oceanic conveyor belt[126] (Figure 1.3). In this concept the Agulhas Current does indeed play a key role. This conveyor belt consists of a deep and a surface part.

Global thermohaline overturning

The surface waters of the northern Atlantic Ocean are more saline by on average of 2.00 to 3.00 than those of their counterparts in, for instance, the North Pacific[126] due to an excess of vapour loss from this ocean region. On being cooled by the cold winter air masses from the Arctic, these overly saline waters sink in the vicinity of Iceland and the Labrador Sea and spread southward in the depths of the Atlantic Ocean as the North Atlantic Deep Water (NADW) with a mean temperature of 2 °C and salinity of 34.93[66]. This wedge of highly saline water, with a low nutrient content, centred at about 2000 m depth, overrides the northward moving Antarctic Bottom Water.

On moving into the Southern Ocean at a latitude of about 35° S (Figure 1.3) the NADW becomes part of the circulation around the Antarctic continent, the Antarctic Circumpolar Current. Here it is mixed with deep water produced in the Weddell Sea and part of this blend subsequently advects into the Indian and the Pacific Ocean. In these two oceans, considerable upwelling of this deep water occurs, the water of NADW origin is heated and the return part of the conveyor belt at the sea surface commences[66]. Some of the mixture of water in the Southern Ocean is taken further east by the Antarctic Circumpolar Current and re-enters the South Atlantic in the upper layers through the Drake Passage south of South America as a cold, shallow return flow.

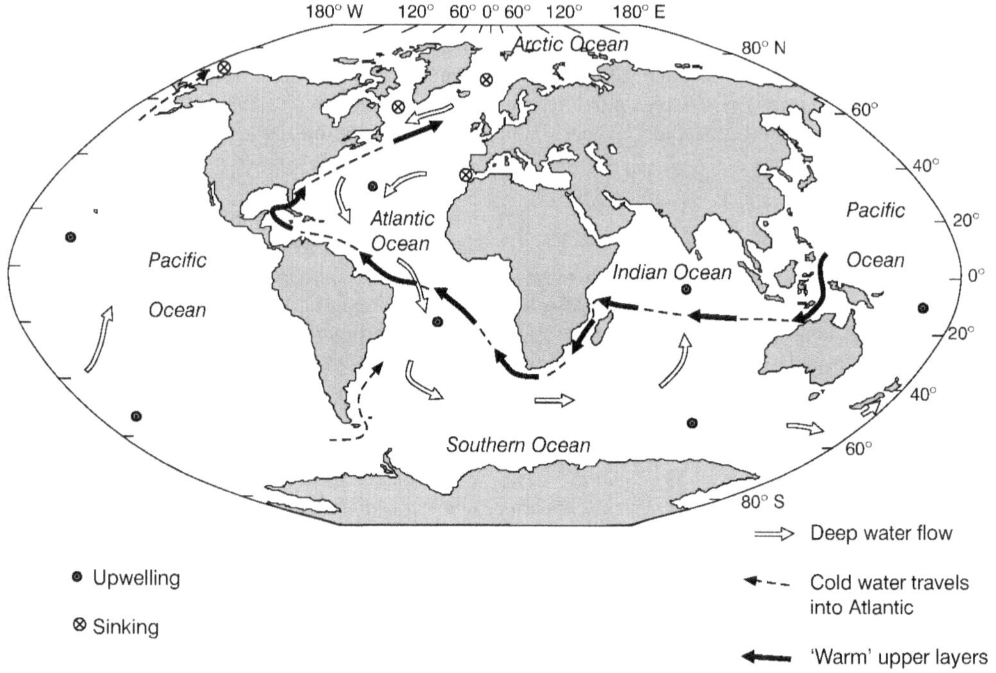

Figure 1.3. Simplified portrayal of the so-called global oceanic conveyor belt, being the global structure of the thermohaline circulation cell associated with North Atlantic Deep Water production. The open arrows describe the route taken by the deep waters; the solid arrows the proposed return path for warmer, surface water[66].

The warm, shallow return flow passes north of Australia through the Timor Sea region[67] and the Indonesian archipelago and then moves zonally across the tropical Indian Ocean at about 10° S latitude. It subsequently passes on both sides of Madagascar to be advected south as part of the Agulhas Current. South of Africa the leakage of Agulhas Current water into the South Atlantic Ocean allows this warm saline water to progress northwards, eventually to cross the equator. In the North Atlantic Ocean this flux of heat is incorporated into the waters of the subtropical gyre[66] of that ocean and carried northward by the Gulf Stream into the ocean regions of Iceland where the cycle can resume by the process induced by high evaporation rates and the cooling of surface waters.

This role of the Agulhas Current in the global thermohaline overturning can be studied most efficaciously with numerical models or by analysing the climatological signal contained in deep-sea sediments. Models have indeed shown[68,668–9] that the overturning circulation of the Atlantic Ocean, a major part of the global thermohaline circulation, is very sensitive to the inter-ocean fluxes of heat and salt from the Indian Ocean. Palaeoceanographic studies in the South Atlantic have in turn shown conclusively that the exchanges between the Indian and Atlantic Oceans have peaked at the end of each of the last five glacial periods[667] and have been coincident with the resumption of thermohaline circulation in the Atlantic Ocean[664] during deglaciation. Tracing the global thermohaline overturning in large-scale ocean circulation models[715] has demonstrated that the absence of inter-ocean leakage south of Africa leads to all manner of abnormal behaviour in the South Atlantic Ocean. This aberrant performance can only be avoided if such models are able to simulate the formation of Agulhas rings.

Inter-ocean leakage by rings

The leakage of Agulhas water into the South Atlantic, although it may represent only a quarter of the return flow[126] of the global conveyor belt, may nonetheless play a critical role in the stability and the variability of the global oceanic circulation. As described above, this leakage occurs predominantly through the process of ring shedding at the Agulhas retroflection[127]. The frequency at which such rings are shed is not known with great statistical precision[74,91], nor is the total inter-ocean flux brought about by these rings[128–9]. Further, the dynamical processes involved in the triggering processes for the formation of such rings remains a tantalising oceanologic problem.

A number of processes that may cause ring spawning events have been suggested; such as perturbation to the fluxes in both the Agulhas and Agulhas Return Currents[91], increases in the mass flux of the Agulhas Current[130] or perturbations caused by the passage of Natal Pulses[131]. Chassignet and Boudra have pinpointed the southward inertia/baroclinicity in the overshooting Agulhas Current[132]. Some, if not all, of these triggering processes may be substantially affected by changes in wind-stress patterns over the South West Indian Ocean[133], thus possibly leading to complicated feedback loops between ocean and atmosphere on the climate time scale in this sensitive part of the global oceanic conveyor belt.

The precise role of the Agulhas Current termination in global climate has therefore as yet not been established, but the preliminary findings to date do strongly suggest that it is considerable.

Influence of the Agulhas Current on local climate and weather

Apart from its role in global climate that is now emerging as possibly a central one, the Agulhas Current also plays an important role in local atmospheric forcing[805]. It would not be unexpected to find that such a large body of distinctly warmer water, maintained by a continuous inflow, has a measurable influence on the weather and the climate of the adjacent continental land mass[806].

Rainfall teleconnections

Statistical studies[134,818] by, e.g., Reason and Godfred-Spenning[698] have in fact demonstrated close associations between inter-annual variability of sea surface temperatures within the Agulhas Current system, including the Agulhas retroflection region, and summer rainfall behaviour over southern Africa. When temperatures are higher in the Agulhas Current region along the east African coast and within the Agulhas retroflection region, to the south of the continent, the summer rainfall has statistically been shown to be higher[136]. These higher sea temperatures precede the higher rainfall[134], suggesting a causative relationship, as well as a potential predictive capability. This also holds true for the winter rainfall region of South Africa, for which it has been shown[699] that higher temperatures in the Agulhas retroflection region may contribute to higher than normal rainfall. There is evidence, however, that these associations between sea surface temperature in the Indian Ocean and rainfall over the southern African subcontinent are not necessarily stable[817].

The geographic distribution of atmospheric pressure cells with extremely low pressures near southern Africa[137] furthermore exhibits a remarkable agreement with that region in the ocean where the Agulhas Return Current is best developed. This supports the general concept of a close association of the greater Agulhas Current system and the atmosphere. Numerous modelling activities have furthermore shown the sensitivity of atmospheric, rain-bringing systems over southern Africa to changes in sea surface temperatures in the Indian Ocean[135,138].

Modelling teleconnections

An anomalous sea surface temperature of 2 °C over the Agulhas retroflection region has, for instance, been shown[139] to influence significantly modelled rainfall over the subcontinent by reducing static stability over the anomaly, thus affecting the structure of westerly troughs in the atmosphere. Even relatively modest anomalies in sea surface temperatures of the South Indian Ocean inserted in models thus simulate appreciable rainfall anomalies over southern Africa[698]. General circulation models show that higher sea surface temperatures in the South West Indian Ocean lead to significantly wetter conditions over eastern and central South Africa[697]. In fact, on an intra-seasonal to an inter-annual scale[691], warmer water in the Agulhas retroflection leads to rainfall enhancement of a significant fraction of the annual total of many regions of southern Africa. A positive anomaly in sea surface temperature imposed on a general circulation model[136] for the central basin of the Indian Ocean by contrast simulated significantly reduced continental rainfall. What mechanisms would be responsible for such changes in rainfall?

Using a general circulation model[693] it has been shown that a dipole in sea surface temperatures in the South Indian Ocean, with a warmer region south of Madagascar and a cooler region west of Australia, will lead to increased evaporation over the warm region, the stimulation of an atmospheric low and strengthened onshore flow of moist marine air for southern Africa as a result. Such a scenario would indeed lead to increased rainfall over southern Africa.

A further study[618] has suggested that the absence of the warm Agulhas Current and its outflow in a model leads to weaker extra-tropical cyclones, with more southerly tracks. This is especially true in winter when the contrast of sea surface temperatures between the Agulhas system and the ambient waters is greatest. How realistic is this? A study of a number of low pressure systems passing over the Agulhas retroflection region during winter has shown[695] that they undergo substantial structural evolution. Those that have passed over the warm surface water of the retroflection exhibit a stronger baroclinic structure, stronger uplift, enhanced zonal winds and higher moisture content compared to those that have not.

The question then arises how these temperature anomalies in the Agulhas system themselves come about. If these rainfall predictors were themselves predictable, their forecasting ability could be substantially extended and so make them much more useful in, for instance, agricultural prognoses.

Some analyses[135,140–1] by Reason and others have now suggested that on time scales up to the inter-annual the anomalies in sea surface temperatures of the South Indian Ocean may be forced largely by atmospheric changes whereas over longer periods, large-scale ocean dynamics may be more important for establishing such anomalies. On decadal to multi-decadal scales a warmer Agulhas Current retroflection will lead, according to some models[141] to changes in surface winds that in turn will lead to modifications to surface fluxes and thus oppose the originally imposed anomaly. Multi-decadal anomalies of sea surface temperature come about due to local atmospheric changes whereas atmospheric changes on multi-decadal scales in general circulation models are manifestations of near-global modes[692]. In fact, some model studies[135] have suggested that the atmospheric response to positive anomalies in sea surface temperatures would be favourable for enhancement of the anomaly, but that at time scales exceeding 20 months dampening of the original anomaly may become more likely[698]. Over the wider Agulhas Current region, models suggest[696] that sea surface temperatures are influenced most by wind anomalies whereas heat fluxes at the ocean-atmosphere interface are more important over the rest of the South Indian Ocean. In order to understand these statistical correlations and teleconnections, the processes involved need to be adequately understood.

The details of the atmospheric linkages between the Agulhas Current system and terrestrial climate over the South African subcontinent therefore still require considerable research, both observationally and in the development of conceptual models. A number of pioneering studies have been carried out to measure directly the influence of the Agulhas Current on the overlying atmosphere.

Ocean–atmosphere interaction

It has, for instance, been known for some time that cumulus clouds tend to form over the Agulhas Current during certain synoptic weather conditions. This is also clearly evident from images in the thermal infrared part of the spectrum produced by satellite remote sensing[142], particularly if wind directions are parallel to the current. Such cloud formation has also been observed over the Gulf Stream[808] and under similarly appropriate wind conditions over Agulhas Current filaments in the southeastern Atlantic Ocean[143]. A dedicated research programme to investigate the influence of the Agulhas Current on the overlying atmosphere[144] has resulted in a better understanding of the prevalent processes.

The atmospheric boundary layer is substantially modified over the current wherever strong fluxes of heat and moisture from the current pertain[145–6]. Under these conditions, thermodynamic considerations adequately account for the formation of cumulus clouds. The vertical extent of these clouds is limited by a subsidence inversion. This suggests that the influence of the warm Agulhas Current extends much deeper into the atmospheric air column than purely the local boundary layer. In the Agulhas retroflection region the effect of strong thermal fronts in the ocean were in fact observed to a height exceeding 2500 m[98]. In general, the heat fluxes from the ocean to the atmosphere are higher in the retroflection region than in any other part of the Agulhas Current system[147], but further observational projects by Jury and Walker[148] have demonstrated the marked changes in atmospheric conditions on crossing the landward border of the Agulhas Current at any location. Observations of boundary layer fluxes that have simultaneously included the Agulhas Current as well as the adjacent land masses[701] show, as would be expected, that surface moisture fluxes from the Agulhas Current to the atmosphere are double those found over cold inshore waters and over the inland desert.

Heat fluxes over components of the southern Agulhas system in general increase on moving polewards. Fluxes of up to 1000 W/m^2 have been observed[496] during storm conditions over an Agulhas eddy adjacent to the Subtropical Convergence. Such fluxes are of course strongly dependent on the weather conditions during which they were measured. Observations of sea-air fluxes over the Agulhas retroflection and over the adjacent Agulhas Return Current[700] give values of total heat flux to the atmosphere (sensible plus latent) of 360 W/m^2 during the presence of a high pressure system and 630 W/m^2 during a cold air outbreak.

The fluxes of heat from the Agulhas Current to the overlying atmosphere are in some respect similar to those of other western boundary currents, but have some notable characteristics of their own. A climatological study[620] has shown that the average latent and sensible heat fluxes from the Gulf Stream are

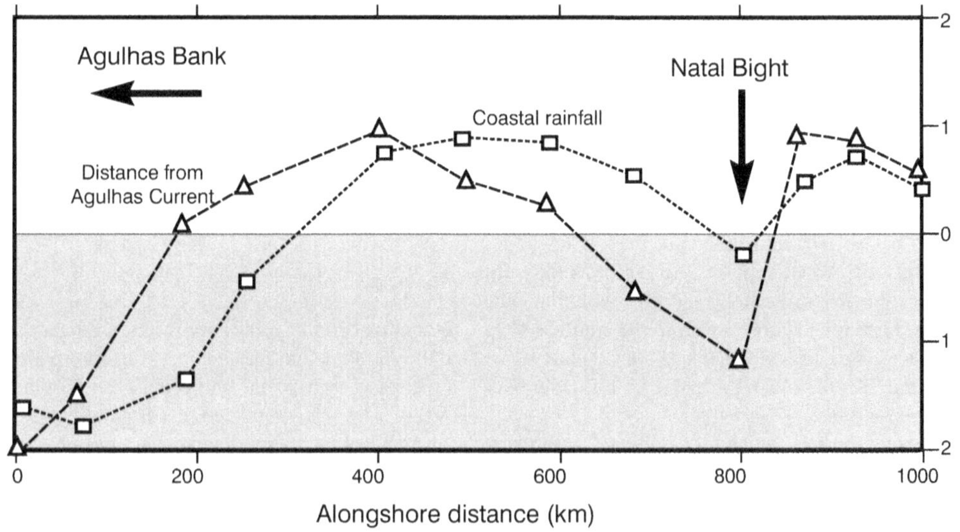

Figure 1.4. The influence of the Agulhas Current on coastal rainfall along the eastern seaboard of South Africa[149]. The abscissa gives distance upstream of Port Elizabeth; the ordinate standardised departures of coastal rainfall (squares) and distance from the thermal core of the Agulhas Current as evident at the sea surface (triangles). The latter has been inverted for comparison. At the Agulhas Bank and at the Natal Bight the current lies farther offshore and the summer rainfall at the coast is commensurably reduced.

200–300 W/m^2 in winter and 50–100 W/m^2 in summer. The corresponding values for the Kuroshio are 50–100 W/m^2 and –15 to 15 W/m^2. Estimated fluxes over the Agulhas Current are by comparison smaller (e.g. 115–180 W/m^2 in winter), but show considerably less seasonal variation[604,620]. These lower values for the Agulhas Current are most probably due to the fact that the air masses that are advected over the Agulhas Current are essentially marine, and are less cold and dry than the continental air moving over the Gulf Stream and the Kuroshio. The values given here are climatological averages; event-scale fluxes may be much larger. As mentioned above, observations of sea-air heat fluxes over an Agulhas eddy[496] have peaked at 1000 W/m^2, comparable to similar measurements made in the Gulf Stream[621] and the Kuroshio[622] during severe storms.

Coastal weather

The Agulhas Current also has a measurable influence on the weather and climate of its closest terrestrial environment, the coastline from Delagoa Bay to Port Elizabeth. It has been shown that summer rainfall along the south-east coast of Africa is influenced by the proximity and temperature of the core of the Agulhas Current[149] (Figure 1.4). Alongshore gradients in rainfall from north-east to south-west mirror gradients in the surface temperatures of the current. It has been surmised that a quasi-stationary mesoscale circulation system over the Agulhas Current interacts with the synoptic weather systems of the region to produce alongshore changes in the rainfall gradient. It has subsequently been shown[701] that during easterly winds the thin marine boundary layer over the Agulhas Current is advected over the adjacent land. Here it interacts with surface continental heating to form convective plumes over the coastal escarpment. The resultant buoyant air with high moisture content feeds developing cloud systems. Any statistically significant changes in the average location or temperature of the core of the Agulhas Current could therefore have a measurable effect on the rainfall over the adjacent coastline[133].

In the cases of comparable western boundary currents it has been noted that the fluxes of heat and moisture from these currents may have an important effect on the intensity of storms. The severity of the infamous *Presidents' Day Storm* of 1979 over the eastern United States has been linked directly to the influence of the Gulf Stream[623] and modelling studies[624] have confirmed the role of the Gulf Stream in the development and potential intensification of winter storms along the eastern seaboard of the US. To date only one, but an instructive, study[619] has been undertaken to investigate the role of the Agulhas Current in the formation of exceptionally severe storms. Such a storm occurred in

December 1998 near the south and near the south-west coasts of South Africa. Rainfall over a two-day period, representing between 150% and 500% of the average rainfall for that calendar month, generated substantial flooding and property damage at two widely separated locations. Tornadoes further contributed to the damage. Using, amongst other data, rainfall estimates from satellite remote sensing[625] and backward trajectories of air parcels it has been demonstrated that the low-level, onshore flow of moisture from the Agulhas Current played a significant role in the storms' evolution. The inland penetration of moisture from the Agulhas Current in fact determines whether subsequent convective rain is limited to the coastal strip or extends wider over the inland land mass[701].

At present the role which the Agulhas Current system plays in determining or influencing the weather and climate of the adjacent subcontinent is still very imperfectly understood. Tantalising suggestions that its effects may be significant will most probably be followed up by further focused investigations to determine this quantitatively. The exact coupling mechanisms that link ocean and atmosphere in this particular system will more than likely be difficult to establish. A better understanding of the Agulhas system itself, its variability and its driving mechanisms, would be an essential prerequisite to rapid advances in our understanding of how it influences local weather and climate.

Forming the physical base for a large proportion of the SouthWest Indian Ocean as well as parts of the SouthEast Atlantic, the Agulhas Current may have an influence that also extends to the ecosystems of the region.

Biological influence of the Agulhas Current

As would be expected from any major current that carries a large body of essentially foreign water into a particular oceanic region, the Agulhas Current has a marked influence on the distribution of a number of pelagic species in the South West Indian and South Atlantic Oceans. This biogeographical imprint may be due to three basic mechanisms, or a combination of these.

Advection

First, the current may carry immotile organisms such as fish larvae[791,800] within its waters thus mechanically skewing their distribution. This process naturally also serves the distribution of flotsam such as plastic debris and seaweeds. It has, for instance, been inferred from a study of plastic particles at the sea surface off the south-western coast of South Africa that the Agulhas Current is a major source of these plastic pollutants[150] The close proximity of this intense current to the coast in places has been invoked[638] as the reason why a number of epipelagic siphonophore species, that generally have poor powers of locomotion, are not found in shelf waters of the northern Agulhas Current, being unable to penetrate against the reigning current.

Marine animals

Secondly, certain species may be evolutionary adapted to the Agulhas Current system, using it purposefully as a means of transportation during their particular life cycles. The marine turtles that lay their eggs on the sandy beaches of Tongaland, KwaZulu-Natal, are a case in point. Leatherback sea turtles (*Dermochelys coriacea*), have been observed to follow the trajectory of the Agulhas Current closely[659] after depositing their eggs. In fact, by fitting these animals with satellite transmitters and comparing their subsequent tracks with satellite altimetry[660] it is clear that they use the current to minimize the energy required to swim. It has previously been surmised[151] that three basic circulatory cells exist in the South Indian Ocean, with increasing periods of circulation depending on their dimensions, and that marine turtles use these cells to circle the South West Indian Ocean so as to arrive back at their nesting sites at the correct time of year. This hypothesis remains to be proven, but there does now exist sufficient evidence that the Agulhas Current plays an important role in the distribution of these animals[152].

According to their movements, monitored by satellite, loggerhead turtles (*Caretta caretta*) on the other hand seem purposefully to avoid the Agulhas Current and to use inshore currents to aid their trips northward to their preferred feeding grounds. This is similar to the movement of displaced African penguins[153] on the Agulhas Bank that also seem able to discriminate and to utilise inshore currents. Apart from such use of the currents, a third possible mechanism for Agulhas Current influence on biota exists.

The hydrographic peculiarities of the Agulhas Current, including water temperature, nutrient availability, vertical stability of the water column as well as advection rate, may play a role in the distribution of organisms that find these characteristics either congenial or not. This holds true, for instance, for the geographic distribution of small cetaceans. Their distribution consists, especially close to the coast, of distinct patterns thought to be due to the wide range of zoogeographical regimes created by the Agulhas system[154]. It has been observed in other systems[641–2] that foraging birds may feed preferentially at the edges of circulatory features such as eddies. This also seems to hold true for Agulhas rings[640] where Leach's storm petrels are found at higher densities at ring edges.

Pelagic species

Considerably more is known on the landward geographic distribution of many pelagic species than on their deep-sea life cycles and how these life cycles may be adapted to the present Agulhas Current system. However, the paleoceanographic evidence presently available from the sedimentary records of the South-West Indian Ocean strongly suggest that a circulatory system much like the present had been established from at least the late Miocene, five million years[155] ago. It would therefore not be unreasonable to expect some pelagic species to have evolutionary adapted their life cycles over this period to this persistent circulatory pattern. Some very tentative hypotheses on the use that juvenile fish and whales may make of coastal counter currents, created by Natal Pulses, have for instance been put forward, but this awaits much further investigation.

Biogeographical distributions of copepods, chaetognaths, tunicates and myctophids all clearly reflect the influence of the Agulhas Current. Surface sampling of copepoda[156] over a total distance of 26 000 nautical miles in the South Atlantic and South Indian Oceans has demonstrated unequivocally that certain species are only found in Agulhas water and may therefore be considered indicator species of this water type. Pelagic tunicates similarly have specific indicator species such as *Thalia democratica* and *Doholum denticulatum*[157]. No chaetognath species have to date been found to be indicator species of particular water masses of the Agulhas system in the somewhat uniform conditions off Durban[158], but in the waters over the Agulhas Bank and along the west coast of southern Africa typical Indo-Pacific species such as *Sagitta enflata* have been found[159] and these have been considered unambiguous indicator species for the presence of Agulhas water.

Gibbons and co-workers have shown[639] that for euphausiid species there is a very tight agreement between the circulation components of the Agulhas Current system and their biogeography. This holds true for a number of siphonophore species as well[638]. These results indicate that the strong gradients in water characteristics that form an inherent part of the greater Agulhas Current system are reflected in the geographic distribution of many species, both pelagic and benthic. This holds especially true for the Agulhas Return Current. Flowing along – or parallel to – the Subtropical Convergence, its path is characterised by intermittent high levels of chlorophyll-*a*[588–9]. It has been assumed that this is due to the juxtapositioning of subtropical waters with high levels of stratification and waters rich in nutrients south of the Subtropical Convergence. It is instructive that the high levels of chlorophyll-*a* observed at this front peter out where the influence of the Agulhas Return Current terminates[96]. This also correlates with an abrupt decrease in levels of mesoscale turbulence. With a decrease in the number of eddies there seems to be a concomitant decrease in chlorophyll-*a* corroborating the concept of eddy-induced enhancement of primary production[690]. Recent combinations of satellite remote sensing and modelling have shown[689] that there is a seasonal element to the observed chlorophyll at the Agulhas Return Current with increased light in summer enhancing the levels of chlorophyll. This seasonality is in opposite phase to the chlorophyll presence found in the extensive subtropical gyre north of the Agulhas Return Current.

The leakage of tropical Indian Ocean water south of Africa, with its attendant pelagic biota, may present an unusual inter-ocean pathway for tropical species from the Indian to the Atlantic Ocean; a pathway that, according to Prell et al.[160], may have been more, or less, accessible over geological time. This inter-ocean exchange mechanism has been surmised, at least for myctophids[161]. Myctophids are namely known to be carried south in the Agulhas Current[162].

Shelf ecosystems

The influence of the warm Agulhas Current on organisms inhabiting the coastline, be they benthic or pelagic, is considerable[803]. This is due for the greater part to the higher temperatures brought about by the current's proximity to the coast. A comparison between the total number of organisms as well as the species composition at the same latitude on the south-east and south-western coastlines of Africa partially demonstrate the coastal influence of the Agulhas Current. The cold, nutrient-rich, upwelled water along the south-western coast naturally also plays a significant role in enhancing this longitudinal contrast.

In two very site-specific regions the effect of the Agulhas Current on inshore ecological systems may be particularly profound. These regions are located at the northern tip of the Natal Bight[163–4], the only area of wide continental shelf off the South African east coast, and at the far eastern part of the Agulhas Bank[165–6], the wide shelf south of Africa. At both locations the passing Agulhas Current is responsible for kinematically driven upwelling[167] that brings cold, nutrient-rich water onto the shelf. It is thought that these upwelling cells are in fact the sources of bottom water with high nutrients for both shelf regions[168–9]. It has been established unequivocally that enhanced primary productivity and higher chlorophyll-*a* values are to be found in both cells[164,170].

The specific role of the Agulhas Current in the ecology of the coastal as well as deep ocean regimes in general is, however, still insufficiently understood[803]. There are two simple but fundamental reasons for this. First, the detailed dynamics and variability of the circulation in the South West Indian Ocean is as yet poorly known and, secondly, the distribution and life cycles of most organisms in this ocean region have also been inadequately studied.

The influence of the Agulhas Current on a particular phenomenon that, in contrast, has received focused research attention and that, as a result, is now considerably better understood, is the effect of the current on the generation of what are popularly known as "monster" waves.

"Giant" waves

Historically the general oceanic vicinity of the southern tip of Africa has been very well known for its extreme storminess and thus its hazards for shipping. The large number of wrecks from the last four centuries along the South African coastline creates an archaeological treasure-trove, but also an eloquent statement on the perils of sailing these waters.

In modern times, with larger and faster-moving vessels, the dangers of giant waves to shipping have also become increasingly apparent. By their nature and dimensions these seemingly unpredictable waves have been able to break large ships in two, to detach whole bow sections and, in the more fortunate cases, only to seriously damage the bows of vessels (see inset on page 105).

Using shipping reports, Mallory[171] has been the first to indicate that these giant waves preferentially occur in the core of the Agulhas Current and particularly on its landward border. Theoretical studies[104,172], measurement programmes and satellite remote sensing[173–4] have all been employed successfully to explore the unique geographical configuration of current and wave generation regions that is responsible for the relationship between the Agulhas Current and these unusual waves[175]. Attempts are even now being made to use real-time wind information from the Southern Ocean and the position of the Agulhas Current to predict the regions in which waves with strongly enhanced amplitudes are to be expected.

Perspective

The African continent has a southerly termination that lies much closer to the equator than does that of South America or Australia; particularly when Tasmania and New Zealand are considered part of the Australian continental land mass. This single fact, more than any other, underlies the exceptional role which the Agulhas Current system plays in global oceanic circulation by linking the waters of the South Indian and South Atlantic subtropical gyres.

This linkage depends predominantly, but not exclusively, on the shedding of Agulhas rings at the western termination of the Agulhas Current. The interbasin flux of heat and salt that comes about through this mechanism has obvious and important climatological implications. Little is as yet known about the factors that control the process of ring shedding, but triggering mechanisms that have been proposed would imply a complex system of sub-mesocale interactions and even – possibly – intricate feed-back loops. This suggests that even relatively small changes in the wind-stress patterns over the Indian Ocean may bring about substantial alterations in the inter-ocean exchange of water and thus conceivably in oceanic climate.

The Agulhas Current has a marked effect on the circulation of the South West Indian Ocean in general and on the continental shelf of south-eastern Africa in particular. This effect extends also to the overlying atmosphere and thus to rainfall over the southern African subcontinent and the south-eastern seaboard. By its very nature the Agulhas Current must additionally play a dominant role in the biogeography of not only the South West Indian Ocean, but also further afield. Since the Agulhas Current would seem to have been in existence for at least four million years, it is highly likely that the life histories of numerous organisms are evolutionary adapted to its present configuration.

All these areas of influence of the Agulhas Current, either singly, but even more so in combination, make this current system worthy of dedicated and focused scientific scrutiny.

2

Large-scale circulation of the South West Indian Ocean

The Agulhas Current system lies imbedded in the larger circulation of the South West Indian Ocean, the South Indian Ocean and, of course, the Indian Ocean as a whole. Even this is not a final, isolated unit but is affected by the inflow of water from the Pacific through the Indonesian archipelago[176] and exchanges with the Southern Ocean to the south. It is probably true to say, in general, that the farther the circulation scales of the regions being considered are from one another, the smaller the direct impact. For instance, the circulation of the North Indian Ocean influences the large-scale circulation of the South Indian Ocean, but its direct impact on detail of the Agulhas Current seems to be small. The behaviour of the greater Agulhas Current system would conceivably be influenced by aspects of the Indian Ocean that are peculiar to it.

The circulation of the Indian Ocean as a whole is unique in this respect that the circulation patterns in its northern and its southern parts are not rough mirror images of one another, as in the Atlantic or the Pacific. The wind-driven circulation of the North Indian Ocean is forced by the seasonal changes in the monsoonal winds and reverses radically with season (Figure 2.1). This is unlike the South Indian Ocean where the conventional anti-cyclonic, wind-driven circulation is persistent throughout the year.

Monsoonal circulation

This unique monsoonal gyre, north of the equator, is driven by the North East Monsoon winds from east to west. This flow starts in the month of November and subsides by April. Off Somalia this flow has to turn south along the African coast and becomes the Equatorial Counter Current (Figure 2.1). This flow along Somalia is strongest in December[178]. The Equatorial Counter Current lies between 3° N to 5° S, but from January to April it moves to 10° S. The general circulation of the North East Monsoon period is rather weakly developed. By contrast, the South West Monsoon circulation reaches its greatest strength during July and is more strongly developed. All flow in its domain is eastward, including the counter current. Most of the water of the South Equatorial Current now flows northward along the African coast[178]. This intense, wind-driven system only starts breaking down again in October. Water motion in this ocean is also driven by thermohaline forcing.

There are three regions of high salinity water formation in the North Indian Ocean: the Arabian Sea, the Red Sea and the Persian Gulf. The formation processes in all three seas are due to an excess of evaporation over precipitation. It is the Red Sea Water which penetrates farthest into the South West Indian Ocean at a depth of about 1100 m.

The northern and southern wind-driven gyres of the Indian Ocean are separated by a strong front at about 10° S[178]. This boundary zone is particularly pronounced below the surface layers. North of the front are waters of low oxygen content and high nutrients; to the south the reverse holds true[178]. At the surface the front is marked by a salinity maximum stretching from Sumatra to Africa.

Subtropical gyre of the South Indian Ocean

The subtropical, wind-driven, anti-cyclonic gyre of the South Indian Ocean consists of the South Equatorial Current, the Agulhas Current system and the South Indian Ocean Current, with little evidence of a northward, eastern boundary current off Australia. This gyre differs substantially from its counterparts elsewhere in that most of the water recirculates in the western and central parts of the ocean basin[101]. The anti-cyclonic flow penetrates to at least 1000 m depth in the South West Indian Ocean sub-gyre, being particularly intense and pronounced here. By contrast the flow on the eastern side of the basin is very weakly developed and,

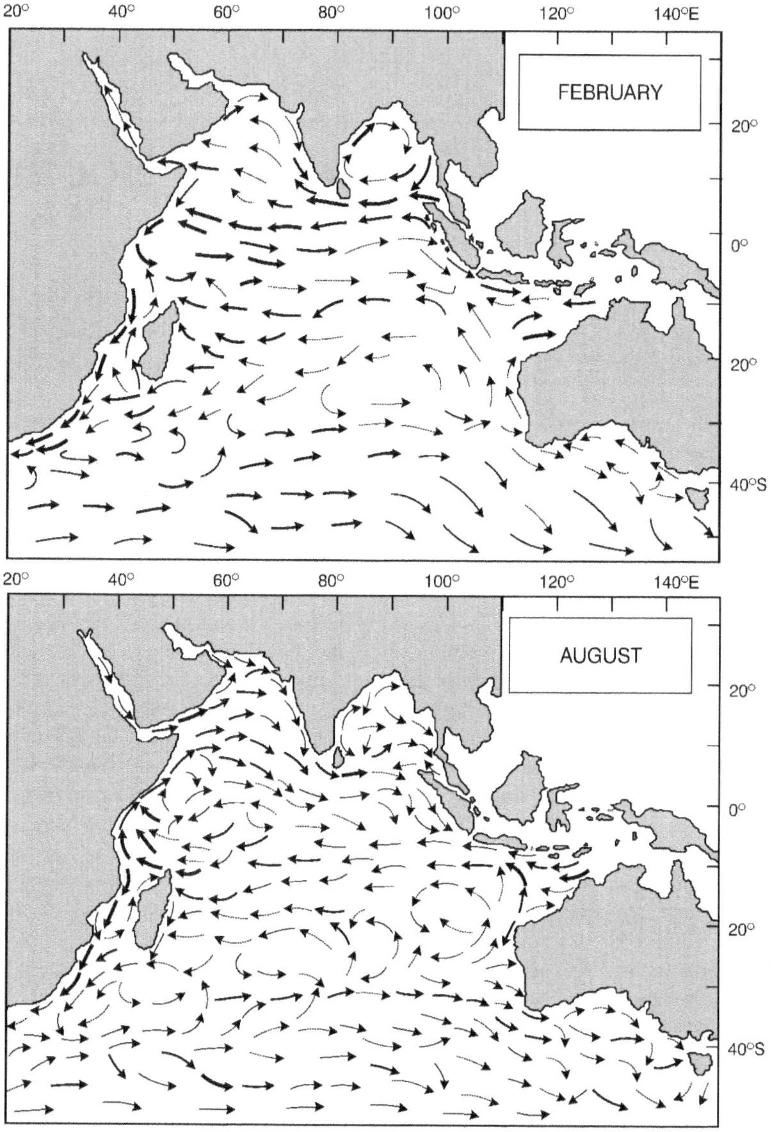

Figure 2.1. The surface circulation of the Indian Ocean during the calendar months February (North East Monsoon season) and August (South West Monsoon season) derived from ships' drift observations[177]. Seasonal changes in the South Equatorial Current – between 10° and 20° S – may have an important effect on the circulation in the South West Indian Ocean.

during the months of January to July, is in fact southward[179–80].

From a biological perspective the subtropical gyre of the South Indian Ocean has very low primary productivity[681–2]. This is similar to most other subtropical systems, since the surface waters of these regions are as a rule nutrient depleted. Modelling efforts by Machu and others[681–3] have shown that there nevertheless is a seasonal signal of primary productivity in the gyre, being higher during autumn and winter. During this season stronger winds enhance deep mixing in the water column and thus the entrainment of deeper water, rich in nutrients to the upper layers. These model simulations agree with what has been observed with ocean colour sensors from satellite[681–2].

The South Equatorial Current forms the northern

Pre-World War II German research on the Agulhas Current

An analysis of the country of origin of publications on the Agulhas Current system shows[196] that over the past century the contributions from individual countries have altered remarkably.

During the first decade of the twentieth century scientists from Britain and Germany were the major scientific participants in building up knowledge on this system. This was the heroic age of extensive exploratory cruises. The results of these undertakings usually were published over periods spanning many years. This was accomplished in the form of numerous volumes of a cruise report, such as, for instance, those by Thomson and Murray[15,197] on the *Challenger* voyage (50 volumes), by Chun on the results of the cruise of the *Valdivia* (24 volumes) and the reports on the research cruise of the *Planet*[21] (five volumes).

From about 1910 to 1940 the research activities on the South West Indian Ocean were concentrated in Germany[196], and more particularly, at the *Institut für Meereskunde* at the University of Berlin. Between 1923 and 1941 at least 28 major papers appeared[196], describing in detail the seasonally changing meteorological[199] and ocean current[44–5] conditions (Figure 3.2), including sea surface temperatures[200–1] and salinities[202]. Interest stretched from the Red Sea[203] to the Subtropical Convergence[204]. Many of the publications of this time[205–207] were based on the results of the *Meteor* cruise in the South Atlantic[33], which had included the region of the southern Agulhas Current. In this heyday of geographical studies of the oceans in Germany a great deal of this material was masterfully combined by Georg Schott in such major works[208] as *Geographie des Indischen und Stillen Ozeans*[49].

Deeper hydrographic measurements were for the first time becoming available in the 1920s and 1930s in quantities that allowed their use for the interpretation of the movement of deep[47,209–10] and bottom waters[211–12]. George Wüst played a major part in these studies. Another person who was very prominent in describing the circulation of the Indian Ocean over its full depth was Lotte Möller[43,46]. Günther Dietrich, who became director of the *Institut für Meereskunde* in Kiel after the buildings of the *Institut für Meereskunde* in Berlin had been destroyed during the war, carried out a descriptive hydrographic analysis of the Agulhas Current for his doctoral dissertation that became the standard reference[40–42] on this current at the time. Little of this work found any resonance in the embryonic marine science establishment in South Africa[213] that worked mainly on fisheries related problems at the time.

The description of the Agulhas Current in the standard oceanographic textbook *The Oceans*[50], used by generations of ocean scientists, was based on Dietrich's work, and on that of many of the other German scientists who had worked in the region.

With very few exceptions, a notable one being Clowes[214] – an Englishman working in South Africa at the time – this benchmark in knowledge on the Agulhas Current system was not referred to again after World War II[198]. Most of the major German papers from this time were not referenced by subsequent investigators[81,215] or in reviews on the Agulhas circulation[216–17]. This may have been due purely to ignorance, as the time interval since the German work had been long; due to inability to read the original texts in German or, possibly, due to a degree of scientific chauvinism[198]. Whatever may have been the reason, most investigations from the 1970s onward started afresh and made no use of the German research of the 1930s. For all intents and purposes it was lost to science.

component of this South Indian Ocean gyre having highest transports during the South West Monsoon. It is strongly affected by the Mascarene Plateau that lies directly in its path at 60° E (viz. Figure 2.2). Even at a depth of 50 m the flow in the South Equatorial Current is constrained to passing through two gaps in the plateau[704] thus severely altering the distribution of the meridional current structure downstream. At the east coast of Madagascar this current splits, about two-thirds going north, one-third proceeding southward along the Madagascarine coastline[178]. The South Equatorial Current thus contributes indirectly to the flow of the Agulhas Current system that forms the pivot of this whole gyre. Water that leaves this system as the South Indian Ocean Current[100] flows eastwards along the Subtropical Convergence at about 41° S latitude, continuously losing water to the gyre to the north along its path. This frontal system is important for other reasons as well.

The Subtropical Convergence

At the Subtropical Convergence a deep mixed layer develops in winter when cooling leads to vertical convection[181]. These thermostads with high oxygen content constitute Subantarctic Mode Water and this subsequently spreads over the greater part of the subtropical gyre. Over the gyre itself there is generally an excess of evaporation over precipitation and subtropical surface water of high salinity is formed. The band of highest salinity lies between 25° and 35° S latitude. The subsurface salinity maximum of this water is found throughout the entire subtropical anti-cyclonic gyre. Near the equator this water is covered by a fresher layer of tropical surface water.

Below the subtropical water a layer of high dissolved oxygen is the product of water formed at the Subtropical Convergence[178]. Both it and the Antarctic Intermediate Water, characterised by a salinity minimum

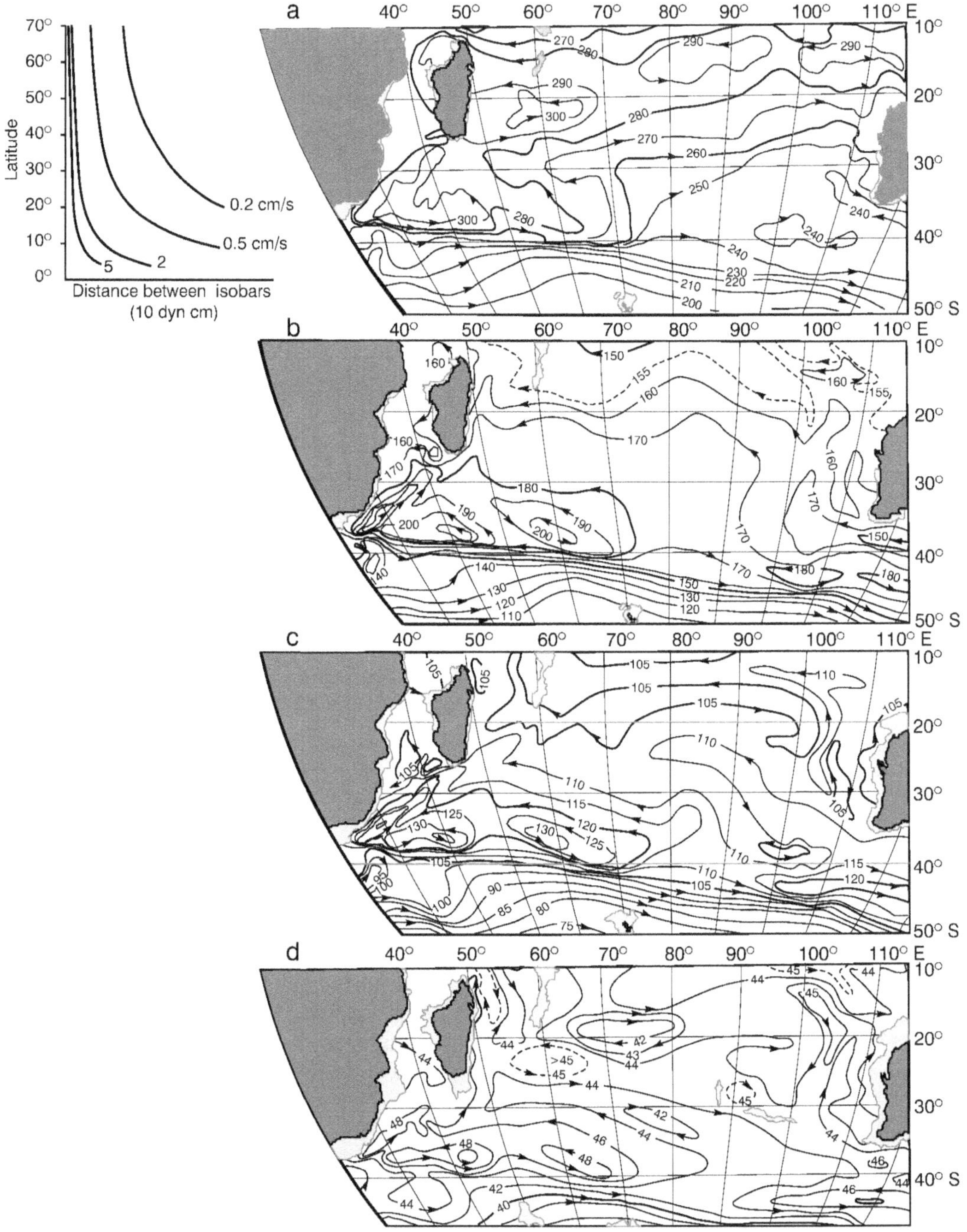

Figure 2.2. The general flow patterns of the South West Indian Ocean[84] for **a**: the sea surface, **b**: 500 decibar, **c**: 1000 decibar and **d**: 2000 decibar; all relative to 3000 decibar. Values are given in dynamic centimetre. Velocities may be estimated by using the insert showing speed versus distance between isobars. The depth penetration of the South West Indian Ocean sub-gyre stands out as one of the main circulatory features.

around 1000 m depth, circulate with the gyre, showing the great depth of this gyre compared to the monsoonal regime. However, the gyre of Antarctic Intermediate Water has a less pronounced concentration to the western side of the basin[182] compared to the gyre of the surface water[101]. Antarctic Intermediate Water in this gyre slowly mixes upwards, north of 32° S. It has been estimated[101] that this upward movement is about 0.6×10^6 m^3/s and is balanced by a downward mixing of about the same magnitude south of 45° S.

Mesoscale variability in the Indian Ocean on the whole is high[183], but highest values are restricted to the southern and south-western regions. Extending from the centre of the Mozambique Channel, through the Agulhas Current system, the high values then follow the core of the Subtropical Convergence eastwards. This mesoscale variability is – as will be seen – important for the large-scale circulation systems and the feedbacks[131] of particularly the Agulhas Current system. It also has biological impacts.

In intriguing contrast to the seasonality in primary productivity in the subtropical gyre of the South Indian Ocean, the highest values for primary productivity in the Subtropical Convergence are found in spring–summer and not in autumn–winter[681]. Using the simulations of a coupled physical–biological model[682], it has been inferred that this inverse seasonality is due to the deeper mixing induced by winter winds. This mixing removes a large proportion of the phytoplankton from the euphotic zone, inhibiting growth.

Limitations

The short description of the main flow regimes of the Indian Ocean that has been given here has a number of serious shortcomings, most stemming from either a lack of information or the geographically inhomogeneous distribution of existing hydrographic observations. In general, the North Indian Ocean and particularly the domain of the Somali Current have been well served with observations. The Agulhas Current has also been fairly well-investigated, but regions such as the centre

Impact of the International Indian Ocean Expedition

The International Geophysical Year (IGY) was the first successful, global geophysical programme that included a large international presence in its conception, planning and its observational phase. Most subsequent international programmes built on its success. One of these was the International Indian Ocean Expedition (IIOE) that took place from 1960 to 1965.

The International Scientific Committee on Oceanic Research (SCOR) considered a proposal for such an expedition in 1957[222], motivated by the fact that of all the ocean basins the Indian had been least observed and was as a result the least known. UNESCO agreed to co-sponsor the expedition and, partially due to this new demand, established its Intergovernmental Oceanographic Commission (IOC). In South Africa a special oceanographic co-ordinating committee of interested parties was appointed under the aegis of the Council for Scientific and Industrial Research.

Early attempts to have a pre-designed grid of stations for the IIOE in order to survey the hydrography of the Indian Ocean most efficaciously[223] failed. Each of the 13 ship-operating countries more or less carried out work of greatest interest to its own scientists. The hydrographic data collected during the IIOE, although therefore quite inhomogeneously distributed, was assembled, analysed and published as a monumental atlas[84], largely achieving the original aims of the project[223,225].

An analyses of the annual number of publications on the physical oceanography of the South West Indian Ocean[196] clearly shows the sizeable effect of the IIOE (see Figure 1, p. ix). From 1958 there was an increase in planning and background documents, relating to the IIOE, that continued up till 1964. In 1964/65 many cruise reports appeared while research publications peaked only in 1967, increasing the annual number of pre-IIOE publications five-fold.

Perhaps of greater importance to research on the Agulhas Current has been the influence this expedition has had by stimulating research in countries, such as South Africa, that previously had shown only limited interest in the deep ocean. In South Africa a number of deep-sea research cruises were carried out, utilising – amongst others – the naval hydrographic vessel S.A.S. *Natal* and the fisheries research vessel, *Africana II*.

It furthermore led to the formation of the predecessor to the South African National Committee for Oceanographic Research (SANCOR), that has played a leading role in guiding and supporting marine science in this country to this day. Shortly after the termination of the observational phase of the IIOE, the first South African National Oceanographic Symposium was held under the auspices of SANCOR[226]. It has taken place at three-year intervals ever since.

The young scientists that became involved in ocean science at this time, or shortly afterwards, were subsequently responsible for the enormous increase in research articles[227] in international journals from South Africa. The bloom in oceanographic research production that resulted was only made manifest in the 1970s and 1980s[228], but may in many ways have originated in the energising effect of the International Indian Ocean Expedition in the early 1960s. South African participation in subsequent large oceanic experiments, such as the WOCE[229] (World Ocean Circulation Experiment), has been politically more difficult, and scientifically less rewarding[814].

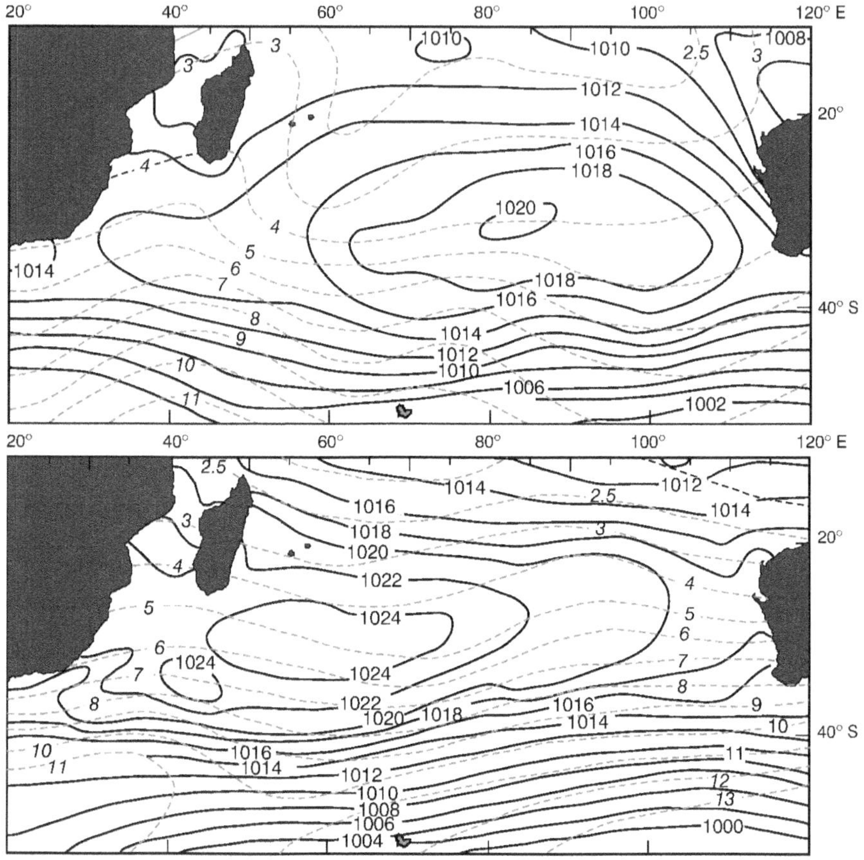

Figure 2.3. The mean atmospheric pressure at sea level for the South Indian Ocean (full lines) with standard deviations (broken lines), both in mbar[186]. The upper panel represents the month of January; the lower panel July. The seasonal shift of the centre of the atmospheric high pressure system substantially changes the wind stress patterns for this ocean region over an annual cycle.

of the gyre, the Mozambique Channel and the region south of Madagascar[184] have remained more or less devoid of an adequate geographic distribution of high-quality, modern measurements for a long time. During the past decade some changes for the better have occurred in this respect[650,658]. Nevertheless, to understand the circulation better, more hydrographic observations, appropriately placed, are required. A more solid knowledge of the wind patterns is also required.

The wind regime of the South West Indian Ocean

By all criteria of wind direction as well as precipitation the South West Indian Ocean falls well outside the monsoonal domain[185]. The general air circulation over this particular oceanic region is anti-cyclonal; stronger in the austral summer than in winter (Figure 2.3) when the centre of the anti-cyclonic high pressure system lies further to the east[185]. Surface winds are at a minimum at about 35° S in summer. This minimum band moves to about 30° S latitude in winter, but wind strengths in the minimum remain about the same, the zonal geostrophic wind at sea level being slightly stronger in winter[186]. From here average climatic wind strengths increase both towards the equator and southwards where they reach a maximum at about 50° S, or at the Antarctic Polar Front. Whereas average zonal winds at 35° S in January are 2 m/s, at 50° S they will be 24 m/s[186].

The region of low wind strengths over the South West Indian Ocean also represents an area of low wind steadiness. Compared to the area in the Mozambique Channel and near Madagascar where wind steadiness of 50 to 75 per cent is evident year round, the centre of the South West Indian Ocean has winds with a mean steadiness of 25 per cent or less[185]. South of the Subtropical Convergence the wind steadiness increases to levels between 50 and 75 per cent. With a lighter wind, it may be expected that the wind stress fields will follow the same pattern and this is indeed the case[187].

In both seasons a zone of minimum wind stress extends from Africa to Australia at about 35° S. In the South East Indian Ocean this minimum becomes slightly less pronounced in austral summer, but in the South West Indian Ocean there is little seasonal variation. The wind stress curl is a positive maximum over this region throughout the year[187]. The band of positive wind stress curl for the South Indian Ocean extends from about 20° S to 45° S latitudes. Estimates of the divergence of the Ekman transports, based on the above results, indicates that the South West Indian Ocean should be a region of average downwelling of about 0 to 20 cm/day[187].

Along coastlines the large-scale wind patterns may be considerably modified. This is particularly important for a fuller understanding of the behaviour of western boundary currents, such as the Agulhas Current and the East Madagascar Current. Around the southern African coast, coastally trapped waves are formed in the atmosphere[188] that propagate from west to east as coastal low pressure systems in concert with passing synoptic pressure systems to the south. Coastal winds therefore are aligned with the coast[189–90] for the greater part of southern Africa and the Agulhas Current.

The general wind patterns over the South Indian Ocean drive the major currents. This is evident from the mass transport stream functions, calculated from the Sverdrup relation for this ocean region[187]. The mass transport stream functions clearly demonstrate the wind-driven part of the ocean circulation. Thermohaline circulation creates some of the major water masses involved in the wind-driven circulation of the South West Indian Ocean, but as a process it is not of major importance in the South West Indian Ocean it self. What then are the major circulation patterns in this ocean?

Flow patterns of the South West Indian Ocean

General flow patterns of the full South Indian Ocean, and the South West Indian Ocean in particular, can be found from essentially three sources: drift observations at the sea surface, the historical hydrographic data and, last, information gleaned from satellite observations.

Surface movement

First, the surface flow from statistical analyses of ships' drift has been assembled in atlases such as those produced by the Naval Oceanographic Office[191–3] of the United States, by the German Hydrographic Office[177] or by the hydrographic office of Great Britain[194]. These data and publications have to a large extent formed the basis for the classical portrayal of the currents briefly described above (Figure 2.1).

The second data set consists of the accumulated hydrographic information that can be used to describe the mean geostrophic flow or the movement of specific water mass types along core layers. For the Indian Ocean this has been done in the *Oceanographic Atlas of the International Indian Ocean Expedition*[84]. Unfortunately the geographic data distribution for the presentations in this atlas is very inhomogeneous for the South Indian Ocean. On both the western and the eastern sides of the ocean the density of stations is probably sufficient for a rough characterisation of the mean flow. In the central part of the gyre there is a dearth of stations and the flow can only be described in the coarsest of terms. Nevertheless, a number of consistent flow patterns emerge for the region (Figure 2.2). Subsequent analyses[710] have filled in some details, but have not dramatically altered the key elements of the surface circulation.

The first element that emerges is that the subtropical, anti-cyclonic gyre is not zonal, but shows a distinct tendency to have its axis in a north-east–south-west direction. This is not an artefact of the data distribution, as is clear from Figure 2.2[84]. Secondly, the pivotal point of the circulation lies far to the west, more so than in comparable basins such as the South Atlantic[195]. Thirdly, the circulation shows a distinct tendency for a clear recirculation gyre west of 70° E (viz. Figure 2.4) with a strong sub-gyre west of 45° E. This is evident at the sea surface but becomes increasingly clear at greater depths[710]. At 500 decibar there is little evidence of the anti-cyclonic gyre east of 95° E, at 1000 decibar it has shrunk to 80° E[84]. It also lies increasingly southward with increasing depth so that at 2000 decibar it is only found between 30° and 40° S latitude.

The South West Indian Ocean therefore plays an increasingly important role in the anti-cyclonic circulation of the South Indian Ocean with increasing depth. Between 0 m and 300 m it is roughly responsible for about 40 per cent of the total anti-cyclonic mass transport in the gyre; for the interval 0 m to 3000 m it carries about 50 per cent of the total flux[84].

Hydrographic results

An analysis of carefully selected hydrographic station lines[101] (Figure 2.4) has confirmed many of the above-mentioned results on the flow patterns of the South

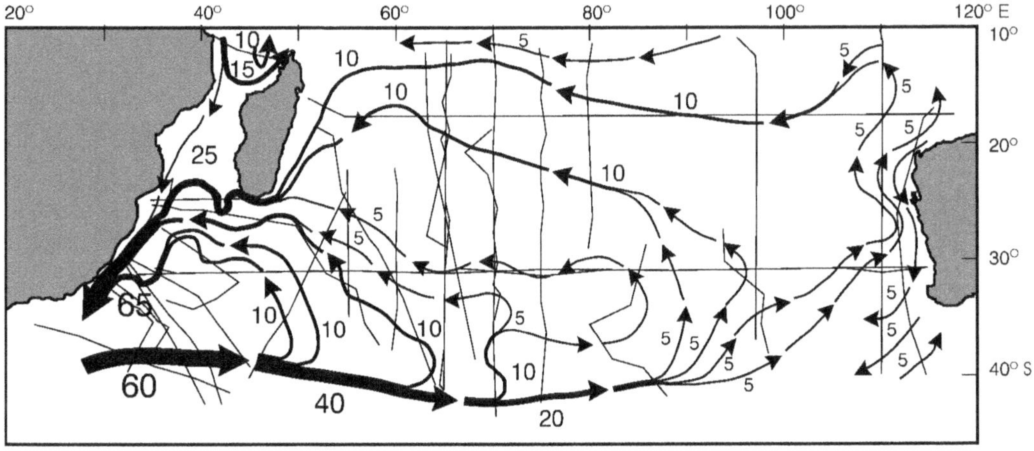

Figure 2.4. The baroclinic volume flux field of the South Indian Ocean[101] over the upper 1000 m. Transport values are in 10^6 m^3/s; thin lines indicate the geographic location of station lines used. Of particular importance here is the concentration of the recirculation west of the 70° E meridian.

West Indian Ocean. Donohue and Toole[708] have used a comprehensive, near-synoptic data set in the general vicinity of Madagascar and derived very comparable results to those shown in Figure 2.4, giving some confidence in the robustness of that portrayal. It demonstrates that for the upper 1000 m the flow in the South Indian Ocean is concentrated in a South West Indian Ocean sub-gyre and that, as a result, the gyral circulation east of 70° E is particularly weak. Flow along the Subtropical Convergence in the Agulhas Return Current is therefore rapidly diminished as its waters move eastwards. This flow forms the poleward border to the subtropical gyre in the South Indian Ocean.

A general flow feature of the South Indian Ocean that becomes clear from Figure 2.2 is the rather invariant location of this southern border to the subtropical gyre. This is seen by a bunching of isolines at about 40° S south of Africa with a southward tendency on proceeding eastwards. South of Perth, on the west coast of Australia, it lies at about 45° S[84]. This geographic location is found for all depths shown in Figure 2.2.

Altimetric results

The third available data set, in addition to those derived from ships' drift and from hydrographic data, is altimetric data from satellite. The results from a number of these latter data sets seem internally consistent[749] and are, furthermore, in substantial agreements with those derived from hydrographic data[218]. Comparisons of the sea level height from altimetric observations with *in situ* pressure gauges has successfully validated the altimetric portrayals[218]. The correlation between float velocities at intermediate depths and geostrophic velocities derived from altimetric observations[747] has been shown to fall between 0.8 and 0.9. All these results give solid evidence for the reliability of altimetric products to accurately depict oceanic motion. The mean dynamic topography for the South Indian Ocean derived from such altimetric data is shown in Figure 2.5.

The broad expanse of the Antarctic Circumpolar Current is clearly evident, its northern border lying between 49° and 45° S latitude, depending on longitude. The well-developed sub-gyre of the South West Indian Ocean is recognisable. Detail of the flow in the Mozambique Channel, east of Madagascar and of the Agulhas Current cannot be easily resolved with these data. An important conclusion from this particular analysis[218] has been that over this 18 month period of altimetric observations there were no significant inter-annual variations of the mean position and magnitude of the major current systems in the South Indian Ocean. Another analysis[218], also based on altimetric data, has presented conflicting results. It suggests that there is a strengthening of the anti-cyclonic gyre in March and a weakening in June and December. The limited contemporaneous hydrographic data on the mass fluxes do not support this latter result directly.

Of substantial importance in assessing the flow patterns of the South West Indian Ocean is an understanding of the geographic distribution of mesoscale turbulence. Reliable estimates of this kind have only become available with the advent of satellite altimetry[724,749]. The distribution of sea height anomalies produced by such measurements give an indication where high variability in ocean circulation can be

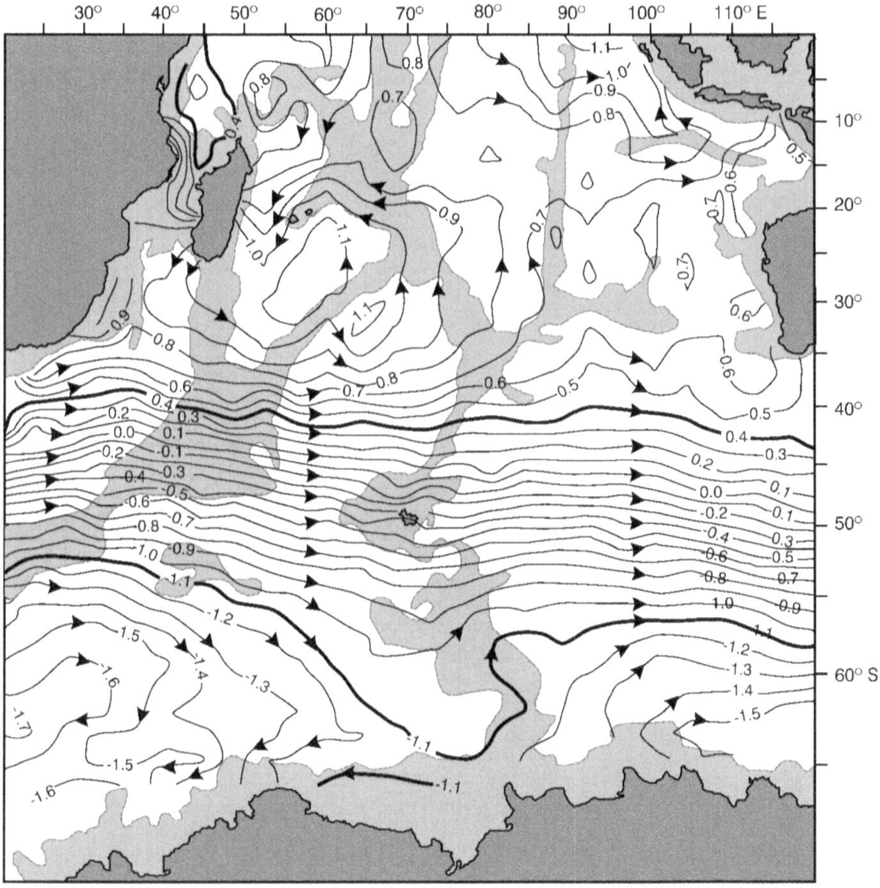

Figure 2.5. The mean dynamic topography for the South Indian Ocean and the Indian sector of the Southern Ocean as derived from altimetric measurements made by the TOPEX/Poseidon satellite[218]. Units are in metre and the two thick, zonal lines represent roughly the outer borders of the Antarctic Circumpolar Current. Regions shallower than 3000 m have been shaded.

expected. The patterns for the Indian Ocean overall are particularly instructive. A region that stands out for its high levels of variability on a global scale is coincident with the location of the Agulhas Current retroflection and the Agulhas Return Current[183,713,724]. Correlation length scales of movement differ within the system[414] and do not necessarily agree with the Rossby radius of deformation[749]. Secondary regions of high variability, evident in satellite altimetry, include most of the South West Indian Ocean and, in particular, the Mozambique Channel[734]. Unique to the Indian Ocean are two zonal bands of slightly higher variability extending across the subtropical Indian Ocean, one at about 15° S and another at about 25° S[724]. It is interesting to note that a high resolution model of the global ocean circulation does not simulate either of these bands, nor the high variability in the Mozambique Channel[724]. The sources of this high variability and their important role in the general circulation of the South West Indian Ocean are discussed in some detail in subsequent chapters.

Putting aside the variability for now, it is clear[218] that the average geostrophic flow patterns derived from both hydrographic and altimetric data (Figure 2.5) show that the anti-cyclonic gyre of the South Indian Ocean is weakly developed in the east and that this is increasingly true at greater depths. The circulation is reduced at depth to a sub-gyre in the south-west. The location of the southern border to the gyre, formed by the adjacent flow of the Antarctic Circumpolar Current, by contrast, remains fairly constant at all depths.

Some of these conclusions are also evident when one inspects a line of hydrographic stations carried out roughly through the core of the subtropical gyre in the Indian Ocean, i.e. at 32° S (Figure 2.6). The dynamic

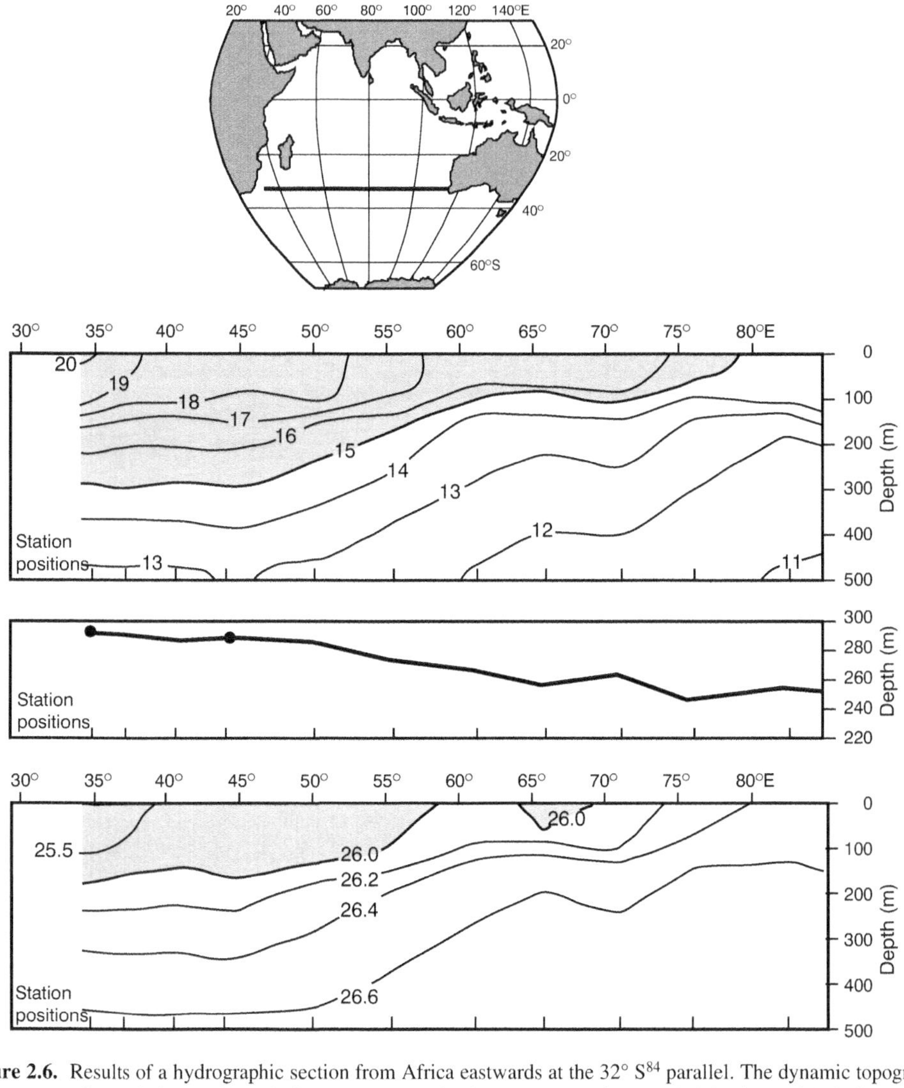

Figure 2.6. Results of a hydrographic section from Africa eastwards at the 32° S[84] parallel. The dynamic topography of the sea surface relative to 3000 decibar (centre panel) shows a strong slope to 65° E after which it flattens off. Potential temperature over the upper 500 m is shown in the upper panel; potential density in the lowest panel. The insert is a locational map of the line of stations.

topography declines steeply to about 65° E after which it decreases considerably more slowly with distance eastward. This represents water in the South West Indian Ocean warmer than 16.5 °C and with a potential density of less than 26. On this latitude water with these characteristics is only found close to the west Australian coastline[84]. This portrayal confirms the concentration of the gyral flow in the western side of the basin. It cannot indicate where this water is formed, what its distribution is or how it is mixed with other water masses.

Water masses of the South West Indian Ocean

The water masses of the South West Indian Ocean and, in particular their volumes, are outlined[84] in Figure 2.7. For temperatures in excess of 15 °C there is a wide range of salinities, representing waters from the subantarctic at 40° S to the tropics at 10° S latitude. These surface waters form a very small part of the total volume. The narrow band in the potential temperature–salinity space between the surface and the intermediate waters, the Central Water, constitutes only 14 per cent

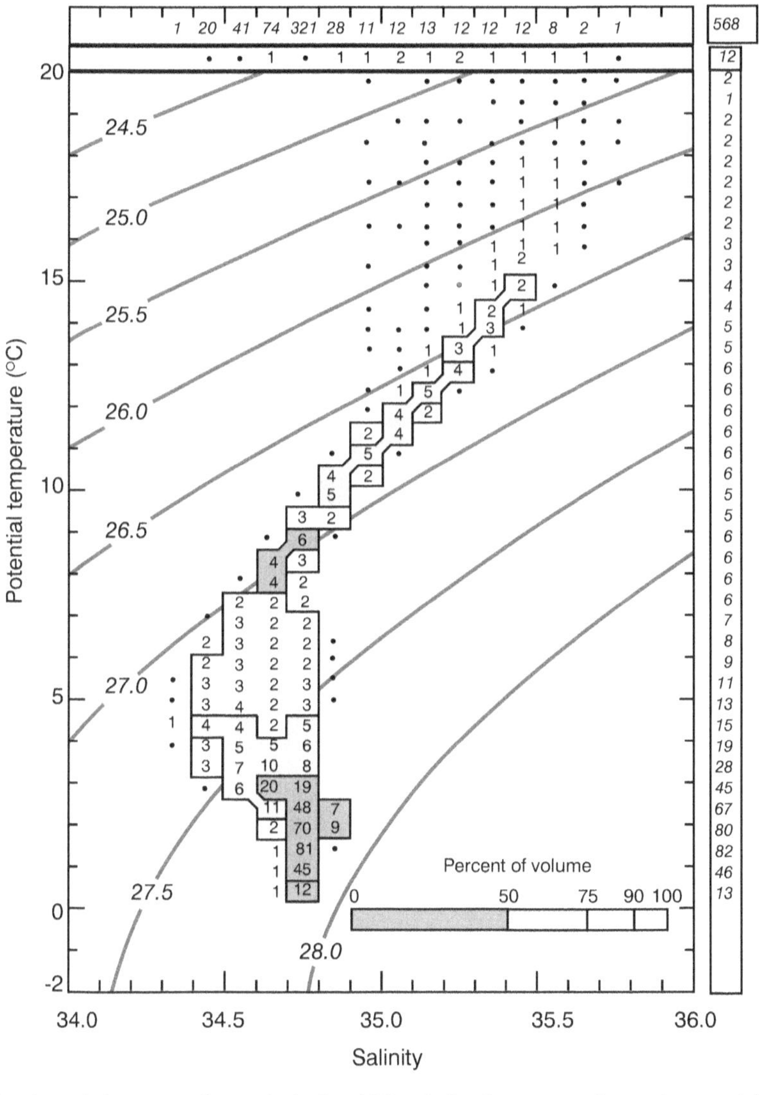

Figure 2.7. A volumetric inventory of water in the South West Indian Ocean, according to the potential temperature–salinity relationships of the water masses[84]. Volumes are in units of 10^5 km^3. The upper row shows the volumes per salinity interval only; the right-hand column per potential temperature only. Isopleths are for sigma-t.

of the total water mass. The various water masses at intermediate depths have a far greater volume, but it is the Deep Waters that form the overwhelming part of the total volume of the South West Indian Ocean. This may be compared to the volumes of similar temperature intervals for the world ocean[220] and the Indian Ocean as a whole[84,221] (Table 2.1).

Compared to the world ocean the Indian Ocean has a higher percentage identifiable bottom water (–2 °C to 2 °C). This may be due to the fact that the South Indian Ocean lies just downstream of the Weddell Sea where the major part of this water is formed. The South West Indian Ocean, on the other hand, consists of a large region in which the water is shallower than 4000 m due to the presence of many mid-ocean ridges. As a result there is a significant percentage reduction of bottom water here. Lying in temperate to warm climatic zones, it is to be expected that the South West Indian Ocean waters would consist of a higher percentage of water between 10 °C and 30 °C than the world ocean, and this is observed (Table 2.1).

For the purposes of this volumetric calculation the

Table 2.1. The percentage of the total water volume taken up by water masses of which the temperatures fall within certain temperature intervals, for the world ocean as a whole, for the Indian Ocean as a whole and for the South West Indian Ocean[220–21].

Temperature	World Ocean	Indian Ocean	SW Indian Ocean
10°–30°	8%	9%	15%
4°–10°	16%	16%	18%
2°–4°	29%	22%	28%
-2°–2°	47%	53%	39%

Indian Ocean included the Southern Ocean between Africa and Tasmania[84]. This explains the low percentage surface water and the high percentage bottom water given in Table 2.1 for the Indian Ocean as a whole. All comparable estimates[221] have unfortunately also been done in this way. The water masses represented in these volumetric analyses can be identified by their potential temperature–salinity relationships and their potential temperature–oxygen relationships shown in Figures 2.8 and 2.9.

Surface and central waters

Near the equator there is an excess of precipitation over evaporation leading to a freshening of surface water warmer than 24 °C, i.e. north of 20° S longitude in winter[81], thus forming Tropical Surface Water. Subtropical Surface Water with a salinity of 35.5 or greater is found further to the south in a band between about 28° and 38° S in the South West Indian Ocean, except near the east coast of southern Africa where Tropical

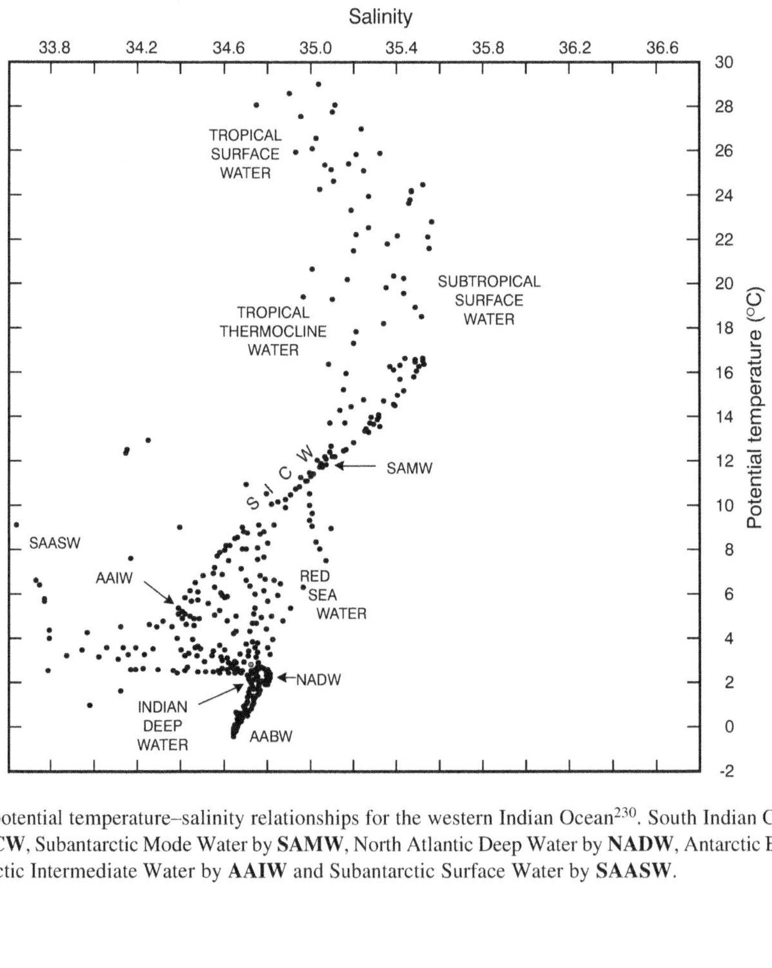

Figure 2.8. The potential temperature–salinity relationships for the western Indian Ocean[230]. South Indian Central Water is identified by **SICW**, Subantarctic Mode Water by **SAMW**, North Atlantic Deep Water by **NADW**, Antarctic Bottom Water by **AABW**, Antarctic Intermediate Water by **AAIW** and Subantarctic Surface Water by **SAASW**.

Surface Water of lower salinity predominates. Both are saturated in dissolved oxygen. Subtropical Surface Water is found as a shallow subsurface salinity maximum in regions to the north and west of its region of formation. Subtropical Underwater is formed by a subduction of saline, winter mixed layer water east of 90° E, but small parcels of it are advected into the South West Indian Ocean as subsurface lenses[731].

Tropical Thermocline Water can be distinguished from Subtropical Surface Water, of the same temperature, by its much lower oxygen content (Figure 2.9). A shallow oxygen minimum is found south of 10° S and is of particular importance to an understanding of the Agulhas Current water movement[231] since it is characteristic of its shoreward side and may be used as a tracer.

Two explanations for the existence of this oxygen minimum have been given. First, it has been suggested[232] that it may be due to the effects of biologically induced oxygen consumption. Secondly, it may be an artifact of a layer of higher oxygen introduced at a slightly greater depth as Subantarctic Mode Water (Figure 2.9). As mentioned above, this water type is formed along the southern edges of the subtropical gyre due to winter cooling and convection[181].

The oxygen, salinity and temperature range (Figure 2.8) of Subantarctic Mode Water falls directly within that of South Indian Central Water. It can only be differentiated from the latter by the characteristic thermostad that it represents. This Central Water forms the mixing product of the surface waters and the Intermediate Waters. Towards its cooler end (Figure 2.8) it is strongly distorted by the presence of Red Sea Water. This is particularly evident in the oxygen distributions since Red Sea Water has a very low oxygen content (Figure 2.9).

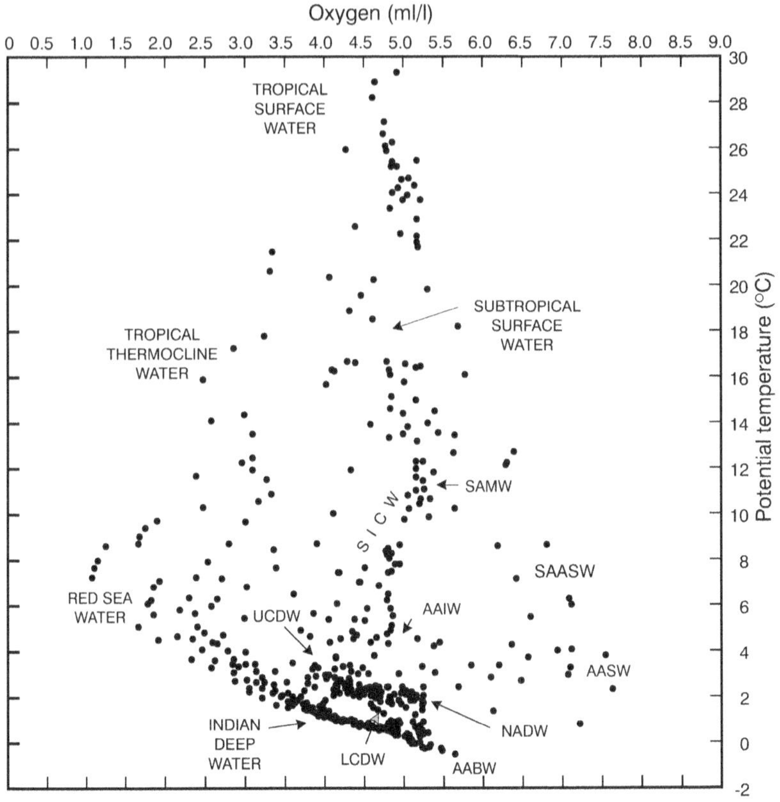

Figure 2.9. The potential temperature–dissolved oxygen relationships for the western Indian Ocean[230]. South Indian Central water is identified by **SICW**, Subantarctic Mode Water by **SAMW**, North Atlantic Deep Water by **NADW**, Antarctic Bottom Water by **AABW**, Antarctic Surface Water by **AASW**, Antarctic Intermediate Water by **AAIW**, Subantarctic Surface Water by **SAASW**, Lower Circumpolar Deep Water by **LCDW** and Upper Circumpolar Deep Water by **UCDW**.

Intermediate waters

This Red Sea Water derives its high salinities and thus densities in the Red Sea and, partially, the Persian Gulf, due to an excess of evaporation over precipitation and moves directly poleward along the east coast of Africa on leaving this region[233]. It forms a salinity maximum at depths of 1500 m to at least 22° S in the Mozambique Channel[234]. A tongue of low oxygen values suggests[84] that it extends even further south. This has been confirmed by more recent measurements in the south-western Indian Ocean[235] and it has subsequently been observed even as far south as the Agulhas Current retroflection[236]. The full geographical extent of its distribution has not yet been determined with any degree of reliability. All present measurements[84] suggest that its influence is felt more specifically on the western side of the South West Indian Ocean. Since it has a different origin, the other water mass found at intermediate depths in this ocean region – Antarctic Intermediate Water – has a very different distribution.

The Antarctic Intermediate Water forms the high oxygen (Figure 2.9) but low salinity member (Figure 2.8) at depths of 1000 m to 2000 m in the south-western Indian Ocean. This water subducts between the Antarctic Polar Front and the Subtropical Convergence. Its thickness is about 500 m but may vary quite substantially from place to place and probably also with time. Both salinity and temperature in this core layer increase slowly as this water mass proceeds northwards. Subantarctic Surface Water may be considered to be its purest end-member (Figures 2.8 and 2.9) with highest oxygen values and lowest salinities and is, of course, found south of the Subtropical Convergence[237]. Below the intermediate waters is the deep water, but it also consists of different components.

Deep waters

Deep Water forming, as shown, by far the major proportion of the water of the South West Indian Ocean by volume, is present as two distinct water types[230], North Atlantic Deep Water and Indian Deep Water. The potential temperature–salinity relationships of these two deep water types differ little (Figure 2.8), but their oxygen characteristics are quite distinct (Figure 2.9). North Atlantic Deep Water enters the South West Indian Ocean around the southern tip of Africa at a depth of between 2500 and 3200 m[178] with a weak salinity maximum[238]. It can be followed northwards in the southern Indian Ocean beyond Madagascar. The salinity and oxygen content decrease northwards, but the already high nutrient content increases. Indian Deep Water, on the other hand, is thought[239] to originate in low and northern latitudes and to flow south. These water masses are not strongly hindered in their movement by the bathymetry. The translation of bottom waters is, by contrast, severely constrained by the bathymetry.

Bottom waters

That part of the South West Indian Ocean that is deep enough accommodates water of Antarctic Bottom Water origin, but in a much diluted form. These water masses are probably not directly involved in the Agulhas Current system per se, but their movement may affect the behaviour of currents higher up in the water column[130] and they are therefore, if not for any other reason, of importance here. It has been estimated[727] that the total northward flow of abyssal waters into the Indian Ocean at a latitude of 32° S is 11.9×10^6 m^3/s. Antarctic Bottom Water is found poleward of the Atlantic–Indian Ridge south of Africa where it has a potential temperature of –0.6 °C, a salinity of less than 34.66[238] (Figure 2.10) and a dissolved oxygen content of more than 5.2 ml/l[84]. Its trajectory partially explains these values.

This water seems to leak northward through a gap in the ridge system at depths greater than 4000 m. This gap lies south-east of Africa. Bottom waters in the basins to the north of here have potential temperatures of 0.1 to 1.1 °C, salinities of 34.70 to 34.80 and oxygen values of 5.0 to 4.8 ml/l[84]. Modern hydrographic data in the southern part of the Agulhas Current system[230,236] show no evidence of bottom water with temperatures below 0 °C. The distinct basins formed by the bathymetry are clearly evident in Figure 2.11. Between the Madagascar Ridge – directly south of Madagascar – and the central Indian ridge at about 70° E, there are three basins: from south to north, the Crozet Basin, the Mascarene Basin and the Somali Basin north of 10° S. The leakage between these different basins has perhaps been evaluated[706,714] best in the Amirante Passage that is the only link at this depth between the Mascarene and the Somali Basins. Estimates from geostrophy for bottom flow in this passage vary from 1 to 4×10^6 m^3/s. These movements through narrow gaps help explain why the Antarctic Bottom Water in the South West Indian Ocean is already in a diluted form. This can clearly be seen from a portrayal of the potential temperature at 4000 m depth in this ocean region (Figure 2.11).

These distributions also give an indication of the time averaged movement of the waters at these depths.

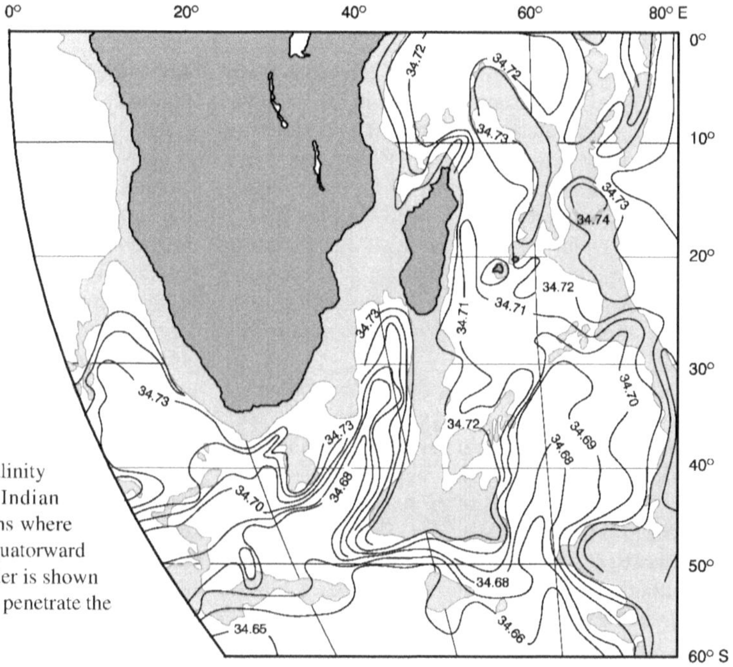

Figure 2.10. The distribution of salinity at the bottom of the South West Indian Ocean[238]. Shading represents regions where the depth is less than 3500 m. The equatorward movement of Antarctic Bottom Water is shown by tongues of low salinity water that penetrate the deep basins in this ocean region.

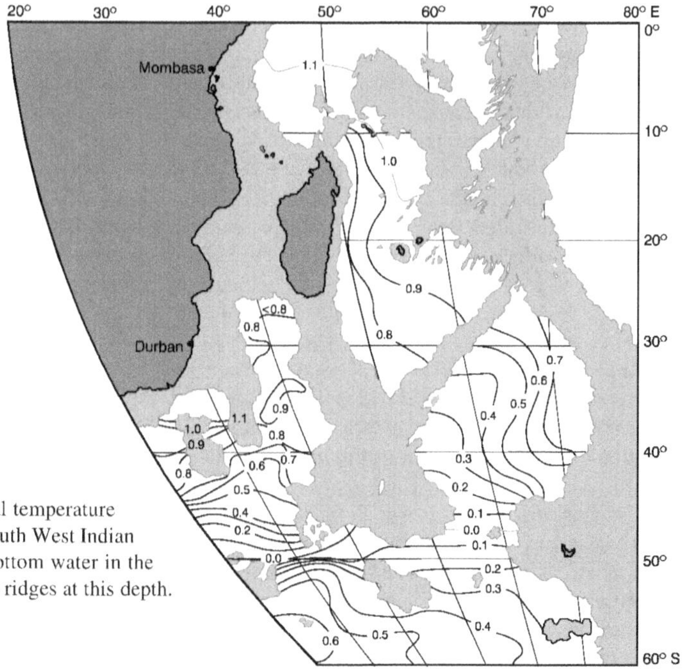

Figure 2.11. Distribution of the potential temperature of water masses at 4000 m depth in the South West Indian Ocean[84]. This shows the movement of bottom water in the relatively small basins found between the ridges at this depth.

Movement of water masses

The Antarctic Bottom Water seems, according to Figures 2.10 and 2.11, to enter the South West Indian Ocean from the south and to be diluted in its northward progression, more so to the west of Madagascar than to the east. This has been confirmed by comparing bottom potential temperatures, turbidity and sea-floor photography[240] showing the major northward flow passing through the gap between the Crozet and Kerguelen Islands. This conclusion is entirely consistent with the interpretation of modern hydrographic data for the region[241].

West of the mid-ocean ridge the coldest bottom water is found south-east of the Agulhas Plateau[215]. There is some indication of a gyral movement, as yet unsubstantiated, of this water in this region. Clearly at these depths the configurations of the basins, as formed by mid-ocean ridges, are a determining factor for water movement[242].

The bottom water in the Mozambique Basin, south of the Mozambique Channel, in fact consists of two distinct types, both originating in the Enderby Basin. Their different hydrographic characteristics are solely a function of the different routes they have taken through the maze of bottom obstructions at that depth[243]. One type reaches the Mozambique Basin via the eastern flank of the Mozambique Plateau as well as by a contribution that leaks northward along the eastern side of the Agulhas Plateau, getting more diluted during this advective process. The other type enters the Mozambique Basin directly from the Enderby Basin through the Prince Edward Fracture Zone. Warren[244] has also demonstrated the leakage of bottom water through fractures in the South West Indian Ridge. Even the smaller-scale characteristics of the ridges therefore play an important role.

Along the ridges, and along the continental shelves, the theory of deep circulation predicts the presence of a western boundary current at depths of 3000 m to 3500 m. Off the east coast of Madagascar such a current has in fact been observed[244] on more than one occasion[245]. The Madagascar Ridge seems to form an impenetrable barrier to bottom flow[244] so that the circulation to the west and east of this ridge are unconnected. The deep boundary current along the east coast of Madagascar seems to derive all its water from that which leaks through the South West Indian Ridge and then moves across the Madagascar Basin[246] towards Madagascar. A result of the existence of these boundary currents is that the bottom water, as inferred from geostrophic calculations[710], has a cyclonic circulation in the Agulhas Basin, in the Crozet Basin and in the Mascarene Basin east of Madagascar.

Deep waters

Above the bottom waters are the very voluminous (viz. Figure 2.7) deep waters. As has been seen, the major water type of this kind in the South West Indian Ocean is the North Atlantic Deep Water, formed in the North Atlantic and moving predominantly southwards along the east coast of South America[195,247]. From there it progresses eastwards and rounds the tip of Africa at a depth of about 2800 m (Figure 2.12). A box model using a perturbed inverse method and that has included organic-decomposition[248] suggests an inflow of $11(\pm 8) \times 10^6$ m^3/s for North Atlantic Deep Water. An investigation[707] using modern hydrographic data gives a value of $11(\pm 4) \times 10^6$ m^3/s. Most of this water can be traced back to passages south of 28° S in the Walvis Ridge from where it moves across the Cape Basin. A substantial part of this water moves in a boundary current against the western continental slope of Africa (Figure 2.13), but the intense deep flow of the Agulhas Current takes its toll, detaches the flow from the boundary and carries most of this water with it in the Agulhas Return Current. A remnant carries on as a counter current below the Agulhas Current itself (Figure 2.13). This portrayal of water movement at depths of 3000 m to 3500 m has been found by a number of other investigators[195,711] as well and is therefore probably a fairly accurate represention. The salinity maximum and dissolved oxygen of the North Atlantic Deep Water are slowly eroded and become increasingly difficult to distinguish the farther eastwards one cares to look in the South Indian Ocean[84], as well as northwards. One may observe it north of Madagascar[237].

Using a zonal line of closely spaced hydrographic stations at 32° S, Toole and Warren[249] have shown that North Atlantic Deep Water is almost entirely confined to the Mozambique and Natal Valley, i.e. west of the Madagascar Ridge and south of the Mozambique Channel[247]. The water that flows northward to the east of here is considered to be Circumpolar Deep Water, not directly derived from the North Atlantic. Others[220] have even considered all the deep water of the Indian Ocean to consist of Circumpolar Deep Water. About 2×10^6 m^3/s from the upper half of the North Atlantic Deep Water appears[716] to flow across the sill in the Mozambique Channel (viz. Figure 2.12) into the Somali Basin to the north.

A second type of deep water, Indian Deep Water, has a much lower oxygen content (Figure 2.9). It does not seem to penetrate far south in the Mozambique Channel[247], but it is found much further south, at depths of 2500 m[232] along the east coast of Madagascar and even along the eastern side of the Madagascar

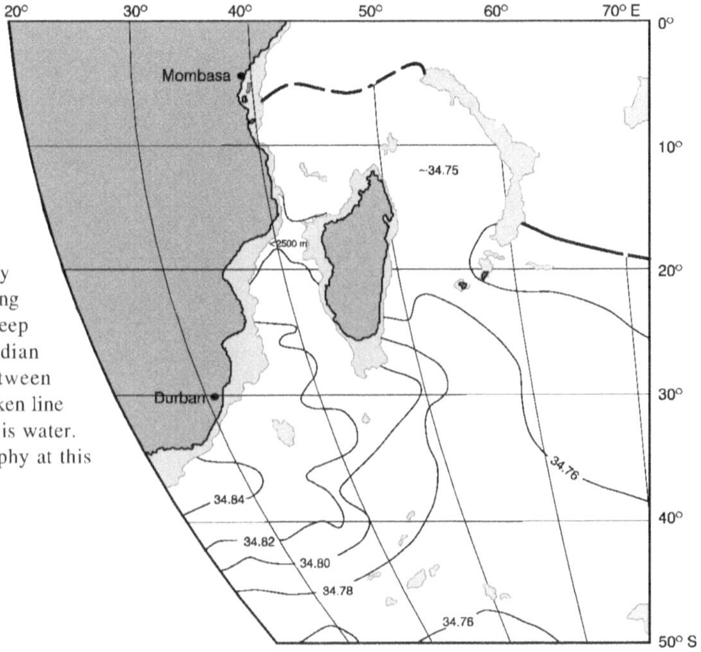

Figure 2.12. Distribution of salinity in the deep salinity maximum showing the movement of North Atlantic Deep Water into the South West Indian Ocean[84]. This is at a depth of between 2750 m and 3000 m. The thick, broken line denotes the equatorward limit of this water. The outline of the bottom topography at this depth is shown.

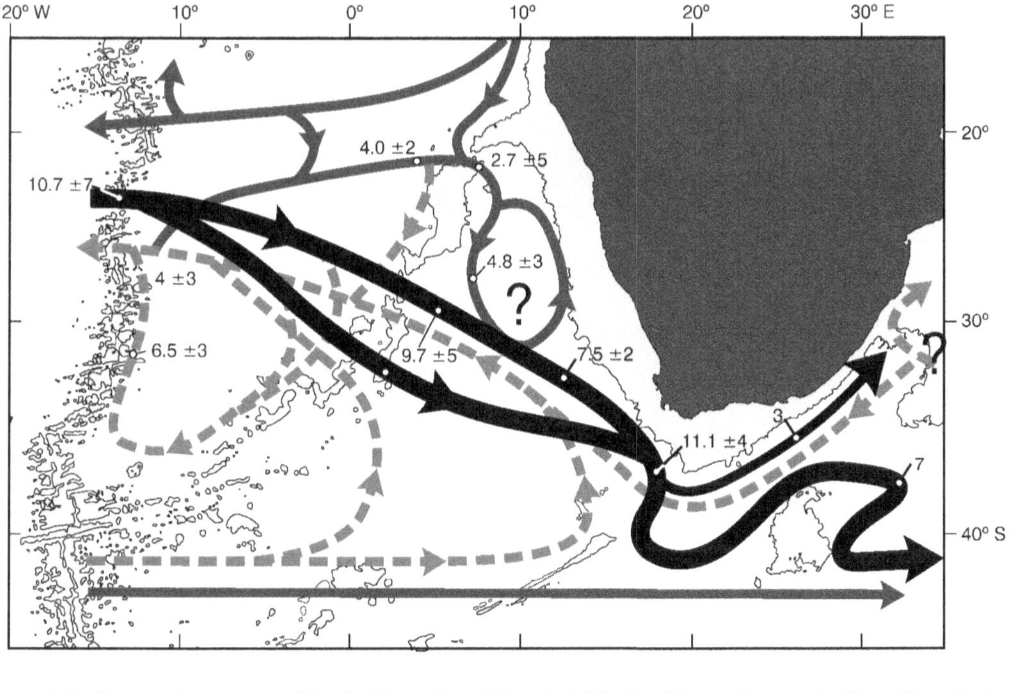

Figure 2.13. Suggested movement of North Atlantic Deep Water (solid line) and Upper Circumpolar Deep Water (broken line) in the south-eastern Atlantic Ocean, based on hydrographic measurements[707]. The 3000 m isobath encompasses the flow. Estimates of the volume flow in 10^6 m^3/s is shown. Question marks denote regions in which the flow is uncertain.

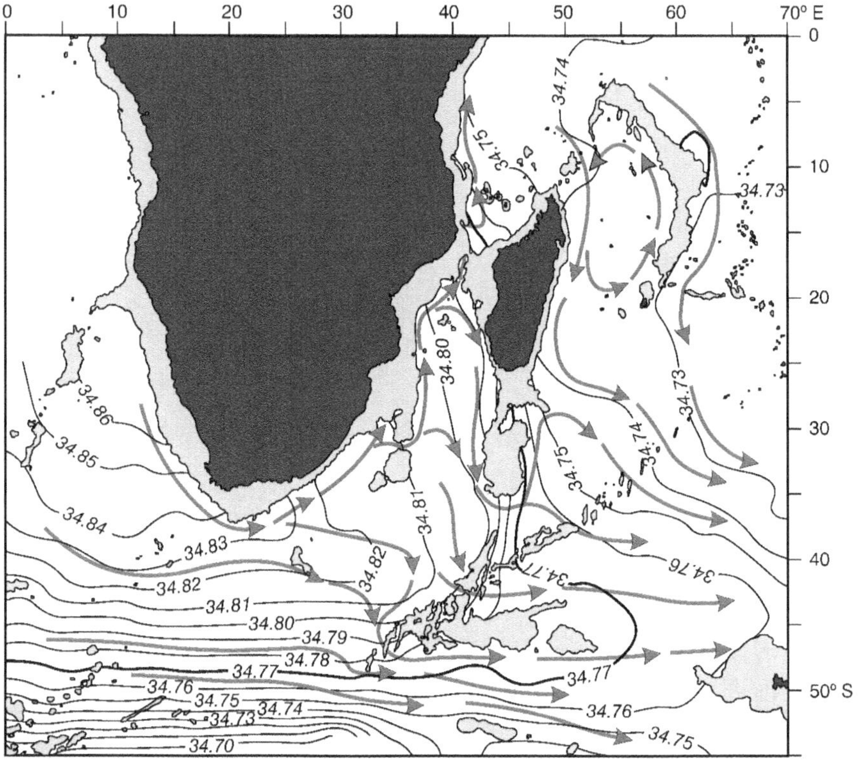

Figure 2.14. A suggested[716] circulation scheme for deep water in the South West Indian Ocean. The 2500 m isobath gives the continental margins at the average depth of this water mass. Superimposed on the arrows is the distribution of the salinity on an appropriate isopycnal surface, the thick isohaline at 34.77 being the approximate boundary between the higher salinity North Atlantic Deep Water and the lower salinity North Indian Deep Water as well as the Lower Circumpolar Deep Water.

Ridge[84]. It leaks southward through the Amirante Passage, the only gap between the Somali and the Mascarene basins, at a rate[714] of about 4–9 × 10^6 m^3/s.

It has been shown[239] that at 18° S, east of Madagascar, Circumpolar Deep Water forms a series of independent gyres in the basins formed by the ridge systems of that part of the Indian Ocean. In the Madagascar and the Mascarene basins the flow of deep water is cyclonic[710]. Deep waters formed at high latitudes tend to move equatorward along the western edge, whereas North Indian Deep Water flows poleward in the basins' interior[708]. This movement is closely analogous to that of the bottom water in these basins[710]. A comprehensive portrayal of the circulation scheme for deep waters of the South West Indian Ocean inferred from both hydrographic data as well as current meter records is shown in Figure 2.14. From this careful analysis by Van Aken et al.[716] it seems clear that it is indeed mostly North Atlantic Deep Water and North Indian Deep Water that are dominant in the South West Indian Ocean. The bottom topography clearly affects the movement of these waters, but does not prevent it from entering any of the basins in this ocean. The general flow pattern portrayed in Figure 2.14 is contradicted by an analysis[722] of hydrographic data on neutral surfaces. According to this study the contribution of North Atlantic Deep Water is largely limited to the South West Indian Ocean. The northward flow of Circumpolar Deep Water from the Southern Ocean is found mainly in the western Crozet and Madagascar basins between 2000 m and 3000 m depth. North Indian Deep Water plays a role in the northern Indian Ocean[721] but has a volume flux that limits its influence. Compared to this relatively simple portrayal of the distribution and movement of the deep waters of the South West Indian Ocean, that of the intermediate waters is considerably more complex (viz. Figure 2.8).

Intermediate waters

To recapitulate, in the South West Indian Ocean there are two major water types in the intermediate depth range that are of importance, namely the Antarctic Intermediate Water and the Red Sea Water. Indonesian Intermediate Water is also present in the Indian Ocean[720], but plays a minor role in the South West Indian Ocean. This is visually apparent in Figure 2.15 where the percentage of each water type on a vertical section along 60° E is shown. Temperature/salinity scatter diagrams show (Figure 2.16) how the characteristic T/S relationships of both the Antarctic Intermediate Water as well as the Red Sea Water are eroded as

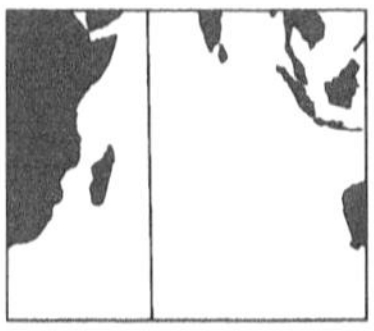

Figure 2.15. The percentage water mass contributions along the 60° E meridian in the western Indian Ocean[720]. The upper panel shows Antarctic Intermediate Water; the central panel Indonesian Intermediate Water and the lower panel the contribution by Red Sea Intermediate Water. Note that the depths shown extend from 500 to 1500 decibar. Black areas denote land masses.

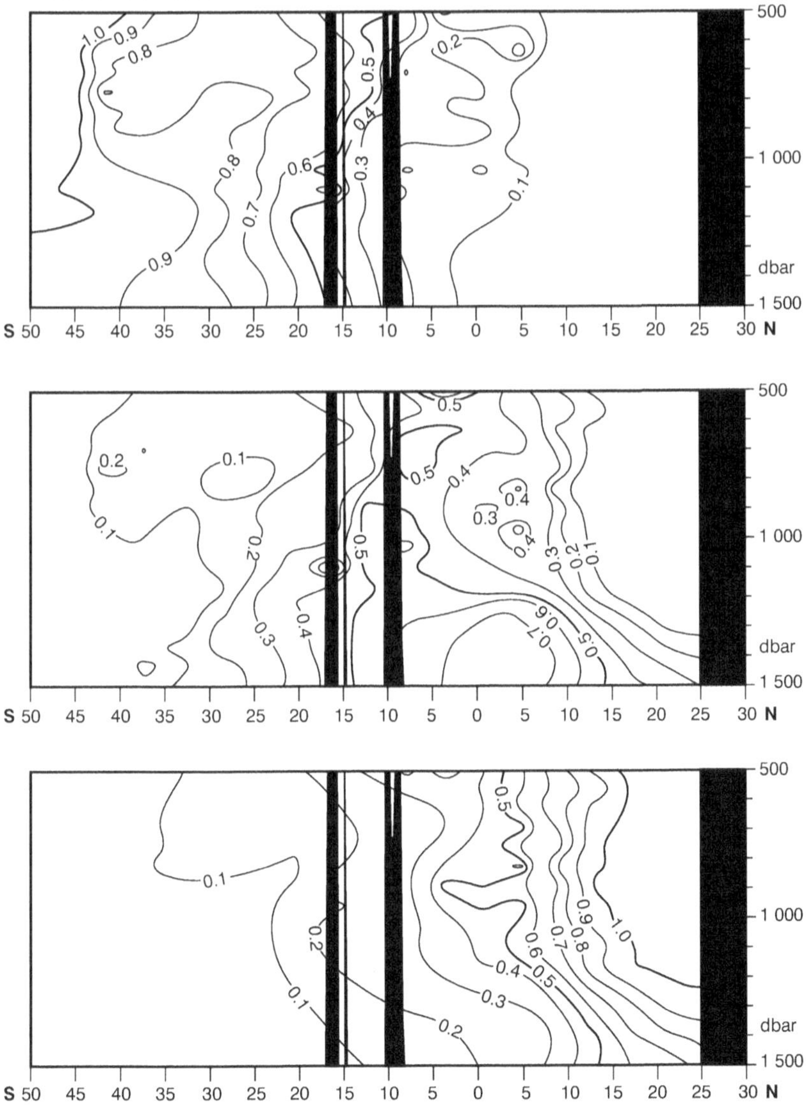

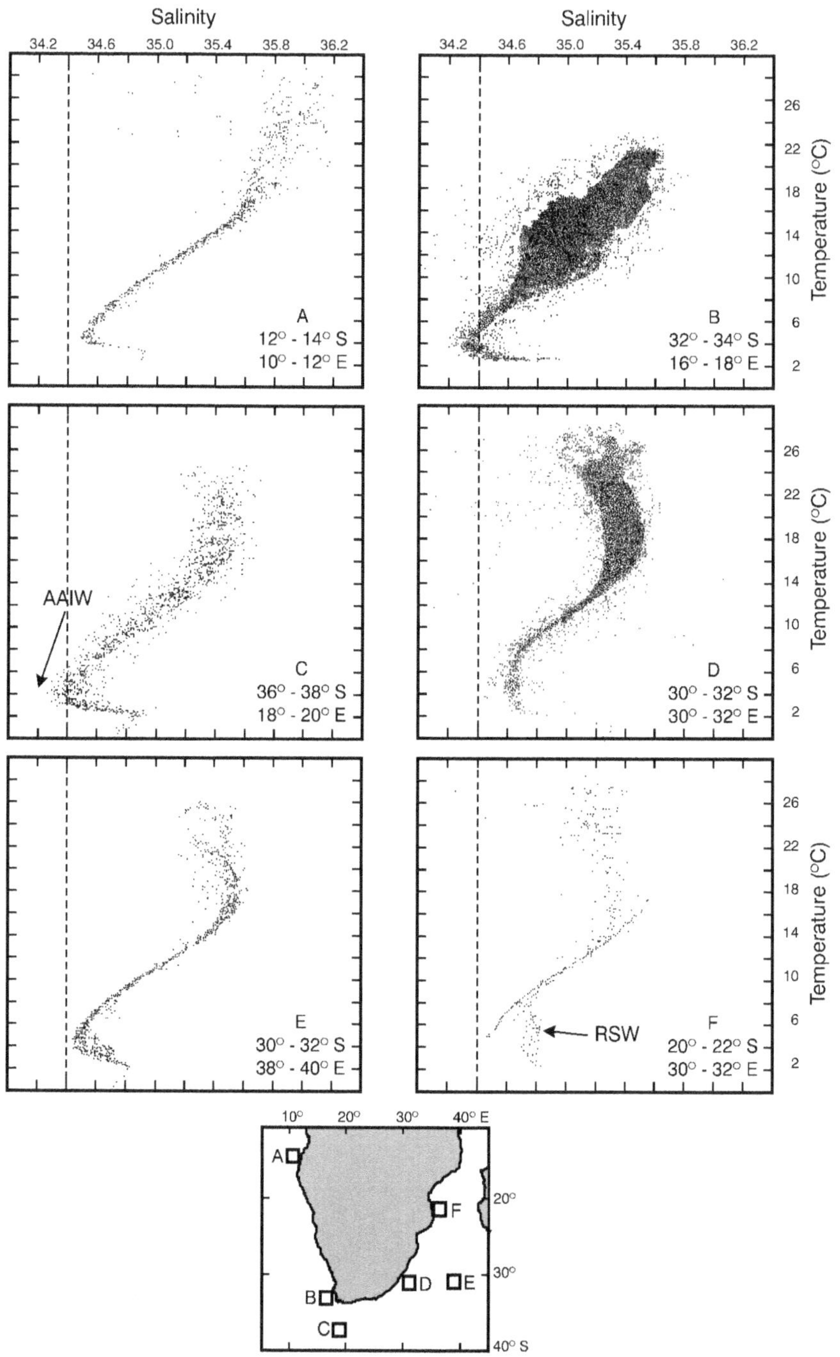

Figure 2.16. Scatter diagrams representing the temperature–salinity characteristics of five regions in the South West Indian Ocean, shown in the map, and one in the tropical South Atlantic Ocean as reference[250]. The greater degree of purity of Antarctic Intermediate Water (**AAIW**) south of Africa (panel **C**) and of Red Sea Water (**RSW**) in the Mozambique Channel (panel **F**) is clearly distinguishable.

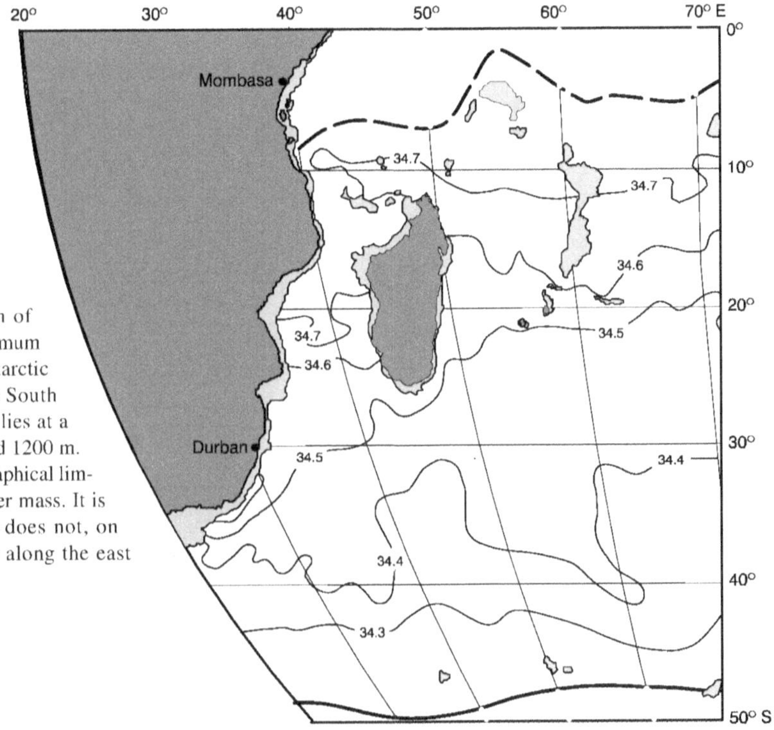

Figure 2.17. Distribution of salinity in the salinity minimum core layer representing Antarctic Intermediate Water in the South West Indian Ocean[84]. This lies at a depth of between 800 m and 1200 m. Heavy lines show the geographical limits to the extent of this water mass. It is clear that this water type does not, on average, penetrate readily along the east coast of southern Africa.

observations are taken increasingly farther away from their origins. Furthermore, the degree of erosion of the salinity minimum that represents Antarctic Intermediate Water is greater along the east coast that the west coast of southern Africa. To this end the T/S relationships at similar latitudes may be compared (Figure 2.16). This would suggest that the Antarctic Intermediate Water does not penetrate the South West Indian Ocean as easily as it does the South East Atlantic Ocean. The inflow of Antarctic Intermediate Water in fact creates scattered lenses of high potential vorticity in the South Indian Ocean[726].

As has been mentioned, recent measurements have shown[235,250] that there may be considerable interleaving, both vertically and laterally, between the two water masses found at intermediate depths. To investigate the motion of these water masses on the larger scale, mean distributions on core layers (Figures 2.17 and 2.18) may represent the major modes of distribution. For Red Sea Water this distribution may have a seasonal component since the initial southward spreading of this water type along the African coast is intensified during the winter monsoon[709].

Antarctic Intermediate Water in general lies at a depth of between 800 m and 1200 m in the South West Indian Ocean[84], the greatest depths found in the centre of the South West Indian Ocean sub-gyre (viz. Figures 3.15 and 3.28). This shows that this water is already part of the wind-driven subtropical gyre and this inclusion has also been demonstrated by geostrophic calculations using a selection of hydrographic data[710]. Its penetration into the South Indian Ocean, and the concomitant erosion of its characteristic salinity minimum is zonally similar, except near the coast of Africa where a tongue of higher salinity follows the path of the Agulhas Current. The immediate temptation is to think of this as remnants of Red Sea Water, but a careful analysis of the movement of Antarctic Intermediate Water around southern Africa[164] suggests that this tongue may possibly represent older, more saline water that has circulated in the South West Indian Ocean. In the Mozambique Channel the juxtaposition of Antarctic Intermediate Water and Red Sea Water is dramatic (Figure 2.19). Here Antarctic Intermediate Water is carried equatorward by the Mozambique Undercurrent[728] while Red Sea Water is carried poleward in Mozambique eddies. The depth of Red Sea Water increases[730] from 900 m to 1200 m along the channel. It has been demonstrated[653] that these eddies may eventually be absorbed by the Agulhas Current and that this may thus be one of the possible sources of Red Sea Water in the Agulhas Current.

Hence it should come as no surprise that in hydro-

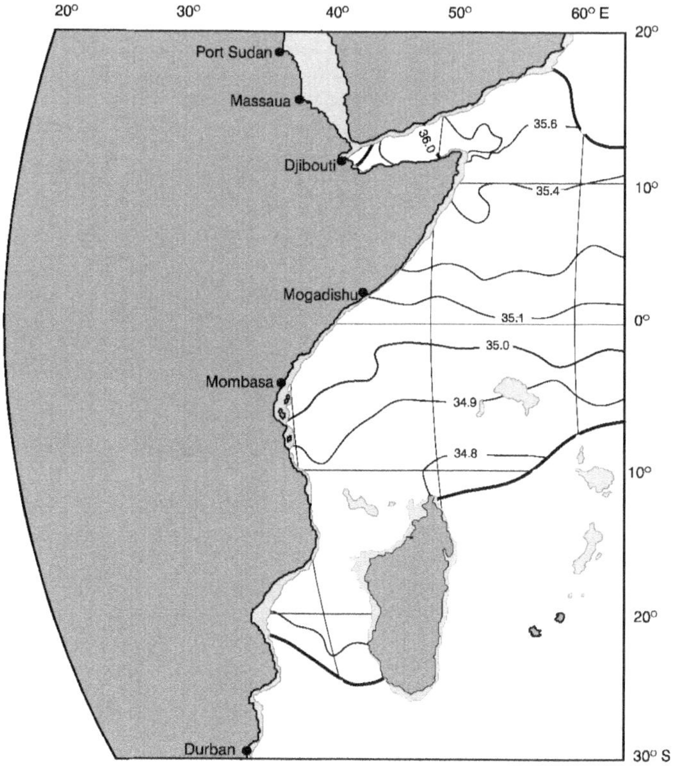

Figure 2.18. The distribution of salinity in the salinity maximum core layer formed at intermediate depths in the South West Indian Ocean by Red Sea Water[84]. This core layer is found at depths of 1000 to 1100 m in this ocean region. The heavy line shows the southern limit to the influence of this particular water mass, much farther poleward in the Mozambique Channel than elsewhere in this ocean region.

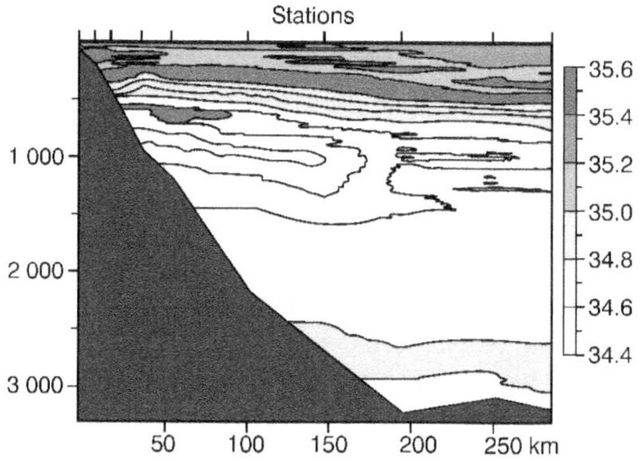

Figure 2.19. The salinity field in a zonal section in the Mozambique Channel at 24 °S[728]. Low salinity (<34.6) Antarctic Intermediate Water at 1000 m depth on the western side of the channel is juxtapositioned to Red Sea Water in the centre of the channel. The Antarctic Intermediate Water is transported equatorward by the Mozambique Undercurrent while an eddy carries the Red Sea Water poleward.

graphic data taken across the poleward flow along the coast of southern Africa (Figure 2.20) there are a sufficient number of higher salinity outliers at the intermediate salinity minimum between 1000 m to 2000 m firmly suggesting the presence of some Red Sea Water[215]. Using all available hydrographic data for the Indian Ocean, Beal et al.[709] have shown that Red Sea Water exhibits a preference for the poleward route along the east coast of Africa (viz. Figure 2.18). Historical hydrographic data[234] gives disparate indications, the salinity showing no penetration of Red Sea Water through the Mozambique Channel beyond 20° S, but the oxygen minimum extending much farther[84,709]. The distribution of salinity on the intermediate salinity maximum that represents the core layer of Red Sea Water (Figure 2.18) extends the indicated influence of this water to the southern mouth of the Mozambique Channel, but no farther south. This also holds true for the oxygen minimum[84].

So what role does the Agulhas Current really play, if at all, in the advection of Red Sea Water? Closely spaced hydrographic sections across the South West

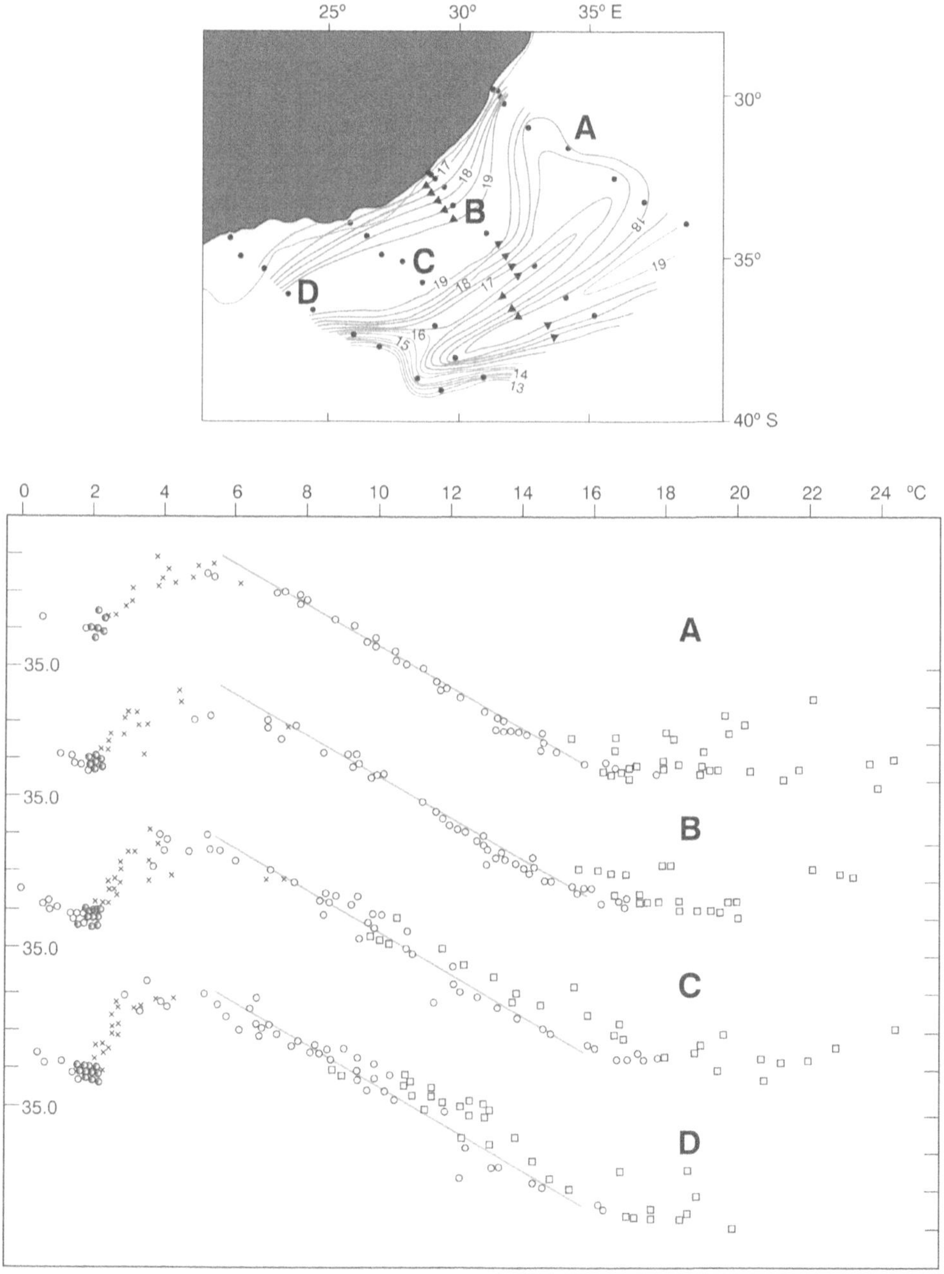

Figure 2.20. Typical temperature–salinity relationships based on four lines of hydrographic stations carried out in the South West Indian Ocean[215] across the Agulhas Current and shown in the upper panel (see also Figure 5.2). Flow directions follow the surface isotherms in °C. Depth ranges are shown by **squares** (50 m to 200 m), **circles** (temperature >2 °C; 200 m to 1000 m), **crosses** (1000 m to 2000 m), **dots** (2000 m to 3000 m) and **circles** (temperatures <2 °C; deeper than 3000 m).

Indian Ocean gyre[249] show that the dissolved oxygen maximum at intermediate depths, as well as the salinity minimum of Antarctic Intermediate Water, substantially disappear close to the south-eastern coast of Africa. Toole and Warren[249] have suggested that this may be owing to the intermittent influx of Red Sea Water.

As mentioned above, Red Sea Water is in fact found in Mozambique eddies[728] in the Mozambique Channel, carrying this water poleward in boluses. The low oxygen concentrations that typify this water decrease poleward[730] throughout the channel. Mozambique eddies have been observed to eventually contribute water to the Agulhas Current itself and thus it has been shown indisputably that the Agulhas Current has carries Red Sea Water southward[709]. A detailed section across the northern Agulhas Current has shown[368] that about 6×10^6 m^3/s of Red Sea Water was transported by the Agulhas Current on this occasion. However, the mechanism of eddy contributions is not necessarily the only one and Red Sea Water is also found[235] in other forms than eddies in the South West Indian Ocean. However, if the main source of Red Sea Water in the Agulhas Current is Mozambique eddies, its presence may be variable, intermittent and a large part of the intermediate water off the coast of south-eastern Africa may turn out to be recycled and substantially mixed Antarctic Intermediate Water.

Investigations of the circulation of Antarctic Intermediate Water in the South Indian Ocean utilising chlorofluorocarbon[252] supports the concept of the Agulhas Current impeding the northward advection of fresh water of this type along the western boundary of this basin. Major inputs are thought to occur farther to the east. From an extensive hydrographic study[709] it has furthermore been concluded that all the Red Sea Water produced in the Indian Ocean may eventually be exported south of Africa in Agulhas rings. This implies that Red Sea Water is the dominant component of the salt budget for intermediate waters. A much more extensive and detailed knowledge of water at intermediate depths around southern Africa is required in order to estimate the inter-basin exchanges of these water masses and their contribution to the global thermohaline circulation. To this end a major, multi-institutional and multinational experiment was launched during the late 1990s[253–4].

Central water

As seen in Figure 2.7, the Central Water – lying above the intermediate waters – has a low volume contribution to the total water masses of the South West Indian Ocean, and the South Indian Central Water forms a linear connection on a potential temperature–salinity diagram of data representing the waters of the South Indian Ocean (Figure 2.8). Although there is little scatter in such diagrams for the central waters, there are, nonetheless, sufficient significant differences in salinity (Figure 2.20) for fixed temperatures to suggest different origins[236] (viz. also Figure 6.12).

Emery and Meincke[220] place South Indian Central Water in the salinity range 34.6 to 35.8 and in the temperature range 8.0 °C to 25.0 °C. South Atlantic Central Water has 34.3 to 35.8 and 5.0 to 18.0 °C, therefore very similar. South Indian Ocean Central Water extends only to 15° S[255] latitude where it meets Indian Equatorial Water. The change in temperature–salinity characteristics across the front formed in this way is quite abrupt[237].

In the South West Indian Ocean the South Indian Central Water is found at depths between 200 m and 1000 m (Figure 2.20). At these depths the sub-gyre in this ocean region is well developed (viz. Figure 2.2) and the water circulates west of about 80° E. Plots of water characteristics at these depths show no influence of central water from the South Atlantic[84]. There have been suggestions[67] that such water may move with the flow along the Subtropical Convergence in a South Atlantic–South Indian Ocean Current continuum and join the circulation of the South West Indian Ocean in this way. Since South Indian Central Water has its origin at the subduction zone of the Subtropical Convergence[256] a continuous flow from the South Atlantic Ocean to the South Indian Ocean would successfully introduce South Atlantic Central Water into the South West Indian Ocean.

A careful analysis of central water types in the Indian Ocean[256] shows that in the South West Indian Ocean the South Indian Central Water dominates, with a minimum percentage volume contribution of 75 per cent east of Madagascar. Red Sea Water's contribution never exceeds 5 per cent, whereas North Indian Central Water may form 20 per cent of the central water in the Mozambique Channel. You has shown[257] that turbulent mixing at thermocline depths leads mainly to downward dianeutral advection in the South Indian Ocean, in contrast to the North Indian Ocean where it leads to mainly upward advection of this type. His analyses are consistent with the concept of Central Water formation at the surface in late winter and its subsequent movement equatorward along isopycnal surfaces. Hence, Indian Central Water is subducted in latitudes 40° to 45° S, advected in the subtropical gyre and then finds its way into the northern Indian Ocean through western boundary currents along the σ_θ 26.7 isopycnal surface.

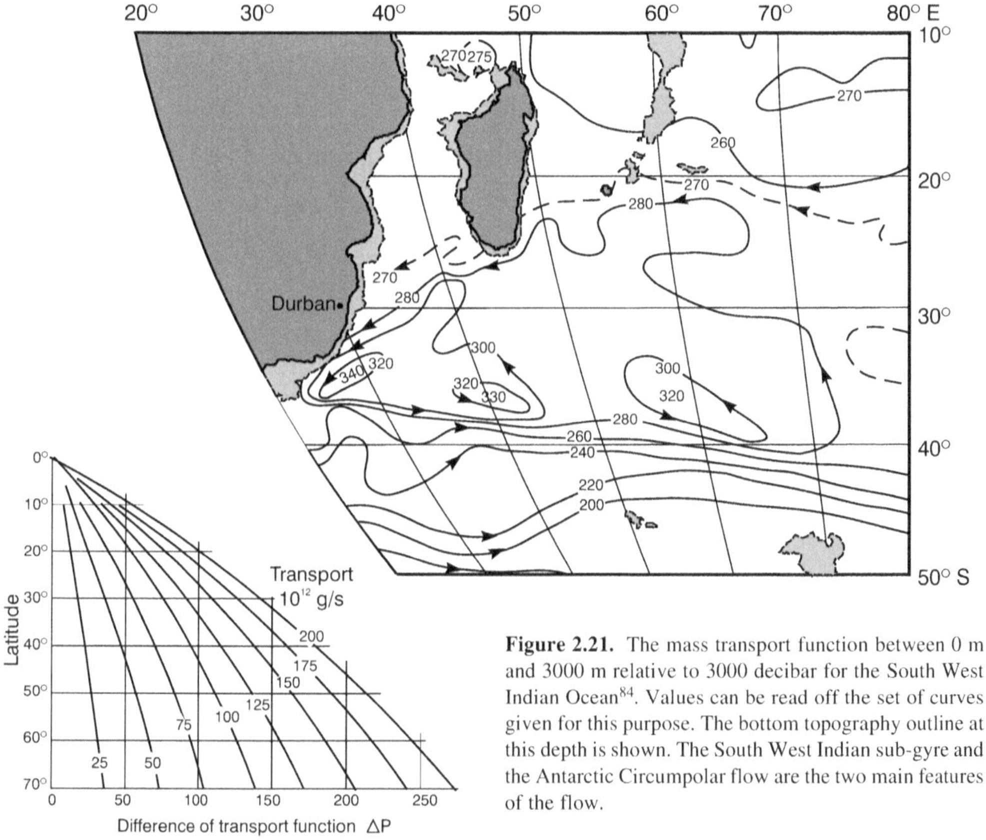

Figure 2.21. The mass transport function between 0 m and 3000 m relative to 3000 decibar for the South West Indian Ocean[84]. Values can be read off the set of curves given for this purpose. The bottom topography outline at this depth is shown. The South West Indian sub-gyre and the Antarctic Circumpolar flow are the two main features of the flow.

Surface waters

Above the central waters lie the surface waters. Both tropical and subtropical types are found in the South West Indian Ocean (Figures 2.8 and 2.9) and the characteristics of both have already been described in some detail. Their motion is largely that of the surface waters of the subtropical sub-gyre (Figure 2.1). A number of different water types can be identified in the main thermocline[719]. In the western Indian Ocean they are Indian Central Water and North Indian Central Water; the latter being an aged version of the former. Indian Central Water has a temperature range of 7°–10° C and a salinity range of 34.7–35.5[729]; North Indian Central Water 7.8°–15.7° C and 34.84–35.10[256]. In Figure 2.15 the percentage distribution of these two water types is given for the summer monsoon along a vertical transect at 60° E. To a large extent it is these surface layers which carry the greater part of the volume fluxes of this ocean region, and this merits some attention.

Transports in the South West Indian Ocean

The mass transport function for the waters in the upper 3000 m of the South West Indian Ocean is shown in Figure 2.21. A volume greater than 65×10^6 m^3/s flows along the Subtropical Convergence. Of this more than half belongs to the circulation of the South West Indian Ocean sub-gyre and recirculates with it, the full amount eventually reaching the Agulhas Current. A more detailed analysis, but only for the baroclinic flow field of the upper 1000 m is given in Figure 2.4. This shows how rapidly the flow along the Subtropical Convergence diminishes with distance eastwards, dropping from 60×10^6 m^3/s at 30° E to 20×10^6 m^3/s at 80° E. Most of this water is drawn off into a South West Indian Ocean sub-gyre and recirculates south of Madagascar[101].

The flow reaching the Agulhas Current directly from the South Equatorial Current is only about 20×10^6 m^3/s, while a trickle estimated at 5×10^6 m^3/s comes from the

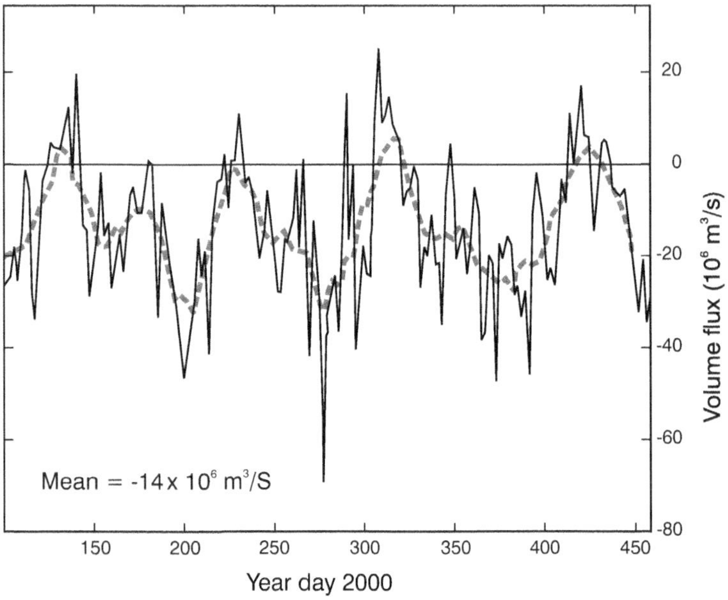

Figure 2.22. The volume transport in 10^6 m^3/s across the narrows of the Mozambique Channel for a full year[733], as estimated from measurements made by a dense array of current meter moorings. Negative values indicate poleward transport. The broken line represents a 20-day running mean. Note the semi-regular cycle in the transport.

Mozambique Channel. There are some conflicting results on the flow through the Mozambique Channel. By employing a hydrographic inverse geostrophic box model, it has been shown[725] that the flow through this channel should be about 14(±6) × 10^6 m^3/s. Others[708] have calculated 18 × 10^6 m^3/s. Using two sections across the Mozambique Channel[730], based on modern hydrographic data, the transport above 2500 m – the sill depth – has been calculated as 29 × 10^6 m^3/s and 6 × 10^6 m^3/s. Needless to say, some enormous variability in the flow has to be invoked to explain these large discrepancies[259]. This will be dealt with in some detail in the next chapter. For now, it is important to note that all the foregoing estimates have been based on geostrophic calculations or on inverse box models. An array of current meter moorings has now given the most reliable and authoritative result[733] to date. It has been shown that the temporal variability in the volume flux through the narrows of the Mozambique Channel is immense, changing rapidly and remarkably regularly from 20 × 10^6 m^3/s equatorward to even 60 × 10^6 m^3/s poleward (Figure 2.22). The mean volume transport over a period of a year was 14 × 10^6 m^3/s. There was no seasonal variability to be found in these observations.

The differences between the flow into the southern Agulhas Current (65 × 10^6 m^3/s) and the outflow in the Agulhas Return Current (60 × 10^6 m^3/s; Figure 2.21) represents the inter-basin leakage south of Africa in this ocean layer. These portrayals integrate the baroclinic flow over the upper 3000 m and 1000 m respectively and do, therefore, not distinguish between the different water masses being carried in the flow, nor the depth distribution of the advection. A more discriminatory portrayal of the flow is given in Figure 2.23.

In this diagram the shaded regions denote flow into the greater Agulhas Current system[75]. Estimates of the geostrophic motion are based on sets of quasi-synoptic hydrographic stations selected to straddle the currents of interest. The South Equatorial Current is shown to be most strongly developed in the upper 300 m. The East African Current (also called the Zanzibar Current) does not extend deeper than 700 m which agrees roughly with findings of a total depth of 500 m for this current[258]. The northern part of the presumed Mozambique Current has a substantially greater flow at depth, as has the East Madagascar Current (Figure 2.23). The Agulhas Current is by far the best developed current of the South West Indian Ocean, extending to the reference level selected. By contrast the westerly flow south of Madagascar is weak, as is that of the Agulhas Return Current. Both estimates, but particularly the one for the Agulhas Return Current, may be weaker than the true values due to the wide station spacing. The Agulhas

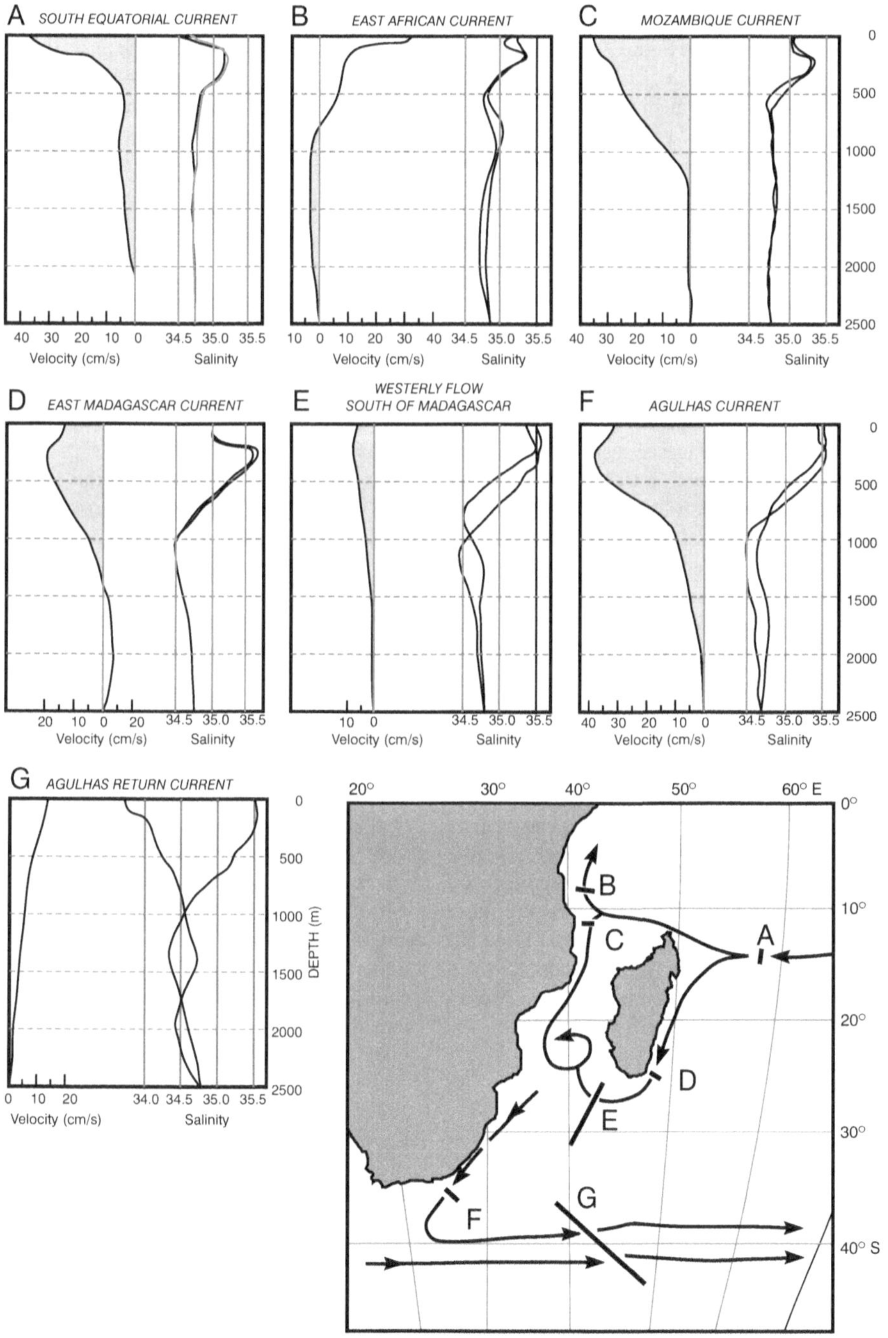

Figure 2.23. The geostrophic velocity, calculated relative to a depth of 2500 m, as well as the vertical profiles of salinity over the same depth interval, for a number of hydrographic stations. These stations were selected for their probable description of key components of the circulatory regime of the South West Indian Ocean[75]. Shaded portions indicate flow into the Agulhas Current region.

Return Current is also fed by the South Atlantic Current that borders the Subtropical Convergence in the South Atlantic Ocean (not shown in Figure 2.23). According to investigations by Mercier et al.[759] this current extends only to the depth of the Intermediate Waters and has a transport across 9° E of 10(±5) × 10^6 m^3/s. Of particular interest are the salinities being carried by these currents and the water masses these characterise (Figure 2.23).

The shallow salinity minimum found in the South Equatorial Current shows the presence of Tropical Surface Water overlying the saltier Subtropical Surface Water. In the East African Current and the flow in the Mozambique Channel the tropical water already has a higher salinity. There are remnants of the Tropical Surface Water in the East Madagascar Current, but in the westerly flow south of Madagascar and in the Agulhas Current these have disappeared, at least on the scale and the resolution used here. In following the flow southwards, the intermediate salinity minimum of Antarctic Intermediate Water is evident in the East Madagascar Current for the first time and then in the Agulhas Current and the Agulhas Return Current.

Details of attempts to quantify the flow in each of these components of the circulation regime of the South West Indian Ocean will be given when discussing these currents individually. It is important to note here that some of these currents, particularly the less well-developed ones, may undergo major changes in their volume transports over time[259]. This means that the portrayals using hydrographic data may be valuable to give indications of the flow characteristics, but should not be used uncritically as fixed values. Not only are there interannual changes, not yet quantified, but there are also seasonal changes that help form the thermal structure of the water column in the South West Indian Ocean.

Thermal structure of the South West Indian Ocean

The thermocline in the South Indian Ocean is ventilated by the central waters[256] discussed above. Above the thermocline the thermal structure of the surface waters varies quite considerably from region to region. Figure 2.24 portrays the characteristic vertical temperature structure over the top 500 m, how it changes seasonally, the depth of the mixed layer as well as the maximum temperature gradient for each region for each month[251].

The waters for most regions are warmest around March except for the northern Mozambique Channel and the Subtropical Convergence where the warmest period is in May. There is evidence[260] that the sea surface temperatures in the southerly regions, between 20° and 40° S, peak in February whereas closer to the equator this occurs in March. The minimum sea surface temperature, on average, is to be found in August. Weeks et al. have demonstrated[261] that seasonal changes in sea surface temperature create stronger horizontal gradients across the Subtropical Convergence in winter than in summer. An average warming of the sea surface of about 0.45 °C from 1910 to 1990 has been observed[260].

The mixed layer depth increases with the onset of winter winds and convection and is usually a maximum in July or August (Figure 2.24). At the Subtropical Convergence the reduced stratification and very severe winds (viz. Figure 2.3) leads to mixed layer depths of up to 300 m[262] thickness. A permanent very thick thermocline, with hardly any seasonal variation is only observed east of Madagascar in the tropics (Figure 2.24). A permanent but shallow thermocline is found in all other parts of the South West Indian Ocean except at the Subtropical Convergence. A summer thermocline is formed throughout the region. An even more detailed analysis of the seasonal temperature variation has been carried out for the region directly seaward of the Agulhas Current.

Gründlingh[263] has shown (Figure 2.25) that at 25° S, assumed to be consistently far away from major advectional influences, the seasonal variation of the sea surface temperature is 5.4 °C, decreasing to 4.7 °C at 35° S. The seasonal thermocline is confined to 100 m to 150 m. Layers deeper than 150 m show variations that are consistent to 400 m depth, but not seasonal. A well-defined mixed layer appears after April when there is a small mean increase in wind speed. A small temperature minimum in the middle of summer may extend to 300 m depth at 28° S, but remains unexplained[263].

The changes in mixed layer depth have some significant biological consequence, particularly in the tropical zone to the east of Madagascar. Satellite observations of ocean colour show a recurring seasonal bloom that starts at the coast of Madagascar and then spreads in an easterly direction during the period from February to May[674]. It has been surmised that this is the result of the annual change in the mixed layer depth that increases from about 30 m to about 70 m depth[676], allowing the entrainment of nutrients into the euphotic zone. Others[675] have modelled this as a plankton wave where diffusion by eddies would play a greater role. Sufficient numbers of *in situ* observations are lacking to give a definitive explanation of this tantalising problem.

As has been seen, a number of other aspects of the

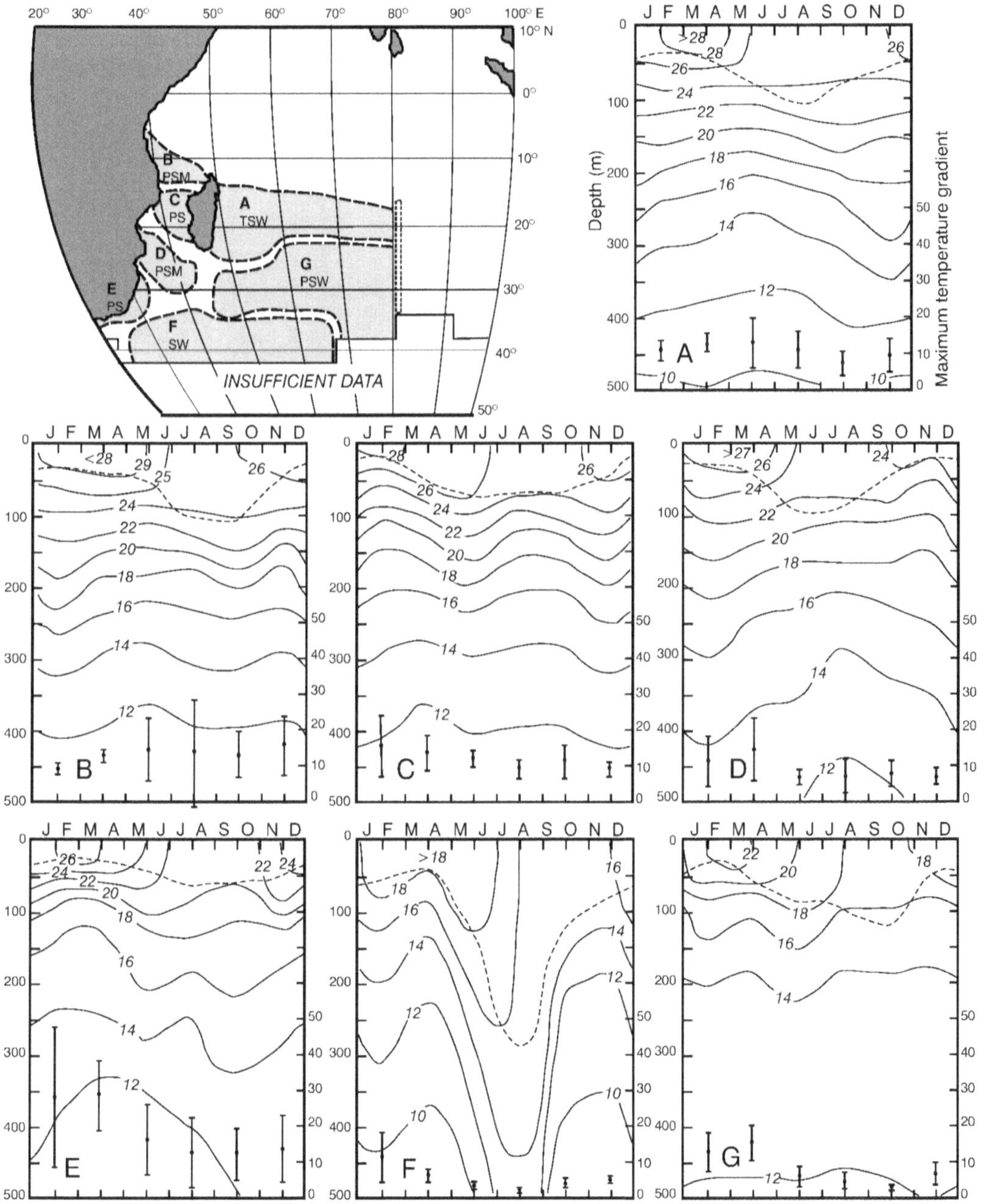

Figure 2.24. Annual variation of the thermal structure of the upper layers of the South West Indian Ocean[251]. Seven provinces with distinct and characteristic stratification types are given in the individual panels; the geographic regions where these types are found are circumscribed in the accompanying map. Vertical bars in the lower part of each panel show the maximum temperature gradient in the vertical for each bi-monthly period with standard deviation. The scale is on the right ordinate. The broken line in each panel shows the mixed-layer depth. The thermal classification is as follows:

P: permanent shallow thermocline
S: summer thermocline
W: deep winter mixed layer (100 m or more).
T: permanent very thick thermocline
M: moderately deep winter mixed layer (80 to 100 m)

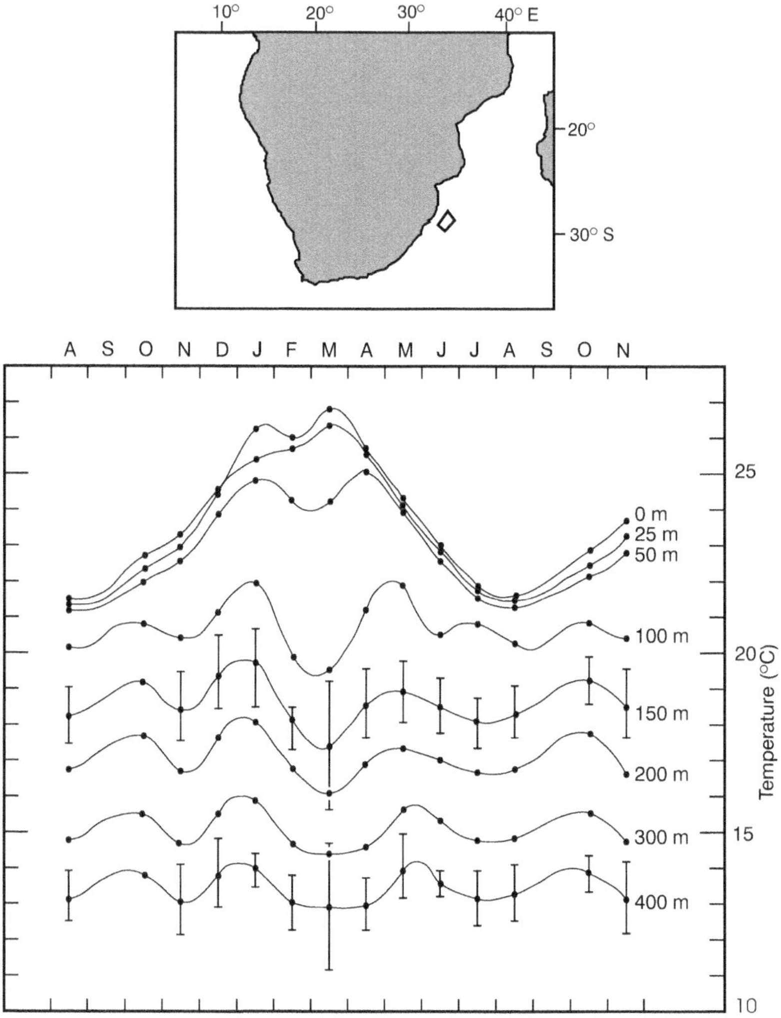

Figure 2.25. Seasonal variations of water temperature for depths between 0 m and 400 m and for the region in the South West Indian Ocean shown in the inset[263]. Standard deviations are given for each month for depths greater than 100 m.

water masses and their movements in the South West Indian Ocean remain as yet unexplained as well. It will be gratifying if numerical models of the flow could be used to suggest solutions to these problems or indicate the optimal locations for hydrographic measurements to resolve them.

Modelling large-scale flows of the South West Indian Ocean

There are only a few numerical models presently in existence that have as their aim the simulation of specifically the movement in the South West Indian Ocean.

Regional models

One quasi-geostrophic, closed-basin model with five layers in the vertical and with a ¼° x ¼° grid resolution[264] has been constructed for this region[265]. This model successfully shows ring and eddy formation in the South West Indian Ocean. It simulates a substantial degree of recirculation in the South West Indian Ocean, strong barotropicity for eddies and Agulhas rings over the top 1500 m and an Agulhas Return or South Indian Ocean Current substantially stronger at depth than the Agulhas Current itself. It gives no further suggestions on the flow of water of different water types.

A non-linear, reduced gravity model driven by observed winds has been used to study the movement of only the surface layers[266] for the region. It accurately simulates the flow of the South Equatorial Current, the southward flowing East Madagascar Current, the East African Coastal Current and indicates a continuous Mozambique Current. Of particular interest is the fact that if the regions shallower than 200 m are taken as impenetrable obstructions, a large current shadow is formed downstream of the Cargados Carojos Shoals that form a shallow part of the Seychelles-Mauritius Ridge[267]. This forces the South Equatorial Current to split at these shoals and not at the East Madagascar coast. As a result, this model shows only a very weak northward component of the East Madagascar Current. The important potential influence of the ocean bottom on this equatorial flow is thus demonstrated[268]. Of growing interest have been observational results suggesting seasonal variability in the overall flow regime[219, 269] of the South West Indian Ocean. Attempts have been made to model this as well[270].

Matano et al. have used a multi-level model with realistic bottom topography and coastlines[270], driven by seasonal wind stress, to investigate the effect of this seasonal forcing on the volume transport of the Agulhas Current. Two distinct geographic modes of variability emerge for the South Indian Ocean, separated by the Madagascar Ridge. On the western side the variability is dominated by a maximum in spring–summer and a minimum during autumn–winter; in the eastern part of the basin the model indicates a marked decrease in circulation in autumn with maxima in late summer and late winter. This is not inconsistent with some recent observations[219].The model indicates that the presumed seasonal adjustment in the South Indian Ocean is accomplished largely by the westward propagation of barotropic planetary waves. This propagation is inhibited in the model by the Madagascar Ridge and the South West Indian Ocean Ridge, isolating the Agulhas Current from seasonal fluctuations in the subtropical gyre farther east of the ridges. The large-scale forcing of the seasonal cycle has subsequently been studied[723] with a model with a ¼° × ¼° grid spacing and realistic forcing fluxes. It has been shown that seasonal changes in the volume transport of the Agulhas Current are linked to the large scale circulation in the Indian Ocean tropics (Figure 2.26). The simulated Agulhas Current in this model[723] exhibits a maximum transport at the start of austral spring and a minimum at the beginning of austral autumn, but its seasonality is not linked to that of the Indonesian throughflow (viz. Figure 2.26). The modelled seasonal cycle in the western Indian Ocean is the result of barotropic modes forced directly by the wind. A comparable seasonal variation has also been observed in a more directed model.

A regional model developed specifically for this ocean regime is that by Biastoch[271]. It consists of a nested system, with flow at the outer boundaries driven by a global circulation model. Within the South Atlantic and South Indian Ocean parts of the model there is an increase in spatial resolution for the greater Agulhas Current to ⅓° × ⅓°. It has 29 layers, with over half of these concentrated in the upper 1500 m. It successfully models both adjacent oceans but is, not unexpectedly since it has higher spatial resolution here, particularly successful in simulating the flow regime of the greater Agulhas Current (Figure 2.27). The flux through the Mozambique Channel is low, whereas the mesoscale variability is high. The Subtropical Convergence, meanders of this front and eddy shedding are particularly well simulated.

Large-scale models

Models covering much larger parts of the world ocean[272], or the whole world ocean[273–4], have in many cases also simulated the circulation in the South West Indian Ocean very successfully. The model by Semtner and Chervin[274] has 20 vertical levels and a ½° latitude by ½° longitude grid spacing. It and other similar models[275] simulate regions of high mesoscale variability, such as in the Agulhas retroflection and along the Subtropical Convergence with a high degree of verisimilitude (viz. Figure 6.3). It also shows the role of the South Equatorial–East Madagascar–Agulhas Current continuum as part of the global thermohaline circulation cell. Because the initial versions of this model did not have a Mozambique Channel the flow from the South Equatorial Current for these versions is highly simplified. This has been corrected in subsequent models in which a seasonal throughflow is indicated for the Mozambique Channel.

The initial model of this series gives no indication of a strong recirculation in a South West Indian Ocean sub-gyre or in a South Indian recirculation at about 70° E[101]. At depths below 3300 m the flow in the region west of the Madagascar Ridge is very vigorous. In the region east of here less so, and along the eastern coast of Madagascar it is very quiescent. Other large-scale models have given somewhat different results.

One of the larger-scale models that has shown substantial success to date in simulating the circulation in the South West Indian Ocean is the FRAM (Fine Resolution Antarctic Model[276]). This is a primitive equation numerical model of the ocean south of 24° S. It has a resolution of ½° longitude by ¼° latitude, 32 vertical

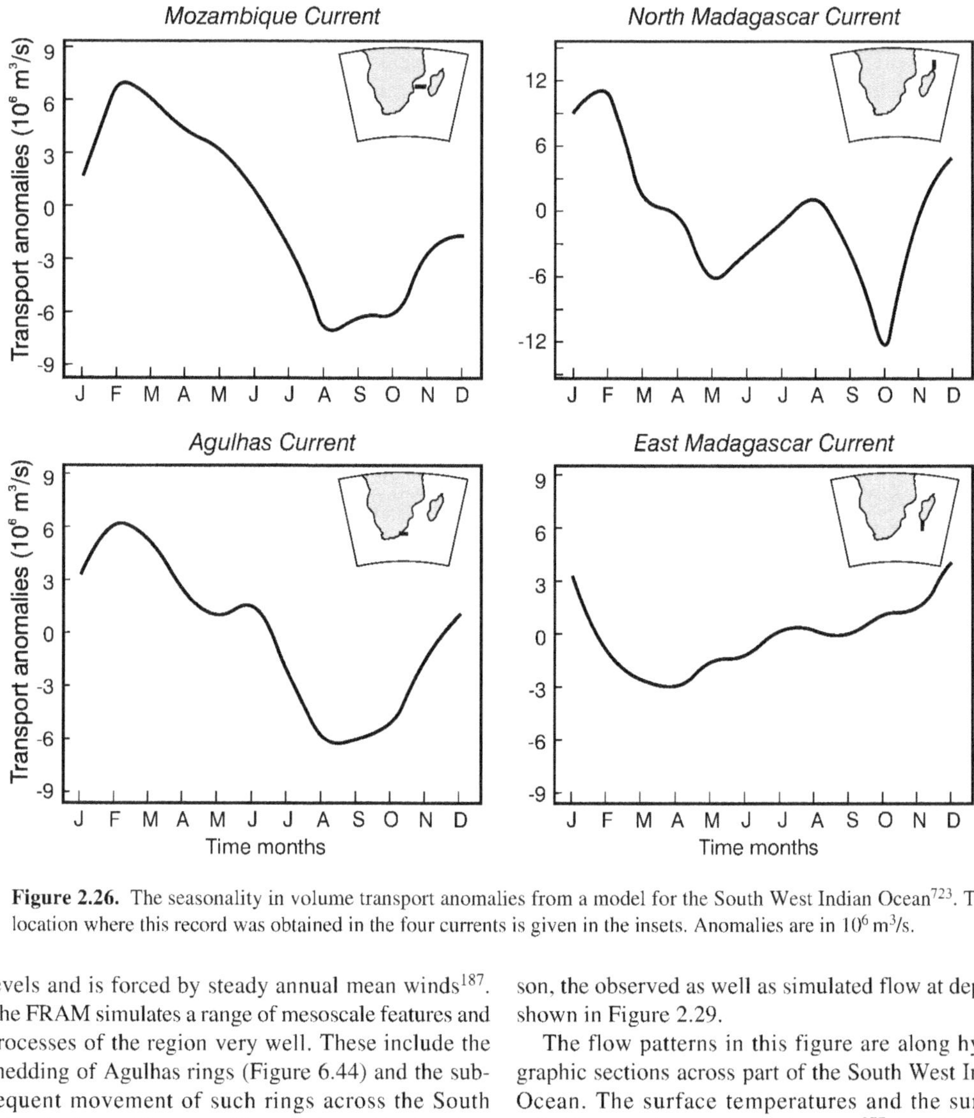

Figure 2.26. The seasonality in volume transport anomalies from a model for the South West Indian Ocean[723]. The location where this record was obtained in the four currents is given in the insets. Anomalies are in 10^6 m^3/s.

levels and is forced by steady annual mean winds[187]. The FRAM simulates a range of mesoscale features and processes of the region very well. These include the shedding of Agulhas rings (Figure 6.44) and the subsequent movement of such rings across the South Atlantic Ocean (Figure 6.45) as well as eddy shedding across the Subtropical Convergence (Figure 7.20). The FRAM also simulates known large-scale components of the general flow regime well.

The recirculation of water in a South West Indian Ocean sub-gyre is shown in Figure 2.28. The wider inflow south of Madagascar, the concentration of flow lines along the south-east coast of Africa and the major zonal outflow eastwards are all portrayed quite adequately. A serious lack of hydrographic data south of Madagascar[184] has forced the drawing of broken lines here (Figure 2.28, lower panel), denoting uncertainty. The model results may in this case actually be more representative of the actual flow. For further comparison, the observed as well as simulated flow at depth is shown in Figure 2.29.

The flow patterns in this figure are along hydrographic sections across part of the South West Indian Ocean. The surface temperatures and the surface salinities are close to those observed[277], but the intensity of the flow, as signified by slopes in the isotherms, is underestimated in the FRAM model. As a result the cross-sectional dimensions of the Agulhas Current are too large and the velocities too low. The salinity maxima at about 1000 m of the Antarctic Intermediate Water as well as the salinity maximum of the North Atlantic Deep Water are both well simulated. Since the FRAM does not extend into the tropics, the salinity minimum representing Antarctic Intermediate Water is found in the simulated Agulhas Current, but no evidence for the presence of Red Sea Water. On the whole the FRAM therefore gives a first indication of the very useful developments in ocean modelling that can be

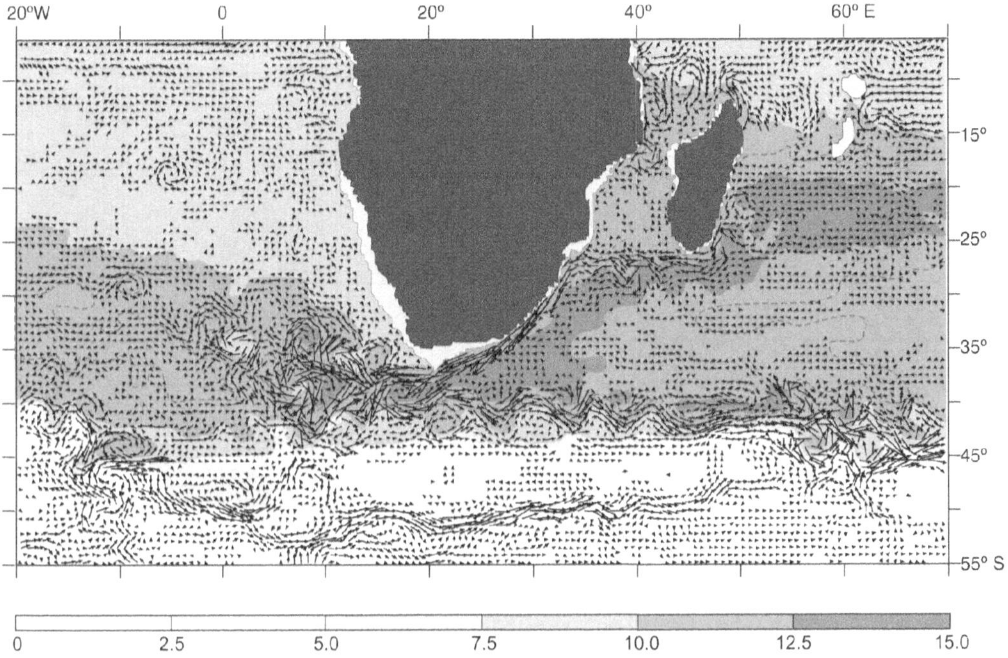

Figure 2.27. An instantaneous view of currents and potential temperature simulated by a numerical model designed especially for the greater Agulhas Current system[271]. Shown are the velocity vectors for all speeds greater than 2 cm/s. Enhanced inflow south of Madagascar, contrasted to hardly any consistent flow through the Mozambique Channel, is evident. Mesoscale turbulence in that channel, in the termination of the Agulhas Current and along the route of the Agulhas Return Current is portrayed realistically.

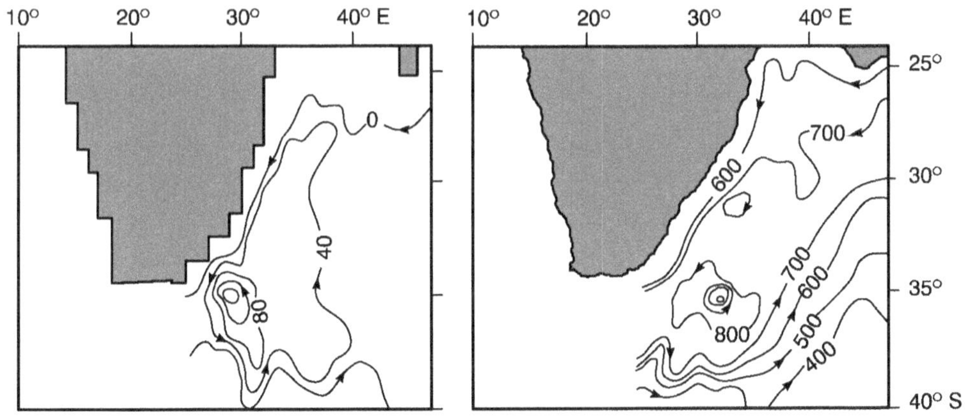

Figure 2.28. The extent and intensity of the sub-gyre in the South West Indian Ocean according to historical hydrographic data (left-hand panel) and according to the Fine Resolution Antarctic Model (FRAM; right-hand panel)[277]. The FRAM simulation shows stream functions, being the vertically integrated baroclinic as well as barotropic volume flux in 10^6 m^3/s. The left-hand panel shows the depth of the 26.80 sigma-t surface for data from six calendar months[82].

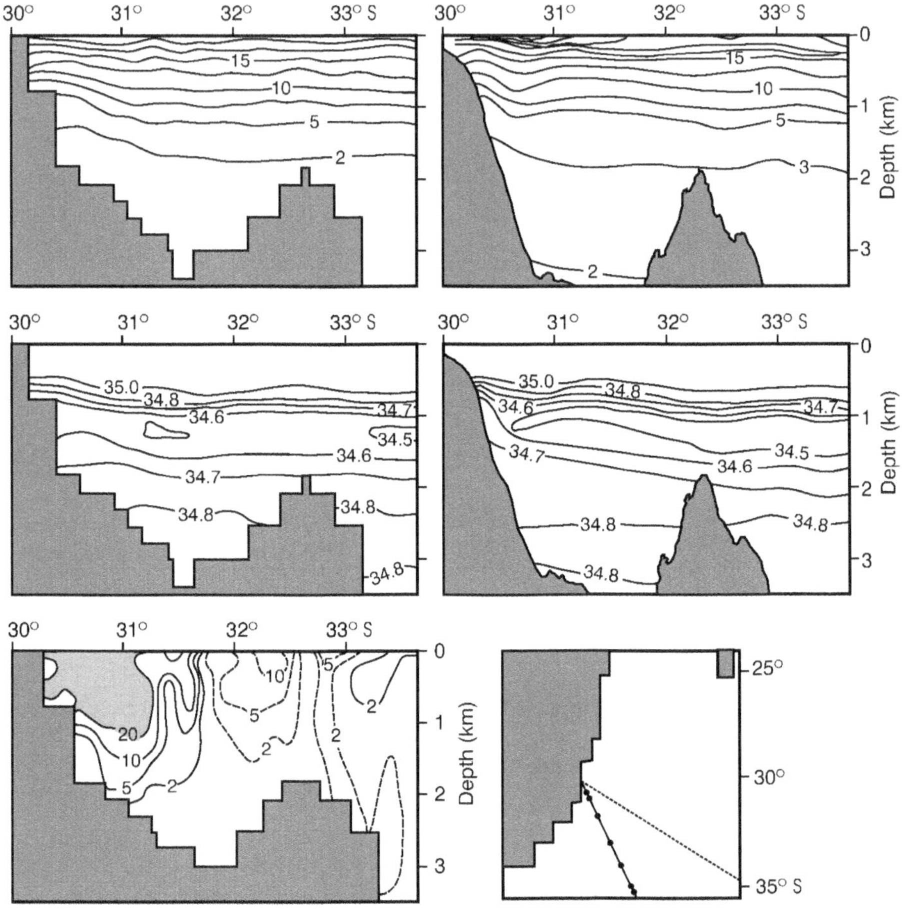

Figure 2.29. A comparison between the results of a line of hydrographic stations from Durban south-eastwards across the Agulhas Current and the simulation for a similar line by the Fine Resolution Antarctic Model (FRAM)[277]. The upper panels show the temperature fields to 3500 m; the model results are on the left. The middle panels show the salinity fields. The lowest panel (left) indicates the velocity component at right angles to the section; solid lines being flow south-westwards, i.e. out of the plane of the drawing. Speeds are in cm/s. The lowest panel (right) indicates the geographic locations of the FRAM section (broken line) and of the hydrographic section with stations[278] (solid line).

expected for the South West Indian Ocean as the quality of model results increase.

Increased understanding of the long-term variations in the flow characteristics of the South West Indian Ocean, due to changes in the atmospheric circulation, is of singular value at a time of climate change. Reliable data sets on the inter-decadal changes in local winds and remote wind forcing are available and these have been used to force a global ocean general circulation model to investigate the sensitivity of the ocean circulation and temperature fields to these changes[279–80]. The dominant effect is shown as changes in the sea surface temperature field in the region of the Agulhas retroflection and outflow zones. These sea surface temperature anomalies in the model are of the correct sign compared to the observations, but are smaller in both magnitude and areal extent. However, the sea surface temperature anomalies are larger, geographically more extensive and bear a greater resemblance to those observed under a scenario where the Indonesian throughflow is increased by stronger winds in the Pacific Ocean. The modulation of the southern Indian Ocean gyre by the Indonesian throughflow is therefore substantial, according to this model[279].

The unremitting increase in spatial resolution in models, their temporal resolution as well as the incorporation of realistic forcing by wind fields observed from satellite is continuously enhancing the accuracy with which numerical models are simulating the flow in the South West Indian Ocean, as elsewhere. The assimilation of altimetric observations as well as the increasingly dense array of profiling float measurements is rapidly changing the field of modelling to an extent unimagined before. It can be expected that this process will continue for some time.

Conclusions

The circulation in the Indian Ocean is unique in a number of respects. First, the monsoonal wind regimes cause a dramatic seasonality in the wind driven currents; a seasonality not seen elsewhere in the world ocean on this scale or of this intensity. Furthermore, there is a significant input of newly formed, very saline water from marginal seas into the North Indian Ocean. One would expect that the unusual wind conditions experienced over this part of the Indian Ocean – and the attendant shifts in current patterns and strengths – would make themselves felt throughout the whole ocean, but this seems not to be the case. In some regions there is a lack of adequate data to identify any possible annual cycle. Notwithstanding this, current information suggests that to some extent the South Indian Ocean seems to be isolated from the North Indian Ocean in the surface layers where most of the seasonality takes place.

By contrast to the North Indian Ocean, the anticyclonic, wind-driven gyre of the South Indian Ocean bears a much stronger resemblance to that of comparable ocean basins elsewhere. There now is ample evidence that this gyre is unusually concentrated in the western part of the basin and that this zonal shift is particularly marked at depth.

The water masses of the South West Indian Ocean are also not unlike those found at similar depths in many other oceans, except for the presence of Red Sea Water. This water mass interacts at intermediate depths with the inflowing Antarctic Intermediate Water in ways that are as yet incompletely understood. Moreover, there is less bottom water in the South West Indian Ocean than in most other basins of its kind. This is a direct consequence of the prevalence of many mid-ocean ridges. These ridges force Antarctic Bottom Water to take very circuitous routes in order to reach certain sub-basins, getting significantly mixed in the process.

The surface circulation of the South West Indian Ocean is similarly affected by the geometry of the region, particularly the presence of the island of Madagascar as a large obstruction in the source regions of the Agulhas Current. This complication is exacerbated by the dearth of good hydrographic data in many ocean sectors around Madagascar that are crucial for a better understanding of the flow patterns in the South West Indian Ocean. Numerical models are increasingly successful in modelling the known aspects of the circulation in the whole region, but cannot be properly verified in many critical regions due to the abovementioned lack of appropriate data.

This is particularly true for the source regions of the Agulhas Current.

3

Sources of the Agulhas Current

Western boundary currents in the ocean are generated by the wind stress patterns over the subtropical basins as a whole and close off the anti-cyclonic circulations of all these basins. In other words, the sources of the energy and motion of these currents are distributed over the whole subtropical gyre. Nonetheless, the origin of the water found in such currents is more specific since it is not an ill-defined mixture of waters from the full gyre that is forced, funnel-like, through the western boundary currents.

The source regions of individual western boundary currents are quite different in their coastal dispositions and morphology. In the case of the Gulf Stream, as was seen above, the water passing through the Florida Straits from the Gulf of Mexico and that passing north-east of the Caribbean islands form the sources of the Gulf Stream. In the South Atlantic, on the other hand, a small southward leakage escapes from the South Equatorial Current where it impinges on the South American continent. This then grows in depth and volume flux along its path to become a fully developed Brazil Current further downstream. In the case of the South Indian Ocean, the presence of the island of Madagascar constitutes a huge obstruction to the circulation in the source regions of the western boundary regime of this ocean. A highly unusual circulation is therefore to be expected for the head waters of the Agulhas Current.

Surface flows

As in all ocean currents, the earliest measurements in the regions of probably tributaries to the Agulhas Current were ships' drifts and sea surface temperatures. These suggested a southward water movement through the Mozambique Channel, a southward drift east of Madagascar and a westward drift south of Madagascar. The source waters of the Agulhas Current were thus easily explained. Unfortunately, these simple and partially erroneous portrayals have persisted in many geography textbooks and atlases, notwithstanding more modern findings. It is therefore important to scrutinise the origins of these early concepts with some extra care in order to judge their validity.

Early portrayals

Early shipping routes crossed two important possible source regions of the Agulhas Current. After rounding the Cape of Good Hope some ships sailed north through the Mozambique Channel on the way to India. Another route lay past the southern tip of Madagascar on the way to the Far East. From a very early stage it was therefore shown in charts of ocean circulation[26] that the currents experienced in the Mozambique Channel are southwards and those south of Madagascar westwards. Berghaus, in his authoritative atlas of 1845[281], has for instance shown a divergence in the westward flow of the tropical Indian Ocean, roughly south of India at 80° E, part of it moving south-eastward past the southern tip of Madagascar, the other branch funnelling water around the northern tip of Madagascar where it is joined by water from the north to flow into the Agulhas Current via a Mozambique Current.

Kerhallet[282] repeated this portrayal, perhaps in a more elegant way: the two tributaries to the Agulhas Current only joining south of Africa. Zimmerman[25] (1865) allowed a more northerly convergence of waters from the two sources, while the influential Krümmel (1882[30], 1911[31]) was the only person to portray the flow of water from east of Madagascar as not joining the Agulhas Current[26]. Petermann, also a producer of influential atlases (e.g. 1850[283]), to some extent combined these somewhat conflicting portrayals (1865[284]) to show only part of the water from east of Madagascar joining the Agulhas Current (Figure 3.1). Nomenclature on the currents was still in a state of flux[285]; the flow from east of Madagascar being variously called the *Branche SO du Courant Equatorial* and the *Madagaskar Strom*[284].

Perhaps of greatest importance in interpreting these portrayals is to realise that both inflows to the Agulhas

Current that are suggested by the nineteenth century charts lie along the main traffic routes of the time. It is highly likely that the inhomogeneous distribution of observations that thus came about substantially influenced and biased the portrayals that were given in the contemporary atlases. Scientifically and statistically more reliable analyses were only carried out during the 1920s and 1930s.

Drift analyses

Using all the possible data sources available, George Michaelis from the *Institut für Meereskunde* of the University of Berlin showed in 1923[44] (Figure 3.2) that the South Equatorial Current would only start to diverge near the islands of Mauritius and Réunion, that a substantial part of the flow rounding southern Madagascar would then flow northwards into the eastern side of the Mozambique Channel and that only part of the northern branch of the South Equatorial Current would feed into the western side of the Mozambique Channel, the rest flowing northwards. Michaelis carried out this analysis for only two calendar months of the year, January and July. Paech[45] did the same for the region around Madagascar, but for every calendar month. Considerable detail was thus added to the portrayal of sea surface currents, but the two major tributaries to the Agulhas Current were still thought to be the Mozambique Current and a flow from east of Madagascar.

Major studies using a large data base of ships' drift were more or less simultaneously undertaken by Barlow[286–91] in the Marine Division of the British Hydrographic Office. Once again, a strong Mozambique Current was inferred as flowing directly into the Agulhas Current and being joined at the latitude of 28° to 30° S by a current which had set south-westwards round the southern end of Madagascar[286]. Of this 500 mile wide current coming from the east, the main part, he stated, proceeds across the ocean to join the Agulhas Current near Durban, a considerable portion breaking off while crossing to Africa and setting southward. Of particular interest are Barlow's attempts to calculate the annual variations in velocity of the various currents and thus to establish their relationships[291].

His results show that the East African Coastal Current has a maximum northward velocity in the month

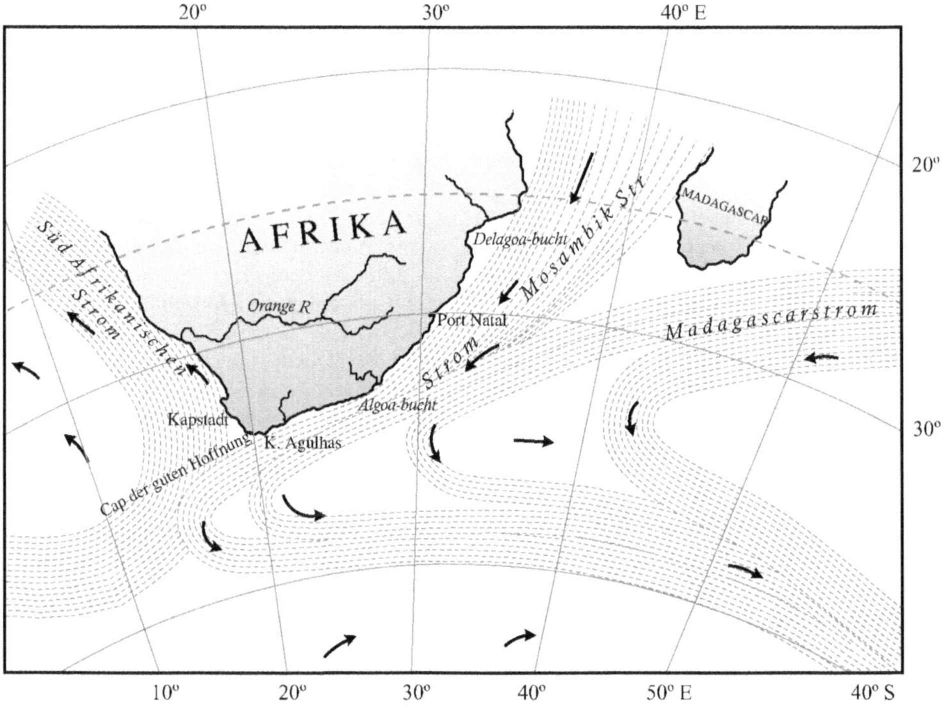

Figure 3.1. An early portrayal of the Agulhas Current system from a 1865 chart by Petermann[284]. It shows two source currents for the Agulhas Current; one from the north – through the Mozambique Channel – and one from the north-east, converging off Algoa Bay. Not all the water from east of Madagascar joins the Agulhas, most rejoining the subtropical gyre before reaching the African coast[285].

of June. This peak precedes a drop in southward velocity of the Mozambique as well as the Agulhas Current by one month. The suggested implication is that more northward flow of Equatorial water in the East African Coastal Current would deprive the Mozambique, and in turn the Agulhas, of its water supply, thus slowing it down. Modern studies have not been able to confirm these results[292].

Recent analyses

Present day atlases of the surface currents[191–3] in this ocean region differ only in detail from the earlier portrayals and atlases. A modern investigation using all available ships' drift observations has shown[293] that the southern branch of the East Madagascar Current extends from 17° S to just south of the southern tip of Madagascar. There is no evidence of any connection between it and the Agulhas Current. The mean speed of the current lies between 0.20 and 0.90 m/s. What can one therefore learn from these results?

Surface drift is wind-driven except in regions where very strong currents, such as the Agulhas Current, dominate the flow. Surface currents also give little information on deeper currents and how they are related by their exchange of water masses. Furthermore, current arrows, portraying average surface drifts over a regular grid, are often linked together in flow portrayals to give a field of isolines then erroneously interpreted as akin to streamlines[45,294]. Nonetheless, in the absence of alternative data, these representations have a strong tendency to become firmly fixed in the general

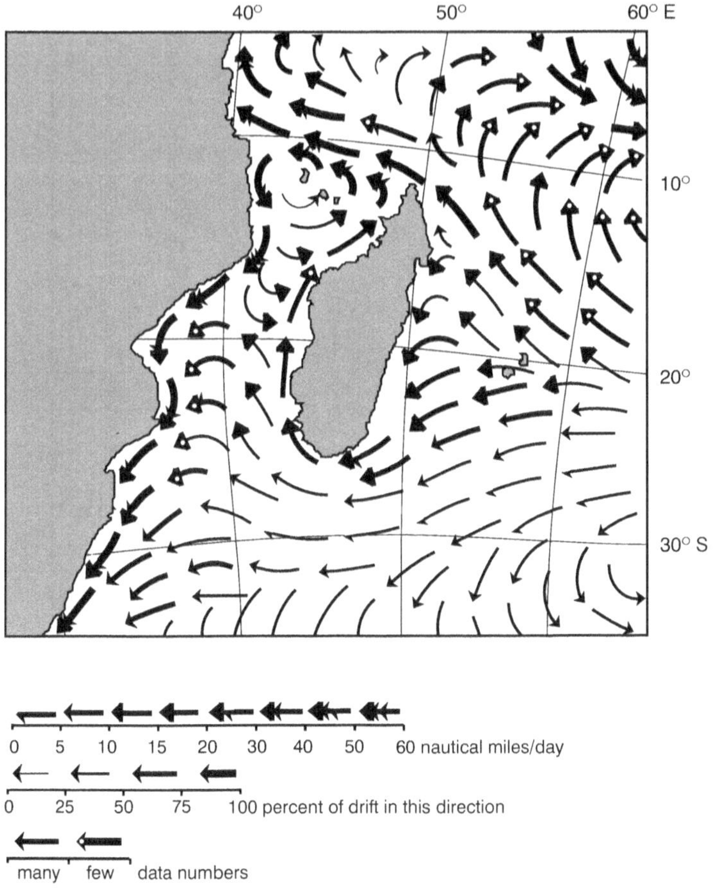

Figure 3.2. A portrayal of sea surface currents in the South Indian Ocean during the month of January. The number of arrow points indicate the speed in nautical miles per day, shaft thicknesses of the arrows the directional stability, and the number of white dots on arrows the decrease in reliability due to scarcity of data[44]. A distinct Mozambique Current has been inferred from these data. The inflow from east of Madagascar is more dispersed.

perception that holds sway in oceanographic thinking. They are only uprooted with difficulty. This phenomenon has especially hampered the case for new concepts on the flow into the Agulhas Current of water from east of Madagascar.

Flow from east of Madagascar

In the atlas based on the results of the International Indian Ocean Expedition[84], Wyrtki has calculated the geopotential topography for this ocean on different levels. However, he has broken up the data set for the upper layers of the ocean into groupings of two calendar months each, leading to a reduction in data that causes enormous differences in data distribution between the periods and thus in contouring. When sufficient data are available for this specific region, the flow of water east of Madagascar and south of 20° S is always southward, but detail of this flow is not clear. Combining these data into three-month bins[75] gives a slightly better data coverage and hence more credible portrayal (Figure 3.3). It suggests a strong and intense southward flow east of Madagascar.

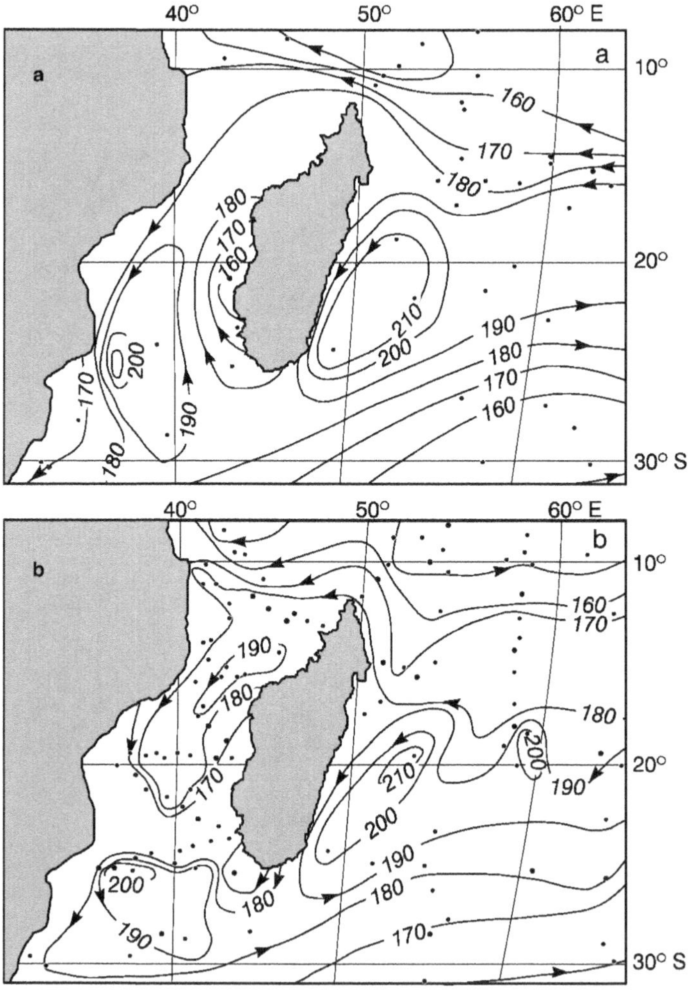

Figure 3.3. The dynamic topography (in dynamic cm) of the sea surface relative to 1000 decibar for the regions surrounding Madagascar. They are for the periods December–January (**a**), and March–May (**b**). Dots indicate station positions[75]. Note that the flow in the Mozambique Channel can be altered dramatically by different, but still legitimate, contouring.

The East Madagascar Current

This detailed analysis of the data set from the International Indian Ocean Expedition[84] shows a transport for the East Madagascar Current of between 20 and 24×10^6 m^3/s relative to 1000 decibar. This represents an estimated 80 per cent of the total flow. Using instead 1170 decibar as a reference level at a synoptic section across the current a value of 20.6×10^6 m^3/s has been found[295]. This value for the geostrophic flow has been closely calibrated with current meter records and is the best estimate of the flux presently available. The section on which this value was measured extended a distance of 110 km offshore. Estimates of the total flux up to a distance of 240 km offshore give a value of about 41×10^6 m^3/s[78]. These results suggest that the current is narrow and that most of the flow is confined to a strip close to the narrow continental shelf. Duncan[75] has in fact predicted that later results would show that the East Madagascar Current forms a small western boundary current system, and this has proven to be the case.

No detailed, dedicated cruise to describe the full extent of the East Madagascar Current has as yet been carried out. By fitting together the results from historic data, a few hydrographic sections and some current meter moorings, a tolerably comprehensive portrayal of this current system may be built up.

Sources of the East Madagascar Current

First, the South Equatorial Current impinges on the coast of eastern Madagascar at about 15° S[82]. This one can roughly estimate from the geostrophic flow patterns portrayed in Figure 3.3 and inferred in Figure 3.15. Large data sets of ships' drift[296] show that surface drifts in excess of 1 m/s may occasionally be found close to this entire coast, but that drifts greater than 1.5 m/s are only observed at the southern and northern ends of this coastline. Results of the movement of floats at intermediate depths[732] give speed values of 8.5 cm/s southward and 7.7 cm/s northward. Speeds past Cape Amber, at the northern tip of Madagascar, were about 11 cm/s. This information does not help in establishing the exact point of divergence between the two branches of the East Madagascar Current, but it does suggest that current speeds increase both southward and northward along the coastline from some central spot. On average the bifurcation point seems to lie at about 17° S[293]. Studies of the movement of intermediate water[732] only suggest a separation point at about 19° S, but do show that this bifurcation extends to at least intermediate depths.

A high-resolution, reduced-gravity model of the wind-driven flow of water[297] (Figure 3.4) suggests that the point of separation should lie at 18° S. This model

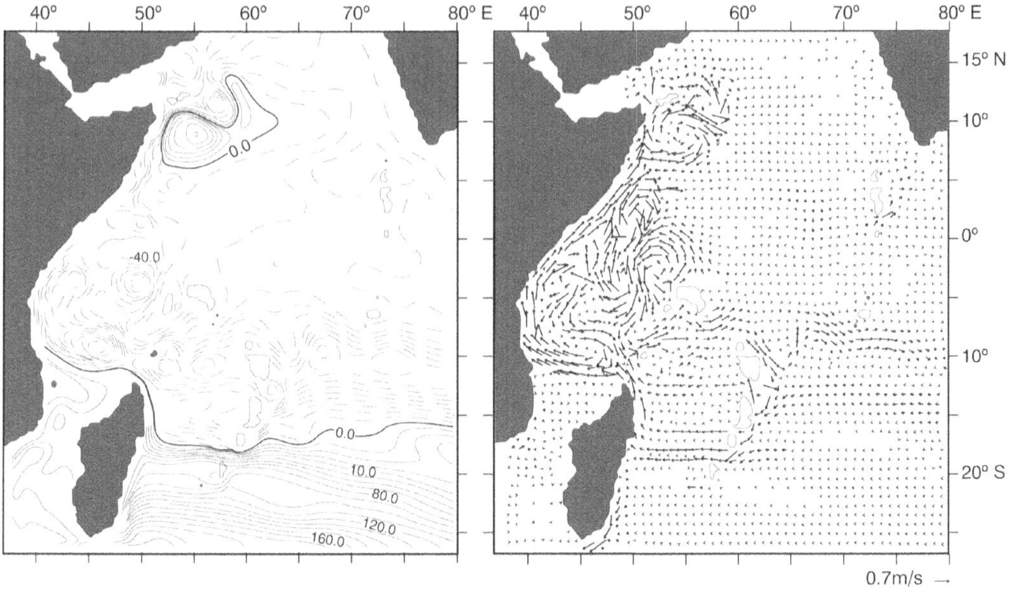

Figure 3.4. Flow and current simulations for part of the western Indian Ocean from a wind-driven, reduced gravity model[297]; **left-hand panel**: deviations of the initial model interface in 10 m intervals; **right-hand panel**: current vectors (the scale is given), vector lengths for speeds greater than 0.7 m/s having been truncated. The South Equatorial Current, the two branches of the East Madagascar Current and the circulation in the Mozambique Channel are clearly circumscribed.

is forced by seasonally changing winds[187] and has been used to calculate the Sverdrup mass transport function from the zonal wind stress curl. A different numerical model[266], with realistic geometry and forced by the same monthly, mean climatological winds, by contrast shows a distinct bifurcation of the flow in the upper 200 m[267] at the Nazareth Bank, well to the east of Madagascar (Figure 3.5). The northern limb of the South Equatorial Current in this model continues zonally westward in the model, past the northern tip of Madagascar. The southern branch also travels zonally to reach the east coast of Madagascar at a latitude of 20° S. In the lee of the Nazareth Bank the flow is nearly totally quiescent. This model suggests that the southern branch of the East Madagascar Current is much stronger than the northern branch. By contrast the global circulation model OCCAM[680] shows that at intermediate depths (Figure 3.6) the point of bifurcation on the east coast of Madagascar undergoes a seasonal shift. In austral summer it lies at 19° S; in winter at 21° S[732].

Ship's drift observed by a research vessel working in the region in 1984[295] implies that the separation point between these two branches of the current might lie at about 17° S. These drift data clearly demonstrate the increase in the speeds of both branches on progressing downstream. Current measurements from on board ship[298] show the fork in the currents to lie at 16° S (Figure 3.16). It has therefore not been established unequivocally where this bifurcation takes place and whether its geographic position changes seasonally. In fact, little seasonality in the overall flow has been observed.

Seasonality

Geostrophic transports based on the historic hydrographic data set[75] show a maximum in autumn, but so few data are available that this conclusion is not very reliable. Modern analyses of ships' drift suggest[293] highest speeds in spring and a minimum in summer. Based on historic surface currents Swallow[295] could find no distinct seasonal pattern, nor was this observed in directly recorded currents below 200 m. In this upper layer other historical data suggest a seasonal amplitude in transport of approximately 0.3×10^6 m^3/s[295], probably well within the experimental error. A significant annual cycle can therefore not be detected in the moored current and transport time series for the East Madagascar Current, despite substantial seasonal variations in the wind forcing over the subtropical Indian Ocean. A reduced-gravity model[295], driven by seasonally varying winds, has also simulated insignificant seasonal variations in the volume transport of the East Madagascar Current, as has a high-resolution primitive

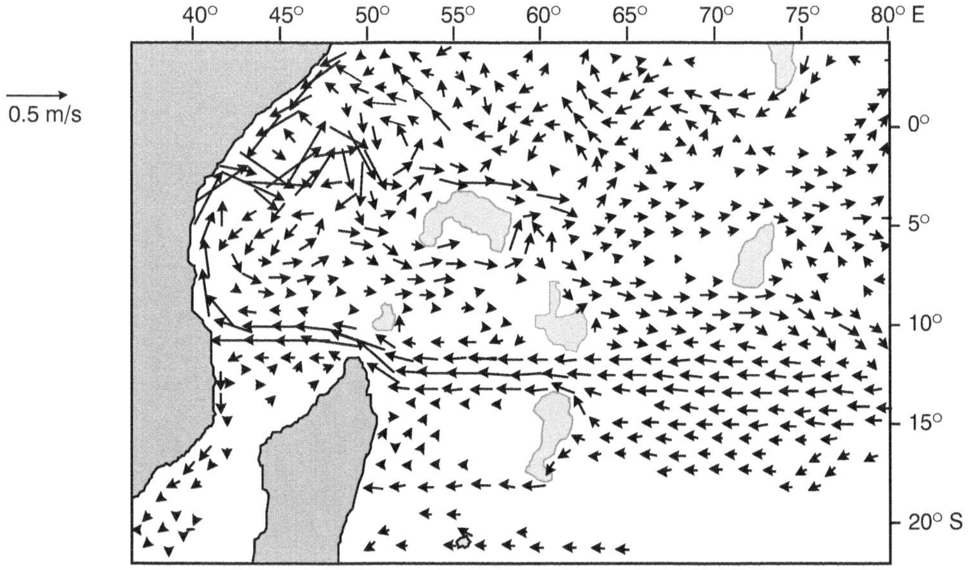

Figure 3.5. The wind-driven ocean currents in the west equatorial Indian Ocean according to a numerical model[267]. Continental shelf regions shallower than 200 m have been darkened; a velocity scale for the current arrows is given. Note the dramatic influence of the Nazareth Bank at 60° E on the flow of the South Equatorial Current, bringing water from the east.

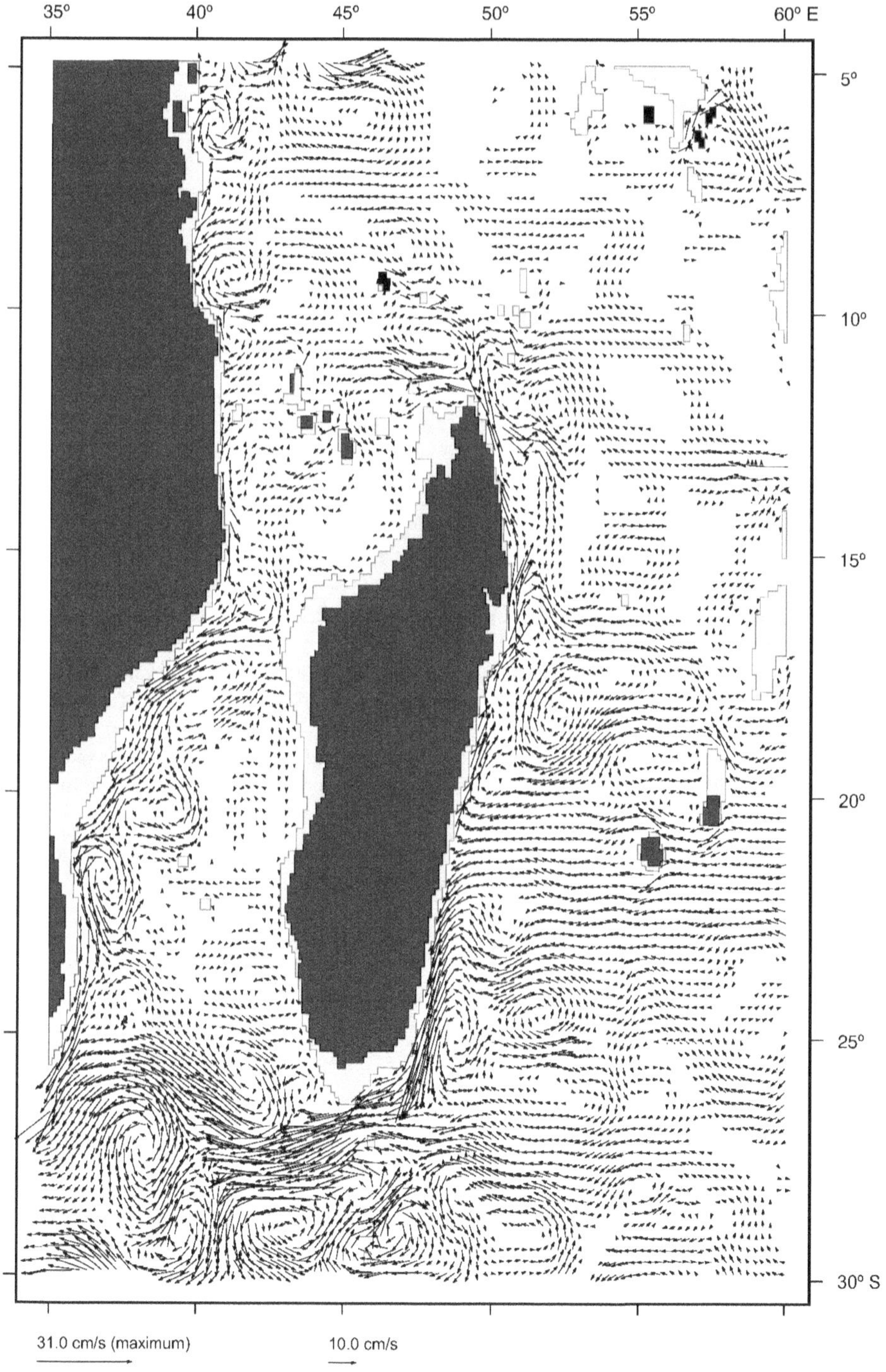

Figure 3.6. The mean currents around Madagascar at a depth of about 850 m in the OCCAM model[732]. This simulated circulation is for the winter months of June to August.

equation model[271], driven by realistic winds[299].

It is of interest that the current records show very energetic transport variations in the 40–55 day period band for the northern branch of the East Madagascar Current. This was measured where this branch passes Cape Amber, the northern tip of Madagascar. The variations in this band contribute about 40 per cent to the total transport variance here. Geostrophic velocities at the sea surface have been estimated to be in excess of 0.6 m/s[300] within the core of the current that itself is about 500 m deep. For the southern branch, the variability was less than 16 per cent. Nevertheless, studies on anomalies in the sea surface height indicate[679] that there are disturbances to the flow of the southern limb of the East Madagascar Current that travel downstream at a range of speeds and durations.

Southern branch of the East Madagascar Current

Nonetheless, most results suggest that the southern branch of the East Madagascar Current has a very stable trajectory and flux. The very narrow continental shelf and steep continental slope would tend strongly to stabilise a boundary current of this kind[87]. All hydrographic sections that have been carried out across this current to date show it to be narrow, intense and, in fact, to hug the continental shelf edge[80,295] (Figure 3.7). An investigation of sea height anomalies[679] has shown that deep sea eddies coming from the east may join the East Madagascar Current at a latitudinal interval between 22° and 24° S. This would be analogous to deep sea eddies joining the Agulhas Current proper[653]. In the latter case such absorption could induce instabilities in the trajectory of the current. To date this has not been observed in the East Madagascar Current except in some models and in one current meter record. In the latter the reversal in current direction at the shelf edge was of short duration.

Some high-resolution models[301,679] have simulated the formation of eddies in the East Madagascar Current. A comparison[271] between the flow variability in such a model (viz. Figure 3.20) and that calculated from altimetric observations indicates that this eddy creation by the model for the upstream part of the East Madagascar Current is almost certainly wrong. The models indicate a band of high variability along the shelf edge of south-eastern Madagascar where these simulated eddies move poleward. The altimetric observations show no such band. With such a narrow shelf and steep shelf slope, the constraints on the path of the current would be substantial[87] giving extra credence to the indication from altimetry that there are no eddies formed.

Swallow et al.[295] have assembled all the hydrographic data collected at 23° S, a latitude downstream of where the southern limb of the East Madagascar Current can be expected to be well developed. They have found an average maximum speed at the sea surface of 0.66 m/s about 50 km from the coast. The

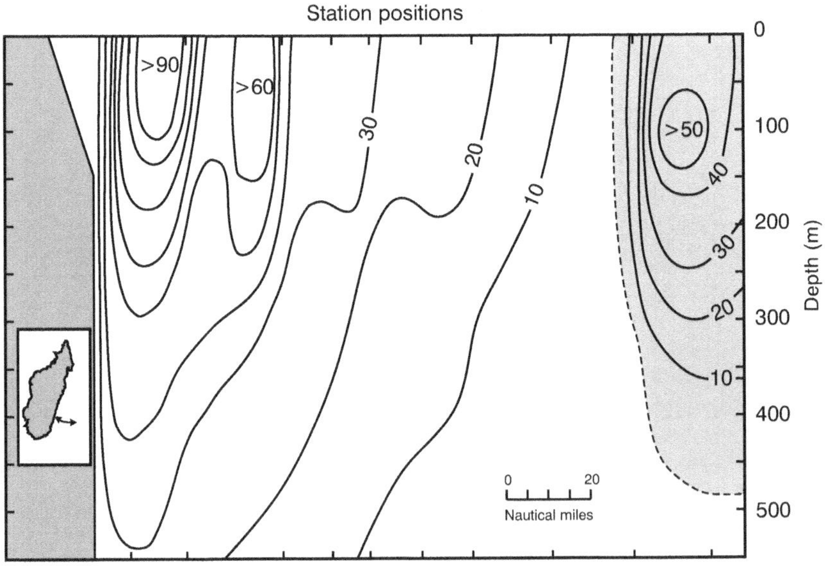

Figure 3.7. Computed geostrophic speeds[80] in cm/s for the East Madagascar Current at 23° S (inset). Shading indicates northward flow. The core of the current is less than 80 km wide and hugs the shelf edge.

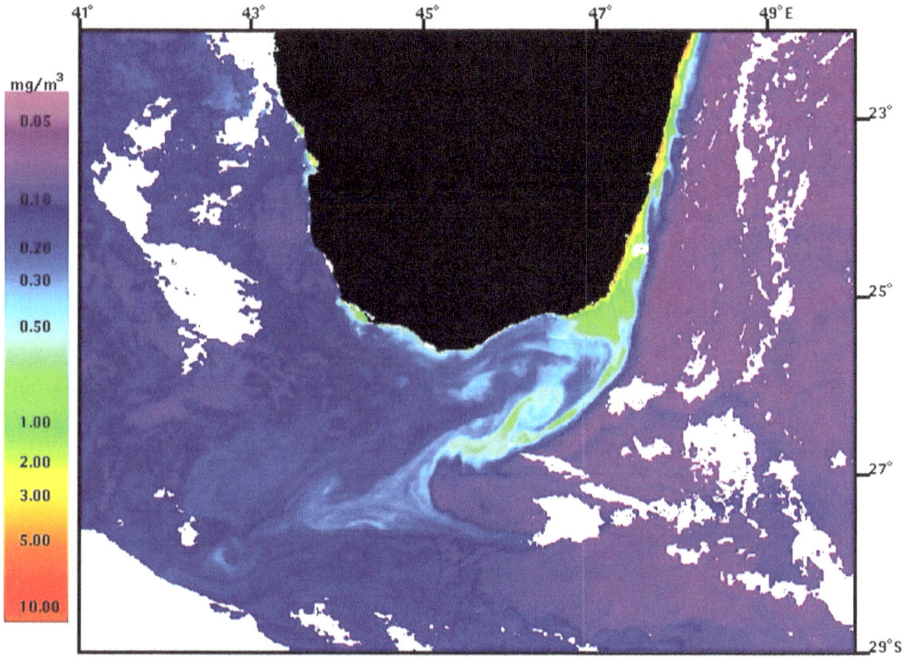

Figure 3.8. The distribution of chlorophyll-*a* at the location where the southern branch of the East Madagascar Current leaves the coast. White patches denote clouds. The location of the inshore upwelling cell off south-eastern Madagascar is clearly delineated, as is the landward border of the East Madagascar Current[304]. The location of the retroflection of the East Madagascar Current on this occasion is also evident from the chlorophyll-*a* that was advected along the edge of the current. This image is from MODIS AQUA for 23 September 2005.

0.50 m/s isotach was confined to within 74 km of the coast and extended no deeper than 200 m. At about 74 km from the coast there is a sharp horizontal gradient in current speeds. Standard deviations to the main geostrophic current field is typically ±12 cm/s in the upper 100 m; ±6 cm/s at 500 m[295]. Mean current profiles from geostrophy, current waters and ADCP (Acoustic Doppler Current Profiler) measurements were all very comparable. As mentioned above, average speeds at intermediate depths of 8.5 cm/s have been observed in this current. These results therefore indicate that this current is indeed very stable in both location and speed.

A line of very closely spaced stations at 23° S[80] (Figure 3.7), to establish the detail of the flow, has also established that the width of the East Madagascar Current as shown by the 0.60 m/s isotach was 79 km. The 0.50 cm/s isotach extended to 300 m depth. These results compare so well with the mean values set out above that one may assume with some confidence that Figure 3.7 gives an accurate portrayal of the current dimension and speeds of the fully developed southern branch of the East Madagascar Current.

Temperature–salinity characteristics

The temperature–salinity characteristics of the water in the current along this coastline are typical for this region. Water in the upper 50 m has a temperature greater than 23 °C and a salinity of less than 35; 34.5 at the sea surface[295]. This is Tropical Surface Water. This is in contrast to adjacent surface water of which the salinity is greater than 35.3, thus more nearly Subtropical Surface Water. In the general region, salinities of 35.75 are usually representative of water at the sea surface[302]. The saline surface characteristics of the current are therefore distinctly different from those of its surroundings.

A subsurface salinity maximum is found at a depth of about 250 m; 200 m in the current itself. Its salinity is greater than 35.6 and therefore consists of Subtropical Surface Water. In the current this maximum does not reach 35.5, and is coincident with a dissolved oxygen minimum of less than 4.0 ml/l, more intense than further offshore. Antarctic Intermediate Water of 4 °C to 6 °C lies between 800 m to 1200 m depth with a salinity of less than 34.6[295]. The water characteristics of the northern branch are similar, with weaker extremes. The contrast to the ambient water masses of the monsoonal regime in general is higher.

What is perhaps of paramount importance here is to realise that the East Madagascar Current is a narrow, intense western boundary current with clearly distinguishable water mass characteristics. Recognising this, the name is no longer applied to the general, wider southward drift in this region.

Upwelling inshore of the southern limb of the East Madagascar Current

The bathymetric disposition of the southern limb of the East Madagascar Current, moving from a narrow shelf east of Madagascar to a much wider one south of Madagascar, is very similar to that of the Agulhas Current flowing past St Lucia (viz. Figure 4.18) and past Port Alfred (viz. Figure 5.12). In both those cases there is consistent evidence for upwelling inshore of the passing current[163,166]. One would therefore expect the same to be evident off south-eastern Madagascar.

Hydrographic observations on the shelf off eastern Madagascar show[645] the temperature/salinity characteristics to be expected in the subtropics of the Indian Ocean (viz. Figure 4.14). Surface temperatures (in June) are 24 °C and surface salinities 35.3. Both Tropical and Subtropical Surface Waters are evident in the temperature/salinity values. The Tropical Surface Water extends from the surface to about 100 m depth. The Subtropical Surface Water, below that, extends to about 250 m. A well-developed subsurface oxygen

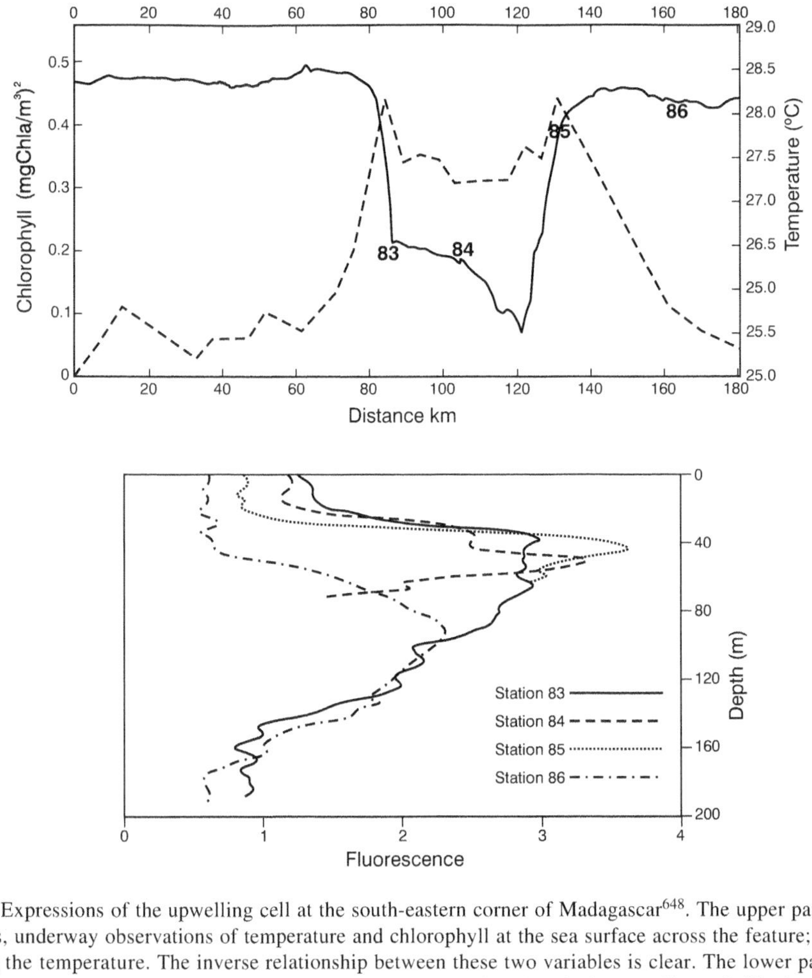

Figure 3.9. Expressions of the upwelling cell at the south-eastern corner of Madagascar[648]. The upper panel shows the simultaneous, underway observations of temperature and chlorophyll at the sea surface across the feature; the solid line representing the temperature. The inverse relationship between these two variables is clear. The lower panel gives the chlorophyll-*a*, as expressed by fluorescence, with depth at a number of stations in the upwelling cell. The locations of these stations relative to the upwelling cell are shown in the upper panel. A chlorophyll maximum is evident at a depth of about 50 m.

minimum is to be found at a depth of 80 m to 200 m. By contrast to the shelf waters off eastern Madagascar, satellite observations of the sea surface temperatures as well as of the chlorophyll-*a* of the waters over the south-eastern part of the shelf show[304,646] persistent patches of colder, greener water (Figure 3.8). Some of this nutrient enhanced water is dragged along the edge of the current (viz. Figure 3.8), but probably has little effect on its flux or on its T–S characteristics.

This image and others like it[647,651] suggests the presence of the expected upwelling cell near the town of Andriamanao, driven by the passing East Madagascar Current. This water, rich in chlorophyll-*a*, has been observed[649] to be drawn off the shelf region by passing eddies in the form of plumes. The question remains whether these remotely sensed suggestions of upwelling may not be due to runoff from land. The few available observations of *in situ* surface salinities and temperatures do all, however, support this contention of an upwelling cell.

During the very first extensive cruise over this shelf region, in June 1983[645], the temperatures at the coast were 2 °C lower than those further offshore. The salinities were up to 35.6, indicating upwelled Subtropical Surface Water. Hydrographic evidence, specifically collected during a cruise[650] dedicated to ascertaining the possible existence of an upwelling cell at this location[647], has recently shown unequivocally that there is indeed upwelling at this location[648]. This was made apparent by lower temperatures and coincident higher values of chlorophyll-*a* (Figure 3.9). Peaks in fluorescence (Figure 3.9) were found[648] between depths of 40 m and 100 m, the shallower peaks found closer to the coast. Concentrations of nitrate, nitrite and phosphate within the upwelling cell were up to six times higher than at the same depth outside the upwelling cell[647] indicating that the water in the upwelling cell had come from a depth of at least 200 m.

At first glance, this upwelling does not seem to be strictly related to wind patterns[304]. However, there is some preliminary evidence that the chlorophyll-*a* concentration at this location has a seasonal pattern with the highest concentrations found in the austral winter and in December, with some inter-annual variability[651]. This seasonal signal is not dominant. Major winds at this location are from the east in winter, from the north-east in summer[332] suggesting winds more favourable for upwelling along the full south coast of Madagascar in winter; at the south-eastern corner in summer. The extent of the water from this upwelling cell also seems to have a seasonal behaviour[649]. During austral winter a plume of green water extends far to the south-west; during summer there is no evidence of such a feature. The question concerning the importance of the wind compared to the current in driving this upwelling therefore remains unanswered.

The biological implications of the existence of this unusual upwelling cell are intriguing, but have to date not been properly quantified. Apart from the remotely sensed chlorophyll-*a*, surveys of fish stocks[645] have shown slightly higher concentrations of demersal fish on the southern shelf than off the adjacent, eastern shelf. Mackerel numbers were higher on the southern shelf, but scad lower. In general fish were found with such a very scattered distribution on this shelf that no firm conclusions can currently be reached on their biogeography.

It is important now to establish what happens to this water once it has passed the end of the meridional part of the continental shelf.

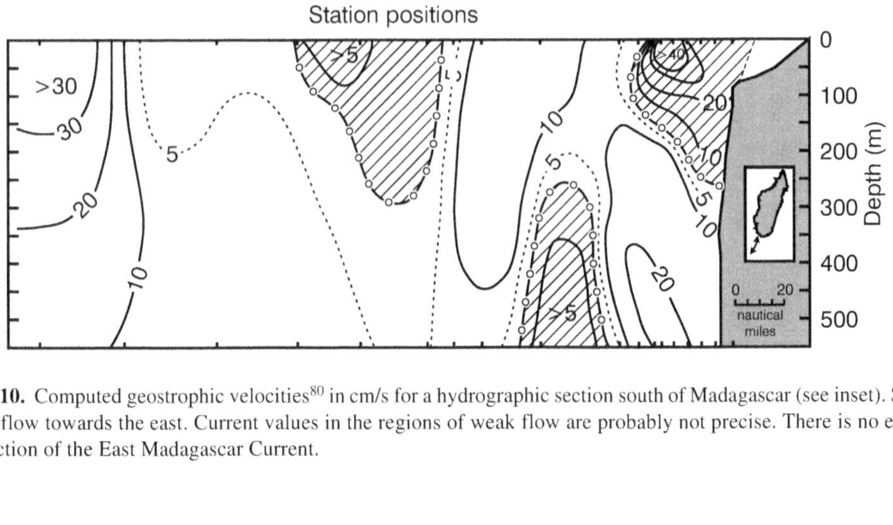

Figure 3.10. Computed geostrophic velocities[80] in cm/s for a hydrographic section south of Madagascar (see inset). Shading indicates flow towards the east. Current values in the regions of weak flow are probably not precise. There is no evidence in this section of the East Madagascar Current.

Termination of the southern limb of the East Madagascar Current

Satellite imagery with high spatial resolution shows[651] that once the East Madagascar Current has left the constraints of the narrow shelf off eastern Madagascar, it rapidly develops shear edge features, characteristic of all western boundary currents. These waves on the western side of the current propagate in the direction of the water movement, reaching a peak velocity about 60–100 km south of the upwelling cell. This is in general agreement with the theory for shear waves[651]. Where does this water in the East Madagascar Current then move to?

One of the few lines of stations that have been carried out to date to clarify this question has shown that south of Madagascar, in the upper 300 m next to the shelf, the flow was in fact eastwards on that occasion[80] (Figure 3.10). Below this there was a very weak flow. Of a strong westward setting branch of the East Madagascar Current there was no sign.

Analysis of historic hydrographic data[82] has likewise shown that water from east of Madagascar may only round Cape St Marie, the southern tip of the island, at a depth of about 500 m. This portrayal is in some contrast to what has been inferred previously from ships' drift (e.g. Figure 3.2) although the dynamic topography, based on historical data, does allow such an interpretation (Figure 3.3). It is interesting to note, however, that modern interpretations of ships' drift[293,296] indicate a current with high speeds off the south-eastern tip of Madagascar, but not extending westward around the island. The dearth of hydrographic data forces one to investigate various possibilities and other data sets to perhaps find the answer to the behaviour of the current south of Madagascar.

Drift tracks of floats at intermediate depths[732] indicate that those that had been part of the southern limb of the East Madagascar Current showed no tendency to move equatorward into the Mozambique Channel. Most moved westwards. Their trajectories also give no clear indication of either a retroflection or direct inflow to the Agulhas Current. Satellite imagery in the thermal infrared has been a very effective tool for interpreting the flow in the Agulhas Current system as a whole[60], but temperature contrasts between the East Madagascar Current and its general environment are small making this method less efficacious for this component

The National Research Institute for Oceanology

The CSIR (Council for Scientific and Industrial Research) was, over a period of many years, one of the key contributors to research on the South West Indian Ocean. Established in South Africa in 1945, it started some marine activities as early as 1954[312]. Work on the South African east coast only came into its own when a Physical Oceanography Division of the then National Physical Research Laboratory[313] was established in Durban. Although research interest initially was mostly coastal[314] and related to marine pollution[315], it soon became apparent that these near-shore investigations could not be carried out in isolation and studies on the adjacent Agulhas Current started[316] in greater earnestness.

By 1974 the marine research activities of the CSIR, distributed over a number of laboratories country-wide, had grown to such an extent that it was decided to amalgamate them all into a new National Research Institute for Oceanology (NRIO), to be located at Stellenbosch near Cape Town. Professor Eric Simpson, renowned marine geologist and later president of SCOR (Scientific Committee on Oceanic Research), was appointed as first director. With its own research vessel[317], an extraordinary large wave flume and an extensive model hall for coastal engineering, a branch laboratory in Durban, well-equipped workshops as well as libraries, the NRIO seemed set to make a major and lasting scientific and coastal engineering contribution.

From a slow start, with some hiccups, this eventually came about[318–19] and the number of published contributions from the NRIO grew apace[288,320]. Information, particularly on the northern Agulhas Current and the waters offshore of KwaZulu-Natal, was probably the single most important oceanographic contribution[227,321] emanating from this research institute during this period.

By 1986 politicians in the central government started questioning the appropriateness of the CSIR carrying out basic research and, shortly after, processes were set in motion to change the culture and functioning of the CSIR to that of a commercial and industrial consultancy[322]. A dramatic transformation of the whole establishment was achieved in record time and as part of this reorganisation the NRIO was dissolved and its workers taken up in a Division for Earth, Marine and Atmospheric Science and Technology (EMATEK), subsequently into an even larger and even more amorphous unit called Environmentek.

For South African research on the Agulhas Current these changes have been little short of disastrous. The research vessel of the CSIR[317] – the *Meiring Naudé* – was sold, active oceanographic research groups broken up and their members redeployed, while a substantial number of oceanographers left the employ of the CSIR. Publications by CSIR scientists based on previously collected data on the Agulhas Current continued to appear for a while, but at a much reduced and ever dwindling rate. By 1994 even this vestigial trickle had dried up[323] and the contribution of the South African CSIR to Agulhas Current research had finally ceased.

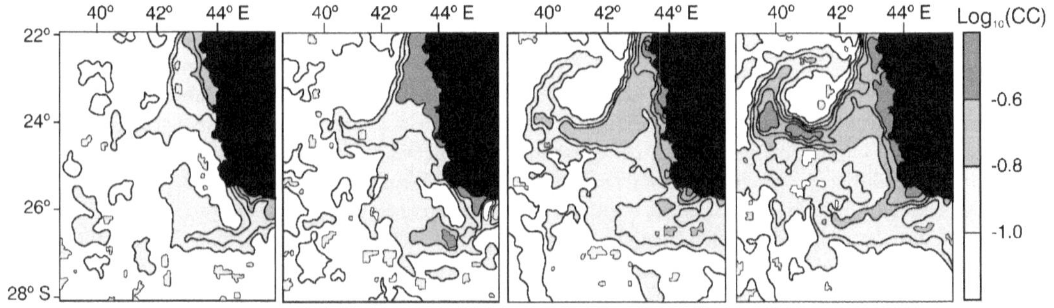

Figure 3.11. The influence of a cyclonic offshore eddy on the waters of the adjacent shelf waters on the eastern side of the Mozambique Channel[649]. The panels show the chlorophyll-*a* distribution as observed by satellite for four consecutive six-day periods starting with 29 November 2000 in the left-most panel. Filaments of shelf water with high concentrations of chlorophyll are drawn around eddies, demarcating the extent of the eddies.

of the system. Nonetheless, an image presented by Harris et al.[59] has shown a filament of slightly warmer water from east of Madagascar drifting into the northern Natal Valley. By contrast, an image obtained at a different period[114] clearly shows the ribbon of warmer water of the East Madagascar Current retroflecting south of Madagascar and subsequently heading eastwards. Analysis of satellite altimetry moreover has suggested[305] that this eastward flow may extend up to 60° E at 28° S latitude. On many occasions the southern part of the current is delineated by a tendril of water with high chlorophyll-*a* content that has been dragged along on its landward side[304,647]. This water characteristic is in strong contrast to that of the waters of the East Madagascar Current itself that has very low chlorophyll-*a* values. This greenish water originates in the Andreamanao upwelling cell discussed above and has the advantage that as a tracer it is much more durable than contrasting sea surface temperatures. A comparison of ocean colour images that suggest retroflection of the southern limb of the East Madagascar Current with concurrent portrayals of anomalies in sea surface height has lead[679] to the concept of no permanent retroflection of this current. Rather, it is thought that the seeming retroflective behaviour is an artifact of water from the East Madagascar Current being dragged around a passing anti-cyclonic eddy leading to strands of chlorophyll-rich waters extending up to 500 km or more eastward. The retroflective behaviour of the current therefore remains somewhat of an enigma.

Eddies from the southern limb of the East Madagascar Current

One of the obvious inferences that could be made from a retroflection of the East Madagascar Current to the south of this island would be that it could shed rings in a way analogous to that occurring at the Agulhas retroflection[91]. This seemingly straightforward deduction has turned out not to be easy to prove.

A number of eddies seem to originate in this region[652], or to pass through here[653] having been generated in other regions. Quartly and Srokosz[649] have used colour images to demonstrate that the ocean region off south-western Madagascar is in fact populated by a substantial number of cyclonic eddies. These eddies become visible when they contain higher concentrations of chlorophyll-*a* or when they draw out filaments of coloured water from the adjacent shelf region (Figure 3.11). A substantial number of these cyclones subsequently move in a westerly direction, across the southern mouth of the Mozambique Channel[649], at a rate of about 8 km/day. Most eddies originating from south of Madagascar subsequently head in a south-westerly direction[670]. The eddies – cyclonic as well as anti-cyclonic – carry about 8×10^6 m^3/s from the East Madagascar Current towards the Agulhas Current. They may be influenced in their movement towards the African coast by the intervening Mozambique Plateau (viz. Figure 1.2). Many of them may move southward along the plateau, hugging its eastern border until they reach its southern tip, where they can then freely move westward. A large number of eddies have in fact been found and studied on the eastern side of the Mozambique Plateau[344,355–9] (viz. Figure 3.32). The characteristics of these eddies are discussed at length below. Based on all the above it has been concluded[679] that the westward flow towards the African mainland is sustained by the westward drift of eddies and not by a steady current. This contrast with the results of most models, such as OCCAM[679], that simulate a much more coherent connection.

Modelling the retroflection of the East Madagascar Current

Model results are of limited help in this regard. None have been designed specifically to investigate the southern termination of the East Madagascar Current. A regional model to study ring formation in the Agulhas system[265] has the East Madagascar Current flowing zonally as a thin jet between Madagascar and Africa, a clear artifact of the boundary conditions of this model. The FRAM model, although extending north to a latitude of 24° S, has its boundaries too close to the region for credible results[276] which then also leads to a purely zonal westward flow for the termination of this current[277]. High viscosity, linear model behaviour and conservation of vorticity may also contribute to the retention of a zonal jet in the model. The early versions of the detailed global model by Semtner and Chervin[273] had no Mozambique Channel and therefore an unrepresentative flow in the region.

However, a more detailed model for the region, with eddy permitting resolution[271] and realistic wind forcing, shows the development of baroclinic, anti-cyclonic eddies east of Madagascar (Figure 3.12) due to baro-

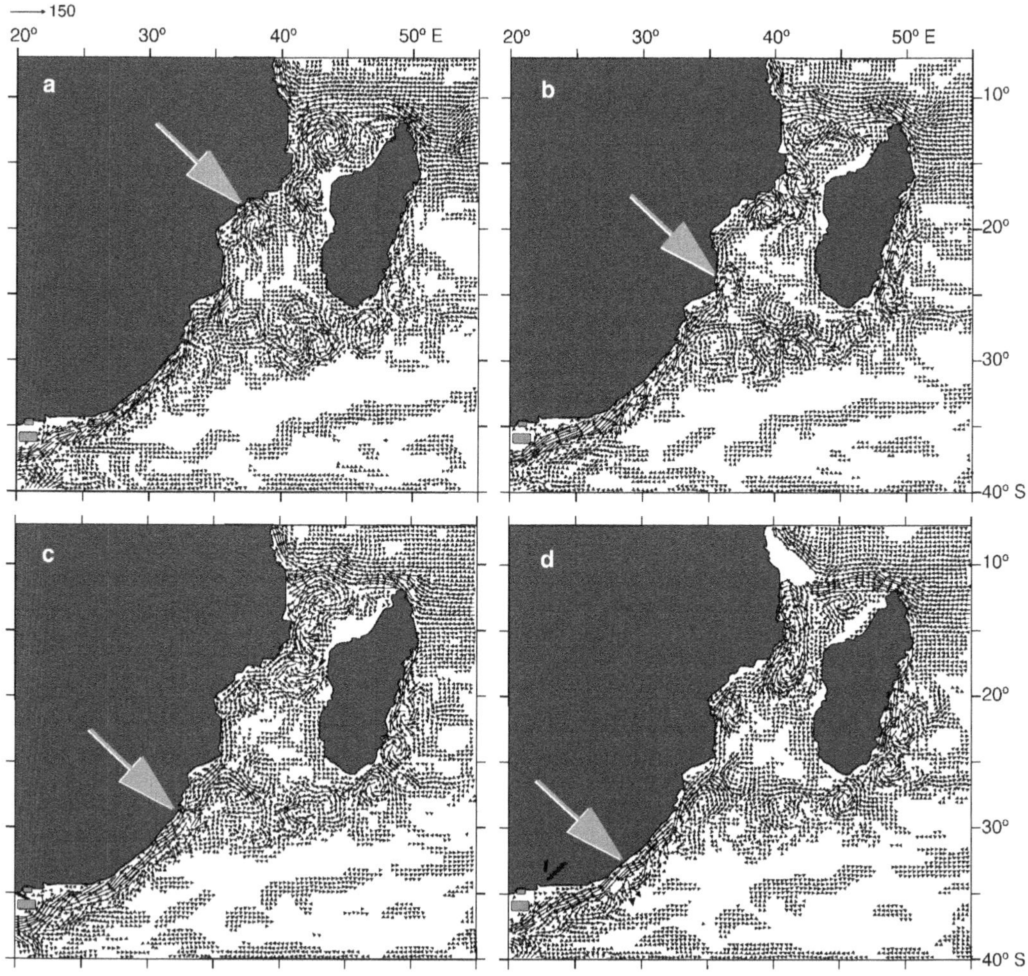

Figure 3.12. Velocity vectors at 41 m depth for the northern part of the greater Agulhas Current system for four model days, roughly a month apart. They come from a primitive equation model[271] with $^1/_3°$ by $^1/_3°$ spatial resolution. Values less than 5 cm/s have been omitted. The progress of an eddy through the Mozambique Channel is followed by arrows. A southern branch of the East Madagascar Current and a well-developed Agulhas Current are simulated. However, no consistent Mozambique Current is evident. Cyclonic eddies, that subsequently join the Agulhas Current, are generated both east of Madagascar and in the Mozambique Channel. The scale arrow indicates 150 cm/s.

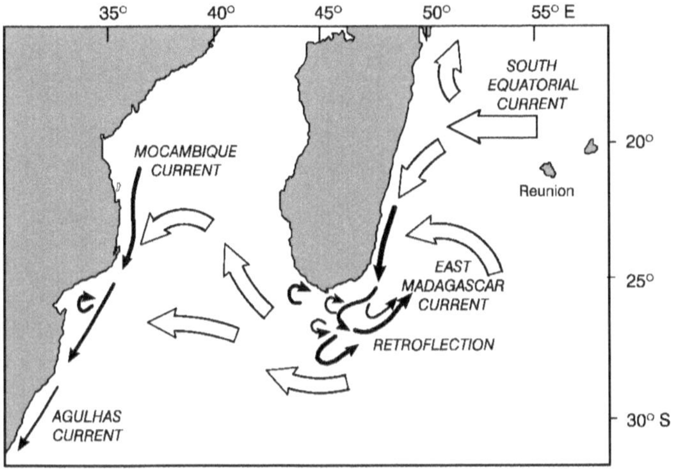

Figure 3.13. A conceptual image of the characteristic flow regime in the source regions of the Agulhas Current[80]. Narrow, intense currents are shown as filled-in arrows. Open arrows represent laterally extensive flows. The rest of the area is believed to be characterised by slow average advection on which mesoscale turbulence is superimposed. The East Madagascar Current is portrayed as a small western boundary current that retroflects south of Madagascar.

clinic instability. These modelled eddies progress westward and are eventually absorbed by the Agulhas Current[301]. The geographic distribution of eddy kinetic energy brought about by these eddies in the modelled East Madagascar Current (Figure 3.20) is however in conflict with that which has been observed[301] (Figure 3.14) and one may therefore assume that these eddies are purely artifacts of this particular model. However, they also occur in other, more detailed global circulation models, such as the OCCAM (Ocean Circulation and Advanced Modelling Project)[680]. A possible reason for this discrepancy may be due to the sensitivity of the simulated current to the bottom topography and the somewhat coarse bathymetry that has been used. These models do, however, simulate a realistic pathway once such disturbances have been shed from the termination of the southern branch of the East Madagascar Current (Figure 3.12). At least one model, a 3.5-layer, reduced-gravity model for the North Indian Ocean, with an embedded mixed layer and driven by realistic winds and thermal forcing[306], does show clear signs of a partial retroflection of the East Madagascar Current south of Madagascar, but with some remnant flow crossing the southern mouth of the Mozambique Channel notwithstanding.

The OCCAM global circulation model[680] (viz. Figure 3.6) indicates a highly variable region south of Madagascar. On average the water movement from the southern termination of the East Madagascar Current is in a westerly direction[732] and joins that of the northern Agulhas Current at about 26° S. Instantaneous portrayals, even for three month periods as shown in Figure 3.6, indicate that the East Madagascar Current may end in a collection of eddies of different sizes and circulation senses. In Figure 3.6 there is an indication that the current may leave the tip of the Madagascarian shelf as a free jet, forming a vortex dipole in the process. This would seem to agree with some of the observations[652] in the region, but this correspondence may be entirely coincidental.

Drifters have turned out to be valuable instruments to trace the Lagrangian motion in this region. They have usually been free-floating, satellite-reporting weather buoys. Some of these have, to date, been placed in the South Equatorial Current and have, fortuitously, drifted into the southern branch of the East Madagascar Current. Those that gave clear signs by their trajectory, speed and temperature, to have been in the current core, all retroflected over the shallow bottom topography south of Madagascar[80]. After this about-turn some drifted into the Mozambique Channel whereas some continued eastwards. A simple, inertial jet model[130] for the East Madagascar Current shows that the current should retroflect over the shallow ridge south of Madagascar, similarly to the Agulhas Current retroflecting south of Africa[130].

The conclusion that can therefore be reached on the flow pattern south of Madagascar, based on all this eclectic evidence, is presented pictorially in Figure 3.13. It shows a narrow, intense boundary current retroflecting south of Madagascar with the possibility of rings, eddies and filaments spilling into the northern

Natal Valley. The general, large-scale, anti-cyclonic circulation is maintained, with the possibility of some water moving northward into the Mozambique Channel. Does this conceptual portrayal (Figure 3.13) imply that no water from east of Madagascar will ever reach the Agulhas Current?

Disposition of water from east of Madagascar

Clearly this is not the case. Surface drift of flotsam[307] and drifters[308] has shown that surface water from this region, but not directly from the East Madagascar Current itself, eventually does reach the coast of south-eastern Africa. Harris[81] has analysed a number of hydrographic sections that had been carried out in the South West Indian Ocean in the spring of 1964 and has come to the conclusion that 35 × 10^6 m^3/s of water from the South Equatorial Current passed south of Madagascar on this occasion to join the Agulhas Current. This constituted 49 per cent of the estimated transport of the Agulhas Current at the time. However, it was not implied that this water came directly via the East Madagascar Current.

Flow variability

The recent availability of altimetric data from orbiting satellites has presented a new instrument to study the region south of Madagascar[70]. The results are unexpected[309]. Portrayals of mesoscale sea height variability[183] based on measurements by the SEASAT satellite show high values in the region south and east of Madagascar filling a large part of the Natal Valley and Agulhas Basin west of the Madagascar Ridge (viz. Figure 1.2).

Greater geographical detail comes from the Geosat[72] and TOPEX/Poseidon[301] altimetric measurements and these exhibit a high degree of mesoscale variability in the assumed retroflection region of the East Madagascar Current. They also show lobes of high values extending eastwards as well as westwards (Figure 3.14). These altimetric portrayals are similar to the geographic distribution of eddy kinetic energy based on variations in ships' drift readings[293]. All these geographic distributions are consistent with the concept of high mesoscale activity in the eastward flow, after the termination of the southern branch of the East Madagascar Current, as well as the westward movement of rings and other retroflection products towards Africa.

More detailed analysis of anomalies in altimetric data that may represent individual eddies has shown[309] such features propagating into the Natal Valley from east of Madagascar. The precise mode of creation of these anomalies could not be established by altimetric means alone, but those features that have been intercepted with hydrographic station lines have turned out

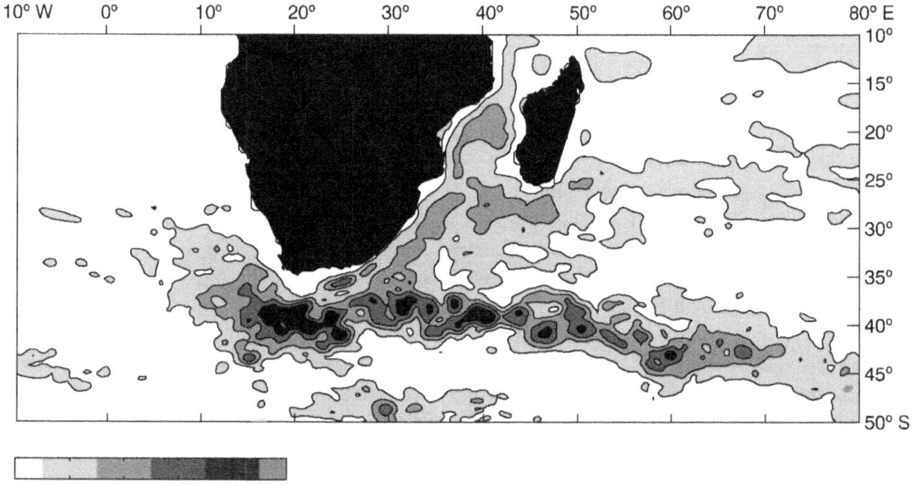

Figure 3.14. The geographic distribution of the variability of the currents in the South West Indian Ocean for the period 1992 to 1998. These values of the standard deviation of sea surface height have been derived from altimetric observations from the TOPEX/Poseidon satellite. Darker hues indicate higher values. The band of very high values centred at 40° S represents the Agulhas Return Current[96] and the high values in the South Atlantic Ocean represent the termination of the Agulhas Current. Note the two strips of high variability originating in the Mozambique Channel and from south of Madagascar.

to be intense, cyclonic eddies. This would indicate a different generating mechanism than simple ring shedding by occlusion of a warm retroflection loop of the East Madagascar Current.

Inspection of large data sets of drifter trajectories gives some further interesting hints[310–11] on the oceanic flow south of Madagascar. Eddy kinetic energy calculated from these observations for the region south of Madagascar is relatively high. What is of particular interest is the fact that the residence time of drifters in this region is also high. This implies that drifters caught in eddies at this location will remain trapped here for a considerable period.

These various data sets give no incontrovertible and conclusive answer on the dynamic behaviour at the termination of the East Madagascar Current or on its effect, if any, on the Agulhas Current. The present data, however, are all consistent, to a greater or smaller degree, with the conceptual portrayal put forward in Figure 3.13. Nonetheless, it is clear that there is a crying need for a concentrated observational programme[184] for this part of the flow regime of the South West Indian Ocean.

Movement at depth

This discussion has dealt almost exclusively with the flow to the east of Madagascar that takes place at shallow and intermediate depths. This is because the bottom topography would most likely preclude deeper water from here reaching the Agulhas Current. Hydrographic measurements have shown[244] that the flow directly beneath the East Madagascar Current, at depths greater than 3000 m to 3500 m, consists of an abyssal boundary current 300 km to 400 km wide, carrying approximately 4 to 5 × 10^6 m^3/s water northwards. This bottom current was observed once again in April 1985[245], but with a slightly reduced transport of 3.6 × 10^6 m^3/s. Hydrographic measurements along the South West Indian Ocean Ridge, that separates the Madagascar Basin east of Madagascar from the Crozet Basin to the south, have shown that sufficient bottom flow takes place through fractures in the ridge to feed this bottom boundary current[324], so that it is estimated that all the water in this current comes from the Crozet Basin.

Having passed past Cape Amber, the northern branch of the East Madagascar Current moves past the Comores in the northern mouth of the Mozambique Channel[45]. Some of its water moves southward into the Mozambique Channel. Similarly, some water of the southern branch spills northward into the channel[80]. The resulting flow through the channel and into the Agulhas Current is therefore also important to an understanding of the sources of the Agulhas Current proper.

Flow through the Mozambique Channel

The classical portrayal of surface drift in and through the Mozambique Channel is given in a very representative way in Figure 3.2. The northern branch of the East Madagascar Current moves directly westward with speeds of about 0.5 m/s and stable directions. At the African coast it bifurcates into a southward Mozambique Current and the northward East African, or Zanzibar, Current. Both have augmented speeds of up to 0.75 m/s. Some modern ships' drift results give values up to 2.1 m/s[296]. Geostrophic calculations, based on modern data[258], show that the East African Current has surface speeds exceeding 1 m/s in northern summer. A volume transport of 20 × 10^6 m^3/s in the upper 500 m, with a current width of 120 km, has been established. Below this the flow is weak and variable. Swallow et al.[258] have put forward the view that most of the surface water of the South Equatorial Current becomes the East African Current, but that its waters below 500 decibar turn southward to feed a Mozambique Current.

The conventional portrayal is of a Mozambique Current that hugs the coastline until it emerges at Maputo, at the southern mouth of the channel, to feed the Agulhas Current. A similarly strong flow, but not quite as rapid, is conceived as moving northward along the west coast of Madagascar (Figure 3.2). The circulation in the central part of the channel seems to consists of an anti-cyclonic gyre in the northern half and an east–west flow that continuously strengthens the Mozambique Current in the southern half. However, modelling this presumed surface drift with a wind-driven, reduced-gravity, numerical model[297] gives a totally different portrayal (e.g. Figure 3.4).

Modelling the flow in the Mozambique Channel

This particular simulation shows no bifurcation at the African coast, but indicates that all of the water of the northern branch of the East Madagascar Current becomes part of the East African Current. There is no strong wind-generated flow on either border of the Mozambique Channel in this model. In the central region of the channel the simulated drift is weak and incoherent (Figure 3.4). A similar model, with realistic geographic configurations[266] and wind-driven, also gives (Figure 3.5) very stagnant flow conditions for the full width of the Mozambique Channel[267] except near the western boundary where a weak current with speeds of only 5 cm/s to 10 cm/s is predicted. However, a

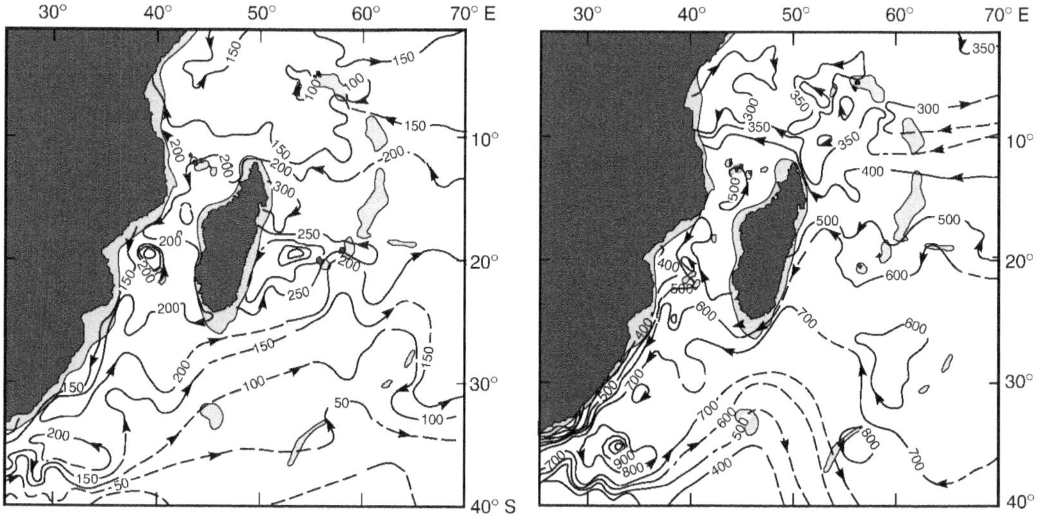

Figure 3.15. The depth of the sigma-t 25.80 surface (left-hand panel) and of the 26.80 surface (right-hand panel), in the South West Indian Ocean according to hydrographic data collected during the north-east monsoon season[82]. Dashed lines denote subjective contouring due to lack of data points. The depth at which substantial amounts of water penetrate through the Mozambique Channel and from the East Madagascar Current is different for these two density surfaces.

model with high spatial resolution designed especially for this region[271], by contrast, shows a substantial flow of water southward through the channel.

This simulated southward flow has strong seasonality[299] as averaged over five years with a maximum of 22×10^6 m^3/s in August. However, it consists almost entirely of anti-cyclonic vortices[301] (viz. Figure 3.12). This simulation of eddies has been repeated in most subsequent modelling efforts[732]. In the global ocean model OCCAM[680], with a $^1/_{12}° \times {}^1/_{12}°$ spatial resolution, the flow at intermediate depths on the western side of the Mozambique Channel is dominated by anticyclonic eddies (Figure 3.6). The flow on the eastern side is in general much weaker, but also shows evidence of the presence of eddies. The circulation in the Comores Basin is weakly anti-cyclonic during most parts of the year[821], particularly on the western side, but may be quite variable[789–90,794]. The flow through the Mozambique Channel should have a marked effect on the flux of the Agulhas Current in most models. This might be seen most clearly in an ocean general circulation model.

A global, high-resolution, primitive-equation model has very similar attributes in the Mozambique Channel[325] compared to the more regional models. It exhibits a strong seasonal cycle with amplitudes exceeding 10×10^6 m^3/s, lagging the seasonality in the Somali Current by about a month. No such seasonality is evident in the simulated Agulhas Current. This has led investigators[325] to conclude that the poleward flow in the Mozambique Channel is not a significant contributor to the Agulhas Current. In general, wind-driven models with no thermohaline component should exhibit no strong flow in the channel. Models with thermohaline circulation could. Nevertheless, the enormous discrepancy between the analyses of ships' drift and the modelled wind-driven surface drift in the Mozambique Channel poses some serious problems.

Discrepancies

The first one of these discrepancies concerns the degree to which water derived from the South Equatorial Current does make its way into the northern Mozambique Channel. The dynamic topography (Figure 3.3) shows this to be a substantial part, estimated at between 20 and 26×10^6 m^3/s in the upper 1000 m[75]. An isentropic analysis[82] also shows substantial southward flow in the upper 200 m (Figure 3.15), but none below about 400 m. Measurements of surface drift by GEK (geomagnetic electrokinetograph) in the region (Figure 3.16) portray the suggested bifurcation at the African coast of the extension of the East Madagascar Current quite clearly, as well as a strong southward flow in the northern Mozambique Channel between 10° and 15° S latitude[326–7].

Such a flow is also evident in the results of the only research cruise to date that has attempted to measure the

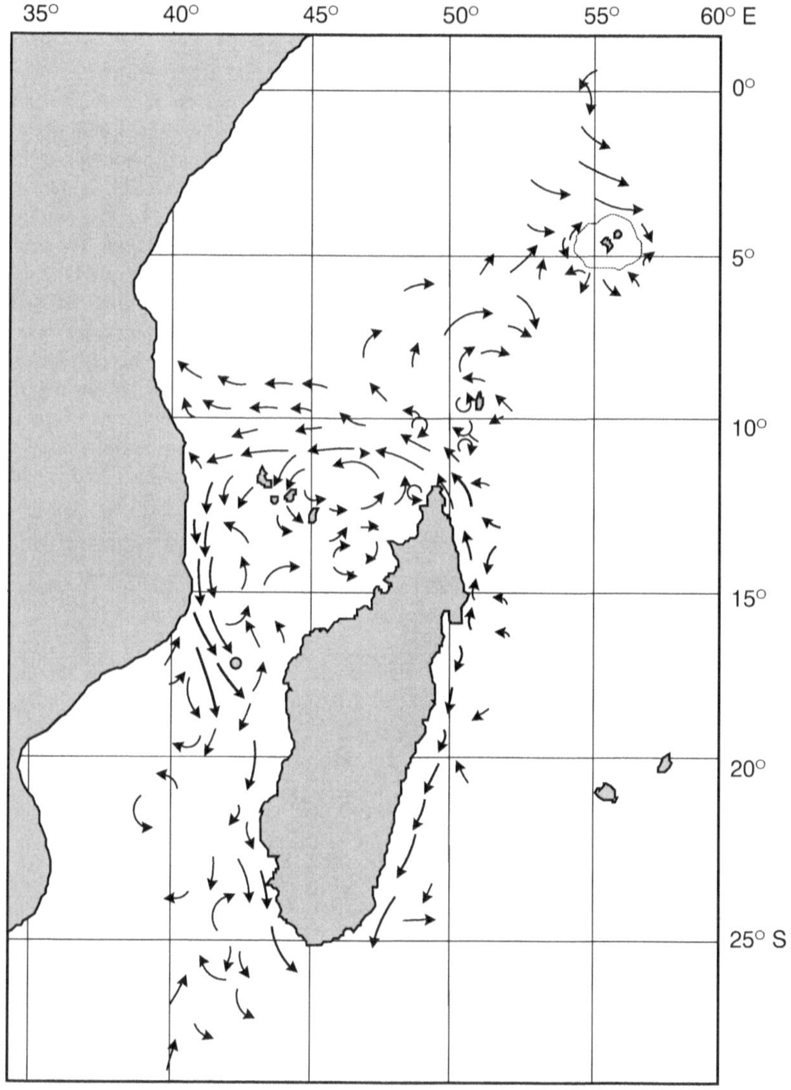

Figure 3.16. Currents in the Mozambique Channel and general environment as measured by GEK[298] (geomagnetic electrokinetograph) over a period of a number of years. There is no consistent northward flow along the eastern side of the channel as suggested from ships' drift observations (viz. Figure 3.2).

hydrography of the whole Mozambique Channel in a quasi-synoptic manner (Figure 3.17). This cruise was undertaken from 11 October to 28 November in 1957 on board the French research vessel *Commandant Robert Giraud*[328]. Unfortunately the station distribution does not include the bifurcation point off eastern Africa, but it does show a strong southward flow along the Mozambican coast at 15° S latitude.

Sætre and Jorge da Silva[79] have gone to great lengths to analyse the data from each individual research cruise that has made observations in the Mozambique Channel up to the time of their investigation. On the whole their results also show a flow southward along the Mozambique coastline in the northern region of the channel (Figure 3.18). Their results, as well as the surface currents measured by GEK (Figure 3.16), the most synoptic dynamic topography (Figure 3.17) and the averaged dynamic topography (Figure 3.3) all show an anti-cyclonic circulation in the northern part of the Mozambique Channel. An analysis of historic hydrographic data (Figure 3.15) suggests that this circulation might extend deeper than 500 m, but it is no longer evident at a depth of about 1200 m.

Flow at the northern mouth of the channel

The results from a model for the region (Figure 3.4) show the northern branch of the East Madagascar Current moving water rapidly past the northern mouth of the channel, nearly blocking it off. Similar models[266] with different internal configurations show much the same. This current branch has a transport, derived from current meter moorings off Cape Amber[297], of about 30×10^6 m^3/s above the 1100 m level and an average velocity at 100 m depth in excess of 0.6 m/s. This high-velocity jet is guided westward by a series of islands and subsurface obstructions[329]. Only three gaps deeper than 2000 m exist in this barrier in the mouth of the Mozambique Channel, the most substantial one at the far western end. It may therefore not be unexpected that this northern branch of the East Madagascar Current has little direct interaction with the waters in the Mozambique Channel. Surface patterns of water mass characteristics support such an interpretation[330].

On the basis of this limited amount of information one might therefore conclude that the flow in the northern part of the Mozambique Channel, the Comores Basin, is variable, but on the whole anti-cyclonic[329]. Some water of east Madagascar origin may on occasion leak southward into the channel along the African coast. One estimate gives this flow as 6×10^6 m^3/s[331] with a geostrophic maximum surface velocity of 0.3 m/s for a southward current with a total width of about 250 km[300]. The persistence of this leakage and its sub-

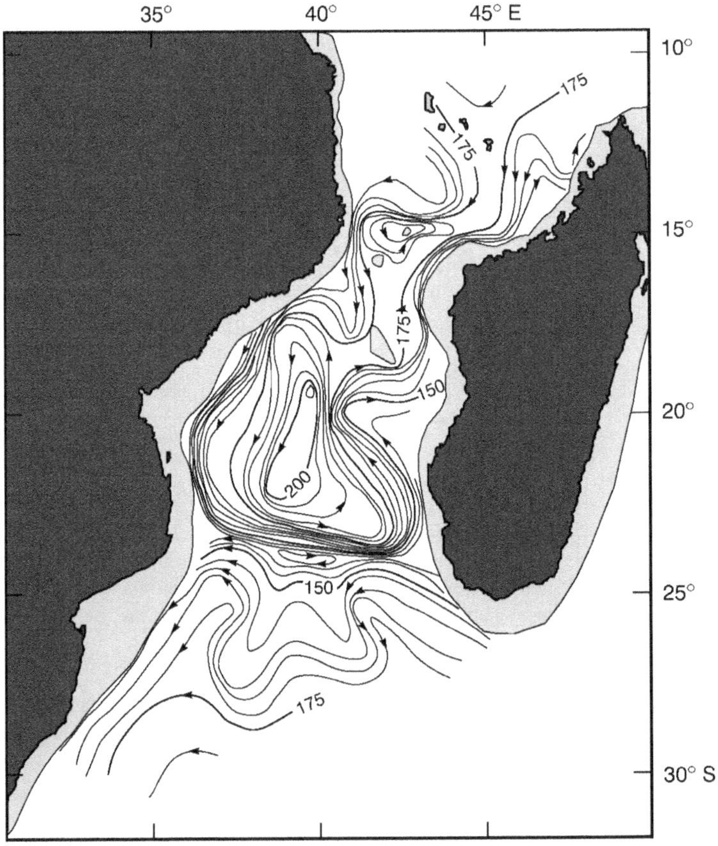

Figure 3.17. Flow patterns in the Mozambique Channel as evident from the dynamic topography of the sea surface relative to 1000 decibar[328]. This is based on a cruise undertaken on board the research vessel *Commandant Robert Giraud* during October–November 1957. The shelf edge is indicated by the 1000 m isobath. Note that different contouring could allow the continuation of the flow along the Mozambican coast.

sequent trajectory have, however, not been unambiguously established.

There are very few direct observations of current profiles in this region. A recent set of current measurements has demonstrated[658] that the surface current along the African coast, in the narrows of the Mozambique Channel, on that occasion was equatorward (Figure 3.19). At 130 m depth this current had nearly disappeared and was replaced by one exhibiting considerable tidal influence. Towards year day 94 (Figure 3.19) the north-setting current was largely destroyed by the advent of cyclone Hudah. These results show that many of the strong surface currents in the Mozambique Channel are restricted to the upper layers and may have their origin in local wind forcing.

Flow in the central part of the channel

South of the narrowest part of the channel at about 17° S latitude, it is shallower than 3000 m, until it again becomes deeper southwards of about 22° S. The circulation in this particular region is not known well. There is considerable evidence from hydrographic data (e.g. Figure 3.3 and 3.17) that a strong northward flow exists along the western coast of Madagascar. Some surface drifters have also moved into the channel along this route[80]. The circulation between this coastal current and the African coastline would then again be anti-cyclonic (Figure 3.17). This is not supported by some observed surface currents (Figure 3.16), and only partially by a careful analysis of all hydrographic information (Figure 3.18). It has in fact been suggested[79] that an inshore counter current moves water southward along the west Madagascar coast. Presumably all these currents would have a marked effect on the shelf currents of the Mozambique Channel.

Flow on the Mozambican shelf

The influence of currents in the off-shelf part of the Mozambique Channel on the water movement over the adjacent shelf of Mozambique has not been adequately studied. First, there is the effect at coastal offsets where lee eddies are prominent. These have been found near the cities of Angoche (viz. Figure 3.25) and Maputo. In the latter case a lee eddy fills the greater part of the Delagoa Bight (viz. Figure 3.26). The biological importance of these eddies may be considerable. The highest concentrations of chlorophyll-*a* measured to date in the whole Mozambique Channel were found in the Angoche eddy[336,809]. These eddies are discussed at greater length below. Based on the limited other observations on the Mozambican shelf, most is known about the Sofala Bank, the shallow continental shelf between 16° and 21° S, to the north of the city of Beira.

This wide shelf is characterized by two contrasting sources of water[822]. First, it is the receptacle for runoff from some major rivers such as the Zambezi. This runoff is highly variable. For instance, during the year 1977–1978 the outflow from the Zambezi was 170 km^3; by contrast during 1982–1983 it was only 50 km^3. After the passage of cyclones this outflow may increase to

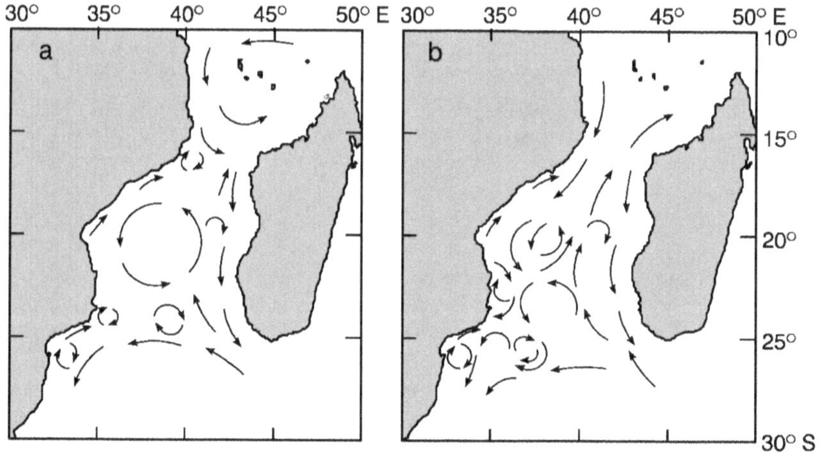

Figure 3.18. Based on a thorough analysis of all available hydrographic data, Sætre and Jorge da Silva[79] have suggested the above circulation patterns for waters in the Mozambique Channel. Panel **a** is for the summer season; panel **b** for the winter season. Note the lack of continuity in the proposed Mozambique Current and the general field of mesoscale eddies in the greater part of the channel.

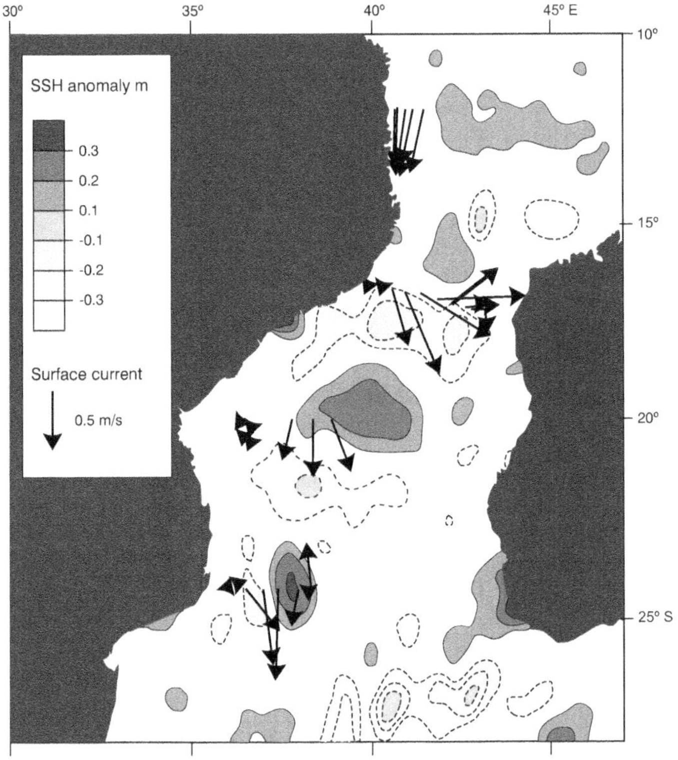

Figure 3.19. Characteristic flow patterns in the Mozambique Channel during March–April 2000[728]. Grey shades show sea surface height anomalies; surface currents were measured by ADCP and averaged over the top 200 m of the water column. Both sets of observations show the dominance of a train of anti-cyclonic eddies in the channel.

such an extent that it would affect the whole shelf region. The downward penetration of this fresh water is, however, highly variable[822]. In certain parts it may be limited to the upper 10 m, while in other parts it may simultaneously extend over the full water column. Near the mouths of the rivers the surface salinity may drop to 20[823] and the fresh water may extend a distance of 50 km offshore. This fresh water influence is counteracted by water from the salt marshes along the southern part of the Sofala Bank[824]. Here sea water may penetrate far inland, particularly during high tide. The salinity of this water is then substantially increased by evaporation and the outflow may be highly saline. Values exceeding 37 have been measured[824] and this very salty water may extend up to 100 km from the coastline.

The circulation on this part of the shelf region as well as in the greater part of the Mozambique Channel remains unclear and may, at best, be considered variable. A careful analysis of ships' drift shows[332] the surface flow along the west coast of Madagascar to be the weakest in the Mozambique Channel and to be highly variable as well.

The Mozambique Current

However, probably the most important question to ask about the movement in the whole Mozambique Channel concerns the existence, persistence and location of a Mozambique Current. Although temperature contrasts at the sea surface are usually weak over the whole channel, Harris et al.[59] have been able to present a thermal infrared satellite image that shows a continuous ribbon of warm water along the path that ships' drift suggest[332] a Mozambique Current would take. These latter data show that surface speeds in excess of 2 m/s are found all along the western border of the channel, but particularly near regions where the African continental shelf is narrowest, such as at 15° S, 24° S and 29° S latitude. It is of interest to note that a simple two-layer, wind-driven model[333] with a flat bottom reproduces these regions of higher flow, and that adding a realistic bottom configuration enhances this effect.

Notwithstanding the complicated internal circulation of the Mozambique Channel[79,81], most – but not all – closely spaced hydrographic station lines perpendicular to the African coast have resolved a southward setting baroclinic current of varying strength[334], usually strong. This current is commonly marked by a subsurface oxygen maximum at about 400 m depth. On some sections across the narrows of the Mozambique Channel this characteristic water mass of the purported Mozambique Current is, however, totally lacking[334]. Zahn[331] has nonetheless estimated the volume transport of the Mozambique Current at 19.2×10^6 m^3/s, suggesting a three-fold increase in its flux from the northern entrance to the channel. The current portrayals have shown that it was usually accompanied by cyclonic offshore eddies, introducing mesoscale variability into the flow regime.

Current variability

Sea height variability as measured by satellite altimetry[183] does show a maximum over the western side of the channel[72,335] (Figure 3.14). This mesoscale variability is much more intense here than in the Agulhas Current, suggesting that a Mozambique Current might by comparison be very variable, meandering in position or shedding eddies. A high-resolution, primitive equation model for the region[271] gives a comparable distribution of the eddy kinetic energy in the channel (Figure 3.20)[301]. This distribution is due to modelled eddies being generated by barotropic instability in the northern part of the channel and moving along the coast (Figure 3.12), eventually being absorbed in the northern Agulhas Current[301]. It may be critical to an understanding of the flow in the Mozambique Channel to note that this model simulates a weak Mozambique Current in the mean drift only, but that in the more synoptic portrayals this mean drift is shown to be constituted by a number of eddies (Figure 3.12). If this is a correct portrayal of the current dynamics in the channel it would go some way to explaining the discrepancy between results from altimetric observations and ship measurements. Some types of eddies have in fact been observed and measured.

Mozambique eddies

In a seminal paper on the flow in the Mozambique Channel, de Ruijter et al.[728] have demonstrated unequivocally that no Mozambique Current exists! During a dedicated cruise[650] to investigate the nature of the purported Mozambique Current, its suggested location adjacent the shelf edge was intersected with eight

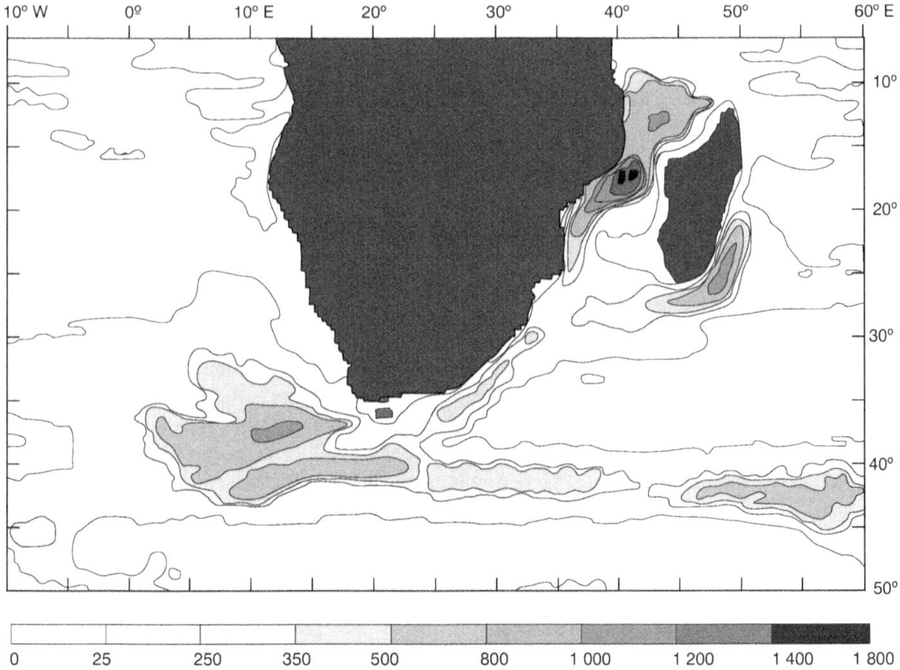

Figure 3.20. The geographic distribution of the mean kinetic energy at 41 m depth for the greater Agulhas Current system, according to a primitive equation model with high spatial resolution[271]. Units are cm^2/s^2. Note in particular the high energy values in the northern Mozambique Channel[301].

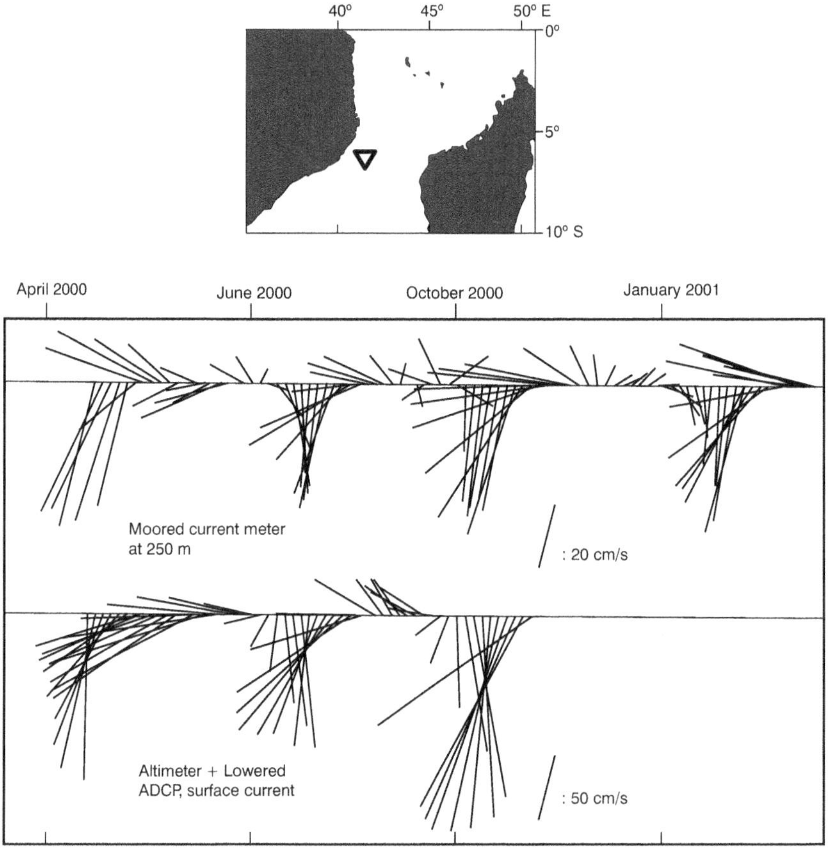

Figure 3.21. Current meter records in the narrows of the Mozambique Channel for a period of 10 months[735]. The map indicates the geographic location of the current meter mooring as well as that of the other observations. The top part of the main panel shows the currents at a depth of 250 m; the lower part shows the geostrophic estimation of the current based on altimetry but referenced to ADCP observations. The agreement is good and shows by the regular current reversals in a clockwise rotation how the western side of anti-cyclonic eddies pass the mooring about four times per year. Note that the scales for the two current records are different and that the current is considerably stronger at the sea surface than at 250 m depth.

hydrographic sections. In no instance was a continuous, persistent western boundary current found. Instead, a number of anti-cyclonic eddies were found, moving poleward as a train of disturbances to the average flow (Figure 3.19). These observational results[650] are in full agreement with the results of the most recent modelling[301] and with portrayals of variability in sea surface height. They also are in total accordance with the current meter records in the narrows of the channel (viz. Figure 3.21) that show the regular creation of eddies[733]. However, even though they do not conflict with results based on ships' drift[293], they are inconsistent with most interpretations of those spot measurements. These erroneous interpretations of the drift in the Mozambique Channel (e.g. Figure 3.2) seem a clear case of confusing Eulerian and Lagrangian results. What are the characteristics of these Mozambique eddies?

First, they exhibit diameters in excess of 300 km, they extend to the ocean floor where their azimuthal speeds are still 10 cm/s, and they cause a net transport of about 15 × 10^6 m^3/s southward through the channel (Figure 3.19). At intermediate depths their azimuthal speeds lie between 10–20 cm/s, a value that has been corroborated by an analysis[732] of the movements of floats. The poleward translation of Mozambique eddies occurs at about 4.5 km/day, but varies with location in the channel[735]. It is about 6 km/day between 12° and 27° S, except between 18° and 21° S where they slow down to 3–4 km/day and south of the channel mouth where they speed up to 8–10 km/day. It is estimated that

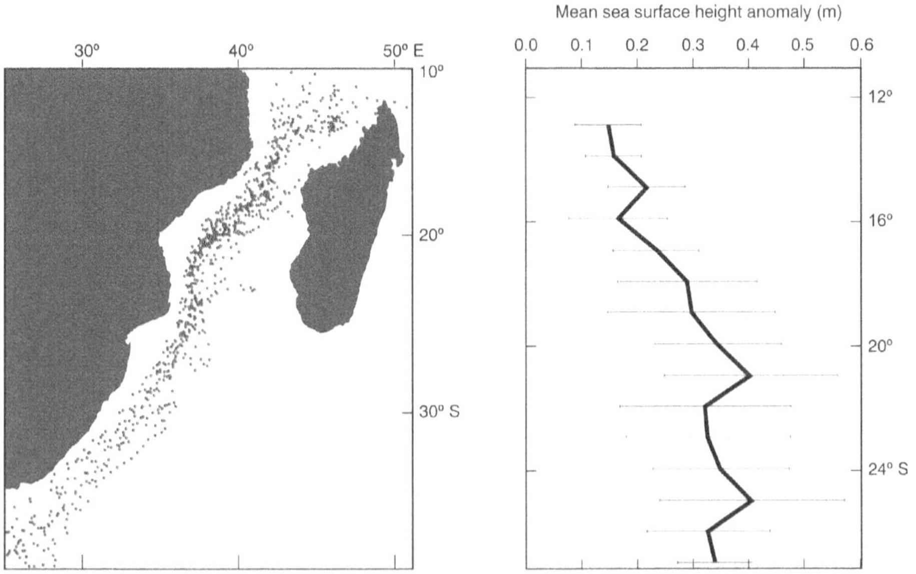

Figure 3.22. The geographic positions of 25 Mozambique eddies as they moved southward from the Mozambique Channel during the period 1995–2000[735] (left-hand panel). A rather distinct corridor is evident. The right-hand panel shows the average increase in energy of the eddies – as expressed by increases in sea surface height – on their way poleward. Error bars give one standard deviation for the measurements of 25 eddies.

about four per year are formed[733] (Figure 3.21) with a marked regular frequency, hence on average three Mozambique eddies should be present in the channel at any one time. As these eddies move past the Mozambican shelf, they sweep coastal waters into mid-channel[649].

It is estimated[728] that these eddies trap water to a depth of 1500 m. This means that they carry Tropical Surface Water, Subtropical Surface Water as well as intermediate water poleward. The intermediate water comes in two well-defined types: Antarctic Intermediate Water and Red Sea Water. The Mozambique eddies are thus effective transporters of Red Sea Water towards the Agulhas Current. In observations of a number of these eddies, as well as elsewhere in the Mozambique Channel, there is evidence of substantial interleaving between these two intermediate water types. Analyses of float trajectories at intermediate depths[732], has shown that these floats spend only short periods in Mozambique eddies before being spun out. This result supports the concept of considerable mixing between the waters in these eddies and the ambient water masses.

From Figure 3.22 some unusual characteristics of Mozambique eddies become apparent. First, it is clear that they follow a very circumscribed path in the western part of the Mozambique Channel and subsequently along the Agulhas Current where they may be absorbed into that current[728] (Figure 3.23). Some move all the way to the Agulhas retroflection. Second, many of these anomalies have their origin in the Comores Basin, equatorward of the narrows. The narrows are therefore probably not the sole source of these features. It has been shown by Ridderinkhof and de Ruijter[733] that the flow through the narrows are represented by two strikingly different regimes. One consists of strong currents across the full longitudinal extent of the narrows, the other with only weak currents. The strong current regime has by far the longer duration. During each strong current period, a Mozambique eddy is formed. The formation of these eddies is directly related to the level of the volume transport through the channel. There is evidence[736] that the formation frequency of Mozambique eddies is remotely controlled by processes in the Indian Ocean as a whole. Perturbations to the oceanic circulation are carried as Rossby waves across the width of the subtropical Indian Ocean where their interaction with Madagascar triggers the formation of Mozambique eddies. This proposed process is described in greater detail later.

The third unusual characteristic of the Mozambique eddies is the geographical energy distribution, as expressed in anomalies of the sea surface height (Figure 3.22). It increases considerably from just north of the narrows to a region close to the southern mouth of the channel[735]. This increase is in agreement with the expected enhancement of eddy intensity due to their lati-

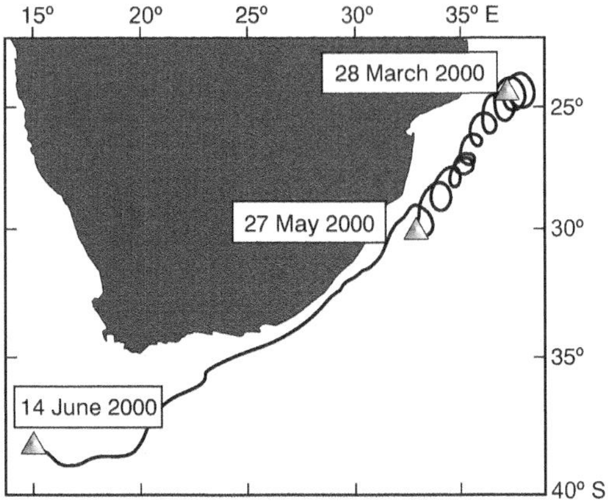

Figure 3.23. Trajectory of a surface drifter released in an eddy in the Mozambique Channel[728]. Dates indicate the location of the drifter on that day. Anti-cyclonic gyrations in the eddy ceased when the eddy got absorbed by the Agulhas Current.

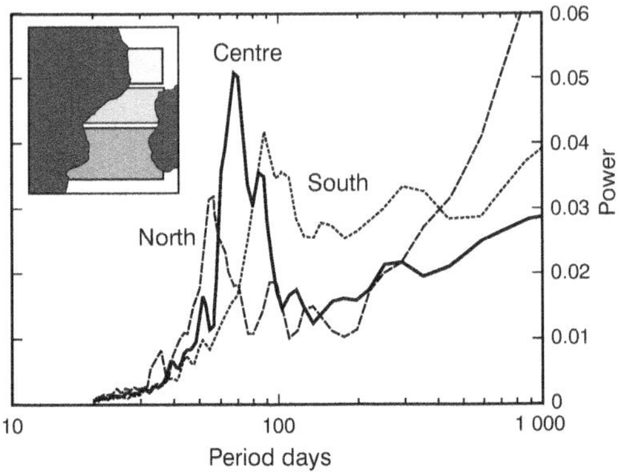

Figure 3.24. The average spectra of sea surface height for three regions in the Mozambique Channel[735], as shown in the inset. The spectra have been normalized by their total variance, thus effectively removing the inter-annual signal. Power peaks shift to longer periods as one proceeds southward in the channel.

tudinal displacement through the planetary vorticity gradient, but this factor is insufficient to account for the entire observed increase[735].

Furthermore, the dominant frequency of variability decreases poleward in the Mozambique Channel (Figure 3.24). The northern region has a peak at a period of 55 days, in agreement with previous observations[297,737] in the South Equatorial Current. In the southern part of the channel this period is increased to about 90 days. Why should seven eddies per year get reduced to about four per year further south? It is possible that a number of Mozambique eddies dissipate at about 20° S and that some others may amalgamate[735], which might cause their energy to increase, as has been observed.

Modelling of Mozambique eddies has been only partially successful. In some models[301] the eddies are too shallow, in others[735] too surface intensified and being shed too frequently. Nevertheless, it is worth noting – as discussed above – that in all modern, eddy-resolving models[274,732] such simulated eddies appear and are the dominant feature of the circulation in the Mozambique Channel.

Of some interest is the influence of these eddies on the local marine ecosystem and in particular the feeding habits of marine predators. An investigation was carried out on the foraging behaviour of great frigatebirds[642] from Europa Island that feed in the centre of the Mozambique Channel. They were tracked using satellite transmitters. It has shown that this top predator tends to avoid the centres of eddies. It feeds preferentially at the edges of Mozambique eddies where the primary productivity is higher.

The Mozambique Undercurrent

One of the most intriguing results of modern, high-quality observations at the continental slope of Mozambique[728,733] has been the discovery of a undercurrent carrying water equatorward. It seems to have a core with maximum velocity at about 2400 m and at 1000 m depth. The maximum velocities at 2400 m were 0.2 m/s. This undercurrent was observed to carry Antarctic Intermediate Water at the lower depth and North Atlantic Deep Water at the greater depth. This was at 24° S. In the narrows of the channel, at about 17° S, the cores of the Mozambique Undercurrent were located[733] at 1500 m and 2500 m. The mean velocity was 4.6 cm/s in both.

Coastal lee eddies

Some of the eddies associated with the flow in the Mozambique Channel may be in the form of trapped, lee eddies[336]. Such a feature is shown in Figure 3.25. The dynamic topography based on these cruise data shows a poleward movement along the whole extent of the coastline where the cruise took place, but intensifying strongly along the headlands at 15° S. Downstream of here it overshoots the shelf edge, driving a cyclonic lee eddy[809–10] with dimensions of about 100 km. Deeper water is being upwelled in this feature[337], shown by an enhanced nutrient content of more than 12 μmol/l nitrate-nitrogen at 75 m depth in the eddy compared to about 2–4 μmol/l in the core of the offshore current[336]. A resultant peak in chlorophyll-*a* concentrations was also observed in this eddy.

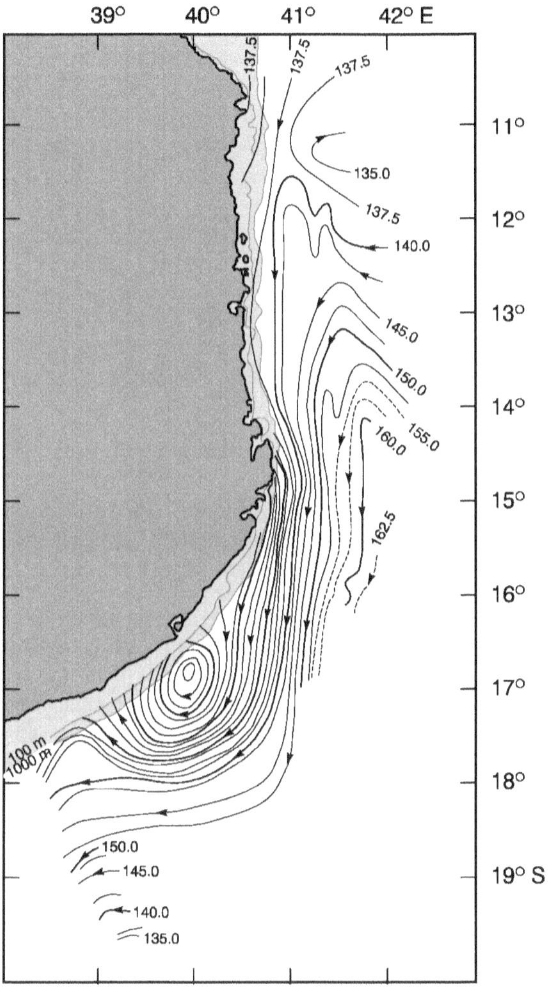

Figure 3.25. A portion of the flow in the northern Mozambique Channel as portrayed by the dynamic topography of the sea surface relative to the 600 decibar surface[336]. Note the evidence for a trapped lee eddy at 17° S latitude. The extent of the continental shelf is indicated by the 100 m and 1000 m isobaths.

These eddies are of considerable importance to an understanding of the current dynamics since in similar locations, such as the Delagoa Bight[338] (Figure 3.26) and the Natal Bight[62] features of the same kind have been observed. Based on all currently available data, the presence of the Delagoa Bight eddy seems to be very persistent. There is evidence[649] that this eddy may be influenced by the changing flow intensities as Mozambique eddies move past the bight. Furthermore, a presumed upwelling cell at the north-eastern corner of the Delagoa Bight can at times be inferred from ocean colour imagery[649]. The occurrence of this particular coastal upwelling seems to be far less consistent than for similar upwelling cells inshore of the southern limb of the East Madagascar Current[647] and inshore of the Agulhas Current[163,166]. The latter ones are driven by a consistent western boundary current; the one in the Delagoa Bight seemingly only by the occasionally passing Mozambique eddies. The intermittency of this upwelling cell may therefore be directly related to its forcing source. The trapped lee eddies may in each case[62,399] become part of a major disturbance to offshore flow. It is possible that the wide shelf between 15° and 20° S latitude in the Mozambique Channel (viz. Figure 3.17) and the weak gradient in the shelf slope may allow barotropic instability in adjacent flow here, thus adding to the major variability visible in the satellite altimetry[335].

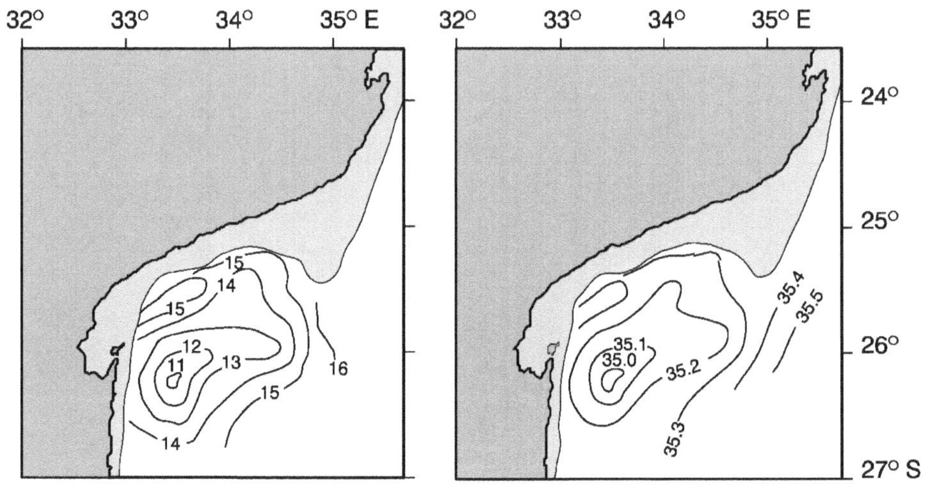

Figure 3.26. The distribution of temperature and salinity respectively at 200 m depth in the Delagoa Bight in the southern Mozambique Channel. An extensive lee eddy is circumscribed[338]. The lightly shaded area indicates the location of the 200 m isobath.

On the basis of the available data it may therefore be concluded that the flow in the Mozambique Channel is dominated by the poleward drift of anti-cyclonic eddies that are largely formed at the narrows of the channel. A continuous, persistent western boundary current in the form of a Mozambique Current, as portrayed in most atlases and textbooks, does not exist. The driving force for such a presumed current always was unclear, since the Sverdrup relationship of zonally integrated wind stress curl shows[187] the Mozambique Channel as a wind-shadow area with therefore no need for a wind-driven western boundary current. These Mozambique eddies move poleward on the western side of the channel. The remaining circulation in the channel seems to consist of weak net flow with a range of eddies and ephemeral circulation elements. Whereas the data in the central and eastern side of channel currently suggest conflicting portrayals of the circulation, this is even more so for the southern mouth of the channel.

Flow though the southern mouth

While most investigators who have worked with the historical hydrographic data set[75,84] have shown a strong flow leaving the western side of the mouth of the channel[82] (e.g. Figure 3.15), the results of the one cruise that has given a good coverage of the whole channel suggests no water leaving the channel southwards[328] (Figure 3.17). The concept of no throughflow has been perpetuated in some oceanographic encyclopaedias[217]. Based on these results Zahn[331] has in fact calculated a net northward flux of about 5 × 10^6 m^3/s through the southern mouth of the channel based on the data collected by Menache[328] in 1957. Recently published satellite imagery for the south-western region of the Mozambique Channel[340] suggests a well-developed current here, at least at the sea surface. A line of closely spaced hydrographic measurements northward into the southern Mozambique Channel[340] shows a weak zonal front at the mouth of the channel with general movement from east to west.

A careful study of the best existing data set (Figure 3.17) shows that the contouring of isolines in this figure between the two most southerly station lines is quite subjective and that these data can be interpreted in a different way. Harris[81] has done this and has derived a volume transport of 25 × 10^6 m^3/s for the presumed Mozambique Current from the same data. Based on the seasonal groupings of historic hydrographic data Duncan[75] has found a highest value of 26 × 10^6 m^3/s above 1000 decibar in winter and a low of 20 × 10^6 m^3/s in summer. Nonetheless, the net flux across the mouth established by Harris[81] is nearly identical to that found by Zahn[331] and is northward.

A closely spaced section across the mouth of the channel carried out in June 1965 on board the research vessel *Atlantis II*[300] shows a well-developed poleward flow about 250 m wide, with a net mass flux southward out of the channel of 6 × 10^6 m^3/s. Based on two hydrographic sections that close off the mouths of the channel, Fu[300] has concluded that the flux through the northern mouth is distributed at depths between the surface to 1500 m depth, whereas the flow through the southern mouth is confined mostly to the upper layers, necessitating considerably internal, non wind-driven upwelling in the channel. By far the most detailed and most accurate hydrographic data across the southern mouth of the Mozambique Channel have been collected during the World Ocean Circulation Experiment in 1995. Based on these data Donohue and Toole[708] have calculate a poleward flow of 18 × 10^6 m^3/s. Keeping in mind the range of temporal variability[259] in the flows observed, this result must remain the most reliable to date. The high variability of this transport may also be influenced by the flow in the eastern part of the Mozambique Channel, about which very little is as yet known, and by the intermittent presence of eddies along the west coast of Madagascar. A study of float trajectories at intermediate depths[732] has shown that most moved southward across the mouth of the channel, except near the south-west coast of Madagascar where there is some evidence for occasional equatorward movement.

Flow along the west coast of Madagascar

If the poleward flow in the Mozambique Channel is indeed well developed at a latitude of 25° S, as shown by hydrographic sections off Maputo[338], and if the throughflow is less than that carried by the Mozambique eddies that constitute this flow, how representative is the northward flow in the eastern part of the channel as suggested by the ships' drift data (e.g. Figure 3.2), one set of direct measurements[820] and some portrayals of the dynamic topography based on historic hydrographic data (Figure 3.3)? Isentropic analysis (Figure 3.15) shows this northward flow, weakly at the surface, but in contrast a true inflow at depth.

Some drifting buoys[80,308] have followed a path analogous to the 500 m depth isoline in the lower panel of Figure 3.15, moving north, then west and then joining the flow southward. Harris[81] has analysed one of the few quasi-synoptic cruises for the Mozambique Channel as a whole, by the research vessel *Almirante Lacerda* in 1964, and found, based on water mass characteristics, a substantial inflow from the south along the eastern side of the channel. Although he has recognised no inflow, Menache[328] (Figure 3.17) has also shown a substantial northward flow on this side.

Modern remote sensing observations have shown (Figure 3.11) that cyclonic eddies are often found in this region and this finding has been supported by some fortuitous hydrographic observations[649]. Such eddies have a characteristic diameter of about 250 km, but may be fairly shallow. They are seen (Figure 3.11) to draw off waters with high chlorophyll from the adjacent shelf and may thus have a substantial local impact on the shelf ecosystem. How do these particular eddies come about? Two possible mechanisms have been proposed[649]. The first is the formation of a cyclonic flow in the lee of the southern tip of Madagascar driven by the passing water of the southern limb of the East Madagascar Current. The second possibility would be that such eddies are formed further to the south and drift into the Mozambique Channel. De Ruijter et al.[652] have shown that vortex dipoles are frequently formed south of Madagascar where the East Madagascar Current separates from the Madagascarian shelf. These dipoles consist of a cyclone on the equatorward side and an anti-cyclone on the poleward side, both usually moving off in a south-westward direction. From an analysis of the hydrographic characteristics of these features it has been inferred[652] that cyclones that form part of the dipole draw their water from the inshore side of the East Madagascar Current, suggesting that they are formed as lee eddies.

The Delagoa Bight eddy

An interpretation of the flow in the southern reaches of the Mozambique Channel, and its influence on the Agulhas Current, may be complicated by the presence of intense mesoscale eddies of unknown origin. First, there is the well-documented cyclonic eddy usually present in the Delagoa Bight[338] (Figure 3.26), mentioned above. This is probably a trapped lee eddy, driven by the rapid flow of the adjacent waters past the substantial promontory. Over a period of 23 years it has been observed 11 times. Sedimentary records in the Delagoa Bight show that an eddy of this kind has occurred here since the Pliocene-Recent, i.e. about 1.8 million years ago[341], and that present sedimentary patterns are determined by the flow configuration of the adjacent flow in the deeper ocean[342]. Water masses in this Delagoa Bight eddy have temperature–salinity characteristics implying substantial upwelling in the core of the eddy from depths of at least 900 m.

The escape of this eddy, causing – or caused by – a huge meander in the downstream path of a presumed Mozambique Current has been described using satellite thermal infrared images[79], but has been observed infrequently. The meander was not observed to move far downstream, dissipating instead over the Mozambique Plateau (viz. Figure 1.2). Such movement has also been observed in satellite altimetry[308].

Altimetric data have moreover shown the southward drift of sea height anomalies southwards from the Mozambique Channel. These eddies are probably for the greater part Mozambique eddies, but a small number may have their origin in the Delagoa Bight. Eddies observed directly south of the channel, over the northern Mozambique Plateau[343,344], but also over the

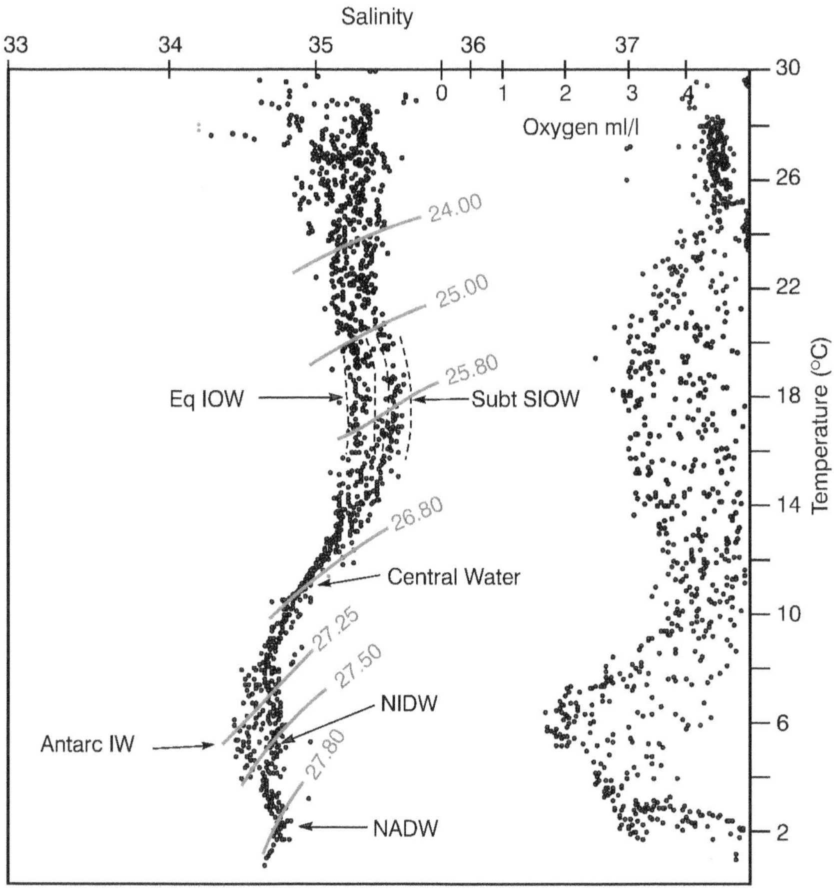

Figure 3.27. The temperature–salinity and temperature–oxygen characteristics of the water masses in the Mozambique Channel[237]. Sigma-t values have been added on the T–S scattergram. Water types that are evident are Equatorial Indian Ocean Water (**Eq IOW**), Subtropical South Indian Ocean Water (**Subt SIOW**), Central Water, Antarctic Intermediate Water (**Antarc IW**), North Indian Deep Water (**NIDW**) and North Atlantic Deep Water (**NADW**).

northern Natal Valley[343] may therefore have their origin either in the channel narrows or the Delagoa Bight. These eddies have diameters of 100 km or more and have an elevation of their 10 °C isotherm of 270 m at the eddy centre. Circular transports of up to 18×10^6 m^3/s have been observed[344]. These values indicate very energetic eddies of a kind usually associated with rings from western boundary currents. The eddies may all contribute to the high mesoscale variability in the region between the Madagascar Ridge and the southern African subcontinent[335] (viz. Figure 3.14) that will be dealt with in more detail later. In order to trace the different water masses that contribute to the Agulhas Current, either directly or by these eddies, a thorough knowledge of the characteristics of the water masses in the Mozambique Channel is required.

Water mass characteristics

A scattergram for the temperature–salinity characteristics of the whole Mozambique Channel is given in Figure 3.27. Sætre and Jorge da Silva[79] have identified three typical T–S traces for waters found in the Mozambique Channel; one representative of the north Indian monsoonal circulation that has very weak, rudimentary remains of the salinity minimum signifying the presence of Antarctic Intermediate Water[237] (viz. Figure 3.27). It also has hardly any sign of a salinity maximum at about 18 °C, showing that at this temperature the water is mostly Equatorial Indian Ocean Water. These water characteristics are only found to the north of the Mozambique Channel with occasional penetration along the western side of the channel[79].

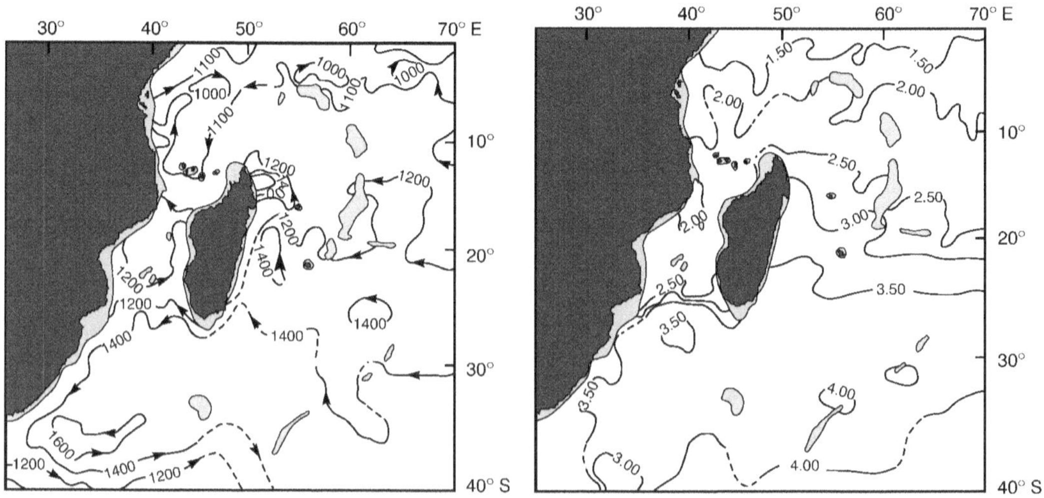

Figure 3.28. The depth of the sigma-t 27.50 surface (left-hand panel) and the distribution of dissolved oxygen content, in ml/l, on the same surface (right-hand panel) in the South West Indian Ocean. This distribution is based on the hydrographic data collected during the north-east monsoon season[82]. Dashed lines are given in regions with sparse data. This particular density surface lies in the Antarctic Intermediate Water and North Indian Deep Water cores (viz. Figure 3.27) and suggests that Antarctic Intermediate Water, coming from the south, does not penetrate far into the Mozambique Channel. Regions shallower than this surface have been lightly shaded.

The second typical temperature–salinity trace for the channel has distinctly more Subtropical South Indian Ocean Water (viz. Figure 3.27), marginally more Antarctic Intermediate Water, but still hardly any sign of North Atlantic Deep Water. It covers about two-thirds of the channel. Only the southernmost third of the channel has clear signs of being much like the rest of the water masses of the South West Indian Ocean[237], with a predominance of Antarctic Intermediate Water, Subtropical South Indian Ocean Water and North Atlantic Deep Water.

One might expect the characteristics of the Equatorial Indian Ocean Water, moving south at about 200 m depth, rapidly to be dissipated by mixing with the ambient Subtropical South Indian Ocean Water. This would not be the case for the deeper water masses. Of particular importance here is to establish the relative degrees of penetration of Antarctic Intermediate Water and North Indian Deep Water. The latter water mass has also been called Red Sea[233, 329], Arabian Sea[345], North Indian Intermediate[178] and Red Sea–Persian Gulf Intermediate Water[220]. (An excellent summary of terminology on water masses for the Mozambique Channel, as used in the scientific literature, has been given by Sætre and Jorge da Silva[79].)

From Figure 3.27 it is clear that both water types are present in the channel, with North Indian Deep Water perhaps having a more dominant presence. Meridional sections through the channel[234,328] show the salinity minimum of Antarctic Intermediate Water petering out between 20° and 18° S in the channel. This water mass does not seem to progress any further north in the channel[79]. An isentropic analysis of the sigma-t surface of this salinity minimum core (viz. Figure 3.27) shows that Antarctic Intermediate Water does not penetrate far into the channel[82], but is taken up by the anti-cyclonic gyre of the South West Indian Ocean (Figure 3.28). Water masses with low values of dissolved oxygen, associated with North Indian Deep Water, fill most of the intermediate depths (Figure 3.28). This is also evident from many meridional sections through the channel[79] using chlorofluorocarbon as a tracer.

At greater depths only North Atlantic Deep Water is found (Figure 3.27), since most of the channel is less than 3000 m deep; the deepest part, to the south, only reaching 4000 m.

Presently available hydrographic data thus characterise the waters of the Mozambique Channel as being a mixture of that found in the monsoonal regime to the north, and the South West Indian Ocean sub-gyre to the south. Water types from the north seem to be dominant, especially in the southward flow along the western side. Penetration from the south seems to be restricted to the surface layer and upper layers of the intermediate depths.

If this is indeed a correct interpretation of the distribution of water mass characteristics in the Mozambique

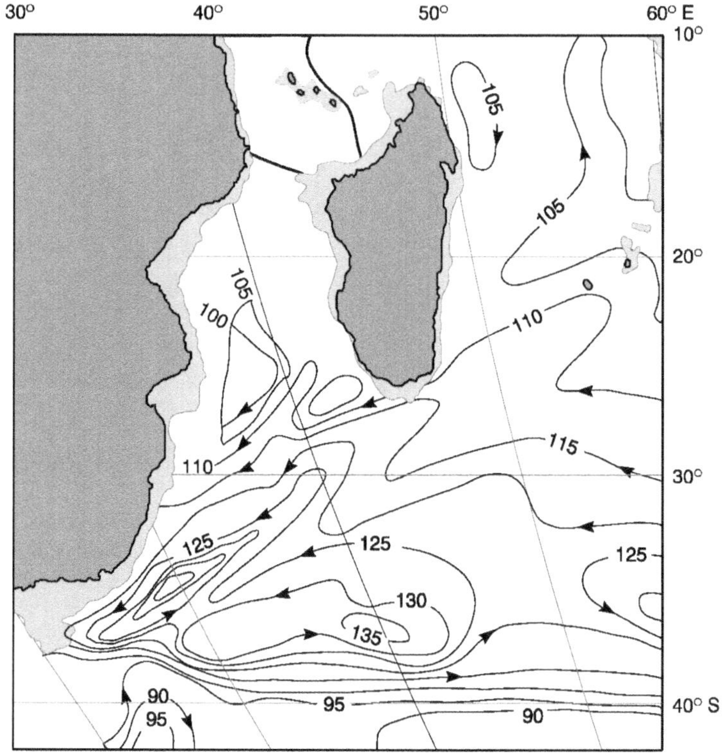

Figure 3.29. Geostrophic circulation in the South West Indian Ocean at 1000 m relative to 3000 m[84]. The 1000 m isobath at the coast is indicated. A tight sub-gyre west of 45° E longitude is evident. This may be compared with the circulation shown in Figures 3.15 and 3.28 based on isentropic analyses.

Channel, it means that water from this channel contributes little to the region further south and, hence, the role of the South West Indian Ocean sub-gyre in feeding the Agulhas Current must be of considerable importance.

Recirculation in a South West Indian Ocean sub-gyre

Although original descriptions of the flow in the South West Indian Ocean give no indications of any perception of a well-developed anti-cyclonic, closed, sub-gyre (viz. Figure 3.1), subsequent portrayals of surface drift do[44]. Closer inspection of the database used for these portrayals unfortunately indicates that most of these were drawn with considerable artistic licence to fill gaps in the data coverage.

Recirculation

Using all the data from the International Indian Ocean Expedition, Duncan[75] has been able to show clear indications of a closed circulation during all three-month periods, but with markedly different lateral dimensions. For this the radically different data distributions between these periods is to blame. The dynamic topography for the sea surface relative to 1000 decibar shows this sub-gyre, but so does the 1000 m surface relative to 3000 m[84] (Figure 3.29). Using an enhanced data set[82] this recirculation cell can be shown to exist throughout the water column (Figure 3.15). All these data sets share a dearth of stations south of Madagascar[184], making it impossible to judge accurately the eastward extent of this sub-gyre.

Recent tracer studies[252] suggest that the circulation of Antarctic Intermediate Water is in fact constrained by the Madagascar Ridge at about 46° E longitude (viz. Figure 1.2). Surface drifters[310–11,346] have not circumscribed this gyre at the sea surface. No drifters to date have moved northwards at a longitude of 40° to 50° E. Using a carefully selected collection of existing hydrographic station lines (Figure 3.30), the best estimates to date of the geographic dimensions as well as volume flux components of the South West Indian Ocean sub-gyre have been made[101].

A South West Indian Ocean sub-gyre

The baroclinic transport in the upper 1000 m shows that about 60 per cent of the flow of the Agulhas Current is returned in the sub-gyre, the eastern limb of the gyre lying between 50° and 70° E. Nevertheless, precise information on the dimensions and flux patterns of the South West Indian Ocean sub-gyre awaits the accumulation of more, and more detailed, hydrographic or altimetric information. It is interesting to note that Feron et al.[347], while developing a method to determine the mean sea surface dynamic topography from satellite altimeter observations, have demonstrated that it is the divergence of the eddy stresses that enhances the formation of a closed sub-gyre in the South West Indian Ocean. It is abundantly evident from these and other analyses that the anti-cyclonic, gyral circulation in the subtropics of the South Indian Ocean is substantially different from that of similar ocean basins such as, for example, the South Atlantic Ocean[101].

The contribution that this sub-gyre makes to the Agulhas Current is of overriding importance. Preliminary analyses[81] have suggested 27 × 10^6 m^3/s, or 40 per cent of the volume transport of the current. Using non-synoptic data, a recirculation flux of between 12 and 18 × 10^6 m^3/s was found[75] based on the surface to 1000 decibar geostrophic transports plus 20 per cent for that below 1000 m. Using a wider-ranging set of existing hydrographic stations[101] (Figure 3.30), this value has been increased to 35 × 10^6 m^3/s, including the top 1000 m only. Notwithstanding the encouraging agreement between these analyses, the underlying database for establishing these figures remains very poor.

Sources of the Agulhas Current

Harris[81] has concluded that the inflow to the Agulhas Current can be established from the gradients in properties across the current. He has considered the inner, shoreward waters as having a Mozambique Channel origin, while those of the outer current are more similar to the waters coming from the east and being recirculated. Using a historical mean data set[348], Gordon[66] has come to the same conclusion. It is of importance to take cognisance of how Harris perceived this recirculation.

The flow of Agulhas Water eastward, after retroflection (Figure 1.2), crosses the Agulhas Plateau and

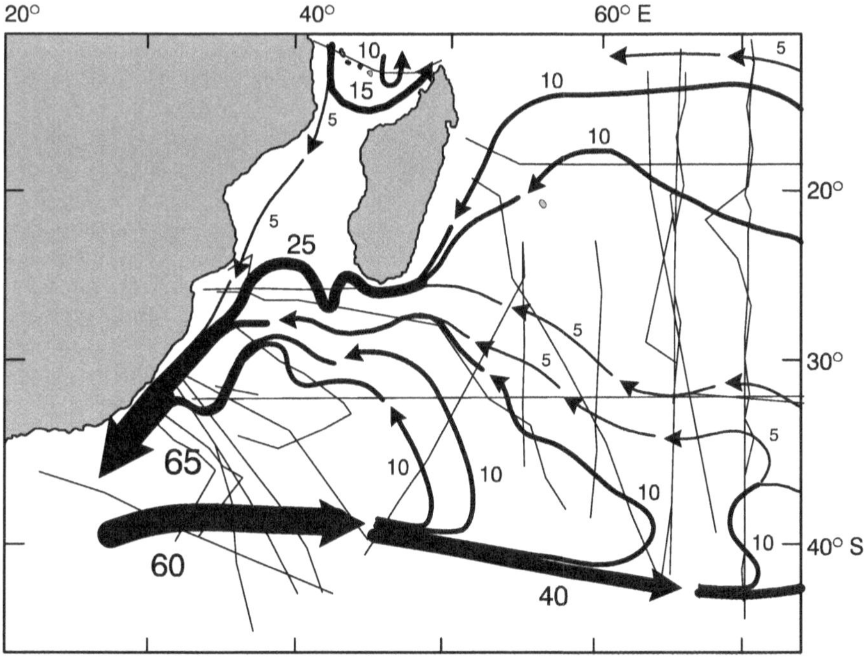

Figure 3.30. A schematic portrayal of the volume transport field of the South West Indian Ocean sub-gyre[101]. Baroclinic transport is for the upper 1000 m and in 10^6 m^3/s. Thin lines indicate the hydrographic station lines used. See also Figure 2.4. The flow through the Mozambique Channel makes a minute contribution to that of the Agulhas Current, but this result is based on a minimal number of data.

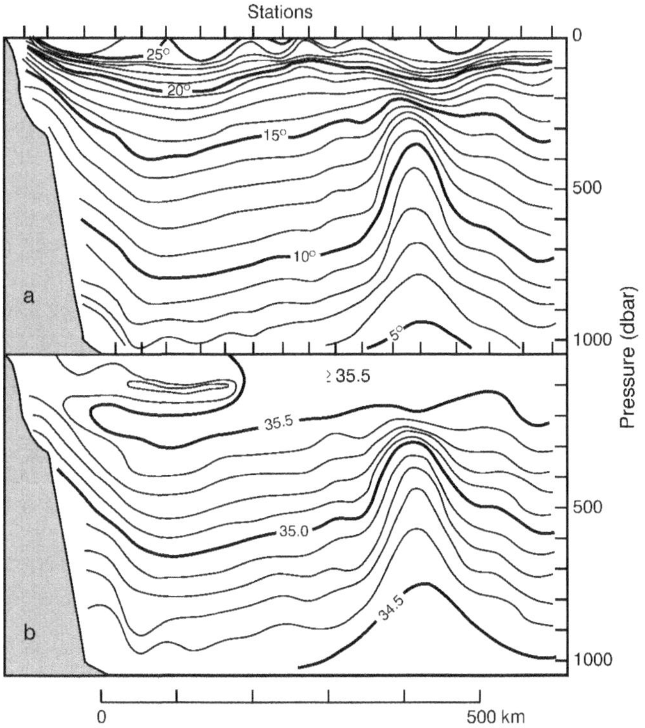

Figure 3.31. A zonal temperature (**a**) and salinity (**b**) section from Durban across the Agulhas Current as well as a typical, cyclonic, deep-sea eddy in the South West Indian Ocean[355]. The geographic locations of the line and the stations are given in Figure 3.32. The slopes in the shown variables are greater for the eddy than for the Agulhas Current.

the Mozambique Plateau extension causing extensive northward excursions in the path of the eastward flow[44,349–50]. These meanders were believed to have a propensity for becoming unstable[98], the eastward flow of the remaining Agulhas Water thus being directed northwards into a closed, anti-cyclonic, recirculation vortex[81]. Drifters that have moved through the region[310,346] have shown no evidence for such an effect. Although the evidence for the existence of a South West Indian Ocean sub-gyre is therefore well established, the deflection of surface water from the Agulhas Return Current northward to join this recirculation at the ridges is not, particularly not so west of about 50° E (Figure 3.29).

Analyses of temporal sea height variability according to altimetric measurements from satellite has shown[73,335] that the centre of this South West Indian Ocean sub-gyre exhibits anomalously high variability values (viz. Figure 3.14). This strongly suggests the passage of intense eddies through this region since no intense, meandering currents are to be expected. As mentioned above, such eddy features have been found and have been studied in detail.

Oceanic eddies

The first major, distinct and well-formed deep-sea eddy of this kind was observed over the northern Mozambique Plateau, at about 28° S, between Maputo and Durban (viz. Figure 1.2) during a cruise in 1977[344]. This discovery led to a decade-long series of cruises that have allowed Gründlingh to establish the locations, dimensions, dynamic characteristics and drifts of many of these deep-sea eddies[355] to a level of detail reached for very few other parts of the Agulhas Current system (viz. Figure 3.31).

The first observed cyclonic eddy[344] had surface speeds of up to 1 m/s, a diameter of about 100 km and a circular transport lying somewhere between 6 and 18×10^6 m^3/s (Figure 3.32). It was characterised by an elevation of the 10 °C isotherm from a depth of 780 m outside the eddy to 480 m at its centre. It furthermore exhibited clear indications of an inverse cap of warmer water in the upper 250 m. Subsequent investigations[356–60] have shown eddies over all parts of the Mozambique Plateau, next to it[357] (see Figure 3.32) and in the adjacent northern Natal Basin to the east of this plateau. Diameters have

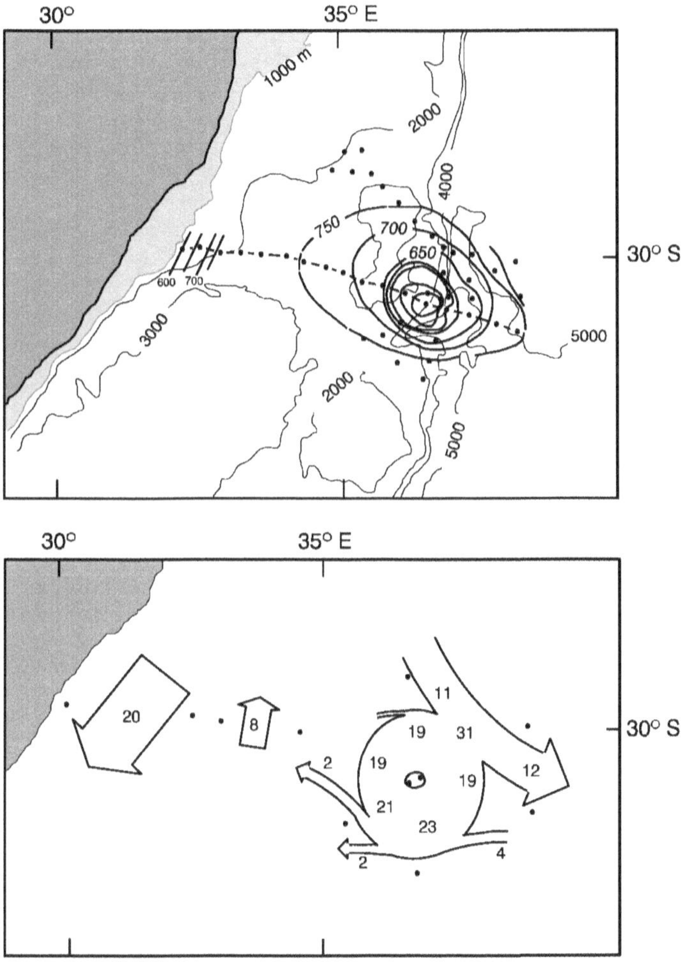

Figure 3.32. Some characteristics of a deep-sea eddy in the South West Indian Ocean[355]. In the top panel the depth of the 10 °C isotherm is given in decibar relative to the station distribution and the bathymetry. The temperature and salinities measured on the near-zonal section from Durban through the eddy are portrayed in Figure 3.31. In the lower panel the distribution of the volume transport is given for the same features relative to 1000 decibar, in units of 10^6 m^3/s. Stations used are shown. The transport around the eddy is the same as that observed for the Agulhas Current at the same time.

varied around 300 km[355-6]. Some were clearly circular[355] while others were elliptical with minor axis 90 km and major axis 240 km[356]. Central elevations of the 10 °C isotherm varied (Figure 3.31) perhaps indicating ageing of these features in eddies where the elevation was smaller. An extreme depth difference between that in the ambient waters (800 m) and in the eddy centre (300 m)[355] was observed in 1982.

Not only cyclonic eddies, but also anti-cyclonic eddies have been observed[358] and although anti-cyclonic eddies seem to be present less frequently[355], satellite altimetry[309] indicates that they are more prevalent than previously thought. A detailed analysis[358] of a well-resolved feature has shown that these anti-cyclonic deep-sea eddies may be at least 2000 m deep, have life spans estimated at one to three years, a potential energy of 7.5×10^{15} J and kinetic energy of 2.6×10^{15} J in the upper 1000 m of their structure.

As discussed above, subsequent analyses[670] have shown that these eddies may move predominantly southward, along the eastern escarpment of the Mozambique Plateau. They may have a tendency to move westward towards the African coast only when they have passed the southern termination of this plateau.

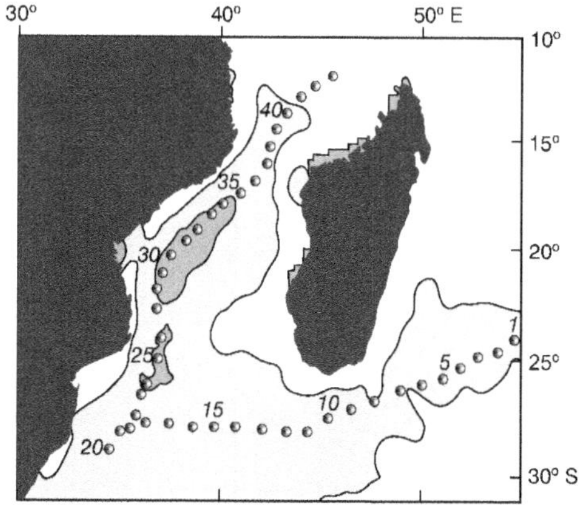

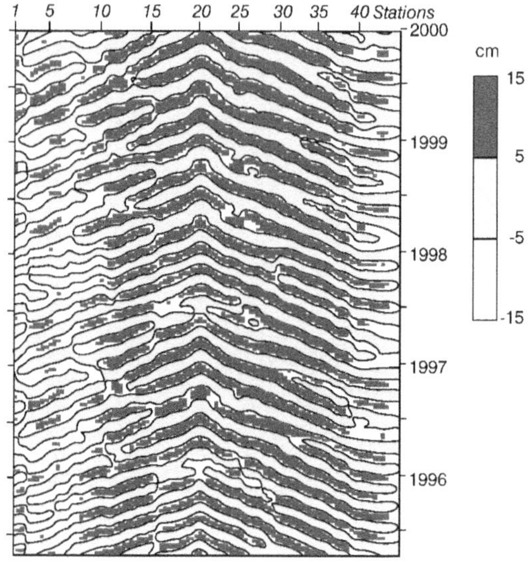

Figure 3.33. A portrayal of the movement of eddies through the Mozambique Channel and from east of Madagascar. The left-hand panel shows the regions of higher mesoscale variability in grey tones. These values are from satellite altimetry[653]. Dots and numbers show the tracks along which anomalous sea surface heights have been measured and portrayed in the right-hand panel. It is clear that anomalies can be followed both through the Mozambique Channel from north to south and from the central Indian Ocean from east to west. Strangely, the two sources of disturbances seem to be synchronized; anomalies from the east and from the north arriving at point 20, off the Natal Bight, at more or less the same time.

Origin of the deep-sea eddies

The source of these unexpectedly strong eddies in what was previously thought to be a region of slow[81-2], rather invariant, anti-cyclonic motion as part of the subtropical gyre initially has led to considerable speculation. Gründlingh[344] has hypothesised that a confluence of water from a South Madagascar Current and the Agulhas/Mozambique Current might lead to such an eddy. Later work[356] has led to the concept of a Mozambique Current, or parts of it, occasionally following the eastern edge of the Mozambique Plateau to form a Mozambique Ridge Current. This current would then be responsible for spawning the rings. Measurements across the eastern edge of the plateau[81,360] have suggested substantial southward flow here, at least on occasion, and models treating a Mozambique Current as an inertial jet are not inconsistent with the concept of a Mozambique Ridge Current. However, no further observations of this hypothesised current have been made. Other possibilities that have been put forward include rings shed from the retroflection of the East Madagascar Current to the south of Madagascar as well as cold eddies shed from the Subtropical Convergence to the south[131]. The existence of such cold eddies has been well established[63,361], they have the correct characteristics and it has been hypothesised that they could drift northward in the time required not to lose their prime characteristics. This ingenious suggestion turns out to be wrong.

Satellite altimetry has shown[362] that all anomalies of the sea surface height field of the South West Indian Ocean that may be cyclonic or anti-cyclonic eddies[309] move westward and southward. Studies of current variability[335] suggest two possible sources (viz. Figure 3.14) for the deep-sea eddies in the Natal Valley. First, in the region east or south of Madagascar meanders in the intense East Madagascar Current may cause it to shed rings. Along the coast of eastern Madagascar this current is, however, expected to have a very stable trajectory because of the shelf morphology. Since some of the anomalies do seem to come from this region[309], this poses serious conceptual problems. Second, altimetric observations from the TOPEX/Poseidon mission show a zone of high mesoscale variability extending along 25° S latitude eastwards from Madagascar. This may hypothetically be the source of these intense eddies.

A third likely source would be the Mozambique Channel. Intense and deep Mozambique eddies do move poleward[728,735], but are almost exclusively anti-cyclonic. At first glance this makes the Mozambique Channel an unlikely source. Some cyclonic features have nevertheless been observed to move southward in the Mozambique Channel[309] and also to escape from the Delagoa Bight where a cyclonic lee eddy has been frequently observed[338]. No adequate and satisfactory

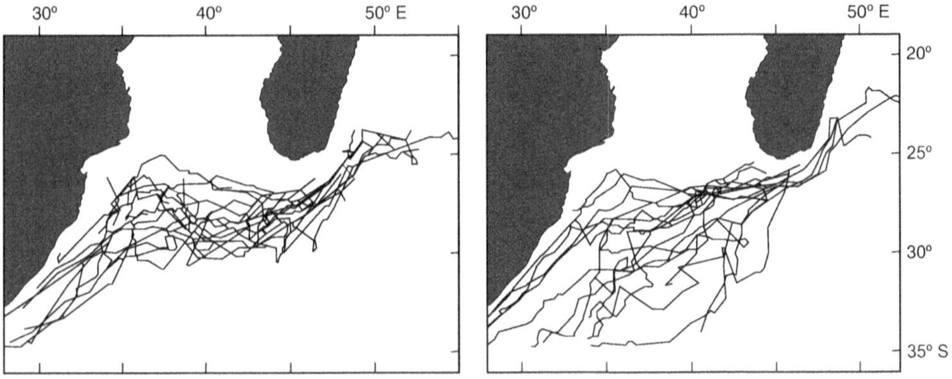

Figure 3.34. The trajectories of a number of cyclones and anti-cyclones that form part of vortex dipoles generated south-east of Madagascar. These circulation features were identified by their anomalous sea surface heights in satellite altimetry[652].

answer to the origin of intense eddies in the Mozambique Basin can therefore be given based on these observations only. At long last, the judicious and detailed analysis of anomalies in sea surface height has made it possible to solve this conundrum (Figure 3.33) satisfactorily.

Schouten et al.[653] have demonstrated that deep-sea eddies in the Mozambique Basin and in the Natal Valley come both from the Mozambique Channel and from the zonal band of disturbances that lies across the South Indian Ocean at latitude 25° S. The influence of these features will be discussed in more detail in a subsequent chapter when dealing with the inter-ocean exchange of waters south of Africa. Important to note here is that both sets of eddies in Figure 3.33 consist of anticyclones, whereas those found to date at the Mozambique Plateau[355] were predominantly cyclonic. Further analyses have suggested why this might be so. Anticyclonic eddies passing the southern tip of Madagascar seem to generate cyclones in the form of lee eddies[652] and these then move together as a vortex pair. Another ostensible source of such vortex dipoles would be an East Madagascar Current that shot off into the Mozambique Basin as a free jet, forcing counter-rotating eddies on either side. Fact is, these vortex pairs seem to be an inherent part of the circulation in this part of the South West Indian Ocean[652]. Their drift patterns seem indicative (Figure 3.34). It would seem that anticyclones avoid the direct vicinity of the East Madagascar Current and then move westward in a somewhat circumscribed zone before moving parallel to the Agulhas Current. By contrast cyclones on the whole seem to be generated close to the Madagascarian shelf and take a much wider range of subsequent trajectories in a westerly, south-westerly direction. There are indications that the paths of both types of eddies are influenced by the Mozambique Plateau, that lies at 35° E. In the Mozambique Basin the eddy pairs behave quite irregularly, with many of them splitting and interacting with other eddies[652]. There is evidence that the formation of such dipole pairs is enhanced during the negative phases of the Indian Ocean Dipole and of El Niño cycles.

Intra-thermocline eddies

Some unusual eddies of another kind have also been observed[731] in the South West Indian Ocean subgyre. These are intra-thermocline eddies consisting of subsurface lenses of water with higher salinities than the ambient water masses (Figure 3.35). Such features have been found east of Madagascar. There cores lay at a depth of 200 m, they were about 200 km in diameter and about 150 m thick. Their azimuthal speeds were between 20 cm/s and 30 cm/s. They exhibited a subsurface salinity maximum of over 35.8 at a potential temperature between 18° and 22° C; hydrographic characteristics that suggests that they are formed in the subtropical southern Indian Ocean east of 90° E where Subtropical Underwater subducts.

Synopsis

From a perusal of the existing literature on the subject, and an analysis of all existing data, it becomes immediately clear that the sources of the Agulhas Current are complex and as yet poorly quantified. The main reason for this ignorance is not hard to find: a serious lack of pertinent data, particularly in the most crucial geographic regions.

The most acceptable conclusion one may currently

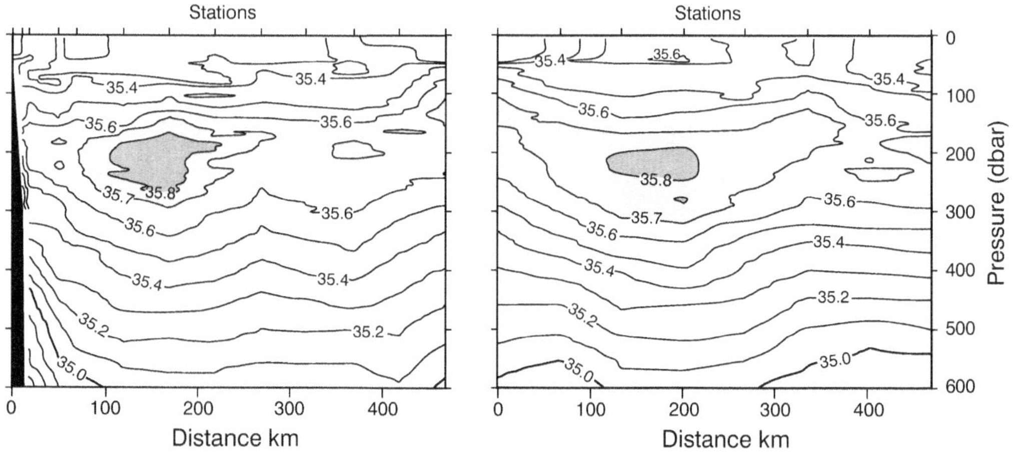

Figure 3.35. Subsurface lenses in the seasonal thermocline to the east of Madagascar[731]. These two vertical salinity sections show anomalous water masses (grey shades) of relatively high salinity centered at depths of about 200 m.

make on the sources of the Agulhas Current is that it is largely fed by recirculation in a South West Indian Ocean sub-gyre. The dimensions of this sub-gyre are zonally limited to the region between the African continent and 70° E and meridionally restricted roughly to the region south of Madagascar. This largely excludes a major contribution to the Agulhas Current from the southern branch of the East Madagascar Current as well as from the Mozambique Channel. The circulation in this channel as a whole remains clouded in a fair degree of ignorance.

The visualisation of a continuous, persistent Mozambique Current as an immediate precursor to the Agulhas Current is a concept that has been firmly held for a long time, but has now been shown to be wrong. Instead the major part of the flow through the Mozambique Channel is in the form of a train of anti-cyclonic eddies. There seems little doubt that the overall contribution to the flux of the Agulhas Current that thus derives from the Mozambique Channel is relatively small. The flow in the rest of the Channel seems to be weak and variable. A band of high flow variability on the western side of the channel demarcates the corridor along which the Mozambique eddies drift southward. Our current understanding of the East Madagascar Current is based on far fewer measurements.

The southern limb of this current seems to represent a miniature western boundary current; fast, narrow and stable. Most current information indicates that at the southern end of Madagascar this current retroflects, with most of its water passing eastwards into the interior of the South Indian gyre. A free jet impinging the waters south of the Mozambique Channel remains another possibility. Rings and vortex dipoles seem to be shed at this termination of the East Madagascar Current and these may subsequently move towards the Agulhas Current.

Both the Mozambique Channel as well as the region east of Madagascar are thus sources of mesoscale variability in the form of intense cyclonic and anti-cyclonic eddies. It seems that this unexpected high level of flow variability in a subtropical gyre is the ultimate consequence of the obstructing presence of the island of Madagascar and of perturbations generated in the North Indian Ocean. The oceanic region off the northern Agulhas Current is thus characterised by enhanced variability. This is somewhat incongruous, since the northern Agulhas Current itself is noted for its high degree of stability.

4

The northern Agulhas Current

The Agulhas Current along the east coast of southern Africa can be considered to be fully constituted somewhere south of Ponto do Ouro, on the border between South Africa and Mozambique. The characteristics of the current downstream from here are such as to allow an indisputable partitioning into a northern and a southern part.

The northern part consists of the upstream section where the current flows along a narrow shelf with a steep shelf slope and where the current trajectory, as a rule, shows little variation. Along this nearly rectilinear path the current grows downstream in both size and strength. At the latitude of Port Elizabeth the continental shelf starts to become much wider, forming the broad Agulhas Bank south of Africa (Figure 1.2). The behaviour of the Agulhas Current also starts changing dramatically here with large lateral meanders forming part of the usual current trajectory downstream from this point. The northern Agulhas Current therefore has rather specific generic boundaries; one at about 27° S in the north, the southern one at about 34° S latitude. These two boundaries are entirely defined by the kinematics of the current.

Kinematics of the northern Agulhas Current

A detailed study of the kinetic energy of the mean flow for the world's oceans as well as the kinetic energy for the variability, or eddy kinetic energy, shows[363] some remarkable expressions for the northern Agulhas Current. In fact, the distribution of these kinetic energies is quite exceptional for the whole Agulhas Current system (Figure 4.1). Mean kinetic energy is high in both branches of the East Madagascar Current and along the western side of the Mozambique Channel. The northern Agulhas Current has an even higher value, 500 cm^2/s^2, that terminates at the border with the southern Agulhas. The high eddy kinetic energy distribution, by contrast, continues well into the southern Agulhas Current and beyond, indicating the greater mean velocity as well as the greater flow stability of the northern Agulhas Current. This somewhat coarse portrayal is considerably improved in other data sets.

Altimetric data show a very similar pattern of kinetic variability[305,335] with high values in the western half of the Mozambique Channel, south of Madagascar, in the Agulhas retroflection[72–3,335] and in the Agulhas Return Current (viz. Figure 3.14). The northern Agulhas Current by comparison exhibits very little variability. The high mean kinetic energy distribution is also very apparent in all other analyses of surface speeds. Observations of ships' drift exceeding 2 m/s are concentrated in the northern Agulhas Current[296] with very few such speeds found either up- or downstream. From all these sets of averaged data it would therefore seem established that the northern Agulhas Current has an exceptionally high mean speed and energy, but little variability, including seasonal variability.

Seasonal variability in current speed

The lack of seasonal variability in the surface movement of the northern Agulhas Current has been documented in considerable detail by Pearce[292] and Gründlingh[86]. Previous analyses of ships' drift[290–1] have suggested that the South Equatorial Current is strongest in the austral winter and that its offshoots exhibit a similar maximum in speeds, but with appropriate time lags. The East African Coastal Current was found to have a maximum velocity in June; the Mozambique and Agulhas Currents a smaller peak in July[291]. Results from global current models with high spatial resolution[325] similarly indicate seasonality in the South Equatorial Current and in a Mozambique Current, but none in the Agulhas Current. In the abovementioned analyses of ships' drift the Agulhas Current was invariably represented by rather large geographic areas. Defining the Agulhas Current instead as only that part of the flow that constitutes a southward jet with speeds exceeding 1 m/s, these data have subsequently been reanalysed[292].

These latter results show that the surface speeds in

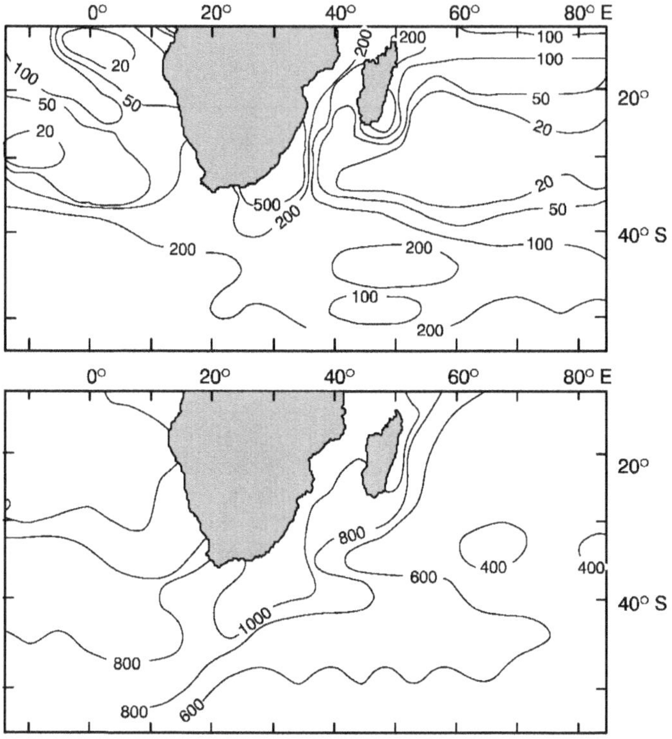

Figure 4.1. The kinetic energy per unit mass of the mean flow (upper panel) and the eddy kinetic energy per unit mass (lower panel) for the greater Agulhas Current system. These results are based on an analysis of surface drift currents[363] measured from merchant ships from 1900 to 1972. Averages for 5° latitude by 5° longitude blocks were used. Units are cm^2/s^2. The overall area of high mean – as well as eddy – kinetic energy for this system far exceeds that of any other western boundary current.

the Agulhas Current exhibit no clear seasonal variations (Figure 4.2). Mean speeds lie between 1.4 m/s and 1.6 m/s, whereas the peak speed from individual months is about 2.6 m/s. The latter variable shows more variability, but also with no clear seasonal signal. These results have been supported by[174] direct current measurements in the northern Agulhas Current[292]. The annual mean for these was 1.5 m/s, with monthly standard deviations of 0.5 m/s. Once again, no distinct seasonal trend is evident. At first glance this seems an unexpected result, since one would expect the dominant monsoonal signal of the North and Equatorial Indian Ocean to penetrate at least to the northern part of the Agulhas Current, as has been suggested by Barlow in the 1930s[288,290].

However, accurate measurements in the southern leg of the East Madagascar Current[297] have found no seasonal variability in this current either while, as was seen previously, the flow in the Mozambique Channel exhibits such large variations[79,259] that no clear seasonal pattern could as yet be established here at all[733]. The strong seasonality in the monsoonal wind regime therefore seems to be restricted to the northern Indian Ocean[292]. If the Agulhas Current is driven predominantly by the Sverdrup relationship of the zonally integrated wind stress curl across the South Indian Ocean basin, the more seasonally stable wind regime over this southern basin would be the main factor in maintaining a steady flow in the current throughout the calendar year. This would furthermore suggest that the connection between the Agulhas Current and the monsoonal regime near the equator is weak, a conclusion also reached in the discussion of the sources of the Agulhas Current in the previous chapter.

Path stability

Even though there may therefore be little seasonality in the speeds of the northern Agulhas Current, as established from these more detailed surface measurements, this steady current could still meander in an irregular fashion, thus bringing about high variability. The fact

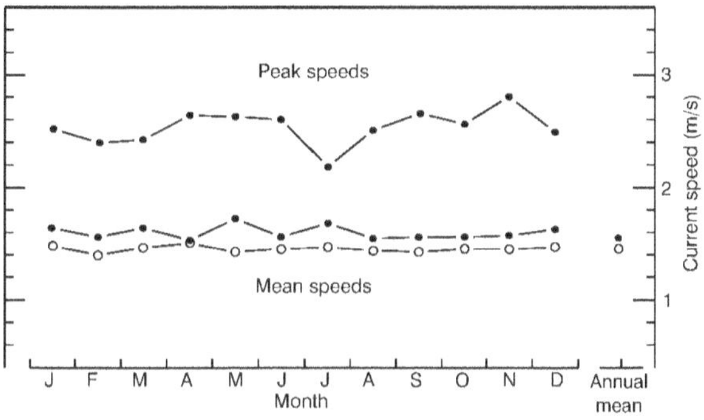

Figure 4.2. Monthly mean as well as individual, peak speeds in the core of the northern Agulhas Current based on selected ships' drift data[292]. Annual averages are shown on the right. Dots are based on published data; circles on as yet unpublished data. Note the lack of any clear seasonality.

that this is a region of low variability therefore implies that the current does not meander. This path stability inferred from ships' drift and altimetric data therefore needs to be confirmed. This has been done admirably by Gründlingh[86]. Based on 237 hydrographic sections across the current core between Cape St Lucia and Port Elizabeth (Figure 4.3) it has been established that on average the current meanders less than 15 km to either side.

The position where the 15 °C isotherm within the northern Agulhas Current intersected the 200 m isobath was used as representing the core of the current in this investigation, although measurements indicate[86] that the highest current speeds at the sea surface may usually be found 20 km shorewards of this indicator. These results were based on a substantial number of hydrographic sections carried out across the current at each position (Figure 4.3). Their number varied between 16 and 41, giving a high level of statistical reliability to the conclusions reached. However, these results represent

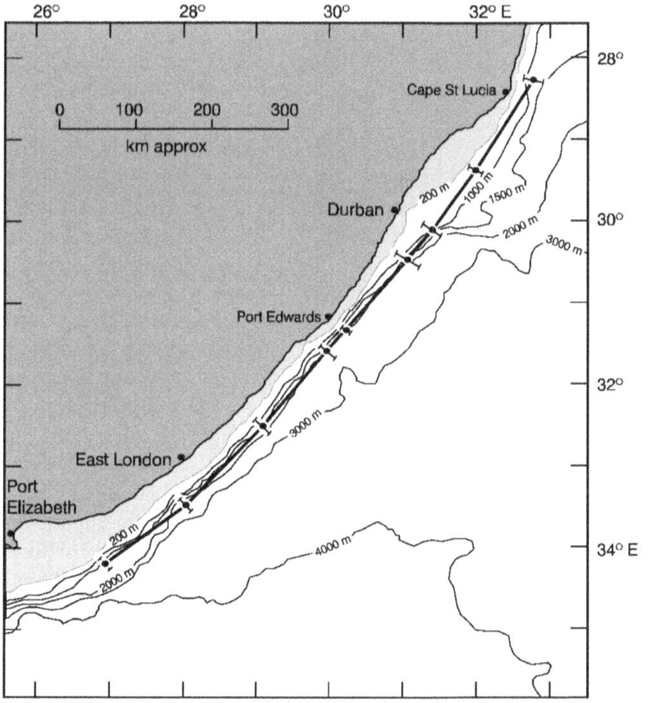

Figure 4.3. The geographic location of the core of the northern Agulhas Current based on a large number of hydrographic sections[86]. The current core is denoted by the heavy line and overlies the slope of the continental shelf, circumscribed by the bunching of isobaths in the figure. Standard deviations in the location of the current core are shown by error bars at nine positions.

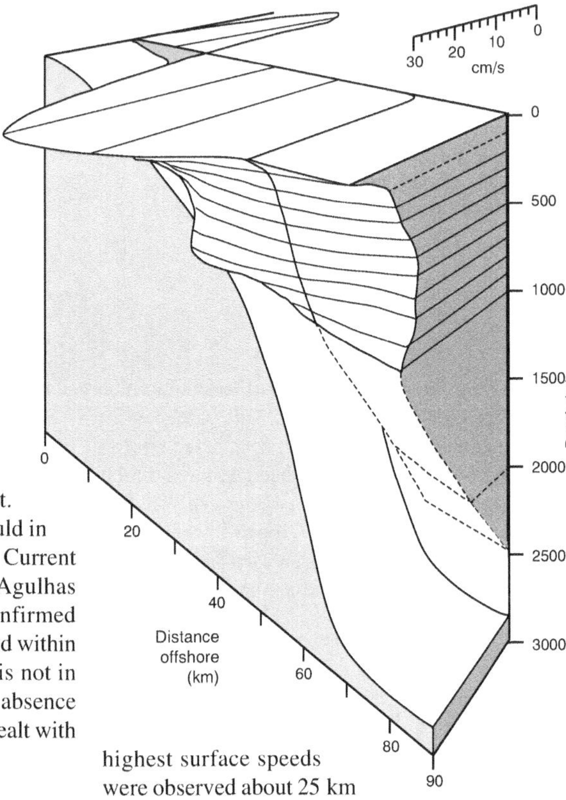

Figure 4.4. The vertical and horizontal structure of the northern Agulhas Current from direct current measurements[75]. This section was carried out on 22 October 1966 downstream of Durban from the vessel *Fraay*.

only 77 per cent of the crossings of the current. On 22 per cent of the crossings, the current could in fact not be found within 100 km of the coast! Current meter moorings placed at 32° S, where the Agulhas Current lies close to the coast[367,740], have confirmed that for 80% of the time the current is to be found within 30 km from the coast. The rest of the time it is not in this usual location. This intriguing occasional absence of the current from its normal location will be dealt with in detail in a later section.

Stabilising influences

The strongly invariant trajectory of the northern Agulhas Current is very unlike that found in comparable western boundary currents such as the Gulf Stream[102] and the Kuroshio[119]. These currents are known to exhibit substantial sideways meanders. The path consistency of the northern Agulhas Current is most likely due to the highly stabilising effect[87] of the strong incline in the continental slope and its nearly linear downstream disposition (viz. Figure 4.3). Modelling the Agulhas Current as an inertial jet[130] has shown the tendency of the current to remain close to the shelf edge. Most other modelling efforts to date[265,273,277] have also simulated the current close to the shelf edge, but probably do not have the spatial resolution adequately to portray meanders if they were to be inherent in the models' physics. In order to be so stabilised by the shelf slope, the northern Agulhas Current must have vertical dimensions of sufficient extent.

Vertical extent of the current

An instructive portrayal of the quasi-two-dimensional structure of the current is given in Figure 4.4. The highest surface speeds were observed about 25 km offshore, over the shelf edge. On most occasions the maximum velocity may be found at an offshore distance of only 15 km[86]. An inshore counter current is evident in Figure 4.4, but at the time this was not considered a permanent feature[364]. Whereas the inshore current edge is sharp and prominent, the offshore boundary is less so and was probably not fully resolved on this occasion, giving a width of only about 60 km for the current as a whole. Flow speeds below 1000 m were estimated. Overall, this portrayal may be considered to be fairly representative, particularly over the top 500 m[365].

More modern data, with much higher spatial and temporal resolution, have modified this portrayal, but also added some important new features. Using three acoustic current profiling techniques it has been shown[366] that south of Durban the core of the Agulhas Current lies about 20 km offshore, is 90 km wide between the 0.50 m/s isotachs[367] and that the current here penetrates to a depth of 2500 m (Figure 4.5). On average the current's influence extends to a distance of 203 km offshore[740]. A maximum surface speed of 1.7 m/s was observed. There was a slope in the current core so that at 900 m depth the highest velocities were found 65 km offshore. It has furthermore been shown[367]

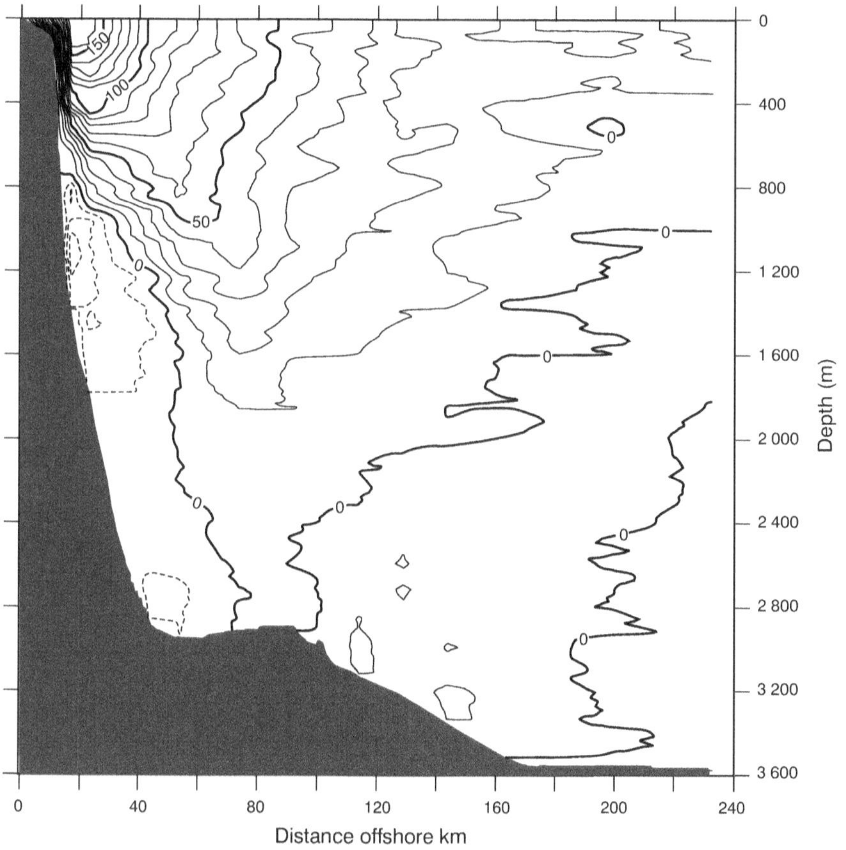

Figure 4.5. Velocity section across the northern Agulhas Current at Port Edward, south of Durban, in February 1995. These measurements were made with an acoustic Doppler current profiler[366–8]. Solid lines indicate movement in a south-westerly direction. Broken lines denote motion in a north-easterly direction. They give the first unambiguous evidence for an Agulhas undercurrent, located at a depth of 1200 m along the shelf slope.

that below 200 m the baroclinic structure of the Agulhas Current is in essence geostrophic. As could be expected, these dimensions are not invariant. A comparison of three velocity sections across the Agulhas Current[738] has shown that on some occasions the current at 32° S extends to the bottom, on other occasions only to a depth of 2300 m. The detailed current observations gathered over a period longer than a year[740] have also shown that over the full depth of the current alongshore velocity variability exhibits time scales of 10.2 days and time scales of 5.4 days for transverse velocity variations.

Most important, Beal and Bryden[368] have presented the first clear evidence for an undercurrent against the continental slope, at a depth of about 1200 m, directly under the surface core of the Agulhas Current. Its volume flux is about 6×10^6 m^3/s, consisting for some part of modified Red Sea Water[366,738]. Over a longer period[740] this equatorward transport was measured at 4.2×10^6 m^3/s. The undercurrent is a noticeable feature of some numerical models[765] that show it at a depth between 300 m and 2500 m at the latitude of Port Elizabeth. Simulated velocities peak at 0.20 m/s. This Agulhas undercurrent is in many respects analogous to that found below the Gulf Stream[369–70] and the Kuroshio[117,371], and exhibits speeds of up to 0.3 m/s.

Having thus established that the northern Agulhas Current is in all respects very stable and that it is sufficiently deep to have this lateral stability induced by the shelf slope morphology, we look now at what is known about the kinematic characteristics of the upper layers of the current.

Kinematic characteristics of the surface layers

Pearce[365] has reported on the main results of a major research programme to study the upper layer of the northern Agulhas Current at Richards Bay, just south of Cape St Lucia, at Durban and at Port Edward (viz. Figure 4.3), and has clearly circumscribed the kinematic characteristics of the current in this region. These are given in Table 4.1. Although considerable inshore and offshore movements were evident over short periods (Figure 4.6), a number of consistent characteristics are nonetheless evident[365].

The northern Agulhas Current consists of three distinguishable cores in its upper layers. First, there is an inshore boundary region of relatively high thermal and velocity gradients, roughly 7 km wide, up to the 1 m/s isotach. A tendency for converging flow, towards the current core, has been observed in this boundary region.

Moving, underwater sand dunes

The high-velocity core of the Agulhas Current lies close to the continental shelf edge and is occasionally observed to lie even partially above the outer shelf. This has some unusual and interesting consequences for the sediments on this shelf region.

Along the whole northern Agulhas Current, from Ponto do Ouro to Algoa Bay, the sediment bedforms on the shelf include extensive dunefields wherever the water is deeper than 50 m. Flemming[372] has shown that these dunes may be up to 8 m high, usually lie about 200 m apart and migrate steadily downstream. Dune fields may extend up to 20 km without interruption. These bedforms imply a water speed of at least 1.3 m/s. It is thought that this conveyor belt of sand is interrupted by submarine canyons that intercise the shelf, thus presenting preferred supply channels for sediment fans that line the continental slopes. Sedimentary processes on the outer shelf off south-eastern Africa are therefore totally dominated by the activity of the Agulhas Current[373].

The discovery of these dune fields was the first documented example of large-scale sediment transport by ocean currents[372]. It is the high velocity of the current and the narrow shelf over which it occasionally transgresses (see e.g. Figure 4.8) that make this phenomenon so clearly discernible here.

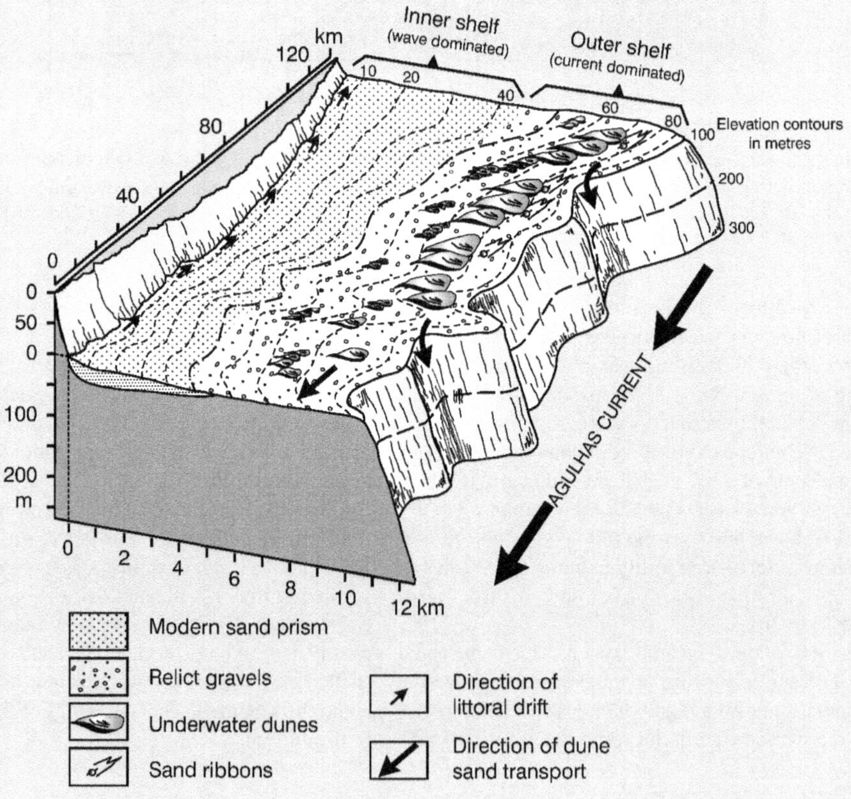

Table 4.1. Kinematic characteristics of the upper layers of the Agulhas Current off Durban[364].

	Mean	Standard Deviation	Maximum
Peak speed in current core (m/s)	1.36	0.30	2.45
Current core offshore distance (km)	52	14	30
Distance offshore 0.5 m/s (km)	35	14	10
Distance offshore 1.0 m/s (km)	42	14	25
Core width, between 1.0 m/s isotachs (km)	34	15	10
Distance offshore temperature maximum (km)	58	20	35
Distance offshore 15 °C/200 m intersection (km)	50	15	25
Distance offshore 35.35 salinity/200 m (km)	47	13	23

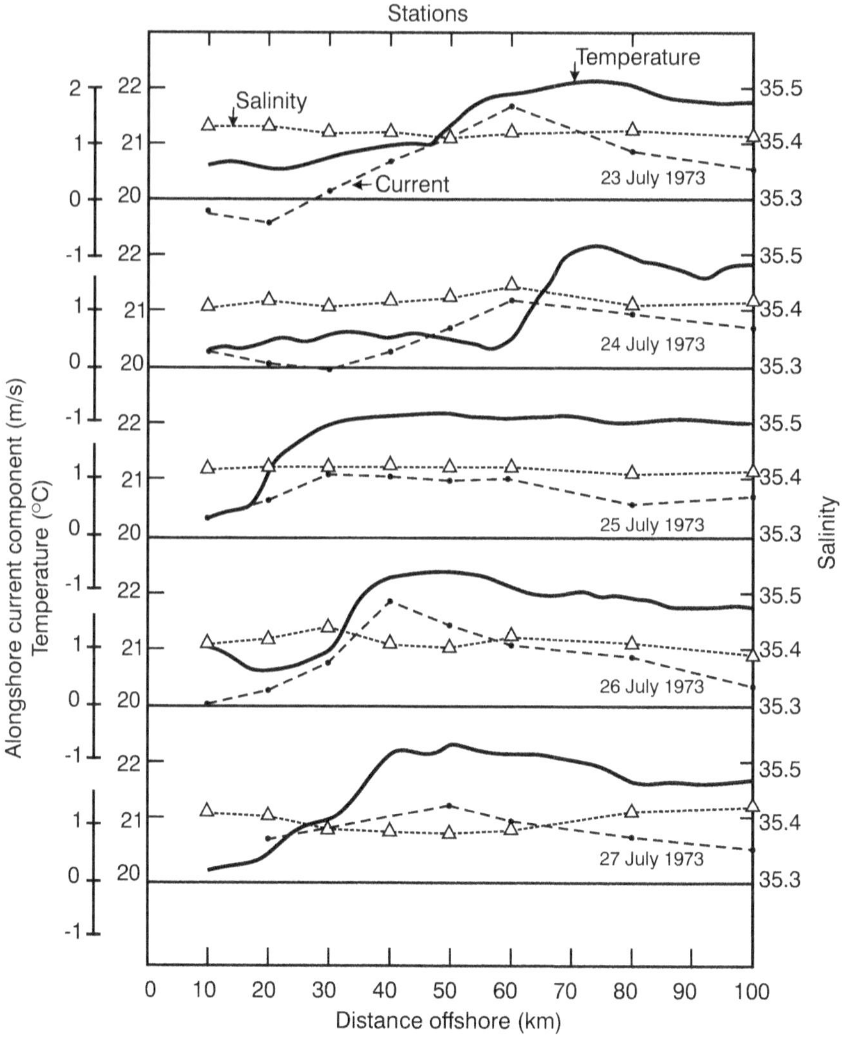

Figure 4.6. Sea surface characteristics of the northern Agulhas Current over a period of five consecutive days[365]. These measurements were made off Durban and demonstrate the variability to be expected in the surface layers of the current. Temperatures and salinities were from 2 m deep and the current velocities parallel to the coast and integrated over the top 100 m (viz. Figure 4.4).

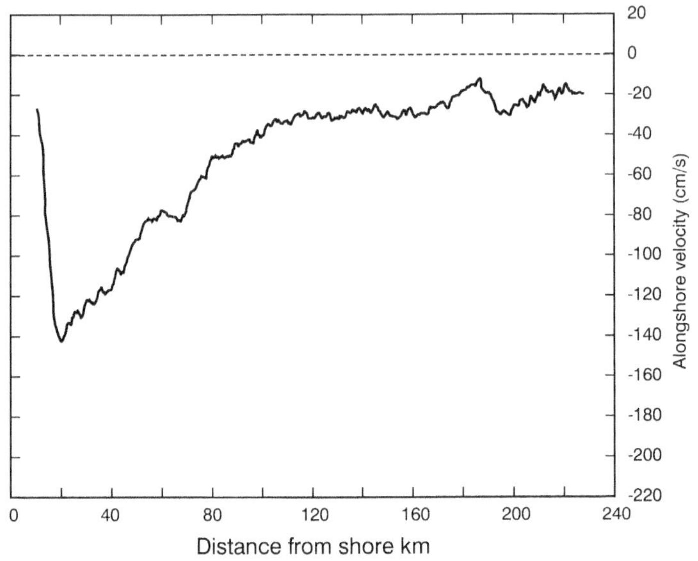

Figure 4.7. A profile of the speed across the Agulhas Current at Port Edward at a depth of roughly 100 m from a ship-mounted acoustic Doppler current profiler[740]. This trace is characteristic for all depths to at least 300 m. The core of highest current speed is at the shoreward edge of the current, about 20 km from the shore.

Individual horizontal temperature gradients may be as high as 1 °C/km at the sea surface, but no consistent salinity gradient is associated with this region (Figure 4.6). The second distinguishable strip is the current core itself where speeds exceeding 1 m/s are found. Its mean width is 34 km (Table 4.1). Peak speeds, averaged over the top 100 m, of 2.45 m/s have been measured[365] in this core. This agrees with advecting speeds observed with free-floating drifters[374]. The third current strip is the anti-cyclonic shear zone seaward of the current core. In this region the horizontal gradients of temperature and current velocity are much less than in the landward border strip (viz. Figure 4.6). The horizontal distance between the 0.5 m/s isotachs in the surface layers of the northern Agulhas Current is about 100 km, which may be considered its width.

A major study of the structure of the northern Agulhas Current using current meter moorings[367,738] has given the best information to date of the average velocity structure across the current[740] (Figure 4.7). Close inshore (10 km offshore) the alongshore speed was about 0.2 m/s at a depth of about 100 m. The peak velocity of 1.4 m/s was found about 20 km offshore. The difference in the peak speed at a depth of 100 m and at 10 m was substantial. At a distance of 200 km offshore the last vestiges of the current had disappeared. Attempts have been made[743] to understand this observed velocity profile in terms of geostrophic balance and simple ideas of eddy viscosity. Standard theories have difficulty to reproduce the large shear near the coast and the slow decay of the current offshore. This profile could only be reproduced using a viscosity that varies linearly with distance from the coast.

How representative are the abovementioned portrayals for the full length of the northern Agulhas Current? Based on repeated measurements made at Richards Bay, Durban and Port Edward[365] the values given in Table 4.1 may be considered an accurate portrayal of the full extent of the current, within the short-term variability evident in Figure 4.6. The vertical current structure of the current in the upper layers is very comparable at the three sites used by Pearce[365] with one, notable, exception. The current core at Richards Bay and at Port Edward lies much closer to the shelf edge (Figure 4.8) than at Durban.

The flow over the shelf is always with the Agulhas Current at the former two sites, whereas at Durban the inshore flow is, on average, a counter current (e.g. Figure 4.4). In the subsurface thermohaline structure this flow is represented by a dome of cold water below 200 m, and about 30 km offshore, suggesting a trapped lee-eddy downstream of the wider shelf of the Natal Bight between Cape St Lucia and Durban (viz. Figure 4.3). The movements over the shelf regions adjacent to the northern Agulhas Current are discussed more fully in a special section below.

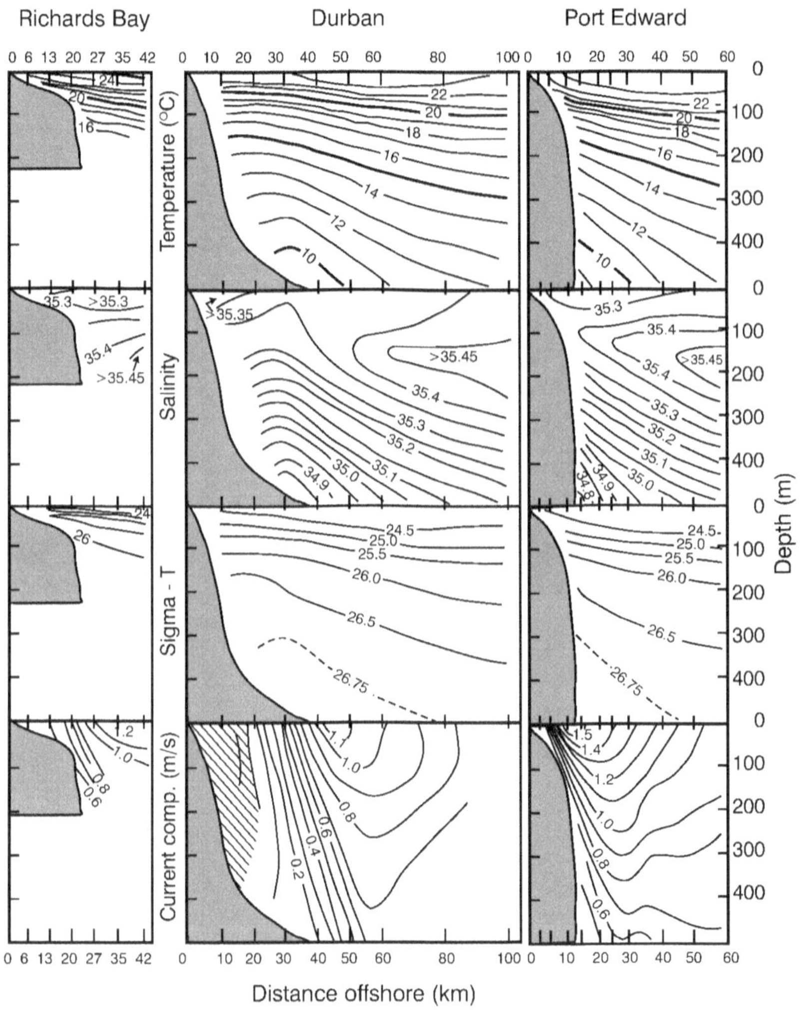

Figure 4.8. Vertical sections across the northern Agulhas Current at Richards Bay, at Durban and at Port Edward for temperature, salinity, density (sigma-t) and the current speed parallel to the coast[365]. A dome of colder, saltier water about 30 km offshore from Durban is reflected in the inshore counter-flow indicated by the hatched region in the lowest panel for Durban. These section portrayals represent averages based on between 8 and 45 individual hydrographic sections each.

Short term surface changes

Having stressed the characteristic stability of the path of the northern Agulhas Current, the shifts in the thermal front at the sea surface that are evident in daily portrayals of surface characteristics (Figure 4.6) come as a bit of a surprise. However, when the location of the core of the flow, as represented by the intersection of the 15 °C isotherm and the 200 m isobath, for sections from Durban (Table 4.1) is inspected[374], it is clear that the current core does not meander here by more than 15 km to either side either. However, the variability at the sea surface is greater, but the thermal front here is not always strongly correlated with the peak in alongshore velocity (Figure 4.6). This has also been found off Richards Bay[375].

An investigation using airborne radiation thermometry as well as current measurements along the edge of the northern Agulhas Current at Richards Bay has shown that there is considerable variability in the location of the surface thermal front as well as in the current directions. Current reversals 6 km off Richards Bay were observed repeatedly. The frequency of these events is very similar to the small meanders encoun-

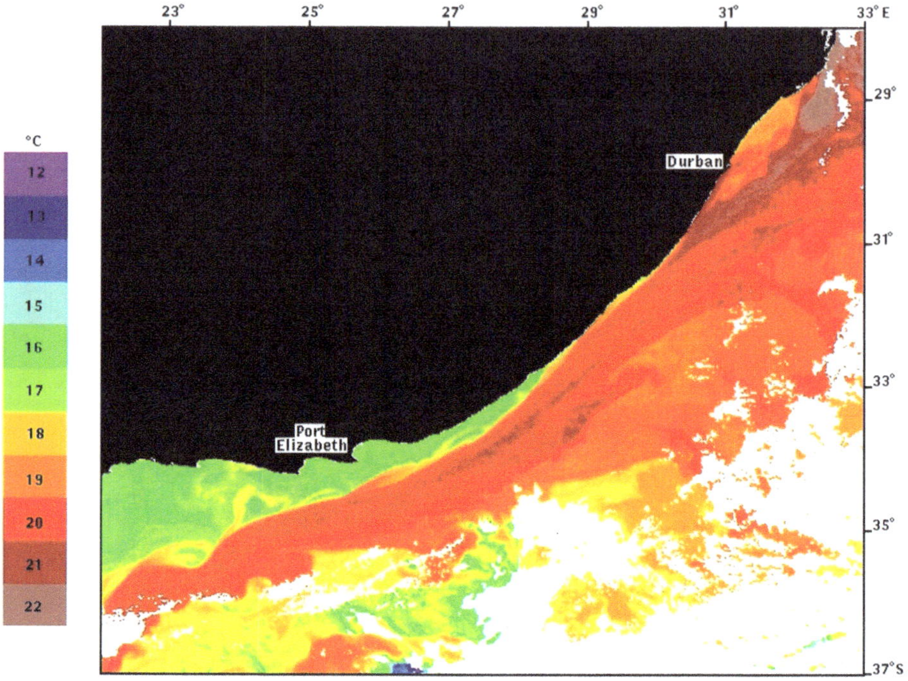

Figure 4.9. The surface thermal expression of the northern Agulhas Current as portrayed in satellite imagery of the thermal infrared radiance. This image is for 26 August 1985 from the NOAA 9 satellite. The rectilinear shoreward boundary of the Agulhas Current as well as cyclonic sheer-edge features over the Natal Bight, upstream of Durban, are clearly visible.

tered at Durban[365]. In both cases attempts have been made to correlate this frontal behaviour with local winds or with the passage of coastal low-pressure cells in the atmosphere[365,375]. Attempts at finding such correlations[316], with different time lags, have not met with any statistical success. Satellite thermal infrared imagery suggests why this may be the case.

The strong thermal contrast between the warm waters of the northern Agulhas Current and the cooler shelf water makes this system ideal for investigation by satellite images in the thermal infrared. A cloud-free image for 26 August 1985 (Figure 4.9) shows that where the shelf is narrow the development of shear-edge features such as eddies and warm water plumes[88] is suppressed, while in the wider Natal Bight, north of Durban, such features are permitted to grow. These shear-edge eddies seem to form a persistent part of the inshore thermal front of the current at the Natal Bight[376] and are also evident in satellite images of suspended particulate matter[377]. Although probably the major factor responsible for the observed variability at the current's edge between Cape St Lucia to south of Durban, these shear edge or border features do not seem to grow laterally to the extent that has been observed on the Agulhas Bank[88] or on the border of, for instance, the Gulf Stream[378] or the Kuroshio[379].

Having explored the small scale kinematics of the surface layers of the northern Agulhas Current, their variability and possible causes for this variability, we now turn to more substantial kinematic characteristics of the current.

Transport of the current

The most representative variable of a current's kinematics may, arguably, be its mass transport since this variable integrates the velocity in both the vertical and the horizontal dimensions. Based on eight sections across the current – one at Durban, six at Port Edward and one at Cape St Lucia – Gründlingh[380] has established a volume transport, extrapolated to the sea floor, of $72(\pm 3) \times 10^6$ m^3/s. Previous estimates have varied[259,278] from 56×10^6 m^3/s to 137×10^6 m^3/s but have been based on geostrophic calculations only. The measurements by Gründlingh were based on geostrophic calculations down to 1000 m, but were matched at the sea

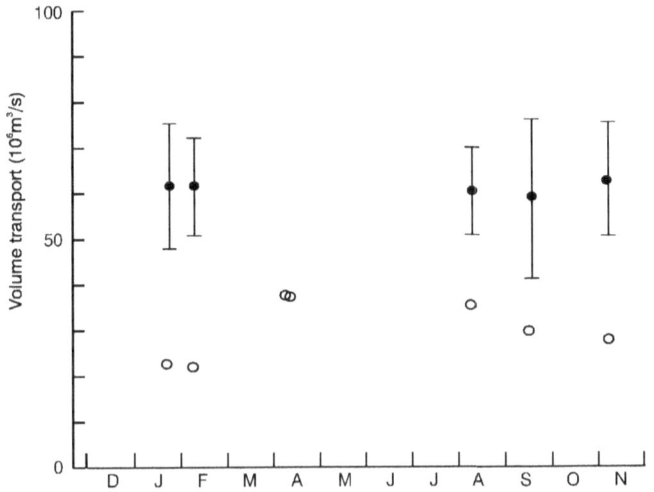

Figure 4.10. Values of the volume transports of the northern Agulhas Current in the upper 1000 m[380] with standard deviations. No clear seasonal pattern is evident. Circles denote averages based on calculated geostrophic flow relative to 1000 decibar; dots the same, but matched with directly measured current velocities over the top 100 m of the water column.

surface with directly measured current velocities. Extrapolation was then done to the sea floor. The results are shown in Figure 4.10. Standard deviations in the mass flux of the current may be considered to be moderate.

Subsequent observations using a set of acoustic current profilers[366,740] have essentially confirmed this result with an estimate of 73×10^6 m^3/s at Port Alfred[367], south of Durban. The average value over a period of one year was $69.7(\pm 4.3) \times 10^6$ m^3/s. This should be considered the best volume transport value to date and clearly establishes the Agulhas Current as the largest western boundary current on the globe. This calculated value is lower than that given by some others[249] (85×10^6 m^3/s) because of the undercurrent that has been resolved for the first time. Of some importance would be an estimate of the volume fluxes for the water mass components of the current. This is given in Figure 4.11.

This diagram gives a representative portrayal of the water masses to be expected in the Agulhas Current. Both Tropical Surface Water and Subtropical Surface Water are present. Tropical Surface Water is found largely between the coast and 100 km offshore, whereas Subtropical Surface Water is dominant beyond 100 km offshore[367]. Red Sea Water is located mostly on the inshore side of the current with some evidence for its presence between 50 and 100 km offshore. By contrast the Antarctic Intermediate Waters are all found between 100 and 200 km from the coast. The cumulative transports substantiate that the offshore current edge lies at about 200 km (Figure 4.11) at this location. At this distance from the coast the poleward transport of all water masses has ceased. It has been shown[742] that this poleward transport in the upper 1000 m is largely compensated by the northward mid-ocean transport in the South Indian Ocean, plus the net evaporation, the Ekman motion and the throughflow from the Pacific at Indonesia. The curves also show that the maximum transport for all water masses lies at the inshore edge of the current, except for Antarctic Intermediate Water. The surface and central waters carry by far the greatest part of the transport. North Atlantic Deep Water flows in the opposite direction to the main flow of the Agulhas Current. The observations on which these measurements are based have extended to the shelf waters[367]. This is important.

A substantial positive contribution to the former values[376] for the volume transport of the Agulhas Current has come from extrapolated near-surface velocities over the continental shelf. This flux contribution from the waters over the shelf is a perennial problem in these type of calculations that has also been encountered by Gründlingh when he attempted to establish the precise seaward border of the current. As was noted above, this border has been recognised as being indistinct[82]. In his study, measurements up to 150 km offshore were carried out in each case. A number of interesting results have emerged.

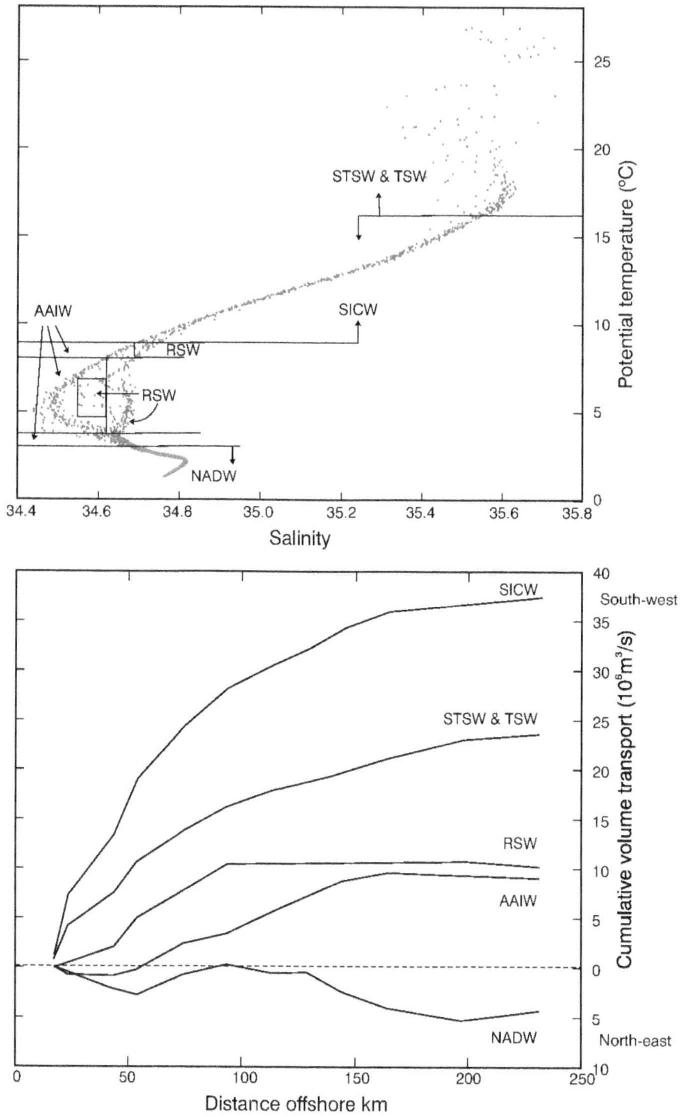

Figure 4.11. A temperature–salinity diagram for the Agulhas Current[367] showing the boundaries for different water masses being carried along (upper panel). They are: **STSW**, Subtropical Surface Water; **TSW**, Tropical Surface Water; **SICW**, South Indian Central Water; **RSW**, Red Sea Water; **AAIW**, Antarctic Intermediate Water and **NADW**, North Atlantic Deep Water. The lower panel gives the cumulative water volume transport across the width of the Agulhas Current for each of these water masses. North Atlantic Deep Water is shown as having a net north-easterly transport.

Variability in the volume transport

First, there is no discernible seasonal variation in the volume transport of the northern Agulhas Current (Figure 4.10). This therefore is in general agreement with the results of analyses of surface currents[292] (Figure 4.2). Suggestions from altimetric studies[219] that the subtropical gyre of the South Indian Ocean exhibits a seasonal, wind-driven cycle could not be substantiated with concurrent volume transport estimates. The latter were all found to be close to the mean values. Nevertheless, careful modelling of the Agulhas Current system[271] has shown[299] that the transport of the current should exhibit a substantial seasonal cycle that origi-

The R.V. *Meiring Naudé*

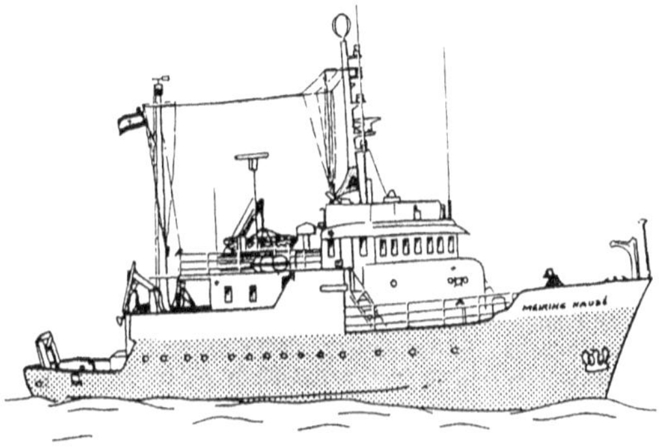

Research on the northern Agulhas Current and, for that matter, on the South West Indian Ocean as a whole, has for a very large part been carried out from one specific research vessel, the *Meiring Naudé*[317].

This 31 m, 364 tonnes coastal research ship was built in 1967 to fill the growing need for an adequate research platform by the CSIR (Council for Scientific and Industrial Research, South Africa). A physical oceanography unit of this body, situated at Durban, had been carrying out research – concentrating mostly on marine pollution and coastal circulation with rented vessels – but increasingly required a dedicated research vessel.

Designed to carry out predominantly coastal investigations, she had a limited endurance, but was nonetheless used for many deep-sea studies, even as far afield as Mauritius, the Mozambique Channel and Saldanha Bay (north-west of Cape Town). For the period in which she was designed and built she was admirably equipped with echo sounders, satellite navigation, winches and hydraulic cranes and even with – novel for the time – the ability to automatically bring a home-designed hydrosonde[374] into a wet laboratory inside the vessel.

For 20 years this dapper little vessel and its crew spent an average of 145 days per year at sea[317], logging more than 300 000 nautical miles in search of data for many different marine disciplines and for the full spectrum of South African ocean establishments. Her crew was a happy one, six having served for more than 14 years; her captain, C.A.E. Foulis, for her whole lifetime as a CSIR research vessel. The greatest part of her activities was concentrated in the northern Agulhas Current.

When the policy decision was taken in the late 1980s to restructure the CSIR into a body concentrating exclusively on applied contract research[322], the fate of the vessel was sealed, and she was sold to private enterprise in 1989. Named after a well-known and well-loved former president of the CSIR, her name has been cropping up in the popular media since for being involved in less dignified activities than oceanic research.

nates in the Mozambique Channel. Changing seasonal wind patterns could force more Tropical Surface Water through the channel in austral winter, thus enhancing the flow in the Agulhas Current. Although not inconsistent with currently available hydrographic data, no evidence is at hand fully to substantiate this suggested seasonality in the volume transport of the current. Accurate results from current meter moorings[367,738,740] have shown that the full volume transport of the northern Agulhas Current does vary considerably, but not on seasonal time scales. Transport values in the poleward direction, over a period of a year[740], varied between 121 and 9×10^6 m^3/s. Comparisons of velocity sections taken across the current at different times[738] – mentioned above – have shown that the current may extend farther offshore on occasions and that its depth of vertical penetration may also change considerably with time.

A second interesting result of repeated sections across the Agulhas Current[380] is that the cross-current profile of the volume transport per unit width has a similar shape to that of surface current velocities (Figure 4.12). However, whereas the peak mean speed is found about 15 km offshore, the highest volume transport lies 40 km from the coast. Estimates[278] of the total volume transport of the southern Agulhas Current give values of 95×10^6 m^3/s south of Port Elizabeth and 137×10^6 m^3/s south of the continent, suggesting an

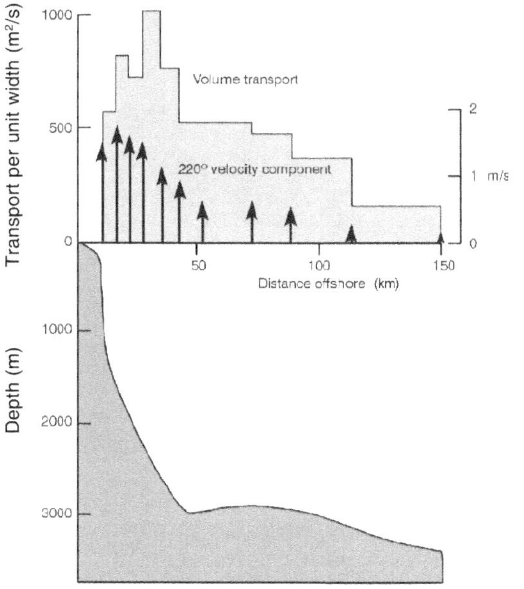

Figure 4.12. The calibrated volume transport in the upper 1000 m of the northern Agulhas Current, the speed parallel to the coast integrated over the top 100 m and a profile of the sea bottom, as observed on 21 and 22 January 1976 off Port Edward[380].

increase of about 6 x 10^6 m^3/s for each 100 km downstream. A detailed model of the greater Agulhas Current system[271] has also suggested an increase of 24 x 10^6 m^3/s over the length of the current. No other measurements of volume transport in the northern Agulhas Current are currently available accurately to confirm this estimate, but there is other circumstantial evidence in support.

Downstream flux increases

Analyses of hydrographic data, albeit a combination of historic hydrographic data[75,82,101], or of individual cruises[75,215,354], all suggest a measure of recirculation in a South West Indian Ocean sub-gyre (viz. Figure 3.30), or at least a shoreward inflow of water into the northern Agulhas Current along its flow path. The location of this inflow varies considerably depending on the data or data combinations used. Based on one of the most extensive cruises on the Agulhas Current undertaken[92] to date, it was established that on that occasion there was a significant deep-sea contribution to the northern Agulhas Current between Durban and East London. The trajectories of two drifting weather buoys have also shown such inflow at this location[374]. These individual measurements may be misleading, but a statistically

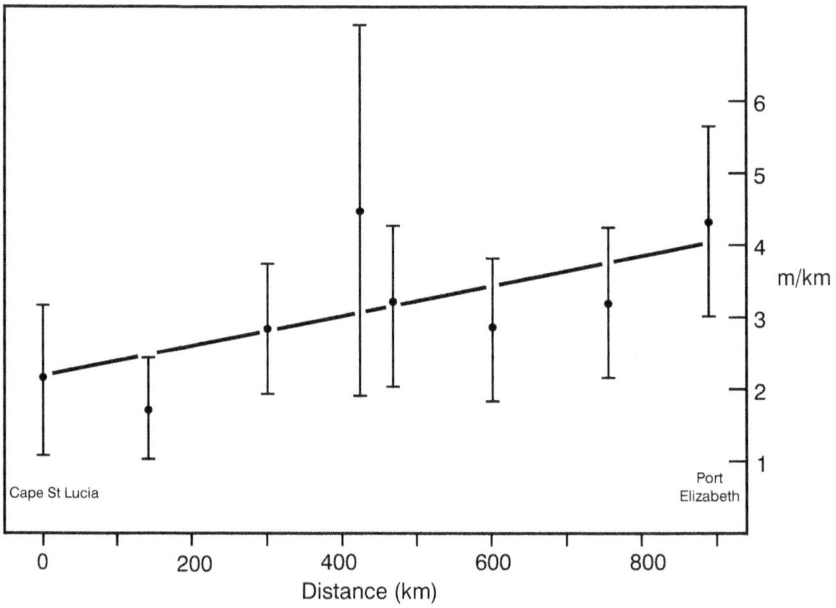

Figure 4.13. The downstream increase in the magnitude of the slope of the 15 °C isotherm, representing the core of the Agulhas Current, between Cape St Lucia (0 km) to Port Elizabeth (900 km) [86]. Standard deviation bars are given for each geographic location shown in Figure 4.3.

more reliable analysis supports the downstream increase in intensity of the northern Agulhas Current.

Gründlingh[86] has established the downstream increase of the slope of the 15 °C isotherm between Cape St Lucia and Port Elizabeth (viz. Figure 4.3) based on a large number of crossings of the current. He has shown that there is an increase along this flow path with a doubling, on average, of the slope over this distance (Figure 4.13). This isotherm slope can be used roughly as a proxy for core speed and transport, therefore demonstrating unambiguously the downstream intensification of the current. Although the precise quantification of the increase in downstream volume transport, and its variations, are still outstanding, these results do show that the downstream growth of the northern Agulhas Current is a real and important part of its kinematic characteristics. Such characteristics are represented by water masses and water types, and for the northern Agulhas Current a considerable amount of information about this hydrography has been accumulated.

Hydrography of the northern Agulhas Current

Surface hydrography

The average temperature fields of the surface waters along the east coast of southern Africa lie more-or-less parallel to the coast, as would be expected on going from a cooler shelf regime to a warm Agulhas Current. The temperature of the surface water in the northern Agulhas Current shows a decline of about 2 °C downstream along its full length, a gradient that persists throughout the year[77]. Maximum average temperatures in the north are 28 °C in February; lowest are 23 °C in July. Off Port Elizabeth, by comparison, they are 25 °C in January; 21 °C in August[77]. Salinities at the sea surface are on average 35.3 along and over the shelf and up to 35.5 in the northern Agulhas Current itself[75], no distinct annual signal being evident.

The water masses of the northern Agulhas Current are naturally part and parcel of those found in the wider

The Agulhas Current and "monster" waves

One of the perennially beguiling aspects of the Agulhas Current is represented by the many anecdotal reports of "freak" or "monster" waves that are thought to form an inherent part of the current's characteristics. Local newspapers are replete with reports of large vessels having been severely damaged, having had their bows removed or even having sunk after a chance encounter with such a "monster" wave. Analyses of the locations where ships had reported extensive damage due to such waves[171] show that they were all steaming downstream on the inshore edge of the Agulhas Current.

In a seminal study, using wave records taken from a research vessel at sea, Schumann[172] was able to demonstrate that a dramatic increase in wave energy can result if a wave group encounters an opposing Agulhas Current. Waves generated at atmospheric low-pressure systems to the south-west of Africa may be refracted into the core of the current, thus further enhancing the local wave energy. In general a substantial part of the wave climate off this coastline[408] derives from the south-west, and the exact subsequent amplification of the heights of these waves is dependent on the location and disposition of the Agulhas Current, its retroflection and even Agulhas rings[409].

With the advent of remote sensing from space the overall relationship between surface waves and the Agulhas Current system could be studied much more efficaciously[173]. Gründlingh has extensively used satellite altimetry[70] to study the wave climate around southern Africa[71]. He has shown that up to 17 per cent of the wave profiles in his investigation that had come from the south-east showed amplification of more than 40 per cent in the Agulhas Current[175], relative to the background wave height. The origin of these waves in storm pressure systems in the Southern Ocean has also been clearly established[174]. The inverse of the wave amplification of south-westerly swells in the Agulhas Current has also been demonstrated using altimetric data. North-easterly swells have their amplitudes attenuated in the Agulhas Current[410] as have westerly swells in the eastward flowing Agulhas Return Current[175].

The probability, although low, of encountering a potentially damaging wave in the Agulhas Current system is therefore largely predictable and the South African Weather Bureau have used this determinacy to broadcast wave warnings on its regular weather reports for shipping.

South Indian Ocean that have been discussed in the previous two chapters and mentioned above where their individual contributions to the volume fluxes have been given (Figure 4.11). To summarise here, the Agulhas Current system would be expected to include Tropical Surface Water (<35.0 salinity), derived from the South Equatorial Current[75], Subtropical Surface Water (>35.5 salinity, 25.8 sigma-t) associated with a subsurface salinity maximum at about 200 m depth, South Indian Intermediate Water (<35.6 salinity, 4 °C to 6 °C temperature and Red Sea Water (>34.7, 27.2 sigma-t). All these would overlie the North Atlantic Deep Water and the Antarctic Bottom Water. The mixture between Surface Water and Intermediate Water is called Central Water and usually forms a linear region in temperature–salinity space between the former two. The presence and characteristics of most of these water types is evident from a portrayal of the temperature–salinity envelope based on about 360 hydrographic stations taken in the northern Agulhas Current[365] (Figure 4.14).

At depths shallower than 100 m there is a distinct warming during the summer months and a decrease in salinity, with great interannual variations. The latter may sometimes be associated with a greater river run-off along this coast[381], or possibly with an increase in inflow of Tropical Surface Water (Figure 4.14). Over the adjacent shelf, seasonal heating to a depth of 50 m has been observed[381], from a minimum of 21 °C to 26 °C in February. The surface waters of the northern Agulhas Current may therefore be considered to be a mixture of Tropical and Subtropical Surface Water with purer Subtropical Surface Water intruding as a high-salinity tongue into the Agulhas Current at depths of 150 to 200 m[365]. This is particularly clear in the average salinity portrayals of Figure 4.8.

The upper limit of Central Water may be considered to be the point 14 °C, 35.3 salinity, with a sigma-t of 26.4. This surface lies at 350 m depth seaward of the northern Agulhas Current and rises to 150 m near the coast. Across the northern Agulhas Current itself there is evidence for separate strips of distinctly different water masses[81,92]. Along the coastal edge of the current a mixed surface water of moderately low salinity (<35.3) may usually be found. Some investigators[66,75] have considered this to be a sign of the presence of Tropical Surface Water, with a separate source in the poleward flow in the Mozambique Channel[92]. The reduced salinity might also possibly be due to mixing

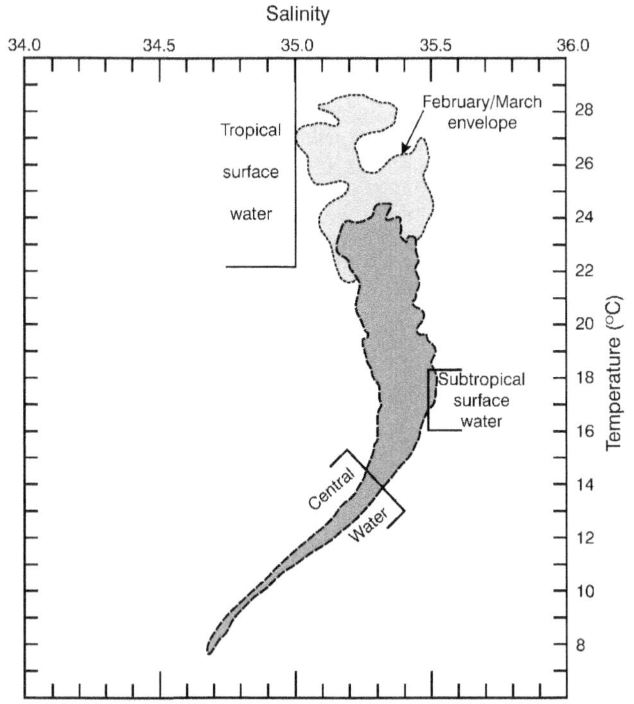

Figure 4.14. The temperature–salinity characteristics of the northern Agulhas Current up to 100 km offshore of Durban[365]. The envelope encloses about 3600 measurements at hydrographic stations occupied from October 1972 to March 1974. The domains of core water types are indicated.

with run-off water from rivers that are found over the shelf. There are, however, further suggestions that this inshore water is associated with a shallow oxygen minimum[230]. These measurements would tend to support the argument for a Tropical Surface Water origin for this low-salinity strip of water.

At a temperature of 17 °C, which corresponds roughly with the core of Subtropical Surface Water[215], a thermostad has been observed, i.e. a layer of unusual vertical extent with a small temperature range. It has been suggested[249] that since thermostads at this temperature are formed by convection in the Agulhas retroflection region[382], farther downstream, this water in the northern Agulhas Current may originate at the retroflection and be recycled in the South West Indian Ocean sub-gyre. It is evident in published hydrographic sections[278]for many parts of this basin, more clearly in some than in others[249].

Subsurface hydrography

Additional measurements across the northern Agulhas Current, but to greater depths[230], show (Figure 4.15, viz. also Figure 4.11) the clear linear extent of the South Indian Central Water and also the presence of both Antarctic Intermediate Water and Red Sea Water at intermediate depths. The presence of Red Sea Water in the northern Agulhas Current should come as no surprise as evidence for intrusions of Red Sea Water into the Mozambique Basin has been presented previously[235] and traces of this water can be followed downstream in the Agulhas Current[230].

In general the subsurface temperature–salinity characteristics of the northern Agulhas Current change very little with distance downstream[215]. South Indian Intermediate Water may be found throughout the northern Agulhas Current region, but at the latitude of Port Elizabeth the salinity minimum associated with this water may be found between 1000 m and 2000 m, as well as between 200 m and 1000 m, whereas farther upstream it occurs only between 100 m and 2000 m[215]. Otherwise few changes except in surface temperature are obvious.

Circulation on the continental shelf

As can clearly be seen in the disposition of isobaths portrayed in Figure 4.3, the continental shelf along which the northern Agulhas Current flows may be

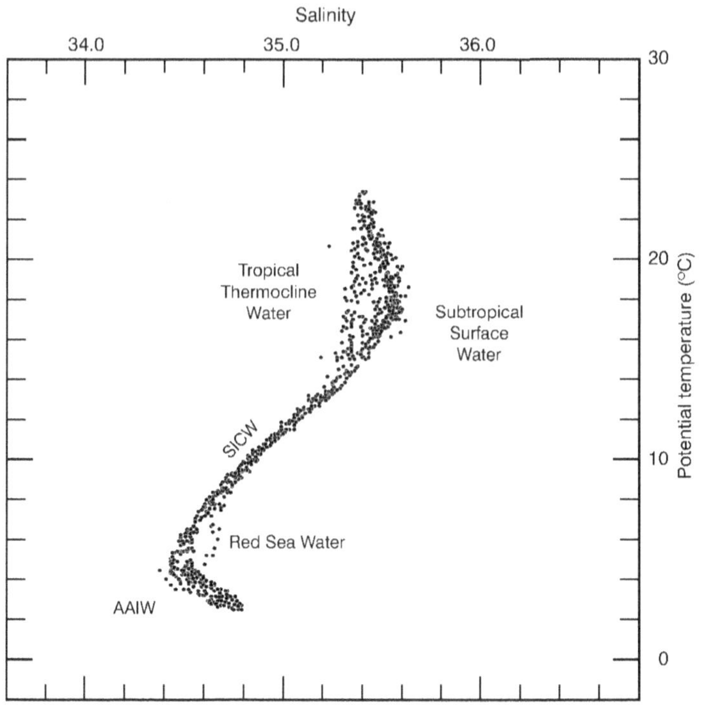

Figure 4.15. The temperature–salinity characteristics[230] of the northern Agulhas Current up to 450 km offshore, from Durban to Port Elizabeth, and to a depth of 2000 m. Distinctly different water types are evident both at the sea surface and at intermediate depths. SICW is South Indian Central Water; AAIW Antarctic Intermediate Water.

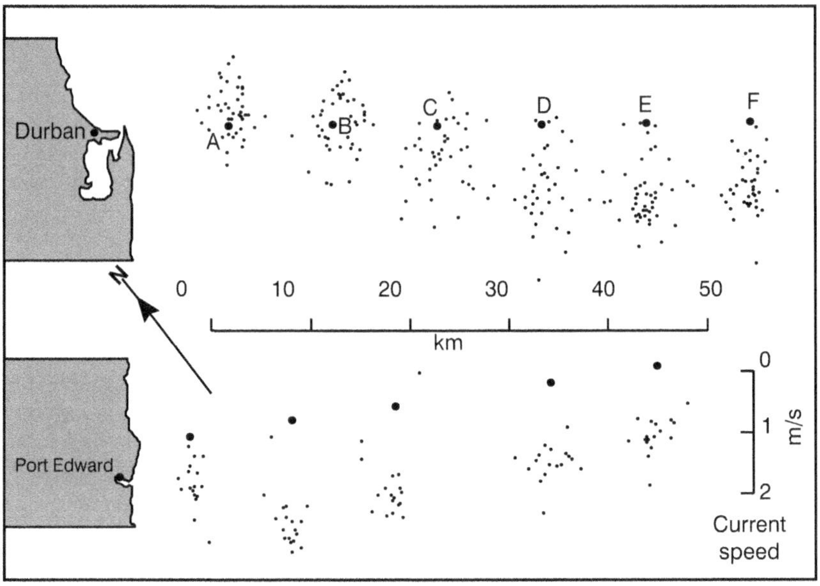

Figure 4.16. Current measurements from on board ship on lines of stations off Durban and Port Edward[364]. Each dot represents the head of a current vector based at the station position. Stations are marked with a large dot. Current measurements were averaged over the top 100 m of the water column. Offshore the south-westward set of the Agulhas Current is apparent; inshore at Durban the current direction is considerably more variable.

divided into two characteristic parts. The southern part, about 75 per cent of the total, is narrow; usually only about 15 km wide. It is found between a point just downstream of Durban and a point just upstream of Port Elizabeth. The other 25 per cent lies to the north, in the Natal Bight, and is much wider, on average 50 km from the coast to the 200 m isobath. Research coverage of this whole shelf region tends to be scattered in both time and space. The circulation on the southern part of this continental shelf will be discussed first.

Narrow shelf circulation

In Figure 4.8 the close proximity of the velocity core of the Agulhas Current to the shelf edge can be seen very clearly for the averaged section at Port Edward, in the centre of the narrow section of the shelf. The mean flow over the shelf here is southerly, in contrast to the section off Durban[365]. Schumann[383] has carried out a series of current measurements at Durban, Port Edward and Richards Bay (viz. Figure 4.3). The currents on the shelf at Port Edward in his results are at all times strongly dominated by the Agulhas Current (e.g. Figure 4.16) and it exhibits low-frequency variations of nine days or less. Local wind-forcing plays a relatively minor role, as do tidal or inertial motions[384]. Nonetheless, temperature variations of up to 5 °C in one day were observed[383]. These may have been due to small shear-edge eddies.

The stability of the shelf current at Port Edward, particularly when compared to the inshore currents measured at Durban, is clearly evident in Figure 4.16. Movement in the very surface layers, i.e. top few metres, is naturally much more variable[385], being driven almost exclusively by the stress of the reigning winds, which may include a considerable land–sea breeze[386–7]. Sediment bedforms show[372] that the dominant direction of water movement along the bottom of the water column over the shelf is with the current and has been deduced to be in excess of 1 m/s in general. Strong counter currents along the bottom have been inferred[388] for that part of the continental shelf 70 km southward from Durban, consistent with the existence of the abovementioned Durban lee eddy.

Clearly there are insufficient measurements[811,812], particularly of sufficient duration, to make any substantial conclusions about the circulation over the whole continental shelf between Durban and Port Elizabeth. Nevertheless, it would probably not be too rash to expect the stable, Agulhas Current-dominated flows at Port Edward adequately to represent the flow along this whole, morphologically very comparable, coastline. The circulatory behaviour over the wider shelf to the north is very different.

Circulation in the Natal Bight

The circulation between Cape St Lucia and a point somewhere downstream of Durban, enclosing the whole Natal Bight, consists of at least four flow features that are not evident over the shelf farther downstream. They consist of a kinematically driven upwelling cell along the northern Natal Bight[163], a general cyclonic circulation in the Natal Bight[377], a persistent lee eddy downstream of Durban[364] and enhanced shear-edge features of the northern Agulhas Current at the shelf edge of the bight. Each of these will be dealt with in turn; first, the upwelling cell.

The St Lucia upwelling cell

The coastal region between Cape St Lucia and Richards Bay has long been known for its relatively low sea-surface temperatures[76,364,381]. From measurements with airborne radiation thermometers a difference in the mean sea surface temperature of 1.5 °C between Agulhas Current core water and coastal water has been observed[375], but in individual flights (e.g. Figure 4.17) differences exceeding 5 °C have been measured. With detailed observations during periods of high temperature contrasts, a clearly circumscribed cell of cold water is evident (Figure 4.17) adjacent to the location where the Agulhas Current passes the broadening shelf. The salinity of this water is also lower than elsewhere on the Natal Bight[390]. Further circumstantial evidence about the nature of this cold-water cell is available.

It has been shown that there exists a negative gradient in nutrient concentrations offshore[164,391,801] of the cell. This conclusion is supported by the distribution of nutrients over the shelf of the Natal Bight acquired during the only single research cruise that has covered the full Natal Bight[164]. It shows (Figure 4.18) a strong concentration in nitrate between Richards Bay and Cape St Lucia. Abrupt changes in water temperature of up to 8 °C on a time scale of days have been observed here[375,392]. Pearce has calculated[390] that the cold water is present about 30 per cent of the time while others[163] have estimated 78 per cent of the time. Plankton production rates and biomass distribution in the area, furthermore, are very patchy[391], with high variability[392]. The St Lucia upwelling cell has been invoked[638] as a replenishment source for certain epipelagic siphonophore species the distribution of which otherwise would be difficult to understand. In Figure 4.18 the chlorophyll-*a* is seen to increase from the core of the

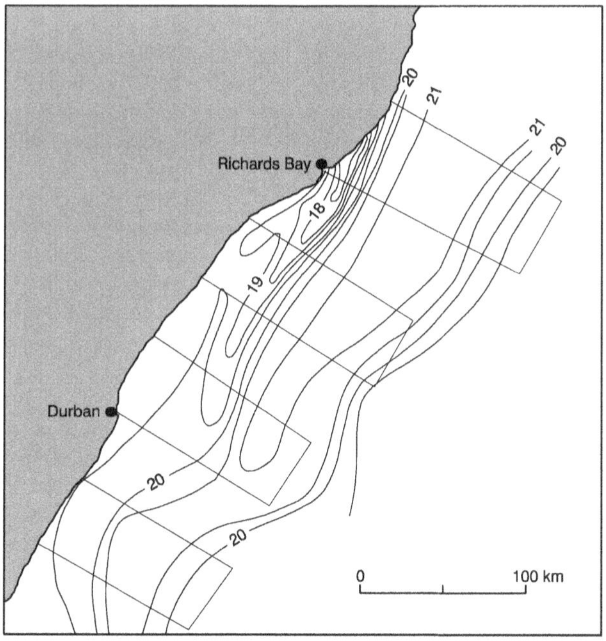

Figure 4.17. The distribution of sea surface temperatures over the Natal Bight and vicinity from airborne radiation thermometry on 9 October 1968[375]. The actual flight track is shown by thin, straight lines. The core of the northern Agulhas Current is shown by 21 °C surface water; upwelled water in an upwelling cell off Richards Bay is 18 °C or less.

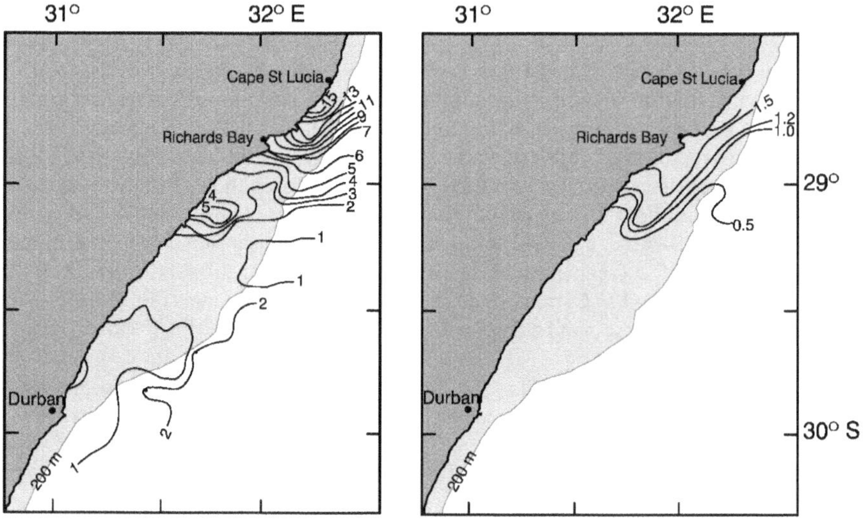

Figure 4.18. The distribution of nitrate (in μmol/kg) at a depth of 10 m over the Natal Bight as observed in July 1989[164] (left panel) and the simultaneous distribution of chlorophyll-*a* at the same depth (right-hand panel). The upwelling cell off Cape St Lucia is clearly circumscribed by both variables. The continental shelf, as defined by the 200 m isobath, is shaded.

upwelling cell southwards and then to decline. This specific distribution is consistent with the delayed biological reaction in an active upwelling cell.

Satellite colour imagery[163] exhibits a high correlation between cold water plumes and green plumes from the Richards Bay region. Measurements of chlorophyll-*a* over the whole bight area[164] show a cell of highest values between Cape St Lucia and Richards Bay (viz. Figure 4.18). All this suggests intermittent and intense, localised upwelling. However, no correlation between these upwelling events and potentially upwelling-inducing winds has been found. This resembles closely the conditions at a similar upwelling cell found at the termination of the East Madagascar Current[299] that has been described above.

It has therefore been concluded[163] that this upwelling cell off St Lucia is kinematically driven by the Agulhas Current where the shelf suddenly widens downstream. Gill and Schumann[167] have demonstrated that, based on theory, shelf topography can cause substantial changes in the structure of a boundary current. A minimum in the shelf width can cause outcropping of cold water near the shore downstream of this point. Furthermore, Ekman veering in the bottom boundary layer under a strong current would tend to carry deeper water towards the sea surface[393]. Such Ekman veering has in fact been observed to take place in the bottom 35 m under the Agulhas Current[394] and may therefore contribute to the upwelling process between Cape St Lucia and Richards Bay.

Cold water with its origin at the upwelling cell has been observed as a cold inshore filament to the Agulhas Current, even as far downstream as south of Durban[163]. Cold, separated bodies of water from the upwelling have also been observed on the shelf of the Natal Bight[389] where they form small cyclonic eddies (Figure 4.19) that may move southward past Durban. A quasi-synoptic, vertical section along the central part of the shelf shows how water rich in nutrients, upwelled at Cape St Lucia, moves along the bottom of the shelf to ventilate this layer (Figure 4.20). This cold, nutrient rich water may have a noted influence on the biological productivity of the Natal Bight and its endemic species[395], a subject that requires further investigation.

Bottom shelf water

Of substantial interest in this regard is the seasonal input of colder, saltier water of offshore origin onto the bottom of the shelf of the Natal Bight in summer[381]. This water is probably Subtropical Surface Water from a depth of 200 m and exhibits an inverse relation to the heating from 21 °C to 26 °C in the upper 50 m of the water column. This seasonal pattern of cold water on the shelf has also been observed for the Agulhas Bank[396], but remains unexplained in the Natal Bight. The second circulatory characteristic of the water on the Natal Bight is its general tendency for cyclonic motion.

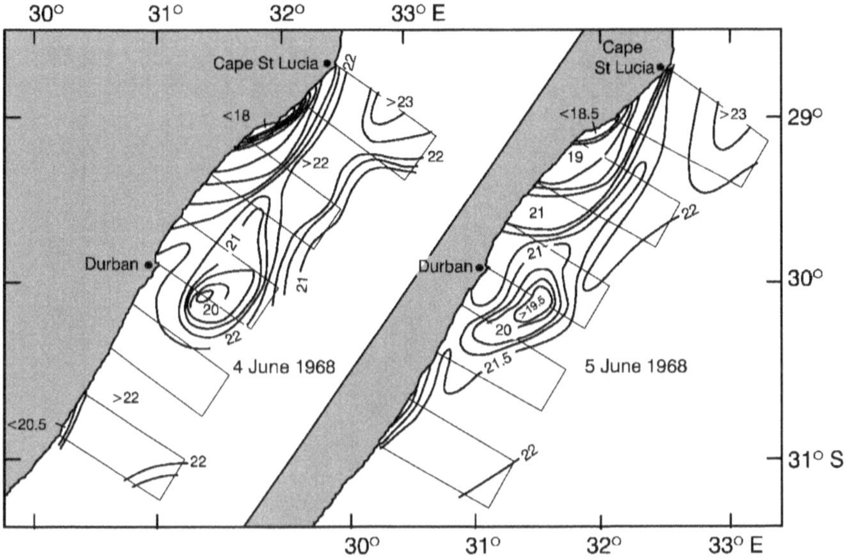

Figure 4.19. Sea surface temperatures observed in the Natal Bight on two consecutive days with an airborne radiation thermometer[389]. Rectangular traces show the flight paths used. Small cyclonic eddies over the shelf and coastal upwelling south of Cape St Lucia are evident.

Lateral circulation

The assumed average circulation over the shelf in the Natal Bight is portrayed in Figure 4.21. This conceptual picture is regrettably based on very few data[397]. Satellite imagery in the visual channels makes it possible to use the sediment outflows from rivers as tracers into the bight. These also show[377] a series of cyclonic eddies.

It is significant that the marked influence of the Agulhas Current on shelf sediments, wherever the shelf is narrow[372], has not been observed over the wider part of the Natal Bight, but only at its north-eastern shelf edge[388]. This implies that high water speeds are found only there. Sediment dispersal and bedform patterns generally support the concept of a few cyclonic eddies[388]. Combined with the little other data available[76, 321] and the logical likelihood that the influence of the Agulhas Current would be to drive a lee eddy, this portrayal (Figure 4.21) is therefore, in general terms, not unlikely.

Plumes of warm Agulhas water have on occasion

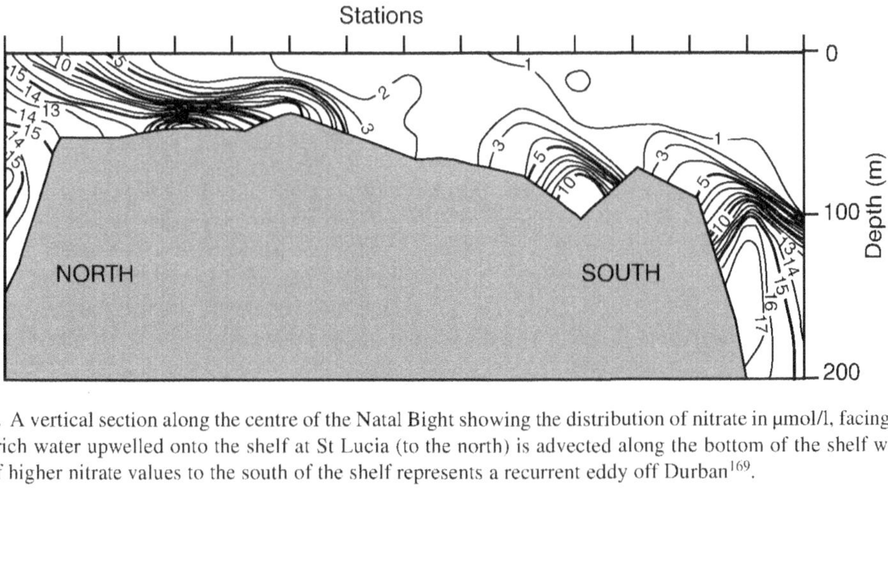

Figure 4.20. A vertical section along the centre of the Natal Bight showing the distribution of nitrate in µmol/l, facing east. The nitrate-rich water upwelled onto the shelf at St Lucia (to the north) is advected along the bottom of the shelf waters. The dome of higher nitrate values to the south of the shelf represents a recurrent eddy off Durban[169].

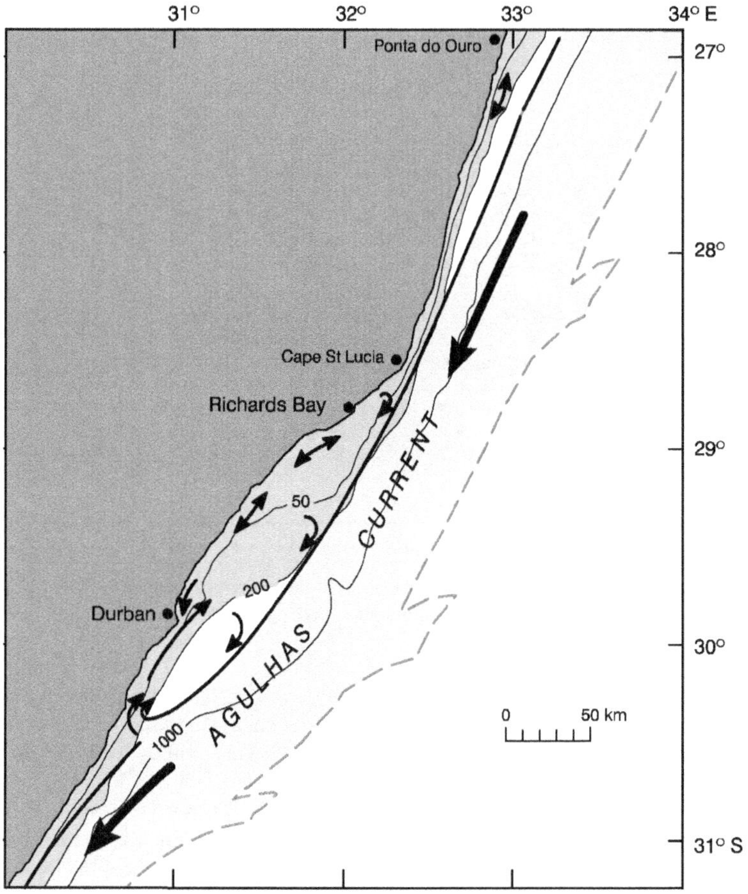

Figure 4.21. A conceptual portrayal of the likely flow distribution over the continental shelf of the Natal Bight[397]. Isobaths are in metres. The circulation in the bight consists of cyclonic eddies and is contained between the offset in the coastline and the Agulhas Current. The latter follows, roughly, the 200 m isobath.

been observed to flow north along the coast[376], in general agreement with this portrayal. Nevertheless, the water over the shelf is so shallow that the influence of the wind must be quite marked[364]. The correlation between the longshore winds and currents at about 16 m depth below the sea surface shows this quite clearly[76] (Figure 4.22). This is probably not the case for the circulation directly downstream of Durban.

The Durban eddy

As was seen previously (e.g. Figure 4.8) a dome of cold, less saline, water at depths below 250 m is regularly present south-west of Durban[76,365]. This feature is reflected in a very persistent coastal counter current (viz. Figures 4.4) and has not been found anywhere else along this coastline. It has also been circumscribed quite clearly by oil slicks caused by the break-up of an oil tanker in 1968[398], and is the major, persistent component of north-eastward flow throughout the water column off Durban[364] (e.g. Figure 4.16). A combination of such current records and Lagrangian drifters show this eddy quite effectively[385]. As has been noted above, sand dunes in the bottom sediments[372] show the effects of a strong counter current here[388] as well, the only place where this has been observed along the flow path of the northern Agulhas Current. Why should a cyclonic vortex have such a recurrent presence here?

The portrayal of the trajectory of the northern Agulhas Current (Figure 4.21) shows that the landward current edge follows the 200 m isobath quite closely upstream of Durban, but overshoots it when the shelf again becomes narrow. The likelihood of a trapped lee eddy in this shelf offset therefore seems high. Theoreti-

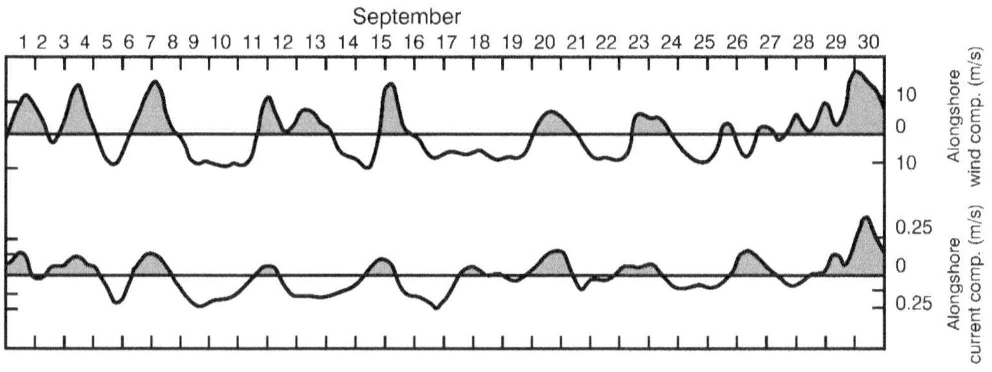

Figure 4.22. Alongshore components of the wind (relative to the inclination of the coastline) and the current (relative to the inclination of the shelf break) at Richards Bay. Shaded areas represent northward movement. Currents were measured 4 m above the bottom in about 20 m deep water[76]. There is a strong correlation between the wind and current speed.

cal considerations[167] also suggest reversed flow upstream of a point where the shelf width becomes narrow. The existence of a specific Durban eddy has therefore – most probably – been demonstrated adequately. Its stability and persistence is as yet unknown. It is not the only shear-edge feature of the region.

Shear-edge phenomena

The last aspect of the circulation at the Natal Bight that is unusual is the occurrence of more extensive shear-edge features[62] here (e.g. Figure 4.9) than elsewhere along the northern Agulhas Current. It has been calculated[376], based on the internal Rossby radius of deformation applicable here, that a typical wavelength of such edge disturbances would be about 80 km and that they are induced by baroclinic instabilities, mainly associated with the upper layer of the current. They have a downstream phase speed of about 50 km/day. Their more extensive growth over the shelf in the Natal Bight may be due merely to the fact that a wider shelf will allow more extensive shoreward penetration. This would contribute water from the upper layers of the Agulhas Current directly onto the shelf, thus modifying the water masses to be found there.

Water types

The waters over the Natal Bight consist of Subtropical Surface Water, probably derived for the greater part from surface water of the Agulhas Current via shear-edge plumes, and Tropical Surface Water, with lower salinity, derived particularly from the inshore edge of the Agulhas Current[164,321]. This is clear from the portrayal of the results of 85 CTD stations undertaken over the Natal Bight in 1989 (Figure 4.23). Considering the shallowness of the shelf and the strong winds experienced during this cruise[164], the lack of mixing between the fresher Tropical Surface Water and the underlying saltier South Indian Subtropical Surface Water is remarkable.

Water in the upwelling cell off Richards Bay consisted of South Indian Subtropical Surface Water with no indication of South Indian Central Water being drawn into it. This is in stark contrast to a similar inshore upwelling cell found farther downstream, at Port Alfred[166], where South Indian Central Water is a major contributor. The shelf at Richards Bay is, however, considerably shallower than at Port Alfred, thus probably preventing deeper water reaching it. Furthermore, South Indian Central Water is found at much reduced depths inshore of the Agulhas Current farther downstream. Overall, it remains unclear what contribution the upwelling near Richards Bay makes to drawing deeper water onto the shelf. By contrast, the contribution by river run-off for the Natal Bight as a whole seems to be relatively modest[164].

Inception of the Natal Pulse

One of the most intriguing aspects of the behaviour of the Agulhas Current has been its occasional absence from the near-coastal location where it normally is found. As mentioned previously, during an intensive investigation on the location of the core of the northern Agulhas Current[86], using hydrographic sections extending from the coast, the current could not be found about 22 per cent of the time. This absence of the current has also been found in current meter records from moorings placed across the current[740]. Pearce[365] has

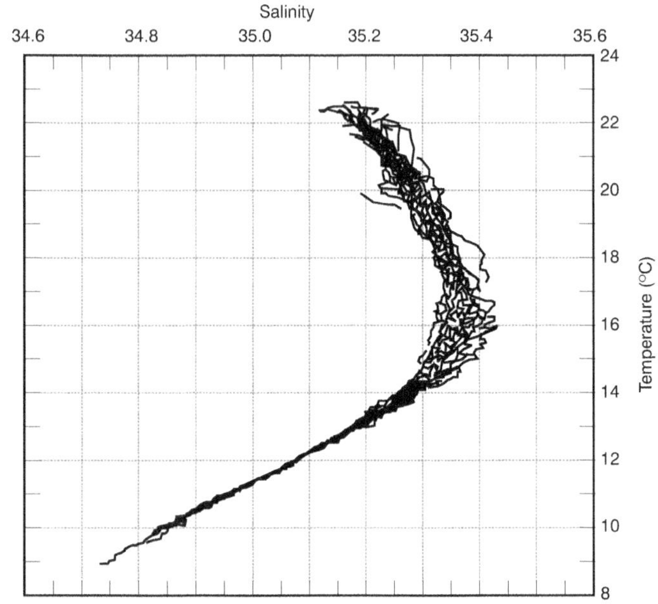

Figure 4.23. The temperature/salinity characteristics for the Natal Bight in July 1989[164].

found the current absent off Durban on 12 per cent of his sections. It was assumed that all these instances were due to very large, but unexplained, meanders in the current.

Occasional meanders

Evidence for the presence of such meanders has first been seen in satellite thermal infrared imagery[59], that has allowed a first estimate of the phase velocity, of about 20 km/day, to be made. An analysis of four-monthly surveys of the northern Agulhas Current between Durban and Port Elizabeth[399] has been successfully used to resolve such a feature hydrographically for the first time. It had a seaward amplitude of 200 km (Figure 4.24), a phase velocity identical to that observed previously in satellite thermometry[59] and there was some indication of cyclonic motion inside the meander itself. Current vectors at 10 m depth taken at the time of the hydrographic surveys are in substantial agreement with an interpretation of a passing meander of the dimensions shown in the hydrography[399]. A later research cruise that successfully intercepted such a meander was able to substantially confirm these results[400].

Subsequent studies using satellite imagery in the thermal infrared have suggested that this feature is a solitary meander that always moves downstream and invariably has its origin in the Natal Bight[60]. A more detailed study of this kind[62], combining satellite imagery and hydrographic data has been able to establish some additional facts about this Natal Pulse (Figure 4.25). First, a Natal Pulse can be seen in some part of the northern Agulhas Current during about 20 per cent of the time, based on five years of observations. It moves downstream at an extremely stable rate of 20 km/day (Figure 4.26). It is shed at the Natal Bight at irregular intervals, and always grows in seaward extent on translating downstream (Figure 4.27) – at least as seen in its surface expression – but not always at the same rate. A very large Natal Pulse has, for instance, been observed off Cape St Lucia[400]. Most of these observations have been either at the sea surface (e.g. Figure 4.27) or relatively shallow (e.g. Figure 4.24). The question on the depth to which these features normally extend has therefore not been adequately addressed with these data.

A complementary set of current meter records, isopycnals floats and altimetric observations has shown[632] that the Natal Pulse extends to the full depth of the Agulhas Current. Distinct temperature decreases and current reversals were observed to a depth of 1960 m as Natal Pulses passed the set of moorings. Close inshore (<50 km) the depth of Natal Pulses was constrained by the presence of the Agulhas Undercurrent. The diameter of a Natal Pulse varied between 90 and 110 km, increasing in extent downstream. The downstream progression of the Pulse observed on this

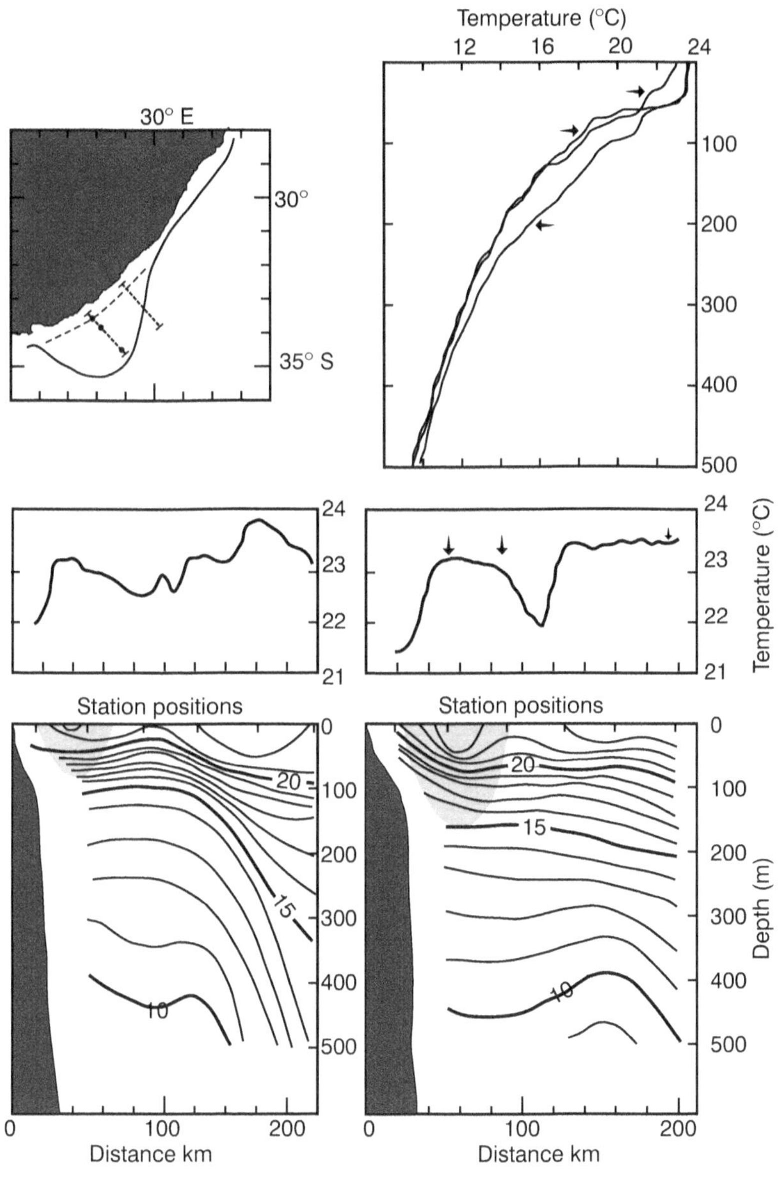

Figure 4.24. Vertical temperature profiles (upper right panel), surface thermograph traces (middle panels) and vertical temperature sections of a Natal Pulse just upstream of Port Elizabeth[399]. The geographical locations of sections and stations, for which the traces are shown, are given in the upper left panel. A dashed line in this panel and shading in the temperature sections indicate a weak inshore current remnant (see also Figure 5.15, page 138).

occasion with the aid of floats was 12 km/day, slower than previous observations. It is instructive to note that eddy-like disturbances on a simulated Agulhas Current in numerical models[301] do move downstream at a very steady rate. Some hydrographic data show the Natal Pulse as having a cyclonic eddy imbedded[62] in the meander. In this specific investigation Natal Pulses could not be followed farther downstream than the Agulhas Bank where there was a suggestion of a rapid deceleration to about 5 km/day where the shelf widened. This decrease in phase speed along the wider part of the shelf has also subsequently been confirmed by satellite altimetry[401].

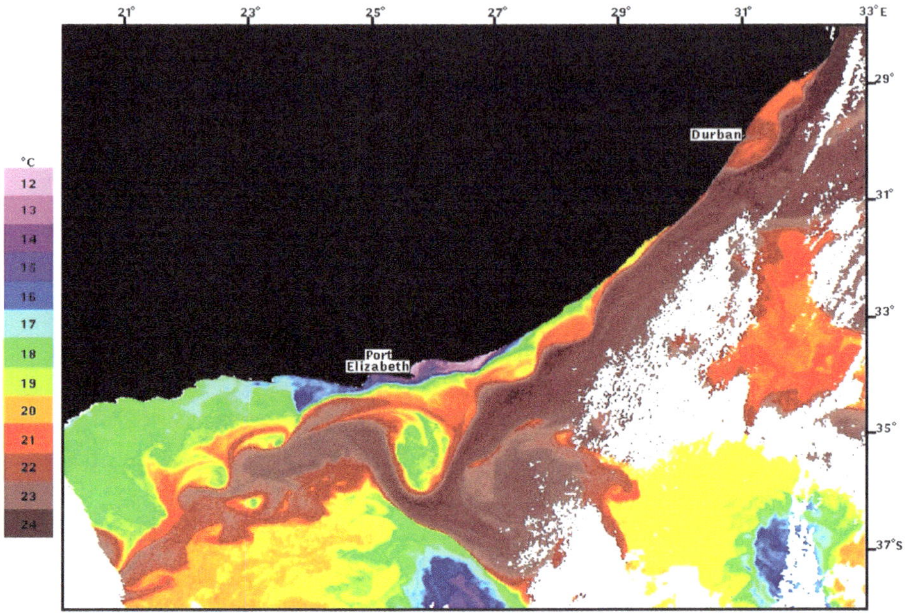

Figure 4.25. Surface temperatures of the Agulhas Current showing a well-developed Natal Pulse directly south of Port Elizabeth[62]. A warm surface expression of a cyclonic eddy attendant on this Natal Pulse is apparent, as is the cold water of the coastal upwelling cell at Port Alfred, just upstream of Port Elizabeth. A possible successor to this Natal Pulse is simultaneously being spawned in the Natal Bight at Durban. Water with higher temperatures is represented by a darker red hue in this image based on data from the radiometer of the NOAA 9 satellite for 27 May 2004. White patches over the sea are clouds. Blue-green patches to the south in this image are cold offshoots from the Suptropical Convergence.

Natal Pulse eddy

The mechanism for the Natal Pulse that has initially been put forward[62], based primarily on the presence of a cyclonic eddy in the meander, was of vortex shedding from the cyclonic circulation in the Natal Bight. This would also be consistent with events of current reversals observed in current records along the coast[364,375]. The anecdotal concept of an inshore counter current to the Agulhas Current that has been particularly persistent in seafaring folklore[402], but that has largely eluded ocean scientists[403], might possibly be explained by this episodic passage of such a cyclonic eddy as part of a Natal Pulse. A study using Lagrangian floats[632,629] has dramatically shown this cyclonic movement inside a Natal Pulse. This movement extends to the full depth measured and shows that the water in the cyclonic eddy is being carried with the Pulse. At a depth of 1200 m the azimuthal speed of the embedded cyclone was 50 cm/s. At 300 m it was 85 cm/s, resulting in a rotational period of six days[629]. Whether this eddy originates as the Durban eddy, mentioned previously, or has been part of the circulation of the shelf waters of the Natal Bight (viz. Figure 4.21) is as yet unclear.

Fortuitous observations of dramatic surface current reversals downstream of Durban, consistent with the passage of a cyclonic eddy, have been accurately predicted using satellite thermal imagery of a Natal Pulse[404]. Such current reversals throughout the water column have also been tied to the passage of a Natal Pulse in strings of deep current water moorings off Port Elizabeth. There is, therefore, some confidence that the kinematic structure of these eddy features is understood. Of passing interest is the fact that the hydrodynamic structure across a Natal Pulse still shows[62,399] that rather weak remnants of the Agulhas Current are retained at the shelf edge during the passage of a Pulse (Figure 4.24). As will emerge below, the mechanism for triggering a Natal Pulse is, however, not vortex shedding.

Delagoa Pulses?

The shelf structure in the Delagoa Bight, upstream of the Natal Bight, bears a strong resemblance, be it on a much larger scale, to that of the Natal Bight. The frequent presence of a deep-reaching, cyclonic eddy in this

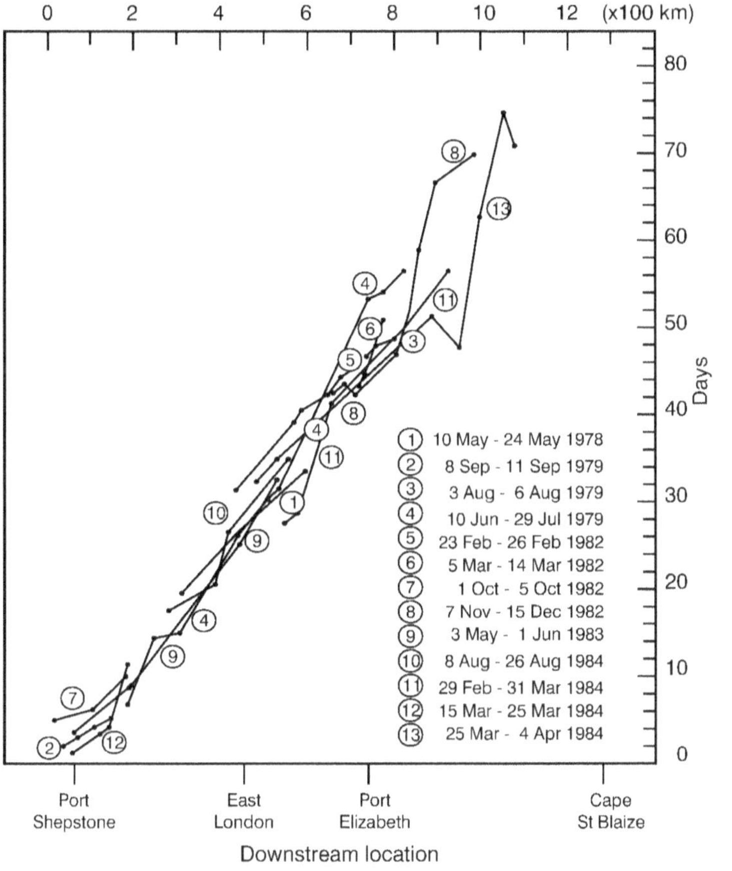

Figure 4.26. The rate of downstream progress of the Natal Pulse along the mean path of the Agulhas Current from 13 series of thermal infrared satellite images[62]. Key geographic reference points along the coast are given along the lower abscissa. A very steady rate of about 20 km/day is maintained, except downstream of Port Elizabeth where there is a suggestion of a marked decrease in phase velocity.

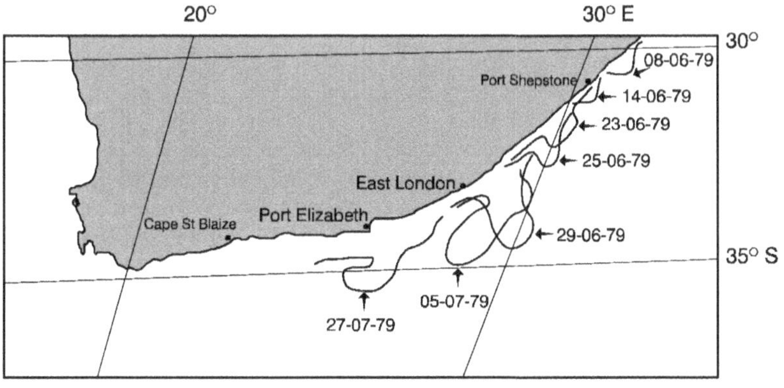

Figure 4.27. The landward thermal edge of the northern Agulhas Current from June to July 1979, showing the downstream passage as well as lateral growth of a Natal Pulse[62].

bight has also been observed[338] and has been discussed previously. This bathymetric and hydrographic similarity might cause one to expect a mechanism of vortex shedding to be evident here as well, similar to that assumed for the Natal Bight. This indeed is the case.

Large meanders reminiscent of those associated with Natal Pulses have been observed to prograde from the Delagoa Bight. This has been based on thermal imagery[398], hydrography[339] (Figure 4.28) as well as altimetry[309]. Large cyclonic eddies have also been observed in this region[360,398], as well as many in the general vicinity of the Mozambique Plateau further to the south[355,356] (Figures 3.31 and 3.32). This is not to imply that all those at the Mozambique Plateau are derived directly from the Delagoa Bight, but only that some may have come from there. An extensive search of satellite images has not been able to find one convincing case of these meander features, that resemble Natal Pulses, advancing southward in the northern Agulhas Current. This may be due to the fact that the Agulhas Current is only found in well developed form farther to the south. This is clear from sediment bedforms along the coast between Durban and Maputo[388]. For the present, until more rigorous data becomes available, we may therefore assume with some confidence that the Natal Pulse on the northern Agulhas Current comes only from the Natal Bight. Why should the Natal Bight be such a special conception area for these features?

Causative mechanism

The first possibility that has been suggested is that this is the only location along this part of the shelf where lee eddies are formed. Such an eddy has initially been thought to be the prime mechanism for the formation of Natal Pulses by a process of vortex shedding, as discussed above. However, De Ruijter and co-workers have carried out an analytical study[87] and demonstrated that barotropic as well as baroclinic instability is impossible in the greater part of the northern Agulhas Current because of the narrowness of the continental shelf and the steepness of the shelf slope. The only exception where this bathymetrically induced constraint might not hold would be the Natal Bight region where the slope gradient is relaxed. The onset of barotropic instability at this spot would then be another possible triggering mechanism for Natal Pulses and not only vortex shedding. For this instability to happen, the velocity profile of the Agulhas Current would have to exceed a certain value. In practice the average velocity profile across the current would not allow barotropic instability, since the mean jet intensity is insufficient[87].

However, about 20 per cent of the hydrographic sections across the current at Durban show a sharpness of the landward front of the current that does exceed this threshold, thus allowing the onset of barotropic instability. Once the current has moved sufficiently far away from the shelf edge as a result of such an instability, the stabilising influence of the slope would be lost and the incipient meander can grow rapidly. The enclosed cyclonic eddy may therefore not play such a principle role in the triggering of Natal Pulses. The next question that then arises as a matter of course concerns the cause for such increases in the jet intensity of the northern Agulhas Current at the Natal Bight.

Triggering Natal Pulses

It is quite possible that natural fluctuations in the transport, or the cross-current velocity distribution of the Agulhas Current, could be responsible[365] for the required increases in jet intensity. Unstable barotropic modes are found[744] inshore of the current with propagation speeds close to those observed. The onset of pulselike behaviour might therefore be found in wave-mean flow interaction of these unstable inshore modes with the current jet. Other possibilities include the adsorption of deep-sea eddies onto the current.

Seaward streamers of warm water from Natal Pulses that have been noticed in satellite imagery[62] (see Figure 4.25) show this as a distinct possibility[339,360]. Absorption of deep-sea eddies into western boundary currents has been well documented for the Gulf Stream[405–6] and is also evident in the Kuroshio[739]. In the latter anti-cyclonic eddies from the west are absorbed by the current, leading to the creation of meanders in the trajectory of the current. Theoretically a deep-sea eddy will be repelled by a current in a stable condition[407] and with comparable strength. It will coalesce only with an unstable jet with vorticity of the same sign. Eddies that might fulfill this function, both cyclonic and anti-cyclonicanti-cyclonic, are present in the Natal Valley[398], over the Mozambique Plateau[344,356], and in the Mozambique Basin[343,360] as was discussed previously.

A suggestion that these eddies may originate at the Subtropical Convergence to the south[131] is, based on altimetric information, incorrect. The modern data indicate[309] that all these deep-sea eddies come from the east or from the north, the latter out of the Mozambique Channel[653,736]. Recent results, efficaciously combining satellite altimetric and thermal infrared data[362], have in fact shown how cyclonic eddies from east of Madagascar have moved west and been adsorbed onto the seaward side of the Agulhas Current. There is positive evidence that at least on one occasion this had striking

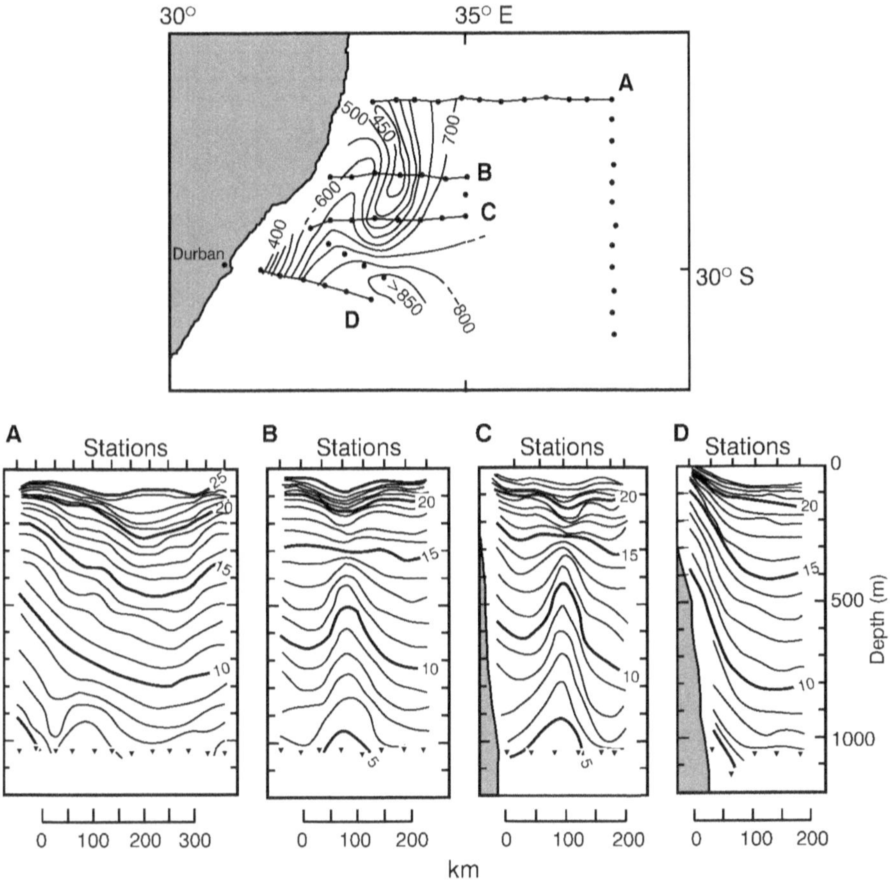

Figure 4.28. Hydrographic characteristics of a solitary meander north of the Natal Bight[339]. The upper panel shows the cruise tracks, station positions and depth of the 10 °C isotherm, a proxy for the dynamic topography. The vertical thermography along four sections across the feature are shown in the other four panels. The bell-shaped isotherms clearly define a cyclonic eddy.

consequences. A drifter placed in a Mozambique eddy[728] in the Mozambique Channel continued its anti-cyclonic gyrations (Figure 3.23) until it reached the latitude of the Agulhas Current. Here it was absorbed by the current, in the process triggering a Natal Pulse[653]. In Figure 3.33 the tracks of such Pulse-inducing eddies were shown graphically. From this figure it would seem evident that both the eddies from the Mozambique Channel as well as from the South Indian perturbation band are headed in the appropriate direction and have the appropriate spin direction to trigger Natal Pulses if they get absorbed in the Agulhas Current at the Natal Bight. Detailed modelling of the system[271] has also shown that instabilities in the flow, generated off southern Madagascar, can move westward to join the Agulhas Current and then progress alongcurrent[301].

Apart from its unusual lateral stability, the northern Agulhas Current is therefore also exceptional for its very large, solitary meanders, the Natal Pulses. It is the combination of shelf morphology at the Natal Bight and the variations in the characteristic dynamics of the northern Agulhas Current that present the required conditions for this special phenomenon. The triggering mechanisms are as yet inadequately understood, but recent results from numerical modelling[273] and from altimetry have shown that each flow perturbation reaching the Natal Bight from the east causes the triggering of a Natal Pulse. It is clear that a significant change in such triggering events could have dramatic consequences for the coastal circulation along southern Africa's eastern seaboard[133].

The influence of Natal Pulses

Natal Pulses are not just dynamic oddities, but have significant and important effects on the general circulation downstream.

It has been noted[411] that drifters placed offshore of the Agulhas Current at Durban may not succeed in passing through the passage between the continent and the Agulhas Plateau (viz. Figure 1.2). Such behaviour would be consistent with some of the flow patterns[355] derived from quasi-synoptic hydrographic data[92]. This implies that the Agulhas Current and the Agulhas Return Current negotiate this narrow gap leaving little room for other water masses to pass through. By implication, any large deflection of any one of these currents from its normal trajectory would lead to a reconfiguration of flow patterns between the Agulhas Bank and the Agulhas Plateau. Thermal infrared imagery has shown that a well-developed Natal Pulse may cause such a path restructuring, causing the Agulhas Current to retroflect upstream of the passage[64]. This is shown diagrammatically in Figure 1.2.

Modelling the current as an inertial jet[64] shows that if the Agulhas Current is forced to overflow the edge of the Agulhas Plateau the shallowing of the bottom will cause this early retroflecting behaviour. Modelling the flow behaviour with a numerical model that is eddy resolving and that has high spatial resolution[277] has also demonstrated that the simulated Agulhas Current is extremely sensitive to the depth of this passage and may be induced to retroflect here without undue forcing. Observations have furthermore shown[412] that this early retroflection may sometimes be only partially completed, with a warm eddy being spun off the seaward side of the current at this position. The depth and extent of the water masses involved in this process of upstream retroflection are at present not known, but are of substantial climatological importance[413], and will be discussed in greater detail later.

Another, rather obvious, question concerning the Natal Pulse is whether such a major meander moving downstream would not have a noticeable effect on the behaviour of the Agulhas retroflection. Would it not precipitate the shedding of Agulhas rings and thus influence inter-ocean leakage? For this to happen the Natal Pulses would have to progress relatively undamped all the way to the termination of the Agulhas Current. There is growing evidence[401,629] that this is precisely what happens. This important effect of the Natal Pulse on the larger-scale circulation will therefore be discussed in greater detail in the next chapters that deal with the southern Agulhas Current and the Agulhas retroflection.

Recapitulation

It is clear from the above discussion that mesoscale processes and sub-mesoscale variability play a substantial role in the larger scale behaviour of the northern Agulhas Current and, most probably, in the wider circulation of the South West Indian and South Atlantic Oceans.

The northern Agulhas Current is first and foremost characterised and defined by its unusual path stability, both in space and time. It is also characterised by the recurrent passage of a large, solitary meander, the Natal Pulse. Both dynamical features play an important role in the circulation of the whole South West Indian Ocean as well as on the circulation of the adjacent shelf regions.

Circulation on these shelves is geographically determined. Over the southern, narrow shelf all motion is parallel and with the Agulhas Current. Over the wider shelf to the north, the Natal Bight, the circulation is by contrast cyclonic. This shelf region also boasts a small, but persistent upwelling cell at Cape St Lucia that is driven by the Agulhas Current. At its southern termination a small lee eddy is often observed to lie off Durban.

In most other respects the northern Agulhas Current resembles many other western boundary currents. These properties include its vertical extent, speed, volume flux, horizontal structure as well as the presence of an opposing undercurrent. The southern Agulhas Current by contrast has some very unusual attributes.

5

The southern Agulhas Current

Once the Agulhas Current waters have moved past Algoa Bay and start following the edge of the ever-widening Agulhas Bank south of Africa, the nature of the current begins to change dramatically. No longer stabilised by a steep continental slope, the path of the current exhibits increasingly larger meanders downstream. The resultant shear-edge perturbations, most clearly evident as plumes of warmer water on its landward side – as well as cyclonic eddies – are no longer inhibited by a narrow continental shelf and extend over considerable parts of the Agulhas Bank. These generic flow characteristics delineate a southern Agulhas Current, clearly distinguishable from a northern Agulhas Current described in the previous chapter.

Whereas the waters over the narrow continental shelf inshore of the northern Agulhas Current have until now been of only moderate scientific interest, and then almost exclusively in relation to what one can learn about the behaviour of the Agulhas Current, this is not so for the Agulhas Bank (viz. Figure 5.16). This widest continental shelf off southern Africa is thought to play a critical role in the ichthyo-ecology of both the west and east coast of the continent. It covers an extensive region and its water masses, currents and stratification are influenced in a complicated way by the Agulhas Current, mostly on the eastern side, and by the Benguela upwelling system, mostly along the western part of the coastline of the Agulhas Bank. These topics have been prominent in research interest in the recent past and have received considerable scientific attention.

The contrasts between the shallow, well-stratified and relatively quiescent waters over the Agulhas Bank and the rapid, deep and turbulent flows of the adjacent Agulhas Current create a range of hydrographic habitats that become especially evident in the distributions of organisms, from the very small to the very large. The disparity in sea surface temperatures and heat content between these contrasting water masses in turn has a substantial impact on the overlying atmosphere through dramatic changes in the atmospheric boundary layer. In many ways the nature, behaviour and influence of the southern part of the Agulhas Current is therefore more complex than that of its northern reaches.

Hydrography of the southern Agulhas Current

When the Agulhas Current waters flow past Algoa Bay, the current is more or less at its peak of development as far as volume flux and hydrographic contrast to its environment is concerned. Surface speeds, as measured by ships' drift[296], show some readings for the southern Agulhas Current in excess of three knots, with a small number greater than four knots. This is in contrast to the northern Agulhas Current where speeds higher than four knots are often observed. It would therefore seem that, at least at the sea surface, the Agulhas Current slows down along the Agulhas Bank. Hydrographic measurements show[230,349] that the mean geographic location of the southern Agulhas Current coincides with the continental shelf break[90,414], but it is perhaps instructive to note that the marked influence of the current on the movement of sand on the shelf ceases abruptly at Algoa Bay[372]. Altimetric data, however, demonstrate[72,183] that the level of mesoscale variability in sea level is higher here than farther upstream. This is most probably due to increased meandering of the current, to be dealt with later.

Volume fluxes

The total, geostrophic volume flux of the current grows by about 6×10^6 m^3/s per 100 km downstream of Port Edward[278] towards the Agulhas Bank. A section from the tip of the Agulhas Bank southwards graphically shows the enormous concentration of this flow (Figure 5.1). The total volume transport between Antarctica and the Agulhas Bank, relative to the deepest common observed level between station pairs along a cross-ocean hydrographic section[278], is 130×10^6 m^3/s. That of the Agulhas Current and the Agulhas Return Current are of nearly identical size. Because these two are in opposite directions they cancel each other out in their

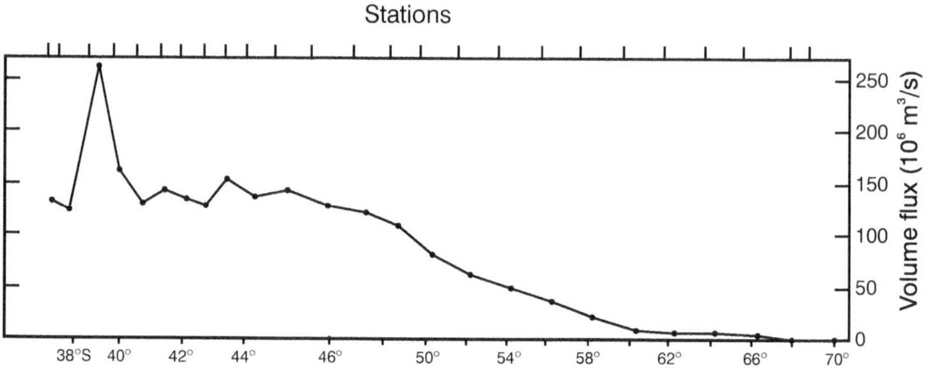

Figure 5.1. The cumulative net volume transport between Antarctica and the tip of the African continent[278]. Station positions as well as latitudes are shown. The largest flow concentrations are at the Antarctic Polar Front, at about 50° S, at the Agulhas Return Current and at the Agulhas Current. This is shown by the strong gradients in the curve at the latitudes of these flow features.

contribution to the total flow between Africa and Antarctica. Purely on the basis of geostrophic approximations, the current here extends to the sea floor.

Further estimates of the total volume flux in the southern Agulhas Current suggest that the steady downstream increase in volume transport from Durban is not maintained past Algoa Bay[230]. The hydrographic data are severely limited here, so that this conclusion can only be treated as indicative at present, particularly since the data may be biased by temporal transport fluctuations of unknown magnitude[259]. The downstream increase in volume transport in the northern Agulhas Current is most probably due to continual additions of water from recirculation in a South West Indian Ocean sub-gyre[101] (viz. Figure 3.30), and deepening of the current. It seems an appealing concept that the inflow from such a sub-gyre would be restricted to a region east of the Agulhas Plateau. Gordon et al.[230] have calculated a total volume flux in the upper 1500 m of the southern Agulhas Current of 70 × 10^6 m^3/s. This is about equivalent to the current's transport to its full depth off KwaZulu-Natal. Geostrophic velocities of the sea surface of 1.10 m/s relative to a depth of 1500 m were calculated at this particular section.

Current kinematics

A set of quasi-synoptic hydrographic sections carried out across the Agulhas Current from Durban, Port St Johns, Port Elizabeth and Cape Agulhas (Figure 5.2) show graphically how the kinematic nature of the current changes downstream[215]. In this case temperatures have been taken as proxies for density; in general this has been shown to be adequate[86,230]. The horizontal gradient in the isotherms on the landward border of the Agulhas Current increases strongly for each consecutive section downstream. The slopes of these isotherms increase also, indicating faster and more concentrated flow. The ribbon of warmest surface water in the current becomes narrower, and possibly deeper. The depths of individual isotherms do not show consistent shifts to greater depths on proceeding downstream (Figure 5.2). What is significant, however, is that the water between the inshore edge of the southern Agulhas Current and the continental slope becomes much colder downstream.

At Port Elizabeth the 10 °C isotherm was found at a depth of about 160 m on this occasion, while farther south (Figure 5.2, section D) it had risen to a depth of only 60 m. This type of upwelling has also been observed in the Gulf Stream[415–6], but has been thought to be due to, and forced by, meanders in that current[417]. This clearly is not the case in the very stable northern Agulhas Current, but may be a contributory factor in the meandering southern Agulhas Current. The increasing closer proximity of the Agulhas Return Current, signified by isotherms sloping in the opposite direction to the south, is evident in subsequent sections of Figure 5.2. It is a very prominent feature, but need not concern us here.

Hydrographic variables

The temperature, salinity and dissolved oxygen characteristics of the southern Agulhas Current are best shown by the results of two stations taken in the core of the current between Port Elizabeth and the southern tip of the Agulhas Bank in 1983[230] (Figure 5.3). Surface water warmer than 16° S, i.e. above 200 m, consists of Tropical Indian Surface Water, with a core of South

Indian Subtropical Surface Water at a depth of about 200 m. The latter is conspicuous for its clear salinity maximum that may be increasingly eroded downstream. Water between 6 °C and 14 °C is South Indian Central Water (viz. Figure 4.15). The cold inshore water being moved onto the shelf (Figure 5.2) is therefore also South Indian Central Water from a depth of about 800 m offshore[215]. The salinity minimum of the Antarctic Intermediate Water, at about 5 °C, shows clear signs of intrusions of Red Sea Water. Below these water masses is the North Atlantic Deep Water. In general this is as expected. The influence of Tropical Thermocline Water, Subantarctic Mode Water[181] and Red Sea Water in the southern Agulhas Current, is more specifically indicated in the temperature–dissolved oxygen relationships (Figure 5.3) but has as yet not been adequately quantified for the southern Agulhas Current[236]. It probably is relatively small by volume.

Current structure

A hydrographic cross-section of the southern Agulhas Current[128] in the concave Agulhas Bight of the eastern shelf edge of the Agulhas Bank, between Algoa Bay

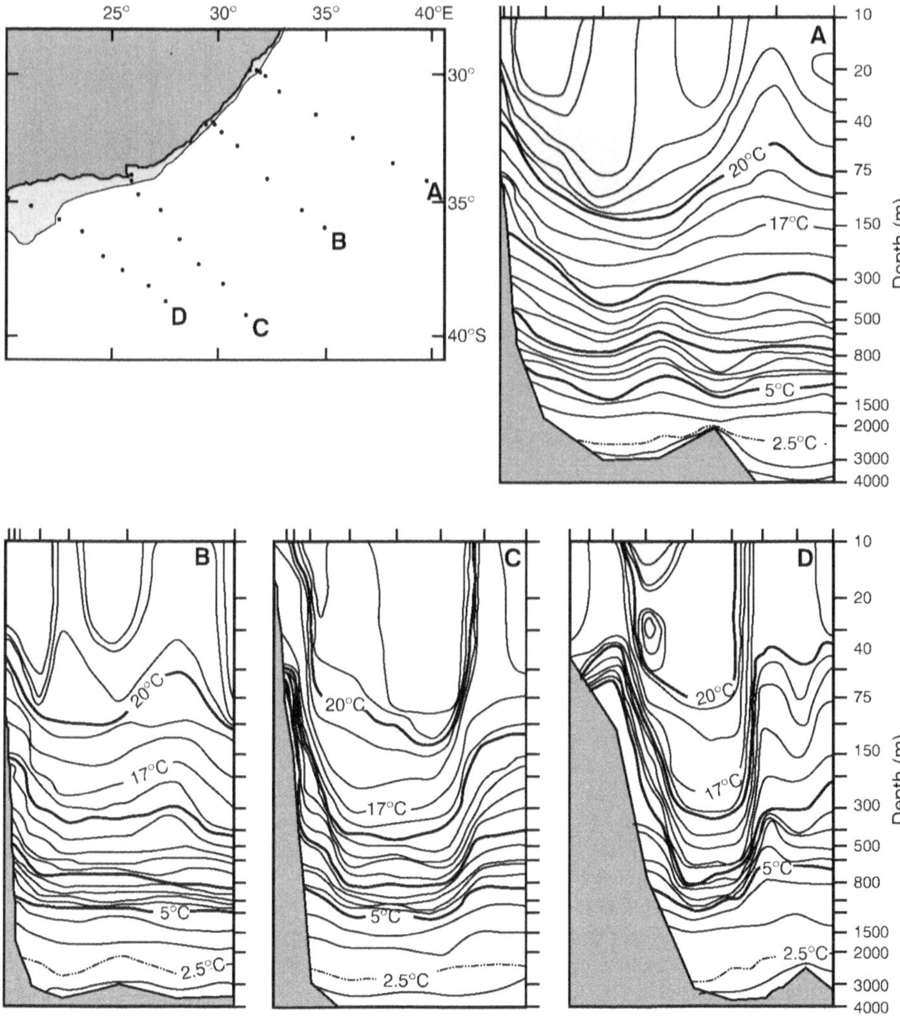

Figure 5.2. Temperature sections across the Agulhas Current from Durban (A), Port St Johns (B), Port Elizabeth (C) and Cape Agulhas (D). Their geographic locations are given in the map[215]. Water warmer than 22 °C that forms part of the Agulhas Current has been shaded. Note that the vertical depth scale is logarithmic. The downstream intensification of the current and the rise of cold water inshore of the current is illustrated.

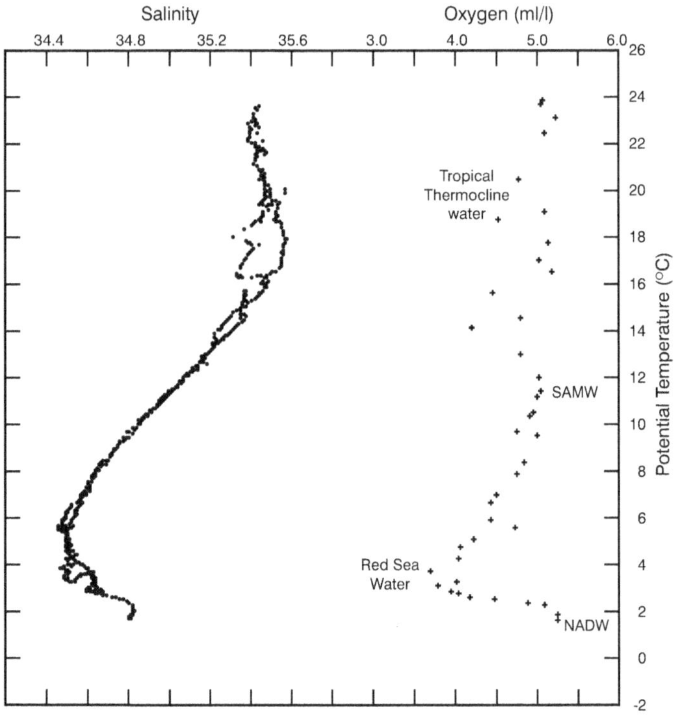

Figure 5.3. The salinity and dissolved oxygen versus potential temperature for two stations in the core of the southern Agulhas Current[230]. The characteristic signatures of Subantarctic Mode Water (SAMW), Tropical Thermocline Water, Red Sea Water and North Atlantic Deep Water (NADW) in the oxygen values are evident.

and the southernmost tip of the Agulhas Bank, shows (Figure 5.4) some interesting internal detail of the southern Agulhas Current. As for the northern Agulhas Current, the landward thermal front at the sea surface is considerably more intense than the seaward front. The ribbon of warmest surface water lies precisely there where the maximum velocities were observed. The current is therefore not symmetrical.

While density surfaces below 400 m at the third station from the left still indicate a flow with the current to the full depth measured, those above give an opposite tendency. Salinities in the upper 200 m of the core of the current indicate Tropical Surface Water with salinities below 35.5, with the salinity maximum of Subtropical Surface Water only evident in the offshore station (Figure 5.4) at 200 m to 300 m depth. This implies either that this division is maintained along the length of the current despite vigorous lateral mixing, or that the water of lower salinity in the core of the current is not Tropical Surface Water, but an admixture of fresher shelf water. The dissolved oxygen distribution suggests that the latter explanation is plausible (Figure 5.4).

Surface characteristics

The temperature of the surface waters of the southern Agulhas Current shows a greater downstream decrease than that found over the same distance in the northern Agulhas Current[77], probably due to greater rates of heat loss to the atmosphere this much farther south[147]. Whereas the alongstream difference in surface temperatures between Ponto do Ouro and Algoa Bay usually is about 2 °C, between Algoa Bay and south of Cape Agulhas it is about 4 °C over slightly more than half the distance. Maximum sea surface temperatures are found in February from 23 °C south of Cape Agulhas to 26 °C adjacent to Algoa Bay. In August these are 17 °C and 21 °C respectively[77].

These averaged seasonal changes in sea surface temperatures may suggest a relatively stable kinematic situation over shorter times. This is not the case. Analyses using hydrographic data[414], satellite thermal infrared imagery[91,418] a well as altimetry[183,335] all show the extremely high mesoscale variability encountered in the Agulhas retroflection region. On closer inspection they also show an enhanced variability for the southern Agulhas Current compared to the northern Agulhas Current. This is due, for the greater part, to the meandering of the southern Agulhas Current mentioned above.

Meanders and shear-edge features

Temperature contrasts at the sea surface between the waters of the southern Agulhas Current and the adjacent Agulhas Bank make the use of satellite thermal infrared images for studying this region very appealing (Figure 5.5) and instructive[59]. Meanders and shear-edge eddies on the current border, that had previously only been inferred[419] from quasi-synoptic hydrographic surveys, can be observed and their behaviour studied with this tool, albeit only their surface expressions. Harris et al.[59] have thus been able to show for the first time that meanders of the current move downstream at a rate of about 23 cm/s and have a crest-to-crest wavelength of about 300 km. Figure 5.5 amply demonstrates the utility of these images.

It shows that shear instabilities only start to grow laterally once the Agulhas water has passed Port Elizabeth. Meanders on the shoreward border grow downstream and each landward crest of these waves leaves behind a plume of warm water. Water in these plumes may eventually disperse over extensive parts of the Agulhas Bank, or may turn current-wards, circumscribing cyclonic border eddies. All these border phenomena are highly time-dependent and during about 35 per cent of the time border perturbations are absent[88] from satellite imagery, the current flowing smoothly past the bank. Modelling of the southern Agulhas Current by two general circulation models[420] has indicated that the meanders in this current are most probably due to baroclinic instabilities.

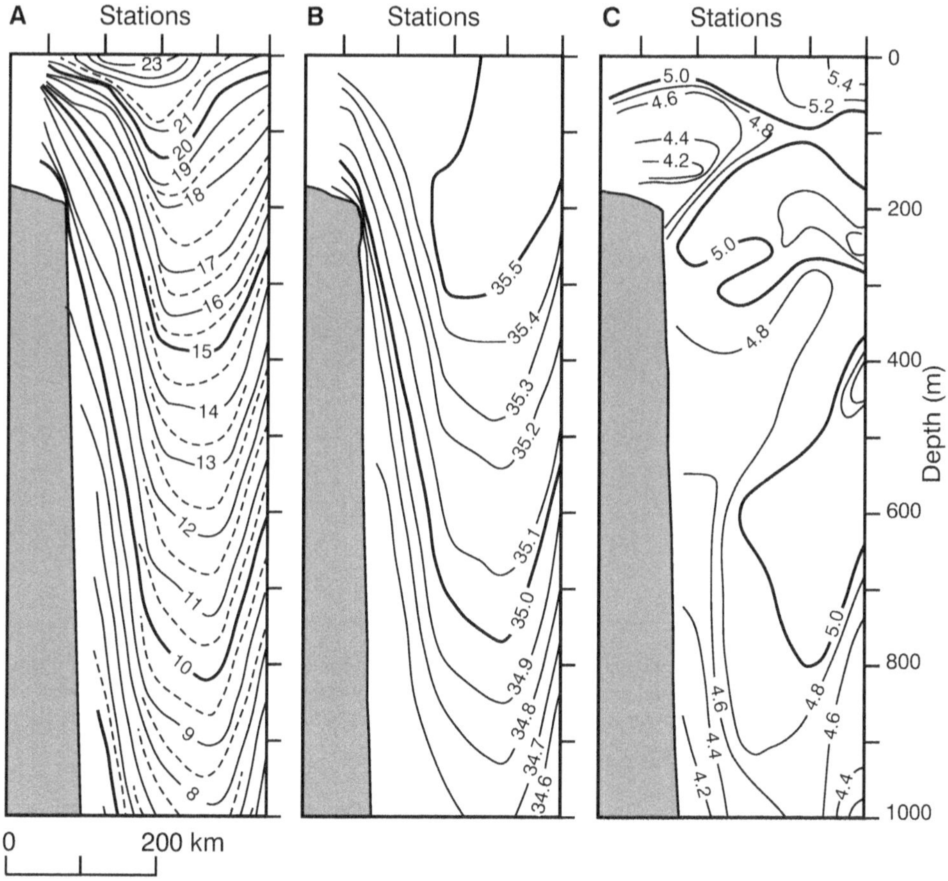

Figure 5.4. A hydrographic section across the southern Agulhas Current[128] showing potential temperature (**A**), salinity (**B**), and dissolved oxygen (ml/l) (**C**). Station positions are indicated.

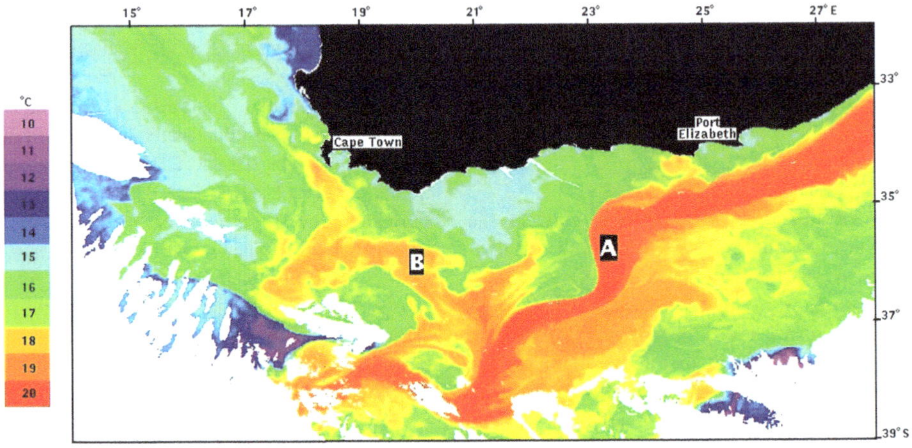

Figure 5.5. A characteristic thermal infrared image for the southern Agulhas Current. The downstream tendency for increasing meandering (A), the inshore prevalence of plumes of warm Agulhas water[88] and the inferred presence of shear-edge eddies, in the lee of meanders, is apparent. The equatorward drift of a warm filament of Agulhas water (B) is also noticeable. In this image of 16 August 1985, which is from the radiometer on board the NOAA 9 satellite, red-yellow hues indicate warm water. The features in this image may be compared with those evident in Figure 4.25.

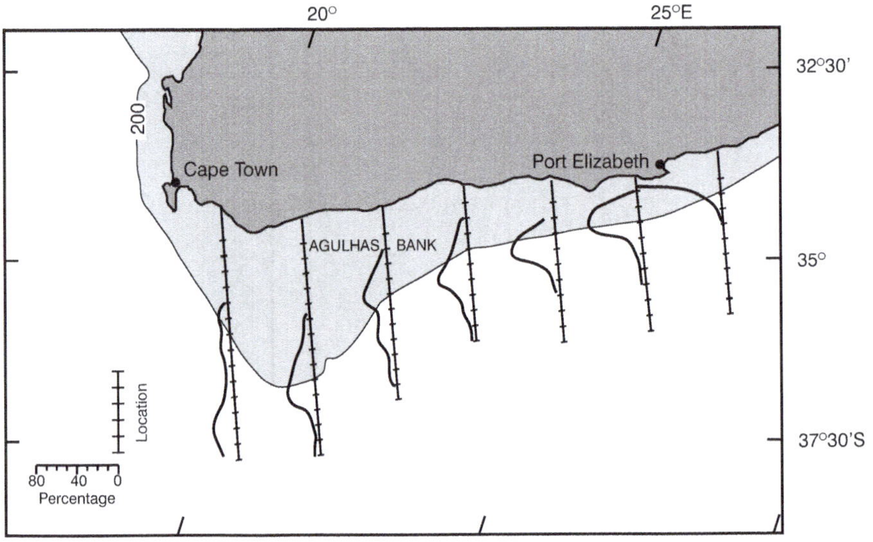

Figure 5.6. The location of the landward border of the southern Agulhas Current relative to the shelf edge[88]. Percentage occurrence in 20 km intervals offshore, according to surface temperatures, shows the increasingly disperse position of this front. This is based on data for a seven year period. The lightly shaded area gives the location of the 200 m isobath that circumscribes the edge of the Agulhas Bank.

Meanders on the current border

Meanders, excluding their attendant plumes, show increasing amplitudes downstream and the geographical location of the landward border of the southern Agulhas Current is thus found over a wider range of latitudes the further one tends to look south-westwards. This is clear from the results portrayed in Figure 5.6. Because of the great lateral stability of its flow path here, the landward border of the Agulhas Current upstream of Port Elizabeth is found in only one 20 km offshore interval during 90 per cent of the time[88]. The remainder of the time the occasional passage of Natal Pulses[62] may move the current offshore.

The most favoured location of the current border relative to the shelf edge, as defined by the 200 m isobath, becomes progressively more diffuse downstream (Figure 5.6). Nonetheless, the current seems to lie slightly seaward in the eastern bight of the Agulhas Bank and to move onto the shelf where the shelf edge turns sharply southwards. Lines of hydrographic measurements in this bight[89] and further downstream[88] are in general agreement with this result. This geographic dispersion is of course due to the increasing amplitude of the meanders. The crest to trough distance for the meanders stays fairly constant at 130 km[88], even once the Agulhas Bank has been left behind. Average plume dimensions by contrast change markedly on proceeding downstream.

Plumes and shear-edge eddies

Plume dimensions include the length of the plume, from where it is attached to the Agulhas Current border; the width of the plume and the distance between the plume and the current. This latter distance may, by inference, describe the diameters of imbedded border eddies. Near Port Elizabeth the mean length of the plumes is 100 km; at the tip of the Agulhas Bank it is 162 km[88]. The width of plumes increases from 27 to 37 km over the same distance, while the size of the enclosed eddies doubles, from an average of 27 km to an average of 51 km. Identifying these shear edge or border eddies by inferring their existence, size and location from warm surface plumes seen in thermal infrared observations naturally is fraught with difficulties[89,421] but at present this is the largest data set available to do this. However, these variables have been confirmed by similar studies[260] not based exclusively on satellite imagery.

The geographic distribution of the border eddies is by no means uniform[88] (Figure 5.7). Directly south of Algoa Bay few have been observed[421-2]. In the eastern bight of the Agulhas Bank they are found clustered on the shelf edge, but downstream of here far fewer are seen, even along the shelf edge. Modelling the circulation along the shelf edge with a high resolution model[631] has been gratifyingly successful with the simulated shear edge eddies bearing a high degree of verisimilitude to those that have been observed at sea. Such models indicate that the main shear edge eddy remains trapped in the shelf bight of the Agulhas Current (viz. Figure 5.8). From there eddies move downstream, representing leakages from the resident shear eddy. This intermittent leakage, as simulated by the model, propagates at about 8 km/day and may trigger the detachment of a cyclonic lee eddy[630] at the tip of the Agulhas Bank. The simulation of this process has been shown to be realistic using floats[629].

Floats were placed in the core of the Agulhas Current near Durban in 1997. On moving downstream, some eventually got trapped in shear edge eddies off the

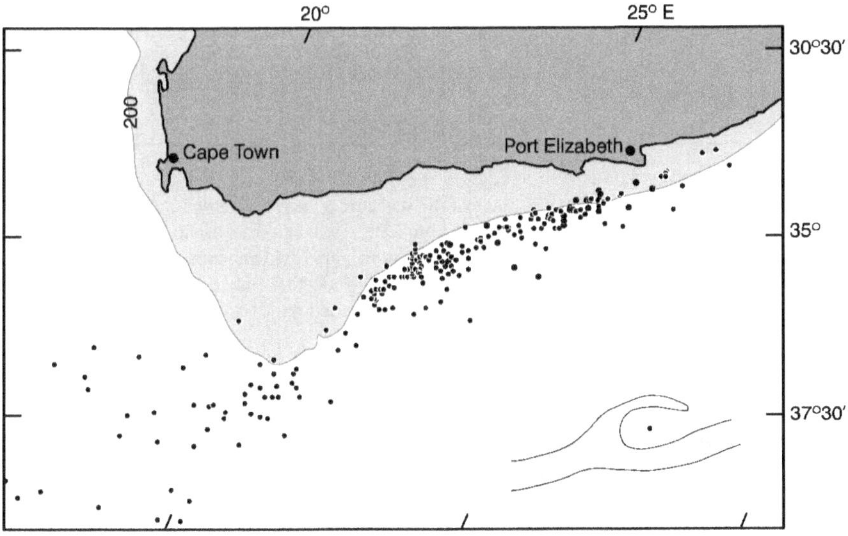

Figure 5.7. The geographic location of shear-edge eddies along the border of the southern Agulhas Current[88]. They are concentrated in the bight of the eastern Agulhas Bank. This distribution is based on satellite infrared imagery for the period 1978 to 1985.

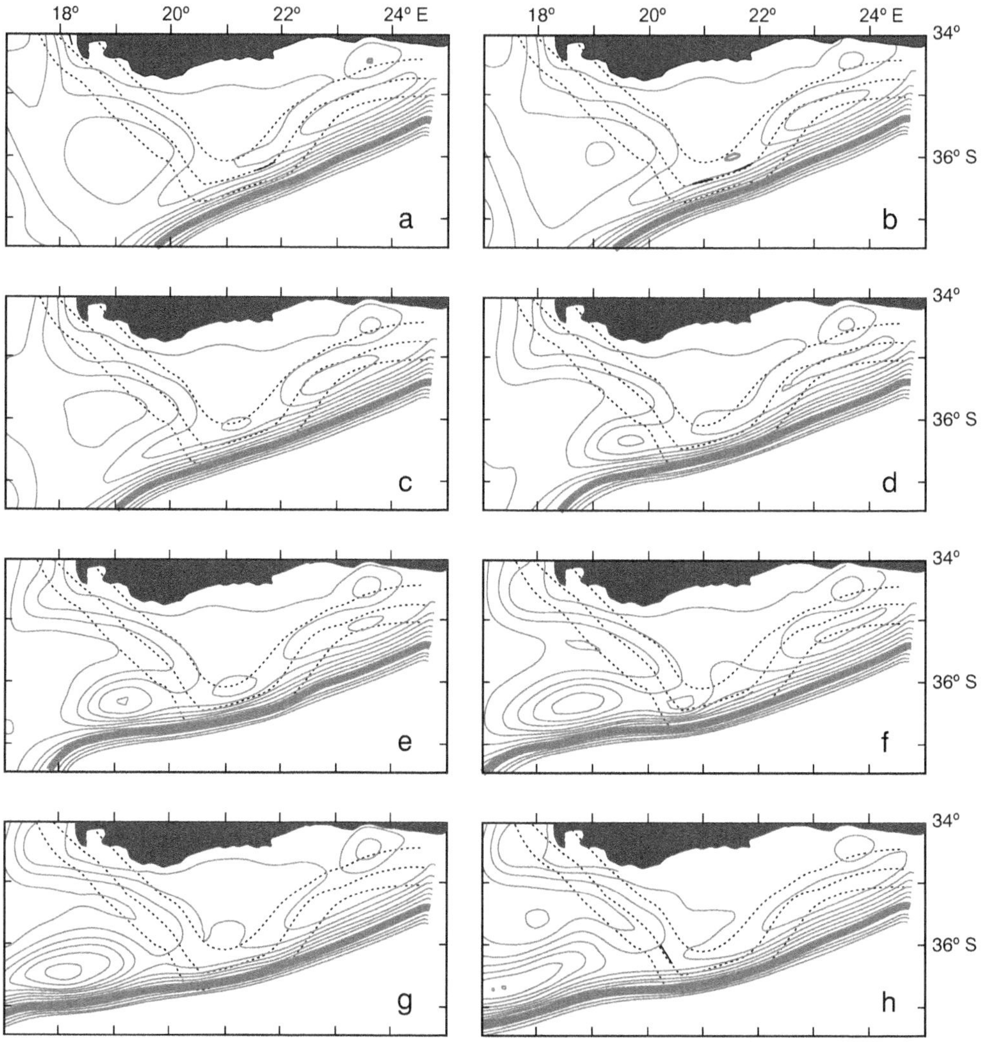

Figure 5.8. The simulated evolution of shear edge eddies along the eastern side of the Agulhas Bank[631]. The broken lines denote the 200 m, 500 m and 1000 m isobaths. The sea surface height is shown in 5 cm intervals with the thick line being the 0 cm isoline. The panels are to be read from left to right and from top to bottom, the temporal intervals being roughly one week in each case. Leakage of sea surface height is evident from shear edge eddies on the eastern side of the Agulhas Bank to a lee eddy on the western side of the bank. This eddy eventually (last two panels) detaches itself from the shelf edge and moves off in a south-westerly direction into the South East Atlantic Ocean.

eastern Agulhas Bank. From there they moved downstream on the seaward side of the current and subsequently were incorporated in the motion of a lee eddy south-west of the Agulhas Bank, very much like the model simulation shown in Figure 5.8. This lee eddy then became detached from the shelf edge and moved westward, cutting through an extensive retroflection loop of the Agulhas Current. This caused the shedding of an Agulhas ring[629]. This sequence of events shows that well-developed, locally formed shear edge eddies may carry out the same function previous reserved for Natal Pulses. These eddies and attendant plumes may also have other functions in the regional ocean dynamics.

Plume dispersion

Of particular importance to an understanding of the stratification of the water column over the Agulhas Bank, is the degree of penetration of warm Agulhas water over the bank through plumes. This is extremely

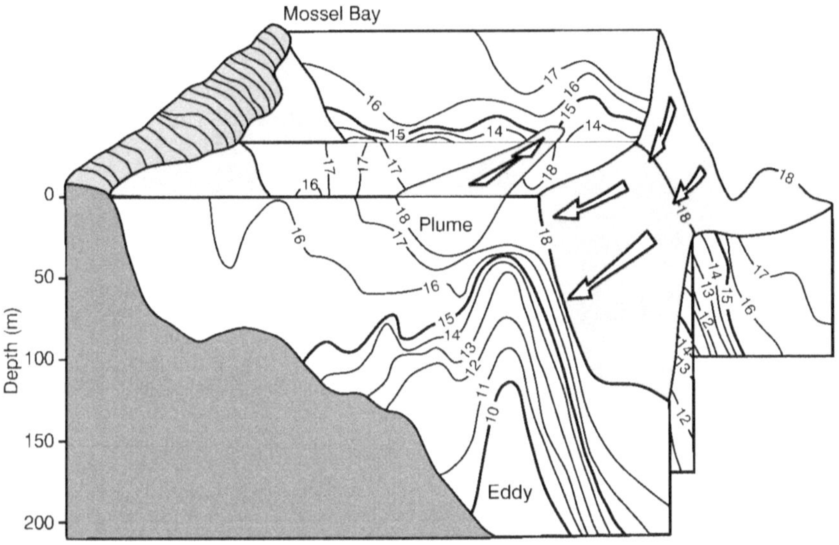

Figure 5.9. The thermal composition of the Agulhas Current, a shear-edge eddy and its associated plume from hydrographic measurements made over the Agulhas Bank in September 1968[88]. The Agulhas Current is defined as lying outside the 18 °C envelope. The plume consists of water warmer than 18 °C at the sea surface while cold, upwelled water is evident in the eddy.

variable and may depend largely on the reigning wind conditions at a specific time[422]. Nonetheless, an analysis of infrared satellite images shows a clear trend of increased spreading away from the core of the Agulhas Current on proceeding downstream. South of Algoa Bay, plumes on average do not carry their warm water farther than 50 km from the current[421]. Goshen and Schumann have shown[422] that under severe wind conditions, however, warm surface water from plumes may penetrate to the coast and even into Algoa Bay, off Port Elizabeth. At the southern tip of the Agulhas Bank the average northward spread increases to about 150 km. On occasion a substantial part of the surface area of the Agulhas Bank may in fact be covered by warm water derived from a number of Agulhas plumes[88]. The shear edge features of the southern Agulhas Current thus play a role in the water profile of the adjacent shelf by inserting warm water via plumes in the upper layers, but they may also contribute by introducing cold water on the bottom of the shelf.

Bottom shelf water

Domes of cold water have been observed on the Agulhas Bank, and inferred to be similar to the domes on the shelf edge that circumscribe the location of the shear-edge eddies (e.g. Figure 5.9). It has therefore been surmised[422] that this cold water on the shelf may have been advected shorewards by shear-edge eddies under the appropriate wind and current conditions. In other parts of the world[748] it has been demonstrated that eddies propagating along the continental shelf alter the structure of the shelf break front. In this way they influence the shelf-slope exchange of biota and water mass properties. This might be true for the Agulhas Bank as well.

This conclusion has been reached[396] based on observations on the Agulhas Bank that demonstrate a prominent seasonality in the water column. It has even been suggested[430] that the very intense seasonal thermoclines over the Agulhas Bank are due not so much to seasonal insolation and wind speeds, but rather to this influx of both warm surface and cold bottom water from the Agulhas Current. As mentioned, it has been suggested that cold water from offshore might conceivably get onto the shelf through shear-edge eddies, or by other upwelling processes. What would be the most likely process by which cold water from greater depths could be upwelled onto the shelf?

An attractive mechanism may be expressed by the shallowing of cold water inshore of the southern Agulhas Current shown in Figure 5.2D. Here water colder than 10 °C, that is usually found at a depth of about 700 m in the South West Indian sub-gyre, is seen as a small dome of cold water at 70 m depth on the edge of the Agulhas Bank. The domed shape of this colder water is significant and suggests a cyclonic eddy. As we have seen, such eddies have for a long time been

How old is the Agulhas Current?

It has been suggested[125] that a reduction in the efficiency of the global thermohaline circulation, in which the Agulhas Current may be an important link, contributed to the cold conditions during the Young Dryas event and that leakage from the Agulhas retroflection into the South Atlantic played a crucial role in the termination of glacial periods[667]. Such intriguing mental images raise questions about the existence of the Agulhas Current in previous epochs. Its former locations, the continuous existence of a retroflection and the possible inhibition of inter-basin water exchange if the Subtropical Convergence ever lay far to the north are questions that come to mind. Reliable information on such major changes in the circulation of the past may indicate possible limits to climate change scenarios for the future.

Various faunal and floral indicators found in the bottom sediments, including oxygen isotope ratios, may give valuable information on the surface temperatures and biological productivity of past periods, thus suggesting changes in the main circulation patterns that occurred during those periods. Paleoceanographic research, based on a limited set of sediment cores, not always located optimally[663], have led to disparate results on the history of the Agulhas Current.

The physiography of sediments underlying the present Agulhas Current[423] suggests the invigoration of the subtropical gyre in the South Indian Ocean near the Eocene/Oligocene boundary, 38 million years before the present, although evidence for the presence of the current goes back as far as the beginning of Tertiary, 65 million years ago[341]. Agulhas Current flow paths very similar to the present were fully established five million years ago[341]. This also holds for the Agulhas Return Current. Seismic studies on the flanks of the Agulhas Plateau[661–2] show that the erosion caused by this current has been continuous since the Lower Oligocene. The south-ward flowing current on the eastern side of this plateau appears to have remained stationary to within a distance of 10 km.

Some results[424] suggest that the Agulhas Current was not a dominant feature during glacial intervals, was weakly present only in summer during these periods and, thus, strongly seasonal. Other interpretations[155] have been of a persistent, but cooler Agulhas Current 18 000 years ago, but considerably less extensive and with a reduced transport. These results are based on sediment cores outside the present position of the current. Winter and Martin[425] have argued that it is unreasonable to expect that the Agulhas Current deviated greatly from its present path in the past since it is strongly controlled by the steep continental slope, the location of which would have changed by less than 10 km during periods of low sea-level during the Pleistocene. They therefore conclude that samples have to be taken underneath the present location of the Agulhas Current.

Sediment surface samples taken directly under the Agulhas Current show that oxygen isotope relationships clearly delineate the narrow band of warm water constituting the current. When these oxygen isotope ratios in the tests of micropaleontological organisms are converted to temperatures[426] they fall clearly within the present temperature range. Bolstered by these positive surface results, analyses of deeper cores show only small temperature fluctuations[425] with time. The sensitivity of these latter results are demonstrated by a 2 °C drop in suggested sea surface temperatures observed in a core taken farther downstream, a result that agrees closely with present downstream temperature decreases in the current[77]. The palaeoceanographic results therefore show no conclusive indication of major shifts in the Agulhas Current position, nor of substantial cooling in the current's water during the last glacial maximum. Its mean position has therefore been stable for at least 150 000 years[425]. It also retroflected in much the same location as today.

This latter result implies that the latitudinal location of the Subtropical Convergence south of Africa had probably not changed much either. This is contrary to some previous findings[427] in which it was suggested that this front moved north by as much as 10° during 65 per cent of the last 540 000 years. Others[160,428] felt that it moved less than 2° of latitude and only marginally[160] during the last glacial maximum, 18 000 years ago. Studies further east in the Indian Ocean[703] have shown that the Subtropical Convergence has been equatorward of its current position for most of the past 500 000 years and has moved poleward during only a few, relatively brief, periods. If this inferred meridional movement is representative also for the region south of Africa, it would imply that a more northerly location of the Subtropical Convergence has not cut off inter-ocean exchange between the Indian and the Atlantic Oceans during this period.

Careful analysis of the incidence of inter-species ratios in cores at the Agulhas Current does suggest marked, but unexplained, changes in biological productivity[425]. These may have come about due to increased meandering of the current[133], leading to more substantial upwelling of nutrient-rich water between the current and the coast[163,429], and thus greater primary productivity than found currently[163,392] at such kinematically driven upwelling cells. The question remains whether the interocean exchanges between the Indian and the Atlantic Oceans – and thus the role of the Agulhas Current in the global thermohaline overturning – changed much over geological epochs.

Analyses of cores taken on the eastern side of the Agulhas Bank[665–6] indicate that the glacial-interglacial cyclicity is well-defined in this region. The influence of Agulhas Current water is enhanced during interglacial periods while the increased presence of subantarctic water is found in glacial periods. Even though the leakage of Indian Ocean water is variable, over the past 450 kyr it has always been present. It becomes most vigorous during the end of glacial periods[667] inducing an abrupt resumption of the interglacial mode of the thermohaline overturning in the Atlantic Ocean[664].

inferred from satellite observations and from hydrographic measurements[419] but very few actual current measurements in these features have been made[89]. These show cyclonic currents to a depth of at least 500 m, when such an eddy is slightly further offshore. As can be expected in a region of such high horizontal shear, the current structure is very turbulent and varies considerably from day to day. Detailed hydrographic observations[88] show (Figure 5.9) that a border eddy will draw up cold South Indian Ocean Central Water. Whether this updrawn water will then spill over onto the shelf is unclear.

The circulation of a shear-edge eddy may cause plumes of warm Agulhas surface water to encircle these features. In the example shown in Figure 5.9 such a plume is only about 50 m deep and consists of water that still has the same temperature as that found in the adjacent surface water of the Agulhas Current at this time. Plumes may on occasion also be much deeper and extend all the way to the shelf floor[88].

Cold water upwelling

Another mechanism by which cold water is upwelled along the edge of the Agulhas Current is through a process of Ekman veering in the bottom boundary layer[193-4]. This process may be responsible for the increasing shallowness of cold water wedged between the Agulhas Current and the shelf edge. This is clearly evident in hydrographic sections across the current taken progressively farther downstream (viz. Figure 5.2). Having, as it were, primed the pump, this cold water is now available to be upwelled[431] in specific upwelling cells, such as that at Port Alfred, or in shear-edge eddies. These border eddies are not in geostrophic balance, but deeper water is drawn upwards in their centres as they are driven by the transfer of cyclonic vorticity from the adjacent Agulhas Current. In regions where the core of the southern Agulhas Current is on average farther away from the shelf edge, i.e. in the eastern bight of the Agulhas Bank, deeper eddies would be more likely to form and to persist (viz. Figure 5.7) than where the current partially overlies the shelf edge.

Other western boundary currents

Both the meandering of the southern Agulhas Current as well as the creation of border eddies and their attendant warm plumes are very similar to that which has been observed in the Gulf Stream. A steady increase in the variance of border location for the Gulf Stream with distance downstream from Florida[432] is comparable to that found in the southern Agulhas Current. The only important exception is the irregular presence of a Natal Pulse in the case of the Agulhas system. A downstream increase in meander amplitudes is also prevalent in the Kuroshio[116]. The phase speed for these meanders in the Gulf Stream decreases downstream[433], something that has not been observed in the Agulhas Current. The downstream distance along which meanders are found on the southern Agulhas Current is, however, much shorter than for the Gulf Stream so that such phase decreases may be difficult to establish reliably here.

Border or shear edge eddies are associated also with meanders on the north wall of the Gulf Stream[434] and are underpinned by an upwelling of cold water[378,435]. They are thought to bring about a subsurface exchange of shelf and slope water[436]. The circumscribing warm plumes are seen in the Gulf Stream as well as in the Kuroshio[379], but those observed in the southern Agulhas Current are usually wider and longer. The driving mechanisms for these features are probably identical, dependent only on the shear parameters of the current boundary and on the continental shelf and shelf-edge configuration.

So, for instance, boundary eddies have been observed to be spun off the edge[417] of the Gulf Stream and subsequently to move independently over the continental shelf. This has not been directly observed for the Agulhas Current and may be due to the limited width of the Agulhas Bank. Cooler water, believed to be associated with such an eddy, has been observed to drift landward[422], but a fully intact feature is still to be observed on the bank. One particular feature that forms part of the border phenomena of the southern Agulhas Current, but that because of its interbasin configuration is unique to this western boundary current, is the Agulhas filament.

Filaments

An example of such a filament, as seen in the surface temperature distributions, may be seen in Figure 5.5. A well-developed plume of warm water, associated with a large meander of the southern Agulhas Current had moved equatorward, past Cape Town, into the south-eastern Atlantic Ocean. Part of it, or a previous one, had been drawn around a cyclonic eddy that itself is not visible from sea surface temperature contrasts, but can only be inferred from the disposition of this filament. From their origin in the Agulhas Current these filaments seem to follow the western edge of the Agulhas Bank[92] closely. They may be advected rapidly northward by the shelf-edge jet[437] along this coastline, enhance the horizontal temperature gradient of the upwelling front found here[438] and in this manner bring about instabili-

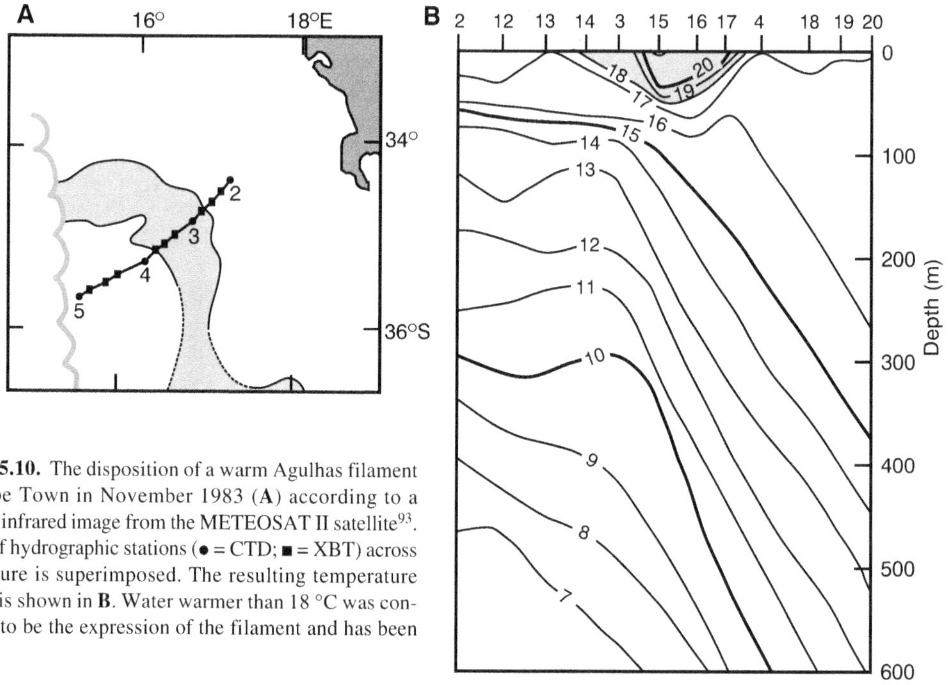

Figure 5.10. The disposition of a warm Agulhas filament off Cape Town in November 1983 (**A**) according to a thermal infrared image from the METEOSAT II satellite[93]. A line of hydrographic stations (● = CTD; ■ = XBT) across the feature is superimposed. The resulting temperature section is shown in **B**. Water warmer than 18 °C was considered to be the expression of the filament and has been shaded.

ties in this front. They may also, infrequently, spread over a large area[439] thus substantially altering the sea surface temperatures, and vertical heat fluxes to the atmosphere, for the region. The obvious question that arises is to what depth these Agulhas filaments penetrate. Are they in any way different to the shear-edge plumes along the eastern side of the Agulhas Bank?

A hydrographic section fortuitously taken across an Agulhas filament is shown in Figure 5.10. It demonstrates, as do a collection of other such crossings[92], that these features are on average 50 km wide and only 50 m deep. This is very similar to the dimensions for plumes along the eastern side of the Agulhas Bank (viz. Figure 5.9). They are present about 60 per cent of the time and most follow the western shelf edge closely. A few are formed farther into the South Atlantic Ocean and then drift northward, parallel, but somewhat distant from the shelf edge. Many get drawn around Agulhas rings[440] that pass through this region and these may make a modest heat and salt contribution to the rings. The annual, inter-basin salt flux accomplished by these Agulhas filaments is estimated to lie between 3 and 9×10^{12} kg/year[92]. Chlorophyll observations show increases in the density of biological material at both inshore and offshore borders of Agulhas filaments[441], but the reason for this, and what local role this phenomenon may play, is not known. It may be due to localised, or edge upwelling. We look next at such upwelling in greater detail and then later on at what role it might play in the ventilation of the water column over the Agulhas Bank.

Shelf-edge upwelling

As discussed above, Ekman veering along the inshore edge of the Agulhas Current may bring deeper, cold and nutrient-rich water closer to the sea surface. This water may subsequently be brought onto the shelf by upwelling in border eddies. As discussed above, this seems to be true for all other, comparable western boundary currents[442–3]. Bringing water onto the shelf by an intense and more-or-less permanent cell of upwelling inshore of the main flow would seem unusual for this type of current. Nevertheless, this seems to occur at the southern Agulhas Current and there is some evidence for a comparable state of affairs off southern Madagascar[304] and even Taiwan[303].

Based on the data collected during the first proper hydrographic coverage of the greater part of the Agulhas Current system[90], Bang[444] has been able to point out a region of what he has called *Agulhas boundary upwelling*, upstream of Algoa Bay. Whereas the surface water in the core of the southern Agulhas Current was in excess of 26 °C on this occasion, the temperature of inshore water was below 15 °C. The cold inshore patch extended about 200 km upstream.

Site-specific upwelling

Subsequent satellite images in the thermal infrared have proven[384,430–1,445] conclusively that this upwelling feature is not uncommon and that it may persist[429] at the sea surface for periods of several weeks. Sea surface temperatures lower than 13 °C have been observed in this upwelling cell[397]. Hydrographic sections carried out from Port Alfred (Figure 5.11) demonstrate how warm surface water from the Agulhas Current may overlie parts of the shelf or may lie offshore of the shelf edge, with cold, upwelled water on the shelf itself[446].

Remote sensing of the Agulhas Current

Over the past three decades remote sensing has been one of the most profitable tools in advancing knowledge on the nature and behaviour of the Agulhas Current system[673]. Because this ocean system exhibits very high horizontal gradients both in sea surface temperature as well as in sea level height, international investigations have frequently used it as a choice case study for the use of new technology[59,72], thus fortuitously aiding the rapid growth in oceanographic knowledge.

The first pioneering studies with remote sensing on the Agulhas Current were carried out by Harris and Stavropoulos[452] who used an airborne radiation thermometer to study sea surface patterns off Natal. This work was extended and continued by Chris Snyman of the National Physical Research Laboratory of the CSIR in Durban, was never adequately published, but results have been included in many subsequent analyses[349,376,397]. Most of these measurements were concentrated in coastal regions[375] due to obvious logistical constraints.

Harris subsequently – then at the University of Cape Town – also pioneered the use of satellite remote sensing, in the thermal infrared[59], for studying the Agulhas Current system over a wider geographical domain. This type of information was subsequently utilised very effectively by a number of investigators and led to the discovery of the process of ring shedding at the Agulhas retroflection[60,91], the Natal Pulse[60,62] (Figure 4.25), the behaviour of shear-edge features[88,339], the retroflection of the East Madagascar Current[114], the drift of Agulhas filaments[92], upwelling on the Agulhas Bank[453–4] and many other circulation features. It has also been increasingly, and efficaciously, used in planning cruises[65,455], in directing underway research cruises[456], in interpreting their hydrographic results[421] and in a range of practical ocean applications[457]. Monitoring the translation of particular circulatory features such as eddies after a cruise[458] and even predicting the behaviour of Natal Pulses[404] has become possible because the satellite data are timeously and readily available. These investigations have largely made use of thermal data gathered by the radiometers on board the polar-orbiting NOAA series of satellites, but also the geostationary METEOSAT series that is particularly well placed for studies of the Agulhas Current, but has a poorer spatial resolution. Both were available to South African scientists at minimal cost from the Satellite Remote Sensing Centre of the CSIR at Hartebeesthoek from as early as 1978. One of the most serious limitations of thermal infrared observations for studying the Agulhas Current system is the obscuring effect of the sea surface by persistent cloud cover. A new satellite system has overcome this debilitating effect by using micro wave radiometry. The TRMM (Tropical Rainfall Measuring Mission) has a multichannel scanning microwave imager that allows the recovery of sea surface temperatures. It has been used to investigate rainfall events[625] and the influence of the Agulhas Current on such events[619]. It has also been used[670] to track features of the greater Agulhas system including eddies from the East Madagascar Current and to investigate various temporal behaviours around southern Africa.

South Africans have also made good use of satellite measurements of ocean colour, particularly of the Coastal Zone Color Scanner (CZCS) on the NIMBUS 7 satellite in 1978 to 1980[459–61] and subsequently of the SeaWiFS satellite in the 1990s[170,304,462]. The South African effort using the CZCS was spearheaded by Shannon and subsequently assembled in an attractive book[463]. These studies demonstrated the feasibility of studying a number of different processes by these means, e.g., the movement of sediment-laden water on the shelf of the Natal Bight[377] and the concentration of chlorophyll-*a* on the inshore edge of the Agulhas Current[441].

With the advent of satellite remote sensing of sea height, i.e. altimetry, aspects of the Agulhas Current system that were previously intractable could be studied[70] and this has occurred fairly rapidly after these data first became available. The otherwise unmonitorable drift of Agulhas rings into the South Atlantic Ocean has received much attention[73,464–5]. They have been followed across the gyre[95] and it has been demonstrated[129] how well the sea height anomalies in altimetric data represent Agulhas rings. Eddy movement south of Madagascar, invisible to thermal infrared instruments, has also been effectively studied using altimetry[87,309,362].

Of long-standing interest has been the prevalence of unusually large surface waves on the Agulhas Current[171], particularly in the northern Agulhas Current. Early attempts to study this phenomenon using synthetic aperture radar on the Space Shuttle[173] have been highly successful, particularly in showing the geographic distribution of such waves. This work has been profitably extended using satellite altimetry[174–5], demonstrating both the amplification of wave height in the Agulhas Current as well as its attenuation in the Agulhas Return Current (see inset on page 105).

A survey of the research publications on the Agulhas Current of the past decades indicates a preponderance of papers using satellite remote sensing. This usage has doubtless played a most important role in the rapid advancement of our understanding of this circulation system over this period.

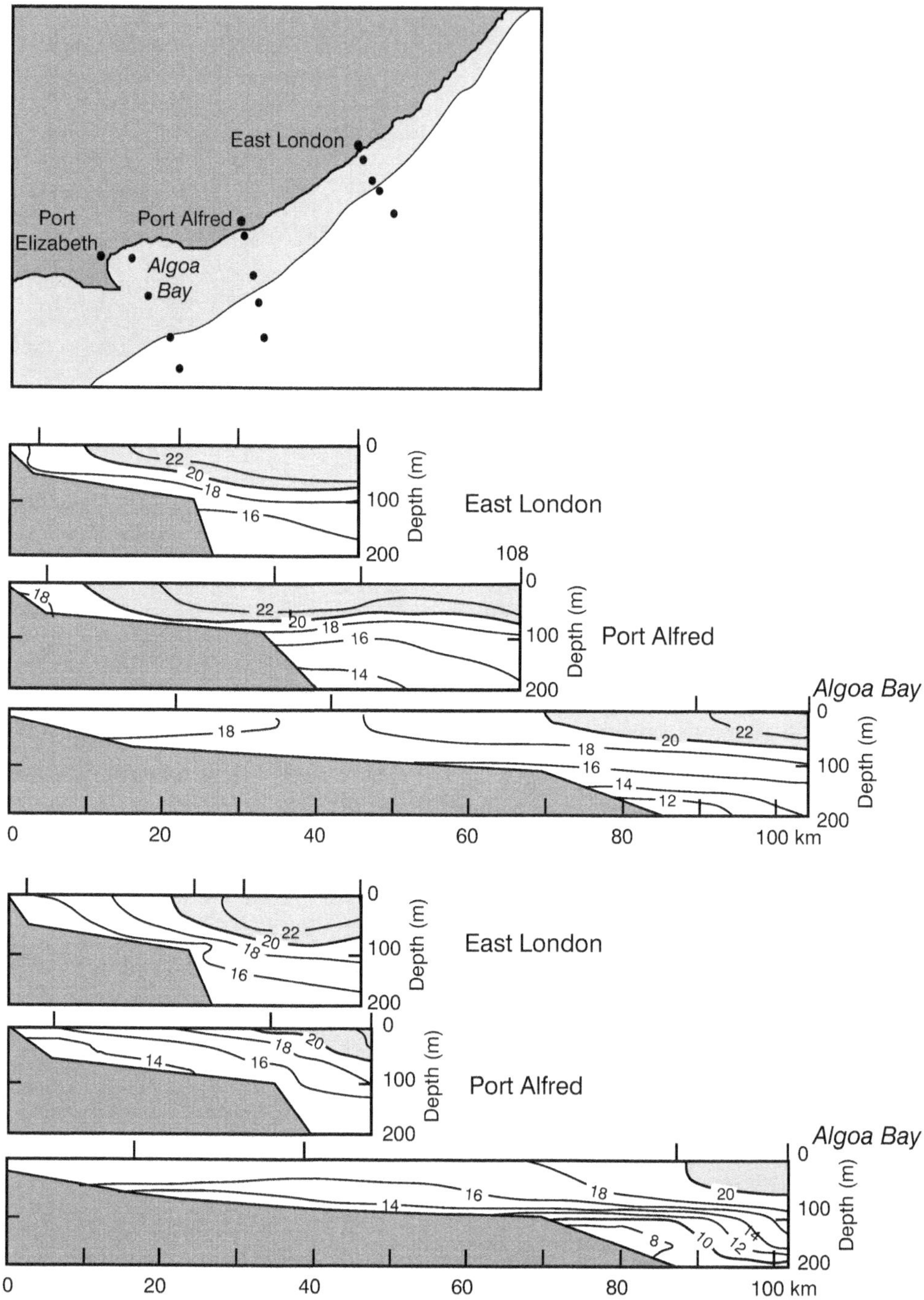

Figure 5.11. Temperature sections across the landward border edge of the Agulhas Current at East London, Port Alfred and at Algoa Bay[446]. The upper panels represent stations carried out in May and June of 1990; the lower panels in October of the same year, during an upwelling event. The geographic locations of the stations are indicated.

Measurements of water movement show[431] that during such an upwelling event the currents are offshore over the width of the shelf and throughout the water column. During this particular event water with temperatures of less than 9 °C formed the bottom water of the shelf while surface temperatures lay between 16 °C and 17 °C. This upwelled water raised the nutrient content of the water considerably. The silicate values, for example, increased from about 3 µmol/l to greater than 9 µmol/l on the bottom and from about 2 to greater than 4 µmol/l at the sea surface. This upwelling cell thus makes a major contribution to the nutrient input onto the shelf[431]. Furthermore, it has a dramatic effect on the heat and moisture fluxes to the overlying atmosphere for this region[146,447]. Rouault et al.[448] have demonstrated that this is particularly effective under conditions of along-current winds.

Three questions about the nature of this particular upwelling regime are important: what is its lateral extent, what is the source of the upwelled water and what influence does this colder water have on the general environment of the Agulhas Bank?

Characteristic of the Port Alfred upwelling cell

The geographic extent, at the sea surface, of water colder than 17 °C, monitored for a summer season with thermal infrared observations from satellite, is shown in Figure 5.12. It is clear that this surface outcropping of cold water is not uniformly distributed along the edge of the Agulhas Current[166], but is restricted to a strip between Mbashe and Port Alfred. During 80 per cent of the time that it has been observed, it is evident at Port Alfred, suggesting that this is the centre of a well-circumscribed upwelling cell.

The source of this upwelled water is in the core of the South Indian Central Water[166]. This is demonstrated by comparing the data collected during two hydrographic sections off Port Alfred (Figure 5.13). These sections represent two sharply contrasting situations. During the first period (Figure 5.13a) most of the water over the shelf was warmer than 20 °C. Offshore stations identified the subsurface salinity maximum characteristic of South Indian Subtropical Surface Water, overlain by less saline, warmer Tropical Indian Surface

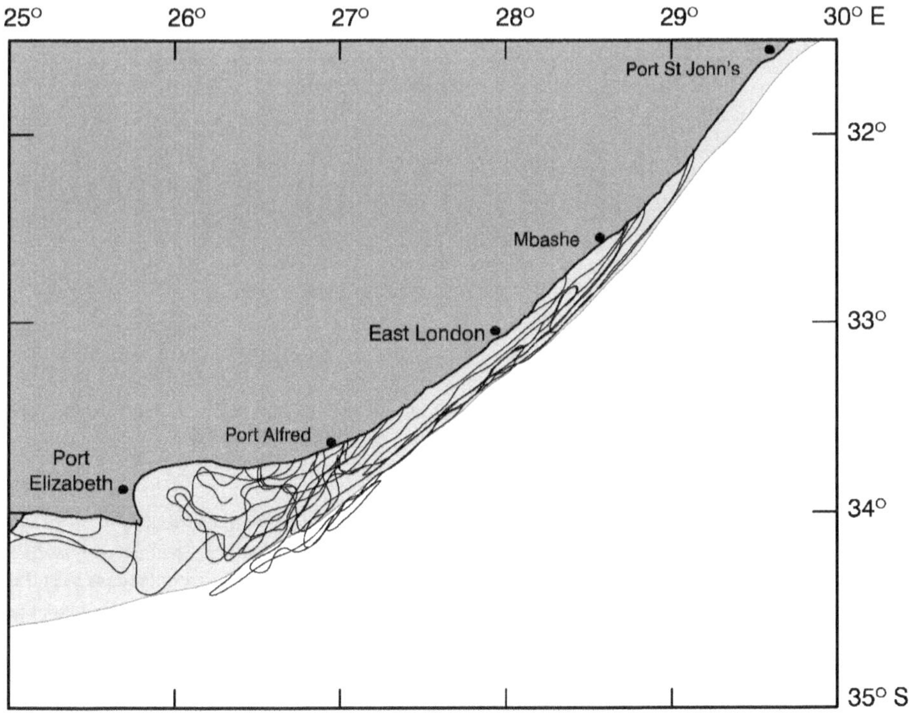

Figure 5.12. An ensemble of outlines of cold upwelling water inshore of the southern Agulhas Current[166]. The 17 °C isotherm in satellite imagery has been used here as representing the border to the upwelling.

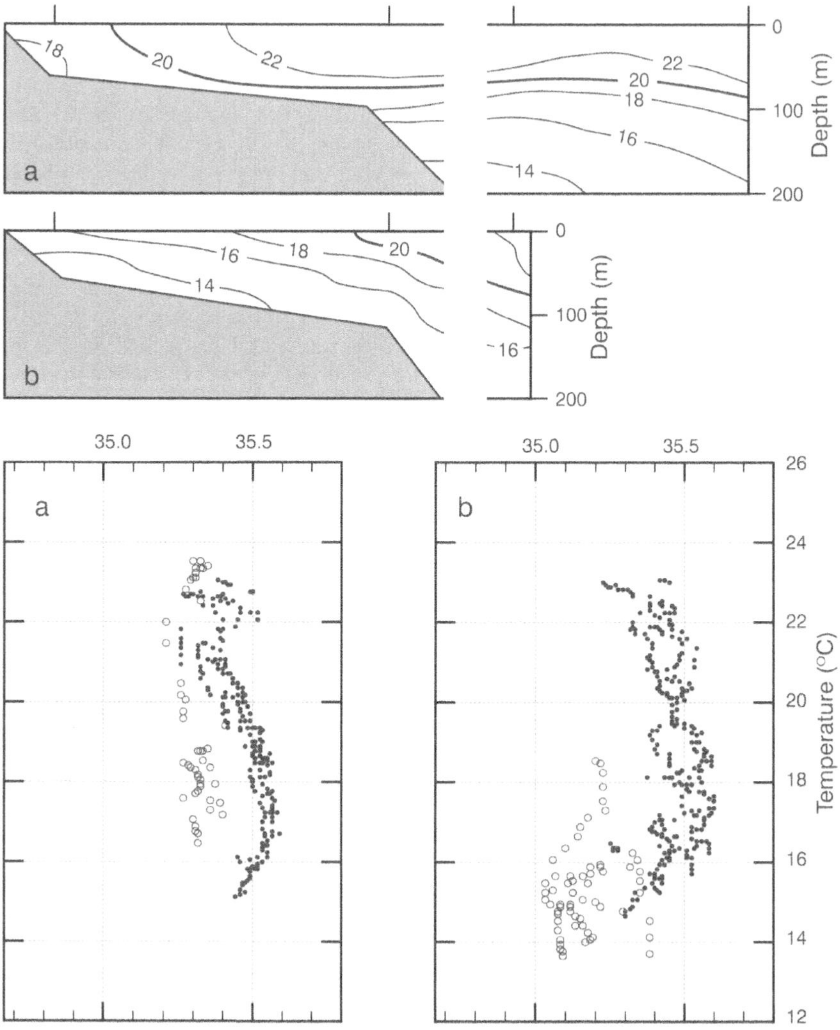

Figure 5.13. Two temperature sections undertaken across the continental shelf at Port Alfred[446] (upper panels). The corresponding temperature/salinity scattergrams for each section are given in the lower panels. Circles represent inshore stations; dots the seaward stations.

Water. This water mass configuration is typical of the inshore edge of the Agulhas Current. Stations taken on the shelf show that the water here consisted of a mixture of these two water types.

During the second period (Figure 5.13b) water warmer than 20 °C was found only at the shelf edge, almost all water on the shelf having been colder than 18 °C. Offshore stations again showed the hydrographic characteristics of the inshore edge of the Agulhas Current, but inshore stations represented much fresher and colder water. Inshore waters were on this occasion decidedly not a mixture of the offshore waters. Comparing their salinity values to those found in the core of the Agulhas Current (Figure 5.3) it is clear that this was slightly warmed South Indian Central Water. This implies that during well-developed upwelling in this cell Central Water displaces mixtures of surface water from the Agulhas Current on the shelf at Port Alfred. Outcropping of such colder water will have a major effect on the ratios of square areas covered by water of different temperatures in this region. This is evident from Figure 5.14.

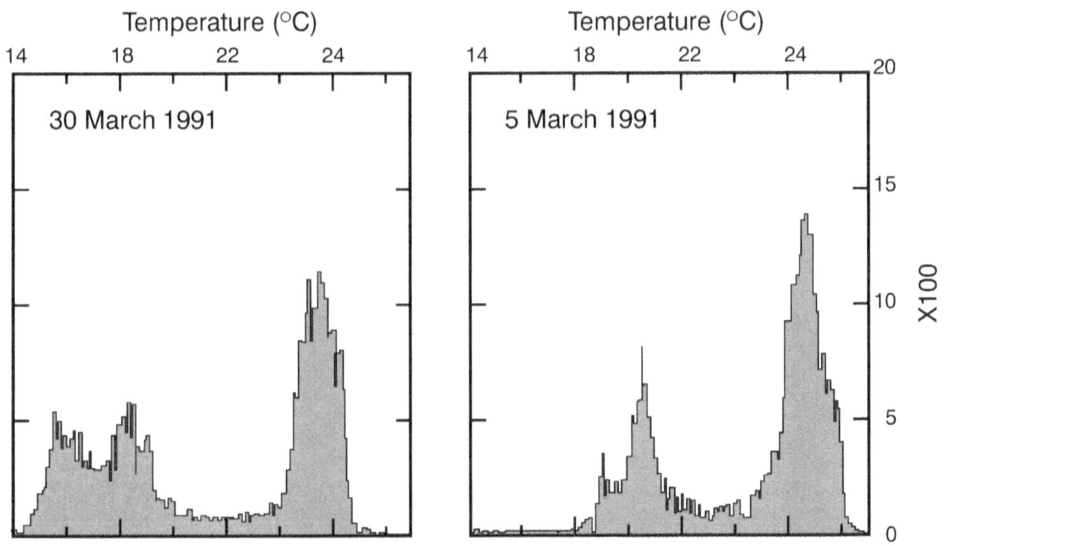

Figure 5.14. The areal distribution, roughly in km^2, for surface temperatures in a region around the Port Alfred upwelling cell during an upwelling outcropping (30 March 1991) and during a period when there was no surface evidence for upwelling (5 March 1991)[166].

The peak of high temperatures at about 25 °C, during a period of no outcropping of cold water, represents surface water of the Agulhas Current. The peak at about 21 °C represents surface water over the shelf. During a major outcropping event (30 March 1991) water of the Agulhas Current had more or less the same temperatures, whereas water on the Agulhas Bank had been cooled to 18 °C to 20 °C, probably due to mixing with cold water advected from the upwelling cell. The major temperature contribution from the upwelling lay in the region 15 °C to 17 °C. At the sea surface this cold water can rapidly be covered by a thin layer of warm Agulhas Current water during winds from the south-west[449].

Local influence of the Port Alfred upwelling cell

These large, wind-induced changes in the sea surface temperature give this shelf region the highest temperature variability along the whole Agulhas Current[166]. It has been shown that the Agulhas Current has a substantial influence on the overlying atmosphere[13,148–9] and also on the rainfall over the adjacent coastal zone[149]. It is therefore of some interest to note that the region of southern Africa that exhibits the greatest variability in vegetative cover is located adjacent to the coastal upwelling cell at Port Alfred[449].

Cold water from this particular upwelling site may be advected at the sea surface over large parts of the Agulhas Bank[445] and into adjacent bays, such as Algoa Bay, where it may contribute substantially to short-term temperature variability[450] of the local waters that have been observed[451] here. A careful study of such an event[429] has shown that one intense upwelling incident may have a dramatic impact on the coastal ecology. On this particular occasion bird numbers feeding in the region increased from only a small number to an estimated 30 000 at the peak of the upwelling, the numbers decreasing rapidly thereafter. Satellite imagery of ocean colour shows this[441] as an area of intermittently high chlorophyll-*a* values, suggesting a region of high primary productivity. Furthermore, the presence of this region of unusually cold water along this coastline is reflected also in the geographic distribution of a number of organisms. A survey of siphonophores along the east coast of South Africa[636] has for instance, shown that the Port Alfred upwelling cell exhibits a distinct aseasonal assemblage of these pelagic organisms, significantly different from that to be found in waters of the Agulhas Current as well as in inshore waters north of here. Certain cowrie species, benthic organisms known to occur in cold water settings only, have been found[637] close to Port Alfred in a very restricted geographical region. These two examples of the biogeography of the region indicate the influence of the Port Alfred upwelling cell also on the local ecosystem. Why should this particular region along the eastern edge of the Agulhas Bank be so especially prone to upwelling? A number of mechanisms have been put forward.

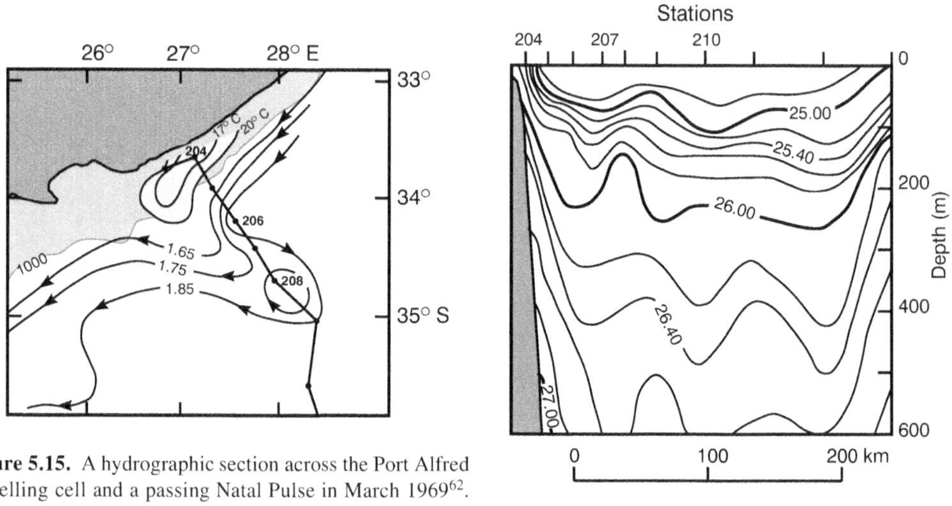

Figure 5.15. A hydrographic section across the Port Alfred upwelling cell and a passing Natal Pulse in March 1969[62]. The isolines in the left panel are for the dynamic height (in dynamic meters) of the sea surface relative to 1000 decibar, showing the flow patterns. The right hand panel shows the sigma-t density surfaces. (See also Figure 4.24, page 115.)

Driving forces of the Port Alfred upwelling

Winds in this region are for the greater part parallel to the coast[189–90]. Easterly winds would cause an offshore movement of surface water in the upper layer due to Ekman drift, causing a local divergence that would bring deeper water to the surface[429]. The fact that the noticeable upwelling only seems to occur at Port Alfred, and not along the whole length of the northern Agulhas Current, presents problems for this hypothesis, but may be explained by the closer proximity to the sea surface of cold water inshore of the current in its southern reaches (viz. Figure 5.2).

A second mechanism could be the occasional presence here of a large, offshore meander in the trajectory of the Agulhas Current[354]. On at least one occasion a hydrographic section has intersected both a well-developed upwelling cell at this location as well as a passing Natal Pulse[62] (Figure 5.15). A statistical comparison of the incidence of upwelling here and the passage of Natal Pulses suggests that Natal Pulses enhance the likelihood of upwelling, but that upwelling occurs much more frequently[429] than the passage of Natal Pulses.

The unusual shelf and current configuration at this location has not been overlooked in attempts to explain the Port Alfred upwelling cell. At the Natal Bight the northern Agulhas Current passes from a narrow to a wider shelf and upwelling is kinematically driven at this point[163]. The geographic arrangement of current and shelf at the Natal Bight is nearly identical to that found between East London and Port Elizabeth. Gill and Schumann[167] have shown that under such conditions currents that can be thought of as jets may become super-critical along narrow shelves, but sub-critical where the shelf widens. This results in a recirculation cell inshore of the current just upstream of the narrow part of the shelf, possibly manifested in the Durban eddy[365,398], as discussed above. Where this jet passes from a narrow to a wide shelf, inshore upwelling is induced, as seen in both the Natal Bight as well as at the Port Alfred upwelling cell.

It is possible that all these processes play a complementary role in the Port Alfred upwelling cell. Ekman veering may bring deeper water closer to the sea surface on the inshore side of the current and this effect may be particularly strong where the shelf starts widening, bringing cold water right to the surface. Winds from the appropriate direction may remove warmer surface layers[449] while any seaward movement of the Agulhas Current might strongly enhance the whole process.

This recurrent upwelling cell between East London and Port Elizabeth may have important hydrographic and biological consequences not only for the immediate surroundings, but for the wider Agulhas Bank as well.

Hydrography of the Agulhas Bank

As continental shelves go, the Agulhas Bank is in many respects highly unusual. Lying between a well-developed western boundary current on the east and an eastern boundary current to the west, an interactive region of high complexity may be expected. Further-

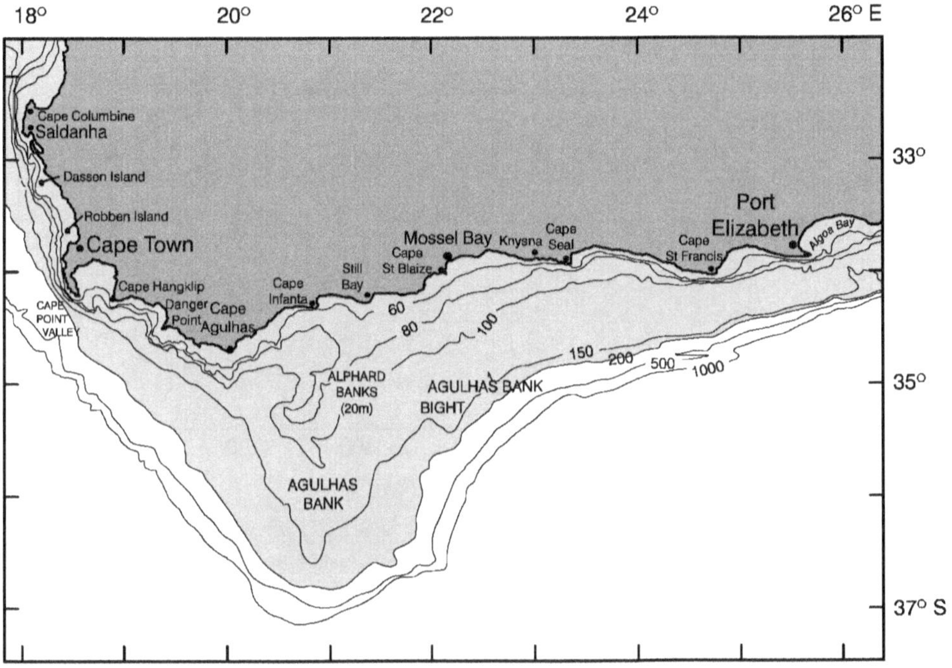

Figure 5.16. The bathymetry of the Agulhas Bank in the intervals shown[466–7].

more, the shelf is bathymetrically divided at about 21° E (see Figure 5.16) between a long, wider shelf with a low shoreward gradient to the east and a narrower, more steeply sloped shelf to the west. The shallowest central part, including the Alphard Banks, forms a division between these two sides and an effective partition for their water masses (Figure 5.17). Since the Agulhas Current plays such a dominant role in the hydrography of the Agulhas Bank, and is in turn influenced by it, the shelf water masses and their dynamics deserve attention.

Influence of the Agulhas Current

As was noted previously, the southern Agulhas Current affects the continental shelf edge of the Agulhas Bank in a variety of ways. It generates shelf-edge upwelling along its whole length and there are indications[441] that very localised increases in chlorophyll-*a* along the inshore edge[468] of the Agulhas Current may, at least partially, be a consequence of this upwelling. It forces the Port Alfred upwelling cell and it creates the border eddies that may help in drawing up deeper water. Plumes of warm surface water replenish the heat and salt of the surface layers over the Agulhas Bank.

The southern Agulhas Current detaches itself from the shelf edge somewhere near 21° E (viz. Figure 5.16), probably dependent on its instantaneous volume flux[469]. An increase in flux may, theoretically, allow it to separate farther upstream. The warm, shallow Agulhas filaments, mentioned previously, advect along the western shelf edge into the Atlantic Ocean. Furthermore, Agulhas rings with diameters in excess of 200 km may lodge against, or move along this shelf edge[440] thus creating high along-shelf velocities. These have as yet not been directly measured and they may occur only intermittently. The oceanic influence is not restricted to the shelf edge, but can be observed over the greater part of the rest of the Agulhas Bank.

Hydrographic provinces

The waters and their kinematics over the shelf may be roughly divided into four hydrographic provinces that reflect the dominant dynamic processes in each. First is the zone close to the coast, then the eastern and western parts of the central shelf and, somewhere between these latter two, a ridge of cold water[185].

The zone close to the coastline is dominated by intermittent, wind-driven upwelling. The coastline from Cape Agulhas westwards (viz. Figure 5.16) may for all intents and purposes be considered to form an inherent part of the general west coast upwelling system. The coastal upwelling between Cape Agulhas and Cape

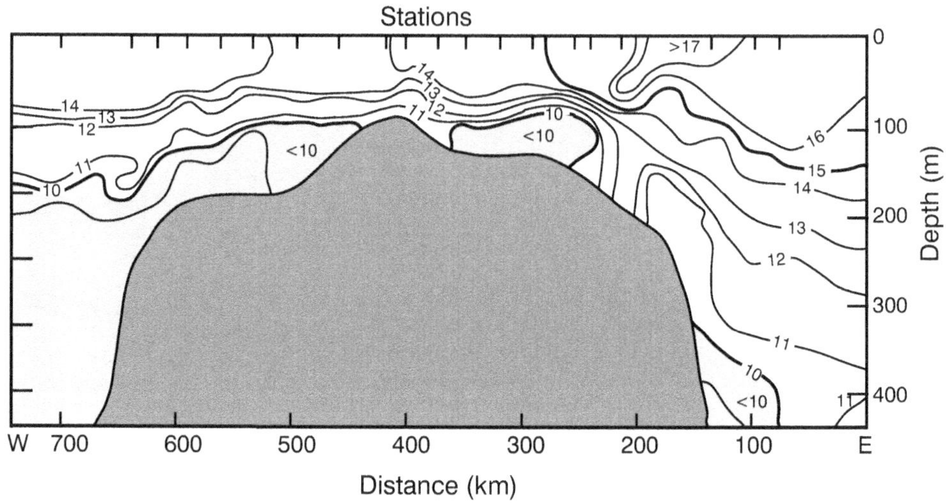

Figure 5.17. Temperatures in a vertical section from west to east across the Agulhas Bank[168], facing north. These data were collected from the H.M.S. *Tiger* in July 1973 along the middle zonal line of Figure 5.22. The shallowest peak in the centre of the shelf is the Alphard Banks. Water colder than 10 °C has been shaded. To the west of the Alphard Banks there is upwelling; to the east the sloping isotherms representing the Agulhas Current. A body of cold water on the eastern Agulhas Bank is not connected to either of the two other regions of lower temperatures.

Point, south of Cape Town, in fact constitutes the southernmost upwelling cell of the Benguela upwelling regime[470]. However, because of the different alignment of the coast south and north of Cape Point, the incidence of upwelling is different. Strong upwelling north of Cape Point has been observed when there was no surface evidence for upwelling east of here[471]. Under special wind conditions upwelling along the western Agulhas Bank is confined to local cells at headlands, or may eventually develop into a wide contiguous upwelling strip. As this cold water is moved offshore at the sea surface it may move past False Bay (viz. Figure 5.16) creating a very noticeable surface thermal front across the mouth of this bay[472]. Cold water has also been observed to enter the bay on its western side forming a meridional front in the bay itself. This bay in some sense forms part of the second hydrographic province, the western part of the Agulhas Bank.

Western Agulhas Bank

Boyd et al. have used observations farther offshore[473] on the western Agulhas Bank to show that a wind-driven seasonal cycle is maintained here as well (viz. Figure 5.18). The results of an investigation using climatological sea surface temperatures[745] has indeed demonstrated the existence of a clear seasonal pattern with average temperatures being 2.5 °C higher in summer. Stratification is much more intense in summer. Central Water moves onto the shelf and is much shallower in summer, allowing a rapid upwelling response, observable at the sea surface, when south-easterly wind events take place. With the onset of strong winter winds mixing to greater depths occurs, evident in higher nutrient concentrations at the sea surface[165]. By carefully comparing high-quality hydrographic data from the western Agulhas Bank, Chapman and Largier[474] have been able to show that the Central Water found here has its origin in the Southeast Atlantic Ocean and fills the lower part of the water column of the whole western Agulhas Bank (viz. Figure 5.17). It is raised onto the shelf by the seasonal lifting of colder water in spring and early summer. Reviewing what is presently known about the region, Largier et al.[475] have come to the conclusion that the western Agulhas Bank may be considered, from its dynamics, as consisting of three subregions. The inner shelf, with its coastal upwelling, is clearly dominated by wind-forcing. The outer shelf is dominated by oceanic forcing, while the mid-shelf is characterised by strong vertical stratification in summer. The oceanic forcing is particularly evident in a shelf-edge front.

Eastern Agulhas Bank

In contrast to the wider upwelling along this western coastline, the coastal upwelling to the east of Cape Agulhas appears to be confined almost exclusively to

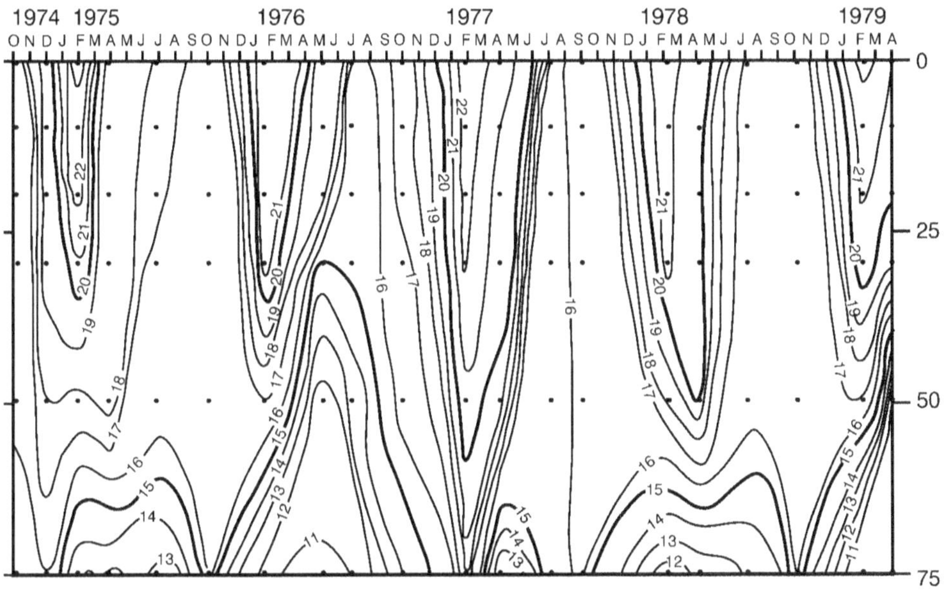

Figure 5.18. Seasonal variations in temperature and stratification over the central part of the Agulhas Bank[165]. Dots denote depths in the water column where observations were made. Seasonal warming from the sea surface and simultaneous cooling from below can be identified, albeit with interannual differences.

prominent capes and headlands[453–4] that seem to create the required coastal morphology for upwelling to occur, mostly during easterly winds. This upwelling is concentrated along the eastern part of this coastline. Upwelling-inducing winds would occur mostly during summer[189–90], but could exhibit considerable inter-annual variability[476–7]as well as site-specific peculiarities, such as at Algoa Bay[815]. Analysis of long records of coastal winds along this coastline and of coastal sea surface temperatures in summer months has shown[477] a significant correlation between stronger easterly winds and lower temperatures[478] (Figure 5.19). This correlation is considerably enhanced when winds exceed a threshold of about 6 m/s. This coastal upwelling, finely tuned to the shape of each particular headland, would naturally lead to localised high variability in temperatures along this coastline[450–1]. The last hydrographic province of the Agulhas Bank is the central region.

Central upwelling on the Agulhas Bank

Apart from the wind-driven, coastal upwelling, the widespread occurrence of cold surface water has also been observed in the central region of the Agulhas Bank[445]. This may be related to the higher chlorophyll-*a* found[745] as an offshore moving plume on the eastern Agulhas Bank, compared to lower average values on the western bank. Both these features have been variously ascribed to the advection of water from the Port Alfred upwelling cell during high easterly winds[430] or to the surface expression of a cold ridge of water in the central region of the bank (Figure 5.20). This subsurface ridge has been observed in many hydrographic sections on the Agulhas Bank[479–80,816], but is not strongly evident in analyses based on mean temperatures[481], thus suggesting that it is intermittent and perhaps does not always occur in the identical location.

This mysterious subsurface feature has been thought to be due to a divergence of surface water supposedly found in the central region of the bank[482], but no supportive evidence has been found for such a divergence. The very surface layer of the shelf's water is moved, as was to be expected, primarily by the reigning wind[483]. Present evidence suggests[467] that on the sea floor cold, upwelled water from the Port Alfred upwelling cell moves parallel to the Agulhas Current and its shear-edge features. It usually forms a dome on the sea floor and only breaks the sea surface under extremely windy conditions[445]. The cold, bottom ridge roughly along the 100 m isobath (Figure 5.22) is therefore probably the main trajectory of cold bottom water from the Port Alfred upwelling cell onto the western and central Agulhas Bank. This figure shows the locations of water colder than 10 °C on the Agulhas Bank based on a number of historical section, one of which is shown in Figure 5.17.

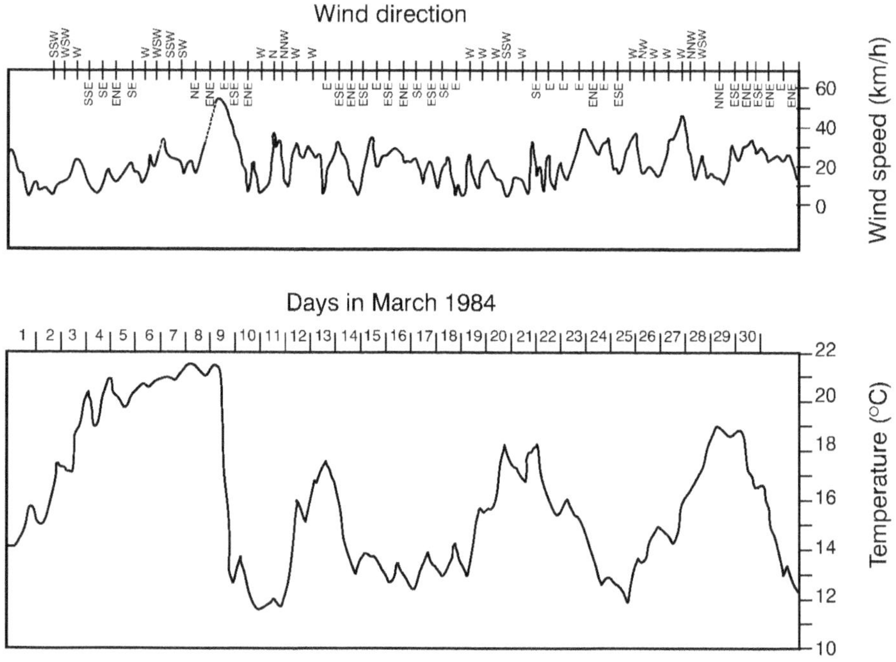

Figure 5.19. Sea surface temperatures and wind speeds at Gouriqua (near Mossel Bay on the south coast of South Africa, see Figure 5.16)[478]. Wind speed is shown as a curve in the upper panel whereas wind direction is shown as letters on the abscissa for the wind speed curve. Temperatures rise and fall at the coastline in direct response to the direction and speed of the local wind.

Cold, bottom ridge

Figure 5.17 demonstrates that the cold water dome on the shelf is not directly connected to that along the shelf edge, except near Port Elizabeth and particularly at Port Alfred. Since it has now become clear that there is a well-defined upwelling cell at Port Alfred and since current measurements near the sea floor in this region all show a net movement parallel to the current[168], this cold-water tongue must indicate water movement away from the Port Alfred upwelling cell. This implies that the bottom water on the Agulhas Bank has its origins in this cell. This portrayal is hampered by the lack of synopticity in the data, but the general result holds for data from one cruise – but over a smaller section of the bank – as well (Figure 5.20). A vertical hydrographic section carried out specifically to study the cold ridge[658] followed the 100 m isobath (Figure 5.21). Water colder than 10 °C was limited to a specific part of the shelf and terminated at about Algoa Bay, not quite Port Alfred. This results is not in immediate agreement with the hypothesized theory set out above, but there is no reason to assume that this cold water would follow the 100 m isobath unfailingly. There seems to be a better connection between offshore water colder than 10 °C and such water on the shelf on the western side of the Agulhas Bank[658,677] (Figure 5.17). This might mean one of two things: either cold water on the shelf cascades into the South Atlantic Ocean here, or cold water is moved onto the shelf from offshore. Current simulations from a high resolution model of the region[633] indicate that water moves in a westerly direction here and moves off the shelf edge on the western side of the Agulhas Bank. This is in disagreement with hydrographic observations that indicate[474] that the bottom water on this side of the bank all comes from the South Atlantic.

This proposed mechanism for bringing South Indian Central Water onto the Agulhas Bank at the Port Alfred upwelling cell would explain the contribution of cold bottom water to the water column stratification (Figure 5.18) as well as the time delay for this cold bottom water to reach the central Agulhas Bank once winter wind-mixing has ceased. It has, after all, to travel along the length of the eastern Agulhas Bank, from Port Alfred. This peculiar cold ridge is, from a biological point of view, considered as being a prominent part, particularly, of the western bank's circulation[468,804]. The origin of this and the other water masses on the Agulhas Bank is therefore of considerable importance.

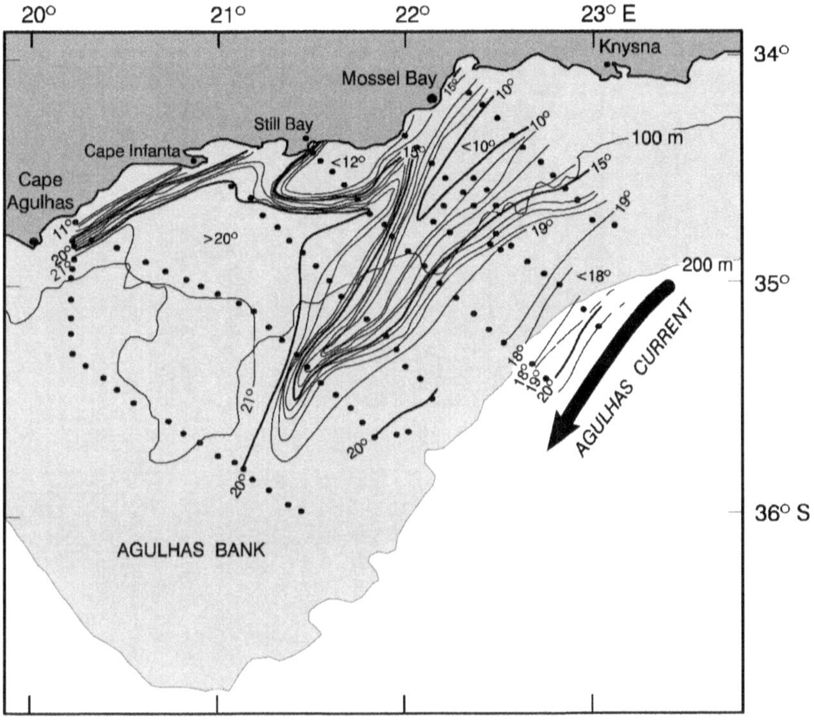

Figure 5.20. The distribution of temperature at 20 m depth in the waters over the Agulhas Bank during March 1986[430]. Station positions are shown. A ridge of cold water, stretching from the coastline in a south-westerly direction, is prominent.

Vertical stratification

The water masses present on the eastern part of the Agulhas Bank all have their origin in the South Indian Ocean[430,474]. This is clearly shown by the temperature–salinity relationships portrayed in Figure 5.23. In summer, the temperature of the surface water is rapidly increased, while an influx of cold South Indian Central Water occurs at the bottom[396] (Figure 5.18). This vertical juxtaposition of warm surface and cold bottom water helps, together with the normal summer insolation and a reduction in wind strengths, to build a very strong seasonal thermocline that lasts until autumn[165]; hence the two thermal groupings for autumn water in Figure 5.23. The enhanced intensity of this thermocline may well play an important role in the distribution of phytoplankton[792–3] and in its retention in the euphotic zone over the Agulhas Bank.

With the onset of strong winds in winter this stratification is largely destroyed and the water mixed through the water column to a depth of at least 75 m. A striking example of the effects of this process is evident in Figure 5.24. It may be noted that in these sections, particularly for the winter, there may be evidence of the role the off-shelf Agulhas Current plays in maintaining and nurturing the shelf stratification by the overspill of warm surface water. In winter a somewhat reduced thermocline is present, but only near the shelf break. That the process shown in these sections is representative for the greater part of the eastern Agulhas Bank can be shown by an analysis of the mean temperature values[481] (Figure 5.25).

In both summer as well as winter the thermal influence of the Agulhas Current is manifested at the sea surface by a strong horizontal temperature gradient. Temperatures in summer naturally are higher. The vertical stratification in the current is rather low in both seasons, but over the eastern shelf it is very strong in summer, but all but disappears in winter (Figure 5.25). The winter mixed layer may well exceed 75 m in depth[396,431]. The summer thermocline, by contrast, may consist of a gradient of 10 °C over a depth of less than 10 m at some locations[479–80]. This seasonal thermocline has been observed to deepen towards the west. This has led to proposals that the control of the stratification on the eastern and on the western sides of the shelf are different.

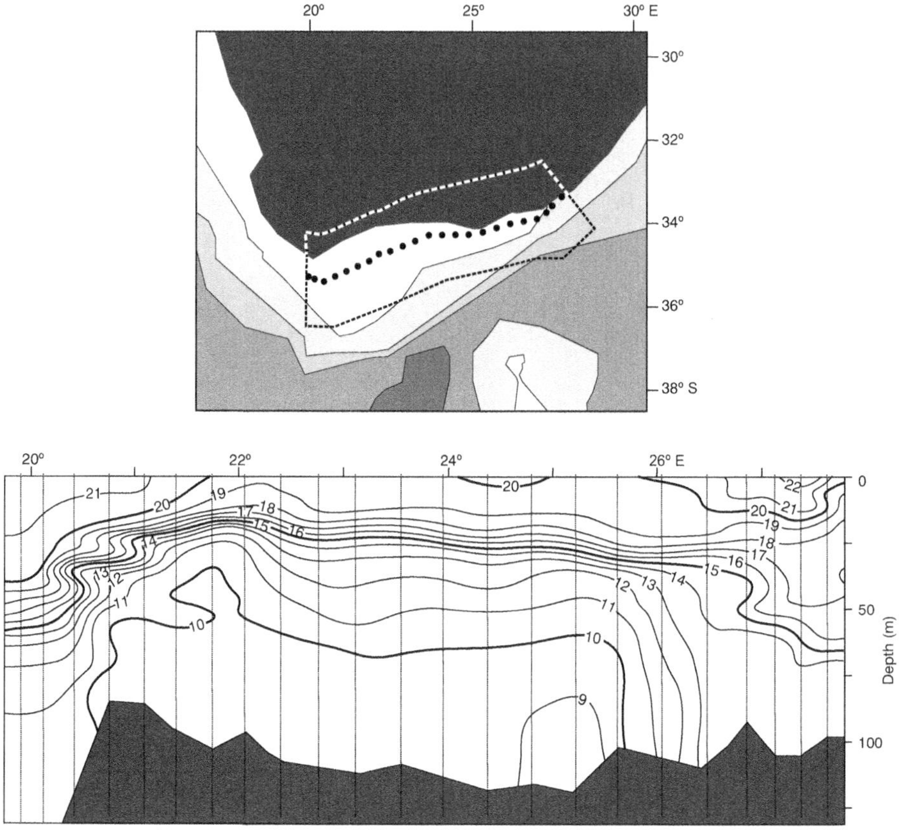

Figure 5.21. A vertical temperature section across the Agulhas Bank south of Africa, roughly along the 100 m isobath. The upper panel indicates the station positions superimposed on the bottom topography. The lower panel gives the temperature to a depth of 130 m. Grey shading represents the ocean floor. Water colder than 10 °C was found only on the Agulhas Bank and not east of Algoa Bay[658].

Stratification control

On the western side the seasonal thermocline would be driven largely by changing atmospheric conditions of wind and insolation[480]. Modelling of the wind-driven behaviour on sea surface temperatures of the Agulhas Bank has shown[746] that they are strictly related to local wind fluctuations. Basal water would come largely from the South Atlantic Ocean and the seasonal inflow should be a function of the seasonal wind-stress patterns. On the eastern side, so the hypothesis goes, advective processes of warm Agulhas plumes at the sea surface and basal water inflows due to the upwelling discussed above would play the major part in maintaining the thermocline. But why would these processes on the eastern side, that are driven by the Agulhas Current, be seasonal, since no seasonality in the flux[380], speed[292] or border eddy formation[88] have been observed?

The inputs by the Agulhas Current are thought, in fact, not to be seasonal, but to be masked by the intense seasonal mixing driven by the winds. The influence of the Agulhas Current is therefore always there, but is only patently evident in summer. In a very analogous situation at the Gulf Stream, it has been suggested that upwelling water intrudes along the shelf bottom preferably in summer when the shelf water has its lowest density[415] and may then displace large amounts of shelf water[485]. Whether this holds true also for the Agulhas Current has still to be established incontrovertibly. However, substantial amounts of cold water are also contributed to the surface layers in summer through coastal upwelling. A thorough quantification of the relative contributions of all these, and other processes, has yet to be attempted. One of the processes that might play an important role in the breakdown of the vertical stratification over the Agulhas Bank, be it atmospherically driven or due mostly to the influence of the Agulhas Current, are currents on the shelf itself.

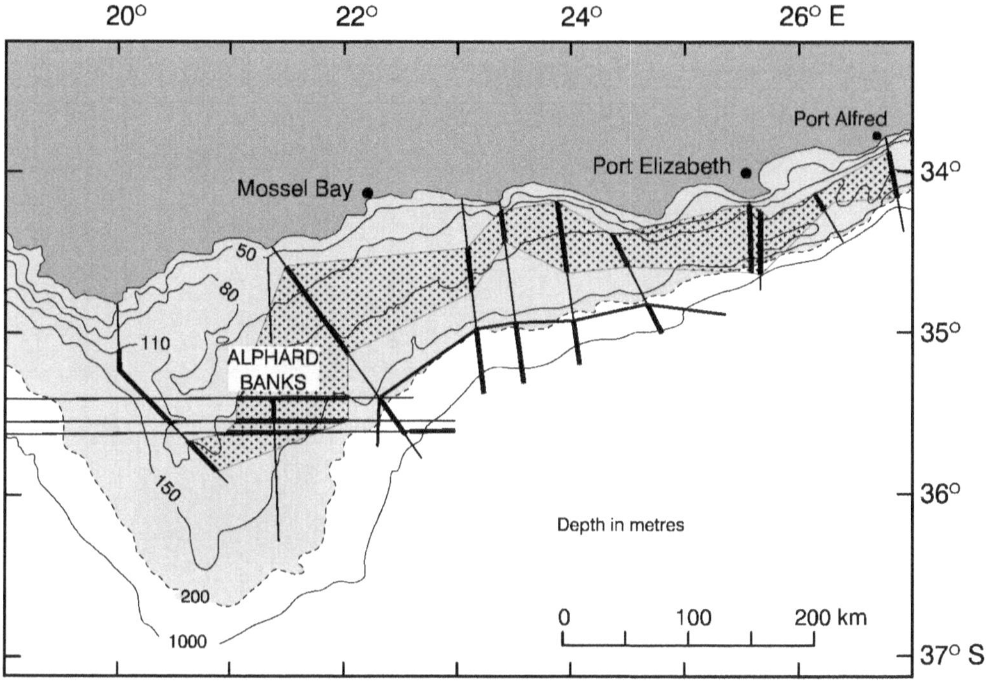

Figure 5.22. The location of water colder than 10 °C on the Agulhas Bank from a collection of historical data[168]. Straight lines indicate the locations of hydrographic sections; heavy lines where water temperatures were below 10 °C and the stippled area connects all the regions where this cold water was found on the shelf. Isobaths are in thinner lines. This distribution suggests a movement of cold water along the bottom of the Agulhas Bank following the 100 m isobath and originating at the Port Alfred upwelling cell.

Shelf currents

It has been estimated that the residence time of water in the surface layer over the Agulhas Bank could be from two to eight weeks or longer[482]. This motion would be largely wind-driven. At the very shelf-edge, internal tides might play an important role in vertical mixing, but this would be dependent on the location. Largier and Swart[480] have shown that the narrow western bank is everywhere subject to large internal tides with a mean amplitude of 6 m and a maximum of 12 m. The outer portion of the central and of the eastern bank only experiences moderate activity, whereas over the inner part of the eastern bank there are internal tides with negligible amplitudes.

Many current measurements have been made over different parts of the Agulhas Bank – to date few have been published. It has been demonstrated[430] that the mean flow on the eastern shelf, at least between the 100 m and the 200 m isobath, is parallel to the bathymetry and, therefore, to the southern Agulhas Current. However, the main movement of water does not lie in the mean or net flow, but in short variations. The results of a frequency spectra analysis carried out for a current record acquired in the Agulhas Bank bight (viz. Figure 5.16) is given in Figure 5.26.

It is clear that there are three main contributions to the current fluctuations observed; namely those occurring at periods longer than three days (i.e. less than 0.014 cycles per hour), diurnal/inertial variations and semi-diurnal fluctuations[384]. The first is probably due to flow changes in the southern Agulhas Current of these periods[89] or longer. The spectra for the tidal and inertial frequencies were found throughout the water column, but with slightly lower energies further down. The tidal fluctuations have been shown to be essentially barotropic, while the inertial fluctuations are baroclinic and highly dependent on the stratification[384].

These measurements are most probably representative of the Agulhas Bank as a whole. Observations on the west coast show even more strongly developed peaks at inertial frequencies[384], so that the western part of the shelf might follow suit. The dominance of inertial motion on the Agulhas Bank makes it very difficult to assess net drifts here[467,486] except if current measurements lasting days are made at each point.

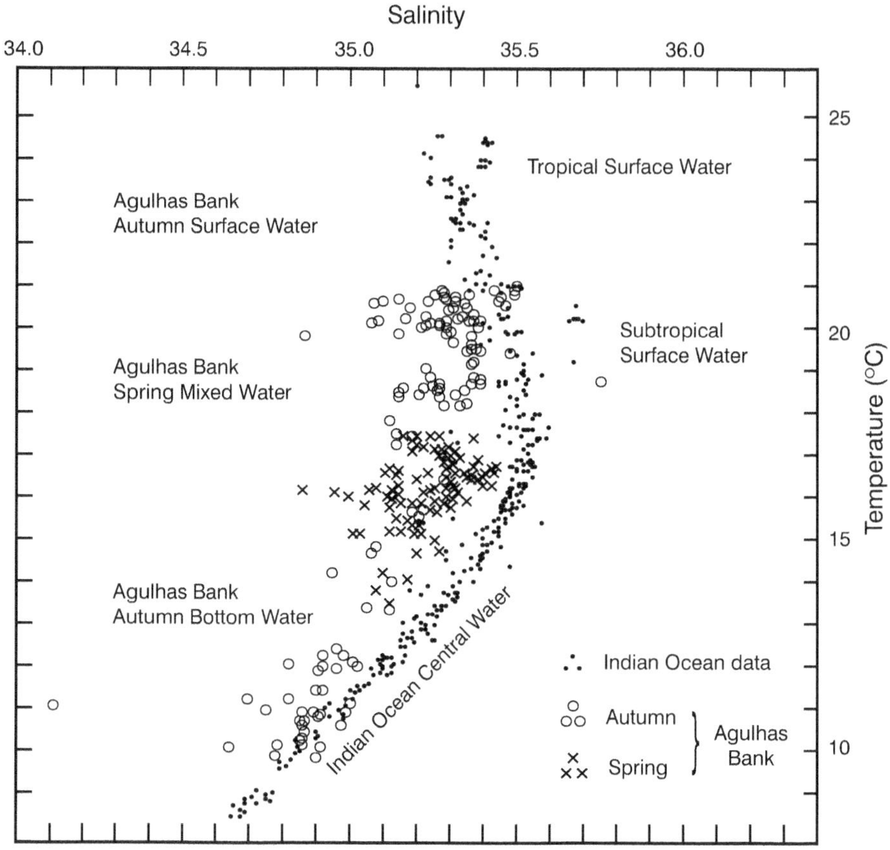

Figure 5.23. Temperature–salinity relationships for waters over the Agulhas Bank (circles and crosses)[396,430] and from the wider South West Indian Ocean[484] (dots).

A part of the Agulhas Bank where knowledge of the currents is of considerable ecological and fisheries importance[799] is the western shelf edge. It has been assumed that a strong current jet is to be found here[437] and that immotile organisms such as fish larvae and weakly swimming fish juveniles will be carried from their spawning grounds on the Agulhas Bank to the upwelling regions off south-western Africa by this jet[797–8]. This hypothesized mechanism would have a crucial role in the spawning success of the main pelagic fish species[796] found on the west coast. There is evidence that such a jet is found here on occasion[795], but it has been shown to be very intermittent and restricted to the northern part of this shelf edge. Along the southern part of the western shelf of the Agulhas Bank a cyclonic lee eddy is prevalent[630–1]. The currents induced along the shelf by this lee eddy are poleward, in opposition to the purported jet.

In summary, the role of the Agulhas Current in controlling the hydrographic structure on the Agulhas Bank seems well established. On the western part of the bank this may be limited, and this shelf may be more logically classified with the west coast upwelling regime. The water masses, stratification and net water movement on the larger eastern bank are all overwhelmingly influenced by the southern Agulhas Current. Cold bottom water and warm surface water are both replenished from the Agulhas Current by processes that take place at its border with the shelf. Wind-driven coastal upwelling plays a role, particularly in the cold cross-shelf ridge, but it is the strong winter winds and atmospheric cooling that lead to the extreme seasonal changes in stratification.

Atmospheric conditions

It is apparent that the southern part of the Agulhas Current, and particularly its interaction with the adjacent shelf waters, is atmospherically influenced to a far greater extent than the northern part. This warrants a

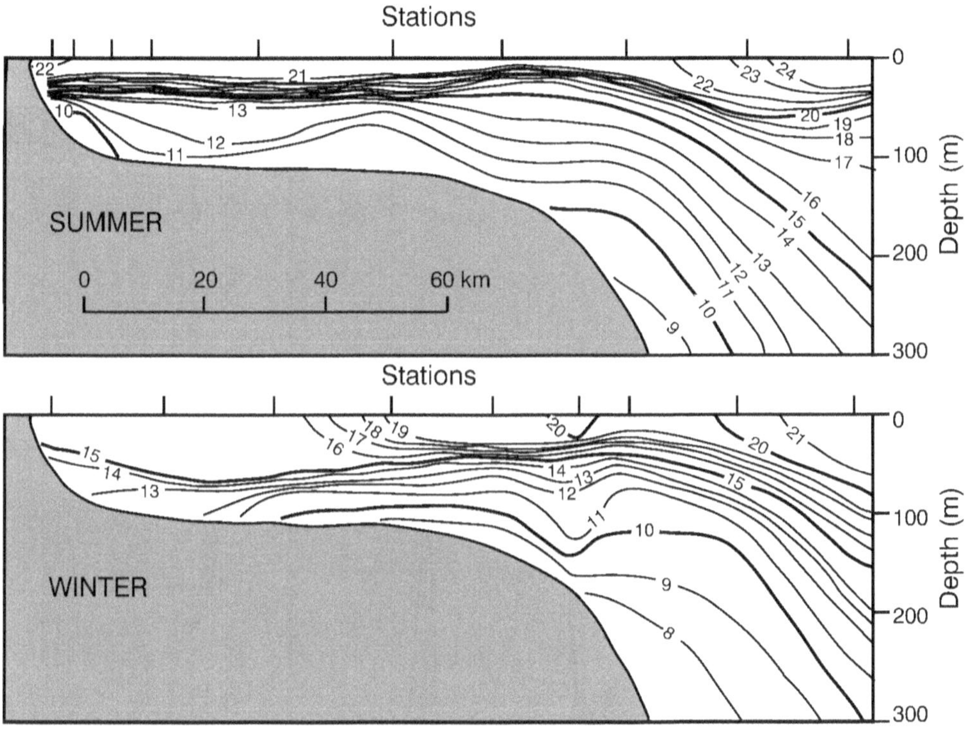

Figure 5.24. Temperature sections across the eastern Agulhas Bank based on hydrographic measurements carried out in summer (February) and winter (August) of 1982[481].

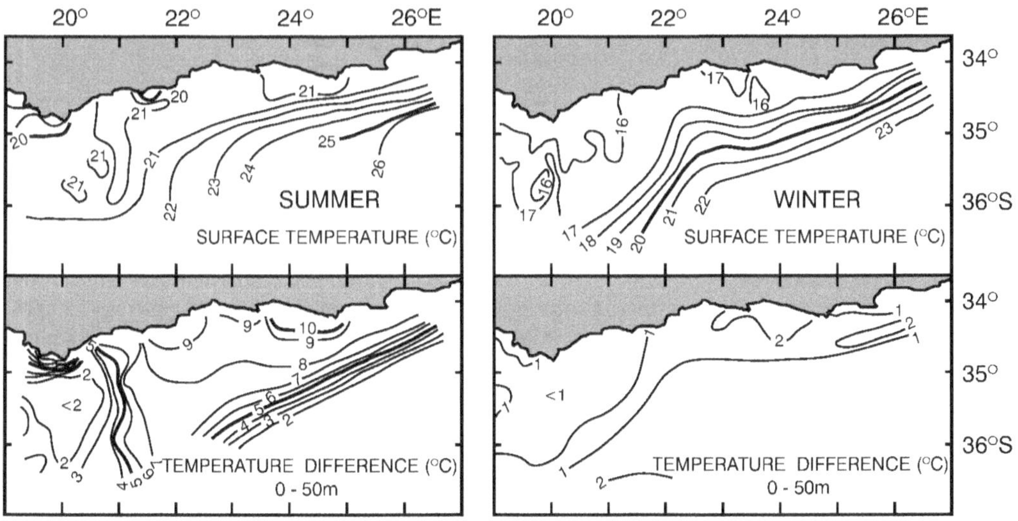

Figure 5.25. Mean sea surface temperatures and vertical stratification of the water overlying the Agulhas Bank in summer and in winter[481]. Winter months used are June to August; representative summer months December to February.

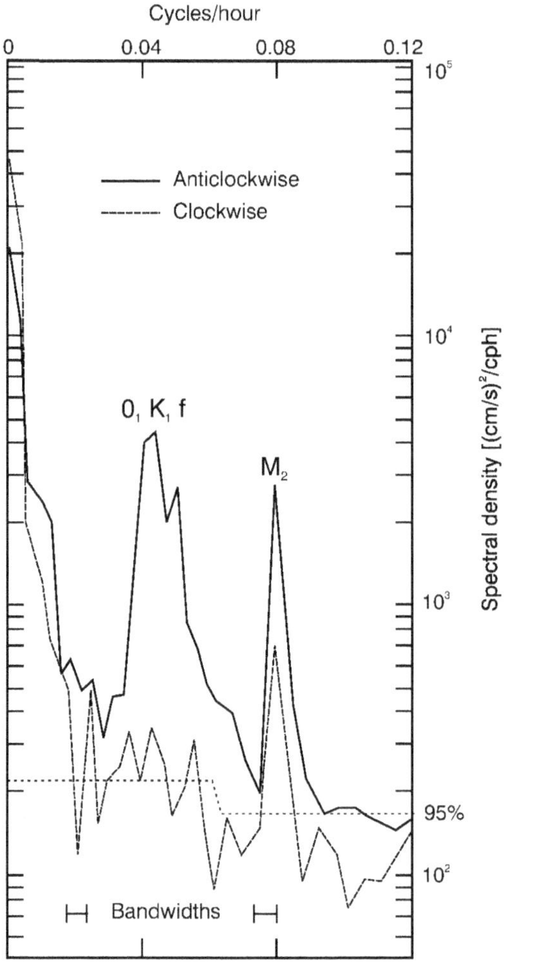

Figure 5.26. Clockwise and anti-clockwise current spectra for a current record obtained at a depth of 38 m in water of about 150 m[384] on the Agulhas Bank. Diurnal tidal peaks are shown by 0_1, K_1; semi-diurnal tidal peaks by M_2. The inertial frequency is indicated by f.

closer look at the atmospheric conditions that pertain here.

The atmospheric conditions associated with the southern Agulhas Current can be grouped into two categories; winds and other atmospheric variables that affect mostly the influence of the current on the adjacent shelf waters and their stratification and, second, the influence of the high horizontal temperature gradients of the current on the overlying air masses.

Winds

As mentioned previously, the large-scale weather over this ocean region is dominated by the high-pressure systems centred at about 30° S, with their seasonal meridional movement of about 6° latitude[487]. These systems move from west to east, but with substantial modifications near the coast. Coastal low-pressure cells are formed as a consequence of the coastal configuration and topography, and are often associated with large-scale cold fronts. Gill[188] has attributed them to coastally trapped waves in the atmosphere. Propagation of these coastal lows is from west to east, following the coastline. Along the western and eastern Agulhas Bank they propagate at about 13.7 and 20.4 m/s respectively[488]. The winds produced by these systems are predominantly in directions parallel to the coast[320, 345]. Winds from the east occur slightly more often in summer; winds from the west dominate during the summer months. It seems that wind strengths increase to the south and to the east over the eastern Agulhas Bank, but that zonal winds increase particularly over the Agulhas Current where the atmospheric turbulence may be very high[308]. A maximum average speed of more than 4 m/s is measured at Port Elizabeth in October and November, while the westerly winds in winter go as high as 4.7 m/s. However, while the strong summer winds have a maximum occurrence frequency of about 45 per cent at their peak in February, the strong winter winds occur about 75 per cent of the time for the months May, June, July as well as August[190]. It is these strong, persistent winds that succeed in breaking down the water column stratification. However, annual averages for these winds show marked variations[476], so that the influence of the wind on the stratification may be very different from one year to the next.

Sea–air interaction

The influence of the Agulhas Current on the overlying atmosphere and thus on weather systems affecting the southern African subcontinent has been shown to be quite marked[139,149,618] and has been described above. Correct estimates of the fluxes of latent and sensible heat above the Agulhas Current are crucial not only for an understanding of local effects, even though they may be dramatic[619], but also for global models of atmospheric circulation. It has for instance been shown[617] that the operational models used by the National Centers for Medium-Range Prediction (NCEP) and the European Centre for Medium Range Weather Forecasts (ECMWF) underestimate this heat transfer. It has been assumed that this is due to the coarse spatial resolution of these models that do not adequately resolve the warmest core of the current. This is particularly true for the southern Agulhas Current[617].

Investigations on the effect of the southern Agulhas Current on the overlying atmosphere have shown that the marine boundary layer is about 400 m thicker over the current[148] than over adjacent waters. Wind speeds may double. Owing to larger temperature differences between ocean and atmosphere, wind stresses and turbulence over the current a latent and sensible heat flux of 340 W/m^2 from ocean to atmosphere has been measured here, compared to 100 W/m^2 over neighbouring water masses. Over the cold shelf waters the latent heat flux[146]may be only 40 W/m^2. The convectively generated turbulence in the atmosphere increases exponentially on crossing the landward border of the Agulhas Current[494]. The border of the current therefore creates an intense atmospheric moisture and thermal front in the atmosphere under alongcurrent wind conditions[447–8]. Under landward wind conditions the air with enhanced moisture content may be advected far into the continental interior[447]. In general it may therefore be concluded that the southern Agulhas Current affects the synoptic pressure field in the overlying atmosphere, allows more moisture to be carried inland and may well enhance the precipitation capacity of passing weather systems.

Of particular interest to local oceanic mixing processes is the observation[489] that a jet of high-speed air movement may be found over the axis of the warmest water in the current. This may be brought about by secondary, vertical circulations due to the enhanced pressure gradients. Such cells have been observed over the edge of the Gulf Stream[495], with ascending motions leading to the development of convective clouds. Such cloud-lines are distinctive features in the cloud patterns over the northern, but sometimes also the southern Agulhas Current[142,145] (Figure 5.27) and even over

Trapped shelf waves off southern Africa

Long, continuous records of sea level at coastal stations, as well as similar current meter records on adjacent continental shelves, have demonstrated the existence of coastal-trapped waves with periods of days to weeks along numerous coastlines of the world's oceans. A number of attempts have been made to identify such features on South African continental shelves as well.

Schumann has carried out the only studies[383,490–1] of this kind on the shelf adjacent to the Agulhas Current. Using data from current meter moorings placed off Natal[383], he was unable to find any coherence in the variability between different mooring sites along this coast, suggesting the absence of such waves here. This has subsequently been confirmed[490] although some signal propagation was observed when the Agulhas Current was at an unusual offshore location.

Further analyses of coastal sea level variations from six sites around South Africa have nonetheless shown[491] that coastal-trapped waves with amplitudes in excess of 50 cm do in fact occur along the west[492] and south coast, but further propagation upstream of Port Elizabeth is inhibited by the proximity of the Agulhas Current. It has been suggested that the wind systems that move along this coastline[189] move at the correct speed to create a resonance condition, thus enhancing the amplitudes of these waves.

These trapped shelf waves are not to be confused with the highly unusual long gravity waves with periods between 12 minutes and 1 hour that have also been observed along the south coast of South Africa[493]. These latter waves are due to oceanic resonance with regular, short-period pulses in atmospheric pressure, occasionally observed along this coastline.

Figure 5.27. Photographs of cloud formation over the Agulhas Current at Port Alfred in April 1995. They were taken[449] during a period of alongshore winds[145]. The left-hand photo shows a view from the ship towards the coast. No clouds of any kind are evident. The next photo shows the simultaneous view towards the ocean from the same spot. A solid bank of cumulus clouds over the Agulhas Current is present.

intrusions of Agulhas water in the South East Atlantic Ocean[143]. In a study at Port Alfred, at the upwelling cell described previously, it was demonstrated[145] that negligible or even downward heat fluxes could occur over the shelf whilst strong heat and moisture fluxes from ocean to atmosphere were prevalent over the adjacent Agulhas Current. A stable boundary layer was in consequence formed over the cold shelf waters whereas sufficiently unstable conditions over the current itself provided the required conditions for cumulus cloud formation. It seems that such conditions may also occasionally be created over other elements of the Agulhas Current system, such as Agulhas Current filaments[143].

Summary

The southern Agulhas Current has a decisively different character to that of its northern counterpart. This is seen in the greater variability in its trajectory and on its more pervasive influence on the adjacent continental shelf. Both aspects seem to be a consequence of the different shelf configuration; the wider shelf and less steep slope allowing greater variegation in the current behaviour.

The path meanders and the attendant shear-edge eddies and warm-water plumes are very similar to those that have been observed in other western boundary currents. Whereas these eddies in the Gulf Stream have been observed to move onto the adjacent shelf, this has not been observed in the southern Agulhas Current. It would therefore appear that it is only the warm water plumes that contribute directly to the water and heat budget of the waters over the Agulhas Bank.

A highly unusual aspect of the southern Agulhas Current is the intense and persistent upwelling cell driven by the current at Port Alfred. This cell has a dramatic influence on the local atmospheric and primary productivity, but – most important – seems to have a crucial role in carrying cold Central Water onto the shelf. From here this water spreads over the whole floor of the Agulhas Bank, enhancing the vertical stratification as well as the supply of nutrients. All these mechanisms seem restricted to the eastern Agulhas Bank.

The southern Agulhas Current leaves the shelf edge at the southern tip of the Agulhas Bank, inducing the intermittent presence of a cyclonic lee eddy. The processes controlling the western part of the shelf seem to be characterised largely by the Benguela Current and wind-driven upwelling. Having become detached from the shelf edge, the Agulhas Current can be thought of as a free jet that terminates in the Agulhas retroflection.

6

The Agulhas Current retroflection

The southern termination of the Agulhas Current is unique for a western boundary current in this respect that it takes place at the meridional extremity of the adjacent continent. This is unlike the continental constraints to which the comparable Gulf Stream, the Kuroshio or the Brazil Currents are subject. Since the African continent separates the Atlantic and Indian basins, the Agulhas Current, at its termination, is also the only western boundary current that lies on the border between two subtropical gyres. This creates unusual conditions for inter-ocean exchanges of water masses, energy and biota between these two gyres with a range of implications for the global oceanic circulation and for biogeographical patterns. Temporal changes in the magnitude of this exchange process may therefore have implications for global water circulation in the ocean and, if such changes are sufficiently large and of sufficient duration, may influence global climate.

Furthermore, the nature of the termination of the Agulhas Current – described below – allows warm tropical and subtropical surface water to remain in the region for a considerably longer time than in comparable western boundary currents. The thus enhanced flux of heat[98,147] and moisture to the atmosphere has a marked effect on the overlying atmosphere[496] of the region. Not unexpectedly, statistical investigations[136,497] have demonstrated that this ocean region has a strong influence on rainfall patterns over southern Africa. Results from ocean–atmosphere models[139] are largely consistent with this view. However, there is consensus that it is the inter-ocean exchanges of water that have the most profound climatic consequences.

The interchange processes that occur in the ocean regions south of Africa are therefore of considerable oceanographic interest and have wide implications. The behaviour of the Agulhas Current must naturally play an important role in these processes. The kinematic nature of the Agulhas Current, once it has passed the southern tip of the African continental shelf, is quite exceptional for a western boundary current. The current turns back on itself in a tight loop, called the Agulhas Current Retroflection, with most of its waters contained in this swift recurvature before they flow back into the South Indian Ocean. The nature and dynamics of this peculiar behaviour have received considerable research attention over the past three decades and are now fairly well understood. This growth in knowledge represents one of the major advances in global oceanography of this period.

Upon closer examination, the scale and the dynamics of the processes in the Agulhas Current retroflection are seen to be of truly monumental proportions[813]. A water mass with an estimated flux of 12 000 cubic kilometre per day, i.e. about 1400 times that of the Amazon River, moving at a rate of 150 km/day, is turned around in a loop with a diameter of about only 400 km to flow directly in the opposite direction. As could be expected, this configuration is highly unstable and global observations of hydrographic, sea surface temperatures and sea surface height have demonstrated that this region is the most intensely variable to be found in the world ocean. The high contrasts in horizontal gradients for a number of ocean variables found here make this area eminently amenable to observation, but the rapid changes that occur severely limit the applicability of a number of standard hydrographic measurement techniques that cannot be used in a quasi-synoptic fashion.

Notwithstanding these serious limitations to observational strategies, brought about by the attributes of the current dynamics itself, much has been learnt about the nature of the Agulhas Current retroflection.

The nature of the Agulhas retroflection

No matter what oceanographic data with a global distribution are used, the extremely high variability south of Africa is always apparent. This result, using modern satellite data[261], has to some extent been adumbrated by analyses of the global[363] and regional[498] distribution of eddy kinetic energy from ships' drift (Figure 4.1) as well as of hydrographic data[499]. Standard deviations of the detrended dynamic height relative to 1000 decibar

Using drifting buoys in the Agulhas Current

After the application of satellite remote sensing, the use of satellite telemetered, drifting buoys has probably contributed most to the rapid increase in knowledge about the Agulhas Current system over the past two decades.

The deployment of drifters in the Agulhas Current was pioneered by Christo Stavropoulos of the CSIR (Council for Scientific and Industrial Research) in Durban, South Africa, when a spar buoy with subsurface drogue was launched 280 km south-east of Durban and tracked successfully by the French Eole satellite for 89 days in 1973[411] (see Figure 7.2). A similar buoy was moored for 315 days on the Mozambique Plateau[521] using the Nimbus VI satellite for positioning and data relaying. With the assistance of the National Aeronautical and Space Administration (NASA, USA) the CSIR subsequently constructed another nine buoys, placed in the Agulhas Current[350,374] and followed over distances of more than 14 000 km. Having booked this substantial success, an additional eight buoys, from the CSIR, NOAA and the National Centre of Atmospheric Research of the US, were then deployed from the South African Antarctic research vessel *RSA* to the south of Africa[522]. Apart from gaining valuable new information on the Agulhas Current system with these Lagrangian drifters that complemented existing hydrographic concepts[92,349], these experiments also demonstrated the longevity, robustness and positional accuracy of these drifters, in particular in the extreme wave and weather conditions of the Southern Ocean.

This South African technical information was effectively used to persuade the international meteorological community and funding agencies that a major endeavour to cover the ocean in the southern hemisphere with drifting meteorological buoys, for a period of at least one year, was technically feasible. It was hoped thus to provide a high-resolution meteorological data set that did not suffer from the large gaps in global coverage due to the small number of weather stations, restricted to land, in the southern hemisphere[523]. This experiment, the First Global GARP Experiment (FGGE, GARP: Global Atmospheric Research Programme), took place from October 1978 to July 1979 with 301 buoys being launched[524]. Twenty-three South African buoys formed part of this international effort[525]. Since South Africa was considered a political pariah at the time, South African participation was not acknowledged in any official FGGE documents, the origin of its buoys usually being designated as "Other"!

The use of the FGGE buoys in the Agulhas Current region has been scientifically very profitable. The role of the East Madagascar Current[80,308], topographic control on the Agulhas Current[130] and the eddy kinetic energy over the wider system[311,346] have all been addressed. Pioneering work in combining drifter tracks with contemporaneous satellite imagery in the thermal infrared[526] has established somewhat of a trend for subsequent investigations[65,458].

The South African Weather Bureau has continued to deploy drifting weather buoys in the South Atlantic[527] on an annual basis, but a substantial proportion of these have been undrogued and are of less use for studying ocean currents. Some more sophisticated buoys have been deployed in the Agulhas Current itself[366], but only on a limited, experimental basis.

The utility of drifters for studying the Agulhas Current is continuing. The possibility of placing subsurface floats to follow specific water masses[253–54] has been realised through a large international programme called KAPEX (Cape of Good Hope Experiment) and has presented many important new results and concepts. The onset of Argo profiling floats[825] is revolutionising the manner in which deep hydrographic information is becoming available. By 2003 more than 80 per cent of such data worldwide came from Argo floats and this percentage is bound to increase. This will make an enormous difference to the study of the South West Indian Ocean in future, but will probably not lessen the need for dedicated research cruises.

show that the region of the Agulhas retroflection has the highest mesoscale variability of any region in the Southern Ocean. A structure function analysis for the Agulhas Current system itself, based on a quasi-synoptic set of cruises[414], has furthermore shown that, within this system, the retroflection component has by far the most intense variability on all spatial scales. A totally independent result could be produced by analysing the Lagrangian movements of surface drifters.

Four separate studies to determine the advective surface flow in the southern hemisphere, by using the trajectories of the large number of satellite-reporting drifters during the years 1978 to 1979 (see box), have been completed[310–11,346]. Up to 300 drifting buoys were placed south of 20° S and in the area of interest more than 2000 hourly measurements were made. Only in the Agulhas retroflection region was a total kinetic energy per unit mass exceeding 4000 cm^2/s^2 found[310]. This extreme is also found for the eddy kinetic energy, the variations being most prominent for fluctuations with periods of months.

The distribution of higher-frequency eddy kinetic energy, i.e. with periods of days and weeks, also show an extreme south of Africa, but distributed in a zonal tongue, from the Agulhas retroflection eastwards[310]. This suggests a dynamical process with longer periods confined to the Agulhas retroflection region. Analysis of current meter records in the Agulhas retroflection by Schmitz and Luyten[500] shows that the normalised frequency distributions of eddy kinetic energy in this region are comparatively depth-independent, peaked at the mesoscale and are the most energetic found in the ocean to date (Figure 6.1). This comparison includes the Gulf Stream and Kuroshio systems. Records from

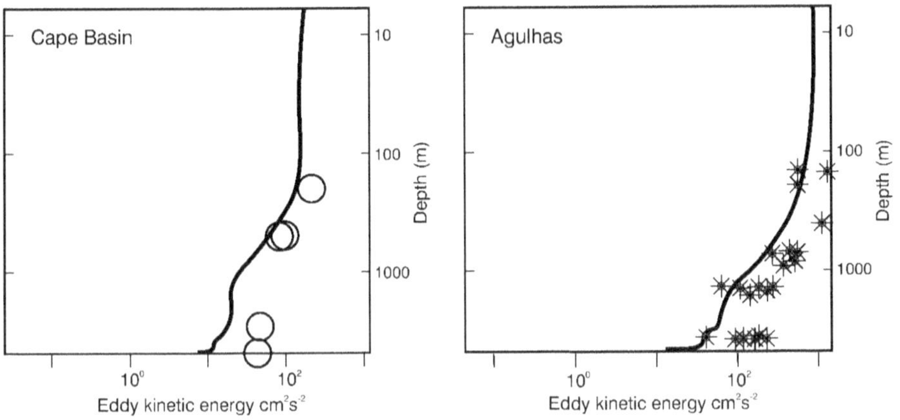

Figure 6.1. The vertical distribution of eddy kinetic energy from current meter records in the Cape Basin (circles) and in the Agulhas Current (stars) compared to calculated profiles from a numerical model[509] for the region, all for the period 1993–1996.

current meter moorings in the Cape Basin and in the Agulhas Current retroflection in general give higher values for the eddy kinetic energy, but with a similar depth distribution.

Sea surface temperatures

Variability in current behaviour can also be gauged from short-term variations in sea surface temperatures, particularly in regions where high horizontal gradients in the sea surface temperatures are known to be present. Such analyses[130,418] show very high values for the Agulhas retroflection, but also a tongue of high variability extending eastwards (Figure 6.3). This tongue is probably a function of meanders in the Subtropical Convergence[131] and eddy-shedding processes associated with the Agulhas Return Current[63]. This is suggested by a compendium of the locations of thermal fronts in the region over

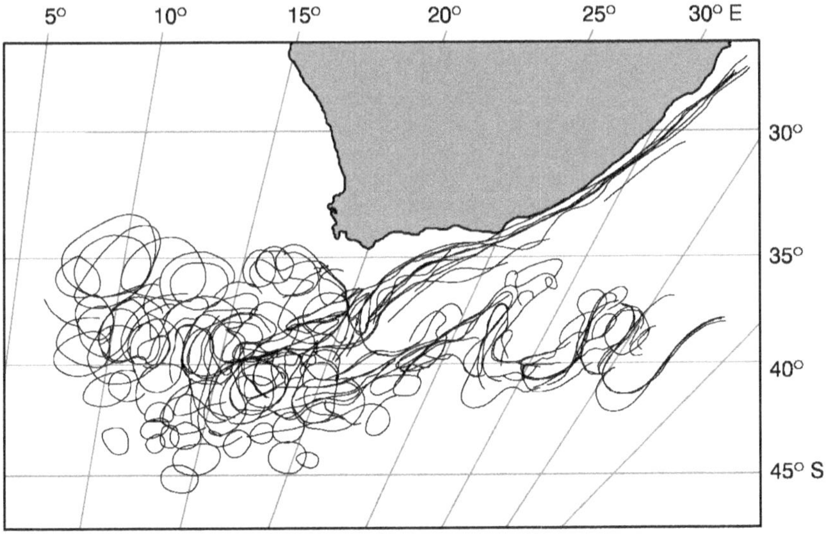

Figure 6.2. Superimposed thermal borders of the southern Agulhas Current, Agulhas retroflection and Agulhas Return Current, for a period of one year[91]. These data are from declouded images in the thermal infrared from the METEOSAT II satellite. The most distinct one for every 12-day period was used. The stability of the Agulhas Current (particularly of its northern part), the relative stability of the Agulhas Return Current and the severe instability and eddy shedding processes of the Agulhas retroflection are all very apparent.

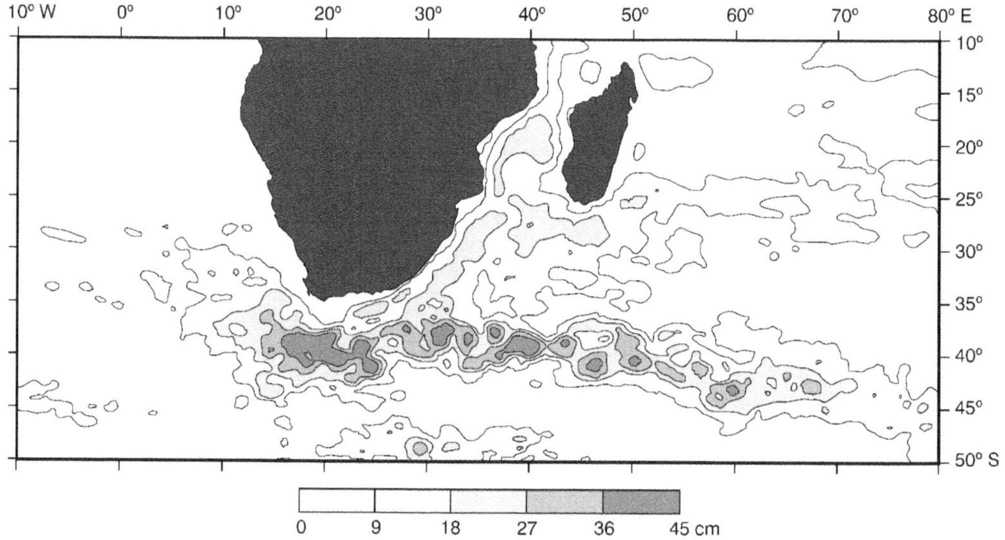

Figure 6.3. Sea surface variability south of Africa for the period 1992–1998. The altimetric data were from the TOPEX/Poseidon satellite. High levels of variability associated with the Agulhas retroflection are particularly prominent. (See also Figure 3.14.)

a period of a full year (Figure 6.2). Eddies, and rings, are seen to be present predominantly around the retroflection region while, downstream, movement in the location of the Subtropical Convergence is probably responsible for most variability there. General variability in sea surface temperatures presents instructive circumstantial data but tells one little about the ocean processes responsible for the variability. Satellite altimetry can potentially do this.

Preliminary analyses of altimetric data have suggested that the Agulhas retroflection is a region of very high sea level variability[305] with the presence of large vortices[501]. The first definitive study of global mesoscale variability based on altimeter data from the SEASAT satellite, by Cheney et al.[183], has confirmed all that had been suggested before with perhaps less reliable data. Not only does the Agulhas Current retroflection represent a large region of high mesoscale variability, in its core the values are higher than anywhere else on the globe. This is amply demonstrated in Figure 6.3.

Similar investigations using subsequent altimetric measurements by other satellites have substantially confirmed these results[501], although the area of high variability and its intensity naturally differ slightly between different periods. At least one study[269] has suggested that there is a seasonal cycle to the variability at the Agulhas Retroflection and that this cycle extends about 30 per cent from the mean. It has a maximum in the austral summer and a minimum during the austral winter, consistent with previous results that were based on another satellite[502]. Based on only slightly more than three full years of data, these interesting results still need to be verified more robustly.

Modelling

Last but by no means least, global, eddy-resolving circulation models[273,277] also successfully simulate this region of particularly high mesoscale variability. Where eddy kinetic energy from the Geosat altimeter data exceeds 1000 cm^2/s^2 in the Agulhas retroflection region, it has been found to be only larger than 500 cm^2/s^2 in a model[503], although the present models do simulate the area of higher variability adequately. Eddy kinetic energy has for instance been modelled[762] with a fair degree of success (Figure 6.1) in a primitive equation model with a $1/3° \times 1/3°$ spatial resolution.

Modelling has also been used extensively in an attempt to understand why the Agulhas Current retroflects and why it does so at this particular location[469,580,583]. It has been shown[580] that the geographic distribution of the wind stress curls is crucial to the behaviour of the termination of the Agulhas Current. Since the line of zero wind stress curl lies well south of the poleward termination of the African continent, the current is a free inertial jet beyond this point and in a purely barotropic model with realistic values for lateral friction will move into the South Atlantic Ocean[580]. Otherwise, the increasing negative relative vorticity of the Agulhas Current overshoot will eventually lead to

Origin of the term and concept *retroflection*

A review of the historical development of concepts on the circulation directly south of Africa[14,285] shows that two flow paradigms have been prominent since the earliest times. The first holds that all of the Agulhas Current's waters flows into the South Atlantic Ocean; the second the opposite, namely that none of it goes westward, but that all is returned to the Indian Ocean. Over the past 150 years these two concepts have battled for supremacy.

Nils D. Bang

It is instructive to note that a recurvature of a major part of the Agulhas Current is inherent to the current portrayal put forward by James Rennell as early as 1832[7]. Subsequent studies[504] that made use of sea surface temperatures[780] as well as ships' drift, undertaken particularly in the Netherlands[10,505] in the 1850s, strongly supported this concept. In fact, in some of these Dutch publications[780] it is explicitly stated that the previous concept of the Agulhas Current rounding the Cape of Good Hope and moving northward in the South Atlantic Ocean is without any doubt wrong. The major work on this subject by K.F.R. Andrau was subsequently seldom referred to directly, more-or-less lost to science, and portrayals of a bifurcation in the Agulhas Current – some water going east, some west – became more fashionable[24,281,284] (Figure 3.1). The quality, quantity and geographic distribution of the data available at the time were such that both interpretations could logically be sustained simultaneously, even when combinations of hydrographic data first allowed a portrayal of the whole water column by Dietrich in 1935[40]. Even he considered the coastal upwelling system along the west coast of southern Africa as representing an Atlantic branch of the Agulhas Current[42]. As late as 1972, Darbyshire[506] still concluded, quite emphatically, that no true return current existed for the Agulhas for three of his four surveys[215].

A comprehensive, quasi-synoptic and detailed set of cruises covering the full Agulhas system[92] was undertaken for the first time only in 1969. When the data from this project were being analysed by Nils Bang[90,444] in the early 1970s, he was particularly struck by the discontinuous nature of the flow, with a host of fronts and mesoscale features. Bang evidently struggled with the interpretation of these data, suggesting a number of alternative explanations for the current's disposition[90]. Searching for a suitably descriptive term that would convey the impression of a dynamic flexing activity instead of a static, geometrically stable process, he came across the term "retroflect", commonly used to describe the turning back on itself of the mammalian gut[14].

This excellent verbal portrayal of the flow regime, suggested by the new data, established a novel conceptual framework for all data gathered before and since. The catchy descriptive terminology aided the acceptance of the recurvature concept by a wider community, particularly once it became clear that all succeeding information fitted it well.

The concept of the Agulhas' retroflection, as well as the new nomenclature, is now firmly entrenched, the term *retroflection* being widely employed in oceanography to describe the behaviour of a number of other currents, such as, for instance, the Brazil Current[507,687].

an eastward turning[581] in both barotropic and baroclinic configurations. Retroflection can also be brought about by increasing the large horizontal friction[582]. Lesser horizontal friction will lead to strongly variable flow. Some[583] have tried to show that under certain circumstances time-dependant phenomena, such as ring shedding, are essential to the existence of a retroflection. To study the requirements for a steady retroflection regime, an investigation has been carried out[778] by modifying the wind forcing, the bottom topography, the lateral friction and the layer depth in a model with steady barotropic flows. It has been found that steady retroflection regimes can be created under a number of conditions, for instance with large friction or with dominant inertial effects when friction values are low. Instabilities in this barotropic steady flow[741] may produce intermonthly and inter-annual variability. Nevertheless, in this barotropic model the frequency of ring formation is set by the physics of the large-scale instabilities and the rectification processes due to these instabilities decrease the degree of retroflection of the mean state. More about wind-driven and other modelling is to be found below under the rubric of the dynamics of the Agulhas retroflection.

Direct measurements

Although the nature of the variability in this retroflection region, as well as its approximate geographic extent, may be estimated from the abovementioned data, only few continuous measurements, such as current observations, have been made here to date. One deployment of current meter moorings, spaced over the full southern Agulhas Current and Agulhas retroflection

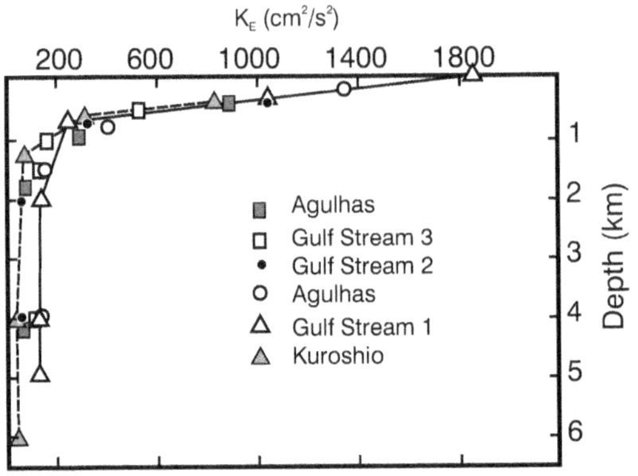

Figure 6.4. The vertical distribution of the current kinetic energy for representative components of the Agulhas Current retroflection[509], a summary of the available data at 55° W longitude (east of the New England seamounts, Gulf Stream 1), a mooring located at 68° W (west of the New England seamounts, Gulf Stream 2), one placed at the New England seamounts (Gulf Stream 3) and the Kuroshio[510]. The energy of the Agulhas Current is high, compared to the other western boundary currents, at all depths.

region[508], has presented data that give some indications of the nature of the current variability. These results may profitably be compared to those found in similar western boundary currents[509] (Figure 6.4).

It shows that the kinetic energy distribution of the currents is very similar, so that in this respect the Agulhas Current is not exceptional. Differences in kinetic energy below 1000 m are mostly due to the dissimilar location of a current meter mooring relative to the current system it was supposed to monitor. Differences between the western boundary currents are of the same order as the differences found between different parts of the same current system (Figure 6.4). The spatial distribution of kinetic energy amongst the different western boundary currents also is very similar, peaking at the mesoscale[509].

There is therefore abundant proof, from a number of totally independent data sets, for the very high mesoscale variability of the Agulhas Current retroflection, implying some continuous process resulting in substantial changes in current structure and position of the main flow elements.

Current predilection

The first description of the southern termination of the Agulhas Current that combined hydrographic data from a number of different deep-sea cruises has already been presented by Dietrich in 1935[40]. It shows a substantial part of the transport, but by no means all[42], flowing back into the South Indian Ocean in a recurvature of the current to the south-west of Cape Agulhas. An analysis of widely spaced hydrographic stations in the region in the early 1960s was the first to demonstrate unequivocally the presence along the Subtropical Convergence of intense eddies[511], and the analysis of the combined data set collected for the International Indian Ocean Expedition (see box) allowed for contouring that also showed some intense eddies here[75]. However, it was only once a full oceanographic project, consisting of three simultaneous research cruises with closely spaced stations over the full region, had been completed that the true nature of the terminal region of the western-most extent of the Agulhas Current became clear[90,444] (Figure 6.5).

In Figure 6.5 the characteristic disposition of the Agulhas Current, based on these data, is demonstrated quite admirably. Having passed by the southern tip of the Agulhas Bank, at about 19° E, the current turns abruptly south carrying its water as far as the 42° S parallel before moving in a north-easterly direction. The neck of the retroflection proper was only about 180 km wide. On this occasion the Agulhas Current, under the influence of a Natal Pulse, was even closer to the Agulhas Return Current in the vicinity of Port Elizabeth, but this is an unusual configuration. It is nonetheless of considerable importance, since this close juxtaposition of opposing currents may occasionally bring about an upstream retroflection here[64,412]. South-east of Cape Town (Figure 6.5) a large, anti-cyclonic eddy is evident in the hydrographic data. The volume transport in this feature relative to 1100 m has been estimated[92] to have been 15 × 10^6 m^3/s, while that of the Agulhas Current itself was 40 × 10^6 m^3/s.

Studying the Agulhas system with large observational programmes

Progress in the understanding of the extended Agulhas system has come about mainly in two ways: by local efforts with a limited geographical scale and by large, usually international, programmes on a much more extensive scale. Investigations using the R.V. *Meiring Naudé* from the CSIR (Council for Scientific and Industrial Research, South Africa) in Durban[317] formed a key component of the former. This is described in an inset of Chapter 4. One of the most important international programmes that stimulated oceanographic research in the region for many years to come, was the International Indian Ocean Expedition[224] of the 1960s. Its effect is described in an inset to Chapter 2. But there have been a number of other sea-going programmes since that have had a decisive influence on the development of our understanding of this current system.

The ARC (Agulhas Retroflection Cruise) took place in 1983 from the US research vessel *Knorr*. Initiated by Dr Arnold Gordon of the then Lamont-Doherty Geological Observatory, it included a number of South African and Dutch participants. It aimed at understanding the inter-ocean exchange of waters at the Agulhas termination[65] and consisted of one of the most extensive cruises in this region up to that time. Many of the results it achieved were seminal[61,458]. It was followed in 1987 by the South African SCARC[787] (Subtropical Convergence and Agulhas Retroflection Cruise) from the R.V. *SA Agulhas*. This multi-disciplinary cruise was one of the first to use satellite remote sensing to guide its sea-going programme[456] and investigated seven Agulhas rings and eddies. It successfully documented one of the most extensive leakages of Subantarctic water[528] across the Subtropical Convergence. These single-cruise projects were followed by a number of multi-cruise programmes.

Arnold L. Gordon

The BEST (Benguela Sources and Transports) was a collaborative programme[535] involving the Woods Hole Oceanographic Institution, the National Oceanic and Atmospheric Administration of the US and the (then) South African Sea Fisheries Research Institute and took place from 1992 to 1993. Its prime aim was to establish the flux of Agulhas water in the South East Atlantic Ocean by the judicious combination of bottom mounted instruments, hydrographic observations and satellite remote sensing. It successfully established that most of the flow in this region was due to Agulhas rings[536] and not the Benguela Current. The geographically largest and most elaborate research programme in the Agulhas system to date has been the KAPEX (Cape of Good Hope Experiment)[677,678,650] undertaken by a number of German, US and South African organizations from 1997 to 1999. It covered the Agulhas system from Durban on the east coast of South Africa to beyond the Walvis Ridge in the South Atlantic Ocean. During the programme a number of sound sources were placed in the region and a large number of RAFOS floats launched to pass through this international array. The results of this highly successful programme filled a special issue[627] of the journal *Deep-Sea Research II*.

A subsequent multi-disciplinary, Dutch-South African programme initiated by Professor Will de Ruijter of Utrecht University consisted of two observational parts: the MARE (Mixing in Agulhas Rings Experiment)[658] and the ACSEX (Agulhas Current Sources Experiment)[650]. The MARE was carried out over a period of one year on three cruises at six-month intervals, starting in 1999. It aimed at studying the slow demise of one particular ring over this period. The ACSEX was carried out during four cruises on the Dutch research vessel *Pelagia* in the Mozambique Channel and in the region south of Madagascar. It has shown that no coherent Mozambique Current exists, but that the flow in the Mozambique Channel is characterized by the regular formation of eddies that subsequently drift poleward[728]. It has been continued by LOCO (Long Term Ocean Climate Observations) in which a current mooring array continues to monitor the flow through the narrows of the Mozambique Channel.

Will P.M. de Ruijter

In later years, the ACE (Agulhas Current Experiment)[788] was funded by the UK to study the fluxes of the Agulhas Current by placing a number of current meter moorings across the current at Port Edward, off South Africa's east coast. It has presented the most accurate estimate of this flux to date and in the process discovered an Agulhas undercurrent[368]. The ASTTEX (Agulhas–South Atlantic Thermohaline Transport Experiment) consists of a similar set of moorings in the South East Atlantic that builds on the results of BEST and will try to quantify the flux of Agulhas water in the Cape Basin.

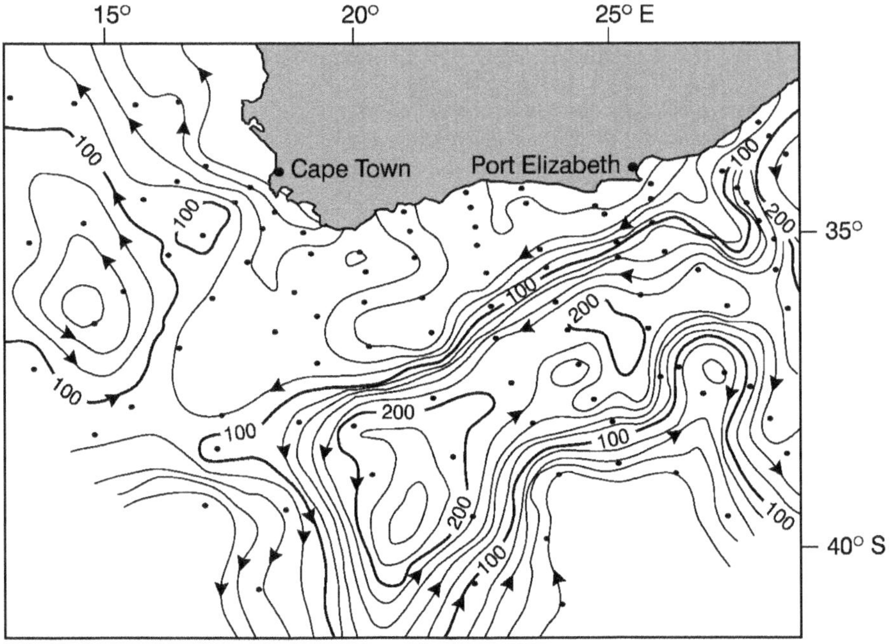

Figure 6.5. The Agulhas retroflection as evident in hydrographic measurements collected in March 1969. Shown here is the depth of the sigma-t 25.80 surface[91]. Dots represent hydrographic stations; arrows inferred directions of movement. The concentration of isobaths identifies the core of the Agulhas and of the Agulhas Return Current. Closer inspection also shows a Natal Pulse off Port Elizabeth (viz. Figure 5.15) and an Agulhas ring west of Cape Town.

Retroflection inconstancy

As might be expected from the multifarious results pertaining to the current's variability (viz. Figure 6.2) this is by no means the only configuration of the Agulhas retroflection. Satellite imagery, particularly in the thermal infrared[59], has shown it in a number of positions[91]. It has in fact been demonstrated that the Agulhas retroflection loop, or its products, may extend continuously as far as 10° E, i.e. outside the Agulhas Basin (viz. Figure 1.2) and well into the Cape Basin[512] west of the Agulhas Plateau. This demonstrates that the Heezen Ridge, which separates these two basins at about 5° E, seems to have no, or limited, restraining effect on the Agulhas retroflection loop.

A representative thermal infrared image from the NOAA 9 satellite for the Agulhas retroflection is given in Figure 6.6. The question arises how representative this one image might be and how much it truly tells one about the movement of water through the system. Using satellite imagery for 623 days it has been shown that the average diameter of the retroflection loop is 341 km (standard deviation 72 km), that anti-cyclonic shear-edge features to its north are 307 km (±89 km) and eddies to its north-west (viz. Figure 6.5) 324 km (±7 km)[414]. Based on about 1000 useful images, the maximum easterly position of the retroflection has been shown to lie at 20°30' E longitude, the westerly position at 9°40' E[91]. In general the Agulhas retroflection seems to lie between 20° E and 15° E, with no preferred locations, as has been surmised previously[59]. Although the range of these features is relatively large, it does demonstrate some consistency in the occurrence and characteristics of these features. Comparing these surface temperature portrayals with the tracks of some drifters has shown that there exists a close and reliable correspondence between them.

A drifting buoy that became entrained in the Agulhas Current south of Port Elizabeth[350] clearly circumscribed a tight retroflection loop at about 15° E longitude before drifting eastwards (viz. Figure 7.3). Other buoys have done the same[513]. Buoys passing through the retroflection show advective rates of more than 1 m/s, very similar to those observed in the Agulhas Current itself. Comparing[349] the main features of all the available drift tracks with the main features of the retroflection, as evident in the results of hydrographic measurements, demonstrates that the portrayals of the nature of the retroflection in satellite imagery are very accurate ones of the water movement through the region. What can

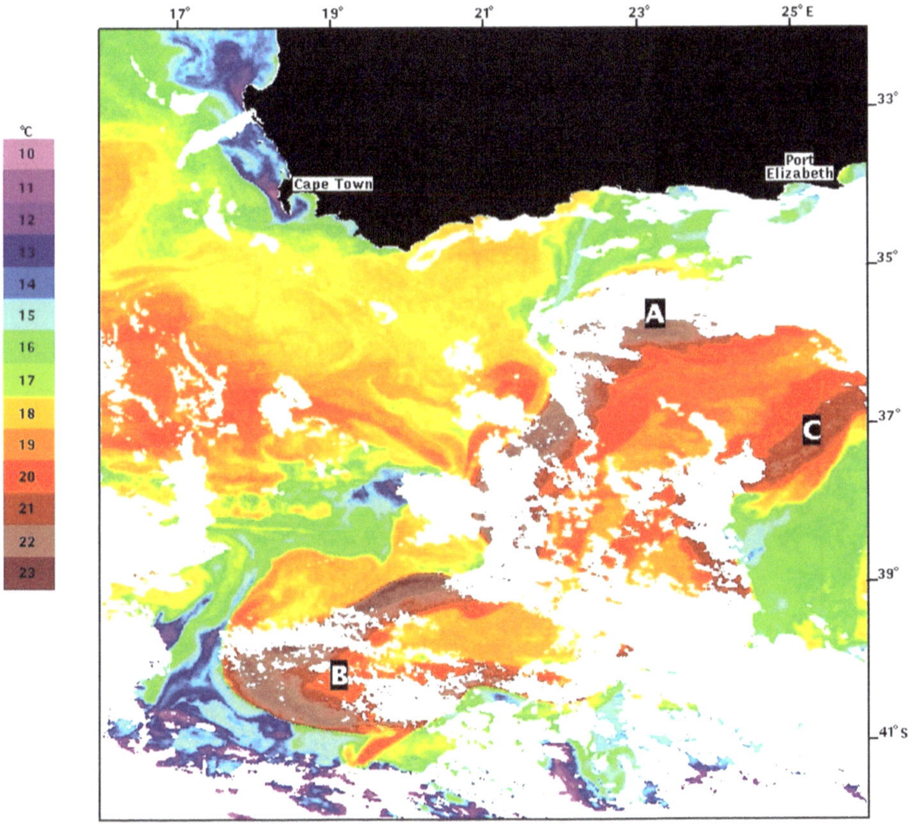

Figure 6.6. The Agulhas retroflection[91] region south of Africa manifested in the sea surface temperatures. This thermal infrared image is from the NOAA series of satellites for 13 November 1985. Red–yellow hues indicate warmer water, black the land and white the clouds. **A** denotes the southern Agulhas Current, **B** the western extent of the retroflection loop and **C** the Agulhas Return Current. Note the cold upwelled water off Cape Town, the cold Subantarctic Surface water at the westernmost extremity of the retroflection loop and the cold coastal water being advected along the Agulhas Bank from the vicinity of Port Elizabeth by the Agulhas Current.

these images then tell us about the transient comportment of the retroflection?

Temporal behaviour

First, they show[91] that the Agulhas retroflection loop normally protrudes farther and farther westwards into the South Atlantic Ocean with time (Figure 6.7). The mean rate of this progradation is about 10 km/day. Sometimes during this process the Agulhas Current and Agulhas Return Current amalgamate, somewhat upstream of the furthermost extent of the loop, and a separate, independently circulating annulus of Agulhas water, an Agulhas ring, is formed. This process was first identified in satellite imagery[60], although the possibility of such a process of loop occlusion had been hypothesised before[444,511], based on the same mechanism already observed in the Gulf Stream at the time[514–15]. Could a major meander in the incipient Agulhas Current trigger or force the occlusion of a ring? Such a meander could be a Natal Pulse[62] that had travelled this far downstream intact. This would constitute a mechanism very different from that found acting in ring shedding events in the Gulf Stream system.

An analysis of the downstream progression of Natal Pulses, using satellite altimetry that had been validated by thermal infrared information[401], has shown that nearly all ring shedding events at the Agulhas retroflection are preceded by the appearance of a Natal Pulse at the Natal Bight, with a lag time of 165 days (Figure 6.8). All the Natal Pulses investigated as part of this particular study proceeded downstream at the previously estimated[62] speeds of about 20 km/day, up to the latitude of Port Elizabeth. Downstream of here their

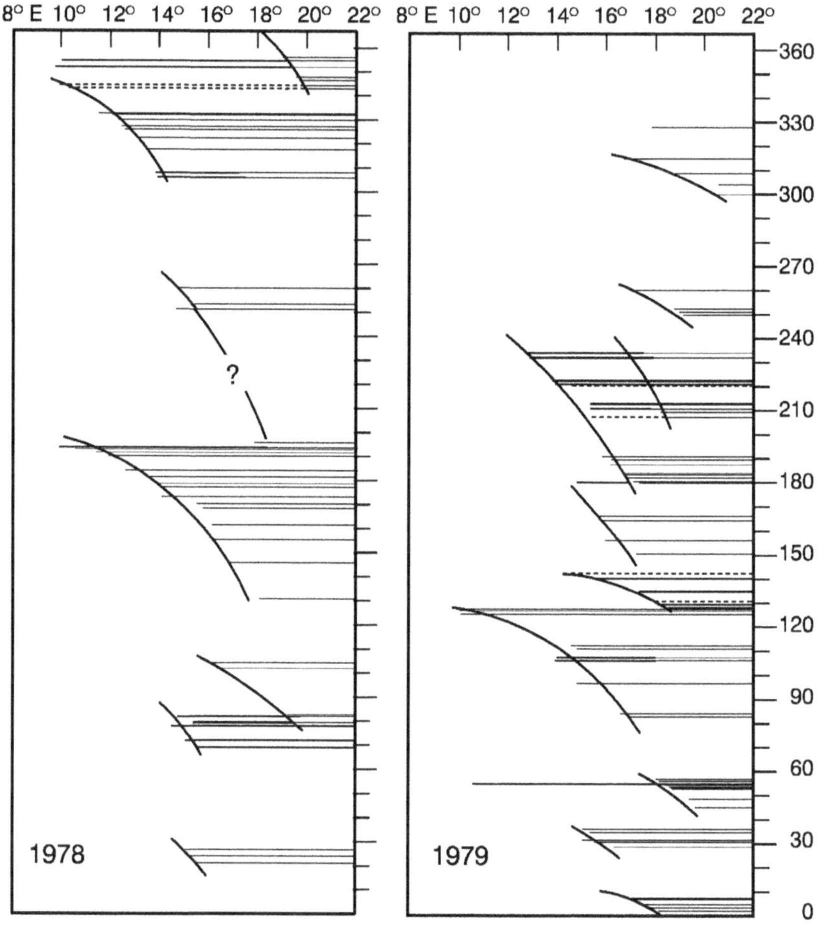

Figure 6.7. The zonal location of the westernmost limit of the Agulhas Current retroflection from thermal infrared imagery from the METEOSAT satellite for 1978 and 1979[91]. Peaks indicate the furthest extent of each progradation event as the Agulhas retroflection loop has pushed into the South Atlantic Ocean. Dotted lines represent less reliable results.

progression slowed to about 5 km/day. This is indicated by the increased slopes of their distance-made-good lines in Figure 6.8 beyond a distance of 800 km from Durban. On these Natal Pulses reaching the retroflection region, a ring was shed in each case.

For this study, extending over more than a year, at least one ring was formed without the intervention of a Natal Pulse. It has therefore been assumed[401,516] that rings will form spontaneously when the retroflection loop has been extended sufficiently, but that the incidence of a Natal Pulse will precipitate such an event. Thus ring shedding may be forced by the Natal Pulse itself or by it interacting with meanders on the Agulhas Return Current[401]. Since the Natal Pulse itself may be triggered by eddies coming from elsewhere[653] this means that the control for the frequency of ring shedding may reside in other parts of the Indian Ocean. It has been shown that monsoonal winds in the Indian Ocean create Kelvin waves that hit Indonesia, then propagates southward along the Indonesian coast and in turn trigger Rossby waves that propagate westward across the subtropics of the Indian Ocean. When they reach Madagascar and the Mozambique Channel they generate eddies which in turn are responsible for the eventual shedding of Agulhas rings. Others[652] have shown that this whole process may be dependant on ENSO cycles and the presence of the Indian Ocean dipole. Inter-annual variability originating in climate modes in the equatorial region of the Indian Ocean may therefore affect the frequency of ring shedding at the Agulhas retroflection[652].

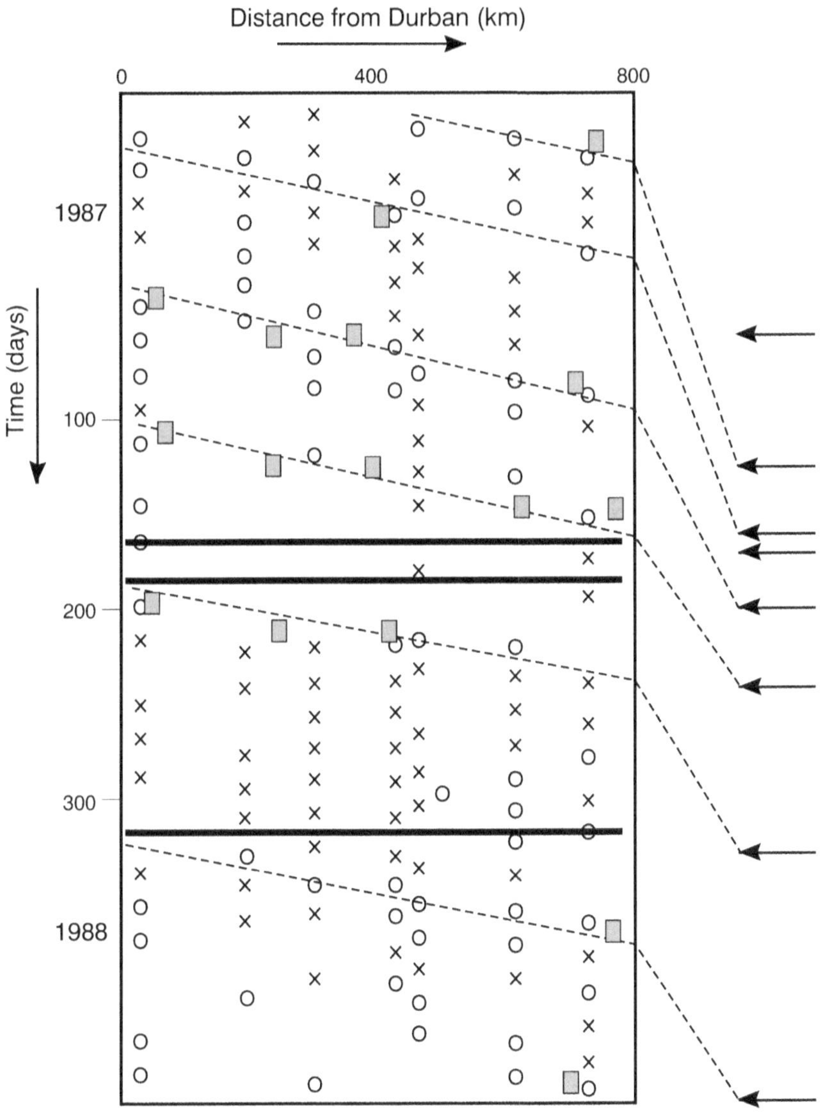

Figure 6.8. A space-time diagram of altimeter (Geosat) and thermal infrared (NOAA) observations along the south-east coast of South Africa starting from 1 November 1986. The ellipses denote Natal Pulses from altimetry; rectangles confirmational sightings in infrared images and crosses observations during which no Natal Pulses were evident. Dashed lines show the assumed tracks of Natal Pulses whereas arrows give observed ring shedding events. The horizontal black bars indicate a cloud-free infrared-image of the full region during which no Natal Pulses were observed. The internal coherence of these independent data sets is impressive[401]. They show that Natal Pulses proceed downstream at a nearly identical and uniform rate until they reach a distance of 800 km from Durban, after which they all slow down. This is along the Agulhas Bank. Shortly after they reach the retroflection region, a ring is shed (arrows) in nearly all cases, suggesting the important role Natal Pulses play in triggering ring shedding events.

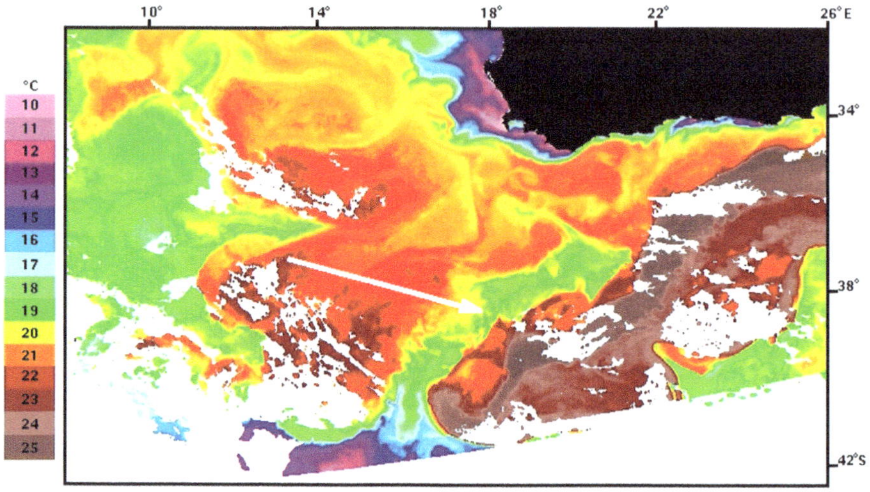

Figure 6.9. Northward penetration of cold Subantarctic Surface Water (blue-green) during the separation of an Agulhas ring. These thermal data are from the NOAA 14 satellite and show the characteristic development of such an event on 16 to 17 December 2000. A similar occurence may be seen in Figure 6.6.

Spawning events

Normally, each of the progradation events of the Agulhas retroflection loop that terminates in the shedding of an Agulhas ring lasts about 40 days[91]. Within each event, westward penetration of the retroflection loop shows an increasing rate of progress until abrupt ring spawning occurs. This event duration may, however, be quite variable. There seems to be no clear periodicity and for long periods there may be no spawning events at all[517].

First results have suggested an annual production of six to nine rings. Other investigators have estimated only four to five ring shedding events per year[74,464,518]. Garzoli et al. have monitored the movement of Agulhas rings past a line of inverted echo-sounders placed along 30° S latitude in the south-eastern Atlantic Ocean[519] and have determined that a minimum of four to six Agulhas rings per year entered this region during the period from 1992 to 1993[520]. In such an extremely variable system it would be highly unlikely that the frequency of shedding events would be identical for each year, although the probable average seems to be stable, about one every two months[413]. This may be compared[755] to the shedding of rings from the southern Brazil Current that exhibits quasi-periodic ring formation roughly every 150 days and the East Australian Current with 130 days.

At least one of these events at the Agulhas retroflection has been hydrographically observed and measured at sea[61]. The newly formed ring essentially retains all the kinematic characteristics of its parental Agulhas Current, at least initially. It extends to the same depth, has the same velocity and temperature structure, but starts cooling very rapidly at the sea surface[262].

Accompanying flows

These ring-shedding events are accompanied by a number of significant, secondary circulations. One of these is the equatorward penetration of a cold wedge of Subantarctic Surface Water, between the newly formed ring and the new retroflection loop (Figure 6.9). This seems to be an inherent part of the dynamics of the ring-shedding process. Usually the width of this throughflow remains relatively modest[91] with the cold water spreading laterally only to the north of the gap between the retroflection loop and the newly spawned ring. However, on occasion it has been observed to be wider than 150 km[512]. Shannon et al.[528] have described an event in which such cold water extended as far north as 33° S latitude, a distance of 1000 km, and was observable at the sea surface for a period of two months. This particular intrusion covered an area of 734×10^3 km^2, 5 standard deviations greater than the mean area for such intrusions established from an investigation extending over nine years[657].

On this occasion temperatures of the sea surface were below 17 °C here and salinities below 34.9, and these anomalous water characteristics extended throughout the upper water column, suggesting that this represents true advection of cold water and not only an out-

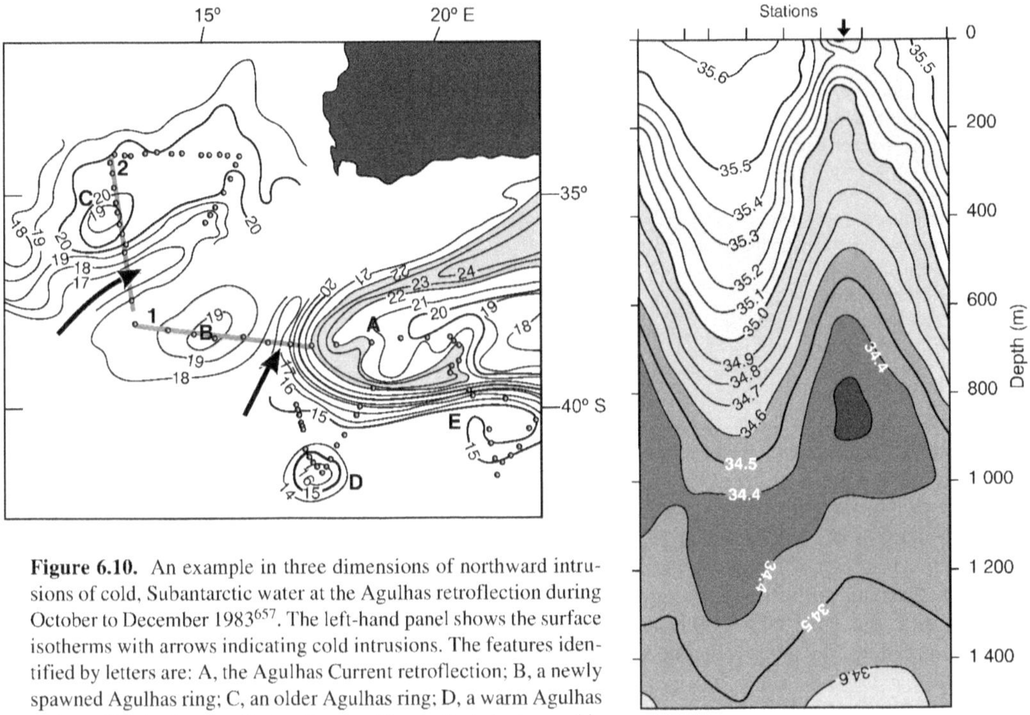

Figure 6.10. An example in three dimensions of northward intrusions of cold, Subantarctic water at the Agulhas retroflection during October to December 1983[657]. The left-hand panel shows the surface isotherms with arrows indicating cold intrusions. The features identified by letters are: A, the Agulhas Current retroflection; B, a newly spawned Agulhas ring; C, an older Agulhas ring; D, a warm Agulhas eddy; and E, the Agulhas Return Current. Dots indicate hydrographic stations. The hydrographic section in the right-hand panel shows a vertical salinity section along line 2. This line is indicated in the left-hand panel. It intersects two Agulhas rings and the section shows the water with lower salinity between these as well as its low salinity surface expression (arrow). Lines on top of this panel show the location of hydrographic stations.

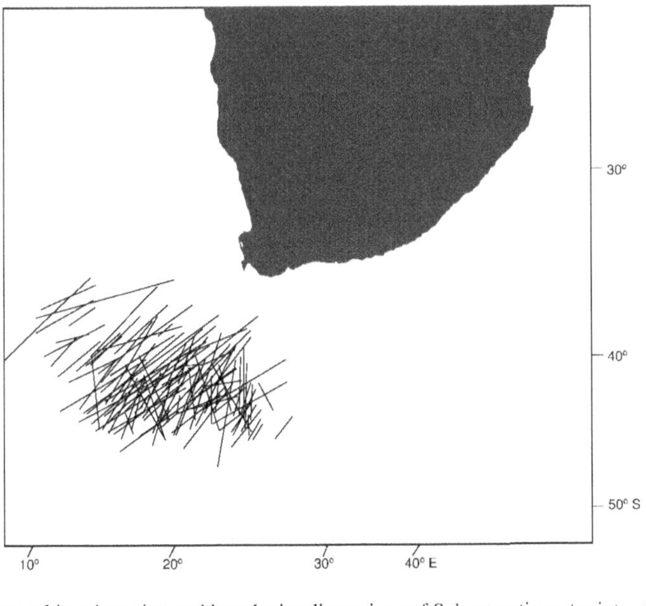

Figure 6.11. The geographic orientation and lengthwise dimensions of Subantarctic water intrusions at the Agulhas retroflection for the period 1981–1990. This portrayal is based on thermal infrared observations from satellite[657].

Table 6.1. Characteristics of the intrusions of Subantarctic Water into the Agulhas retroflection region[657] from thermal infrared observations by satellite.

Zonal distribution	8° E to 22° E
Meridional distribution	Subtropical Convergence ~35° S
Average number per year	5
Temporal occurrence frequency	38 per cent
One intrusion present	21 per cent of times when intrusions are present
Two intrusions present simultaneously	49 per cent of times when intrusions are present
Three intrusions present simultaneously	30 per cent of times when intrusions are present
Average length of intrusions	410(±220) km
Average width of intusions	80(±100) km
Average surface area of intusions	159(±118) × 10^3 km^2
Average duration of intrusions	28 days

cropping as suggested by some numerical[529] models. Other studies[657] have supported these conclusions and demonstrated that these wedges of cold water found between newly shed Agulhas rings and the Agulhas retroflection may extend deeper than 1500 m (Figure 6.10). On this occasion two wedges were evident at the same time. From the vertical sections across these features it is clear that these cold wedges are only weak surface expressions of a much larger body at depth.

Over a period of a decade cold wedges associated with the shedding of Agulhas rings were found to lie between 8° and 22° E, i.e. the expected zonal range for ring shedding events (Figure 6.11). Their general orientation nearly always was in a south-west/north-east direction. Intrusions are evident about 38 per cent of the time (see also Table 6.1). The recurrence of this pulse of cold water, probably carrying a collection of foreign biota, has an as yet unquantified effect on the South East Atlantic Ocean[530]. The converse, i.e. an unusual flux of warm Agulhas water into the South East Atlantic, has also been observed[439].

In this particular instance of an enhanced flow of warm water, the configuration of Agulhas retroflection, Agulhas rings, and winds was conducive to drawing considerable amounts of surface water from the Agulhas Current retroflection, through Agulhas filaments and the like. A large ocean area off Cape Town was covered with warm surface water that was replenished from the Agulhas retroflection for an unusually long period. This exceptional culmination of a number of factors that seem to influence flow of warmer water from the Agulhas coincided with 1986 being the warmest year on record in the South East Atlantic Ocean[439]. But it is not only the surface waters that are influenced by the behaviour of the Agulhas retroflection. It has been demostrated[707] that the deep boundary current along the west coast of southern Africa, consisting mainly of North Atlantic Deep Water (viz. Figure 2.13), is influenced by changes in the circulation at the Agulhas retroflection. This temporal variability causes it to separate from the continental slope on some occasions and to enter the Indian Ocean in the deep return flow. The causal relationship between the behaviour of the Agulhas retroflection loop and changes to the trajectory of the deep boundary current of the South East Atlantic Ocean has not been established.

As one would expect for a region where waters from the Indian Ocean tropics, the South Atlantic gyre and the Subantarctic meet in a dynamic system of extreme variability, the water masses of the Agulhas retroflection are a true mélange of water types.

Water masses

Using all the presently available, high-quality hydrographic data of the Agulhas retroflection region, Valentine et al.[236] have tabulated the water types to be found here and their thermal and saline characteristics (Table 6.2, Figure 6.12).

The pictorial representation (Figure 6.12) exhibits considerable variations in the surface water warmer than 16 °C and in the Antarctic Intermediate Water. In the Central Water, between these two extremes, there is some indication of two preferred temperature–salinity relationships that represent the hydrographic characteristics of South East Atlantic and South West Indian Ocean water respectively. A precise volumetric analysis of the water masses present[236] shows that the warm, saline surface water of the Agulhas Current contributes relatively little to the volume of the upper 1500 m of the region. Pulses of cold Subantarctic Surface Water, with low salinities, make a distinctive, but very small overall contribution to the volume. By volume alone, the North Atlantic Deep Water is the dominant water mass, accounting for 40 per cent of the total volume.

Table 6.2. Thermal and saline characteristics of the principle water masses found in the Agulhas retroflection and its direct vicinity[236].

	Temperature range [°C]	Salinity range [psu]
Surface Water	16.0–26.0	>35.5
Central Water		
South East Atlantic Ocean	6.0–16.0	34.5–35.5
South West Indian Ocean	8.0–15.0	34.6–35.5
Antarctic Intermediate Water		
Characteristic *T/S*	2.2	33.87
South East Atlantic	2.0–6.0	33.8–34.8
South West Indian	2.0–10.0	33.8–34.8
Deep Water		
North Atlantic Deep Water (South-east Atlantic)	1.5–4.0	34.80–35.00
Circumpolar Deep Water (South-west Indian)	0.1–2.0	34.63–34.73
Antarctic Bottom Water	–0.9–1.7	34.64–34.72

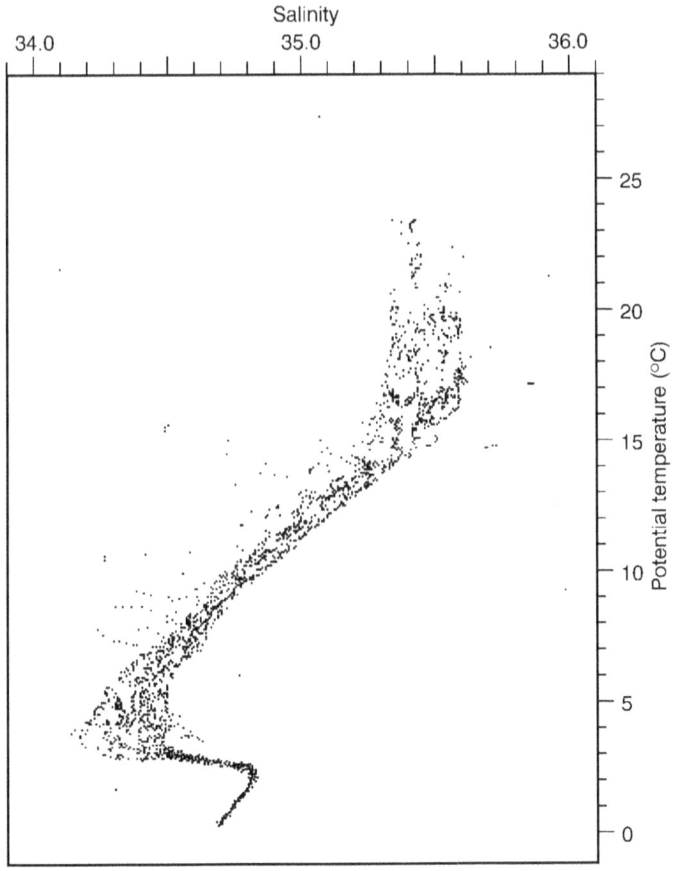

Figure 6.12. A scatter diagram of the potential temperature–salinity relationship of the water masses found in the Agulhas retroflection and its direct environment[236]. These data are all from CTD (conductivity–temperature–depth) measurements taken to the ocean bottom and represent the wide range of water masses to be found in this mixing region.

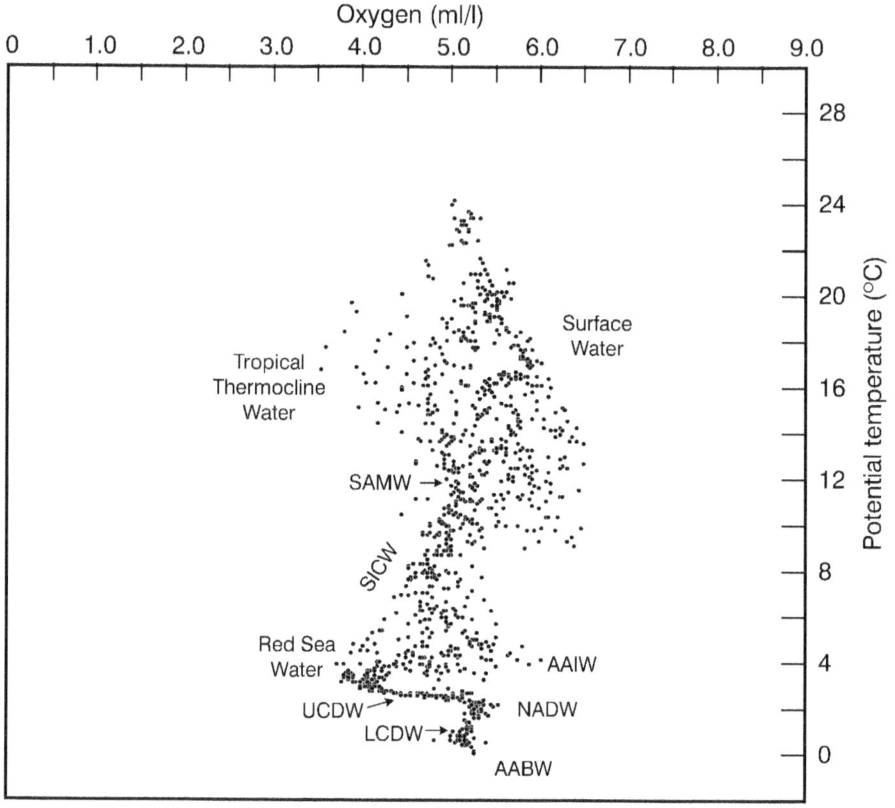

Figure 6.13. The potential temperature–dissolved oxygen characteristics of the water masses in the Agulhas retroflection region[230]. SAMW: Subantarctic Mode Water; SICW: South Indian Central Water; AAIW: Antarctic Intermediate Water; UCDW: Upper Circumpolar Deep Water; NADW: North Atlantic Deep Water; LCDW: Lower Circumpolar Deep Water; AABW: Antarctic Bottom Water. Oxygen content is particularly valuable in distinguishing between different water masses at intermediate depths.

By comparing the temperature–salinities characteristics found in the Agulhas retroflection (Figure 6.12) to those found in the northern Agulhas (Figure 4.15) and southern Agulhas Current (Figure 5.3), it can be seen that the two components of the surface waters of the Agulhas Current, Tropical Thermocline Water and Subtropical Surface Water, arrive in the retroflection region fairly intact. The central and intermediate waters in the retroflection region by contrast show many more outliers towards lower temperatures than they do farther upstream, indicating the influence of the subantarctic waters, understandably not so evident to the north. Of interest in Figure 6.13 is also the presence of Subantarctic Mode Water, made manifest by its deep oxygen maximum[258]. This water is introduced along the southern edge of the subtropical region and, befitting the proximity of the retroflection to the Subtropical Convergence, is much more prominent here than in the South Indian Ocean as a whole (viz. Figure 5.3).

Gordon et al.[230] have shown that the water of Indian Ocean origin introduced into the retroflection region by the Agulhas Current is restricted to the upper 2000 m. They have also shown that a substantial, or at least a very conspicuous, remnant of Red Sea Water is carried downstream as part of the Agulhas Current flow. It is not clear whether this Red Sea Water passes through the Mozambique Channel[234], or whether it comes from east of Madagascar[235]. Its low-oxygen characteristics are clearly seen in potential temperature–dissolved oxygen plots for the region (Figure 6.13).

Shallow oxygen minimum

Of particular relevance to an understanding of the circulation and mixing of water masses in the Agulhas retroflection is the presence of an oxygen minimum[230] found at depths of between 100 to 150 m. It is marked as Tropical Thermocline Water in Figure 6.13, and is associated with the warm surface water of the Agulhas Current. During some cruises that have intersected Agulhas rings[230], this minimum was not found in older rings, suggesting that this particular water mass had been mixed out.

Water in the tropical surface layers in general has significantly lower levels of dissolved oxygen than in the subtropics[84]. On moving southwards, this water is overlain by Subtropical Surface Water with a higher oxygen content and underlain by the deeper oxygen maximum of Subantarctic Mode Water (Figure 6.13), thus creating a shallow oxygen minimum. This layer of low oxygen extends to the south within the western margins of the Indian Ocean. Warren[232], as was noted previously, has suggested that this particular minimum may represent the effects of biological oxygen consumption due to decaying organic matter.

Nonetheless, the layer is characteristic of the core of the Agulhas Current Water. In sequential stations carried out from west to east along an isobath of the Agulhas Bank (Figure 6.14) it can be seen that this tropical signature can be used as a valuable tracer of Agulhas Current Water[531]. Chapman[231], by making use of all appropriate historical data, has been able to show that a layer, with a thickness of between 50 and 150 m, depleted in oxygen by about 1.1 to 1.5 ml/l occurs con-

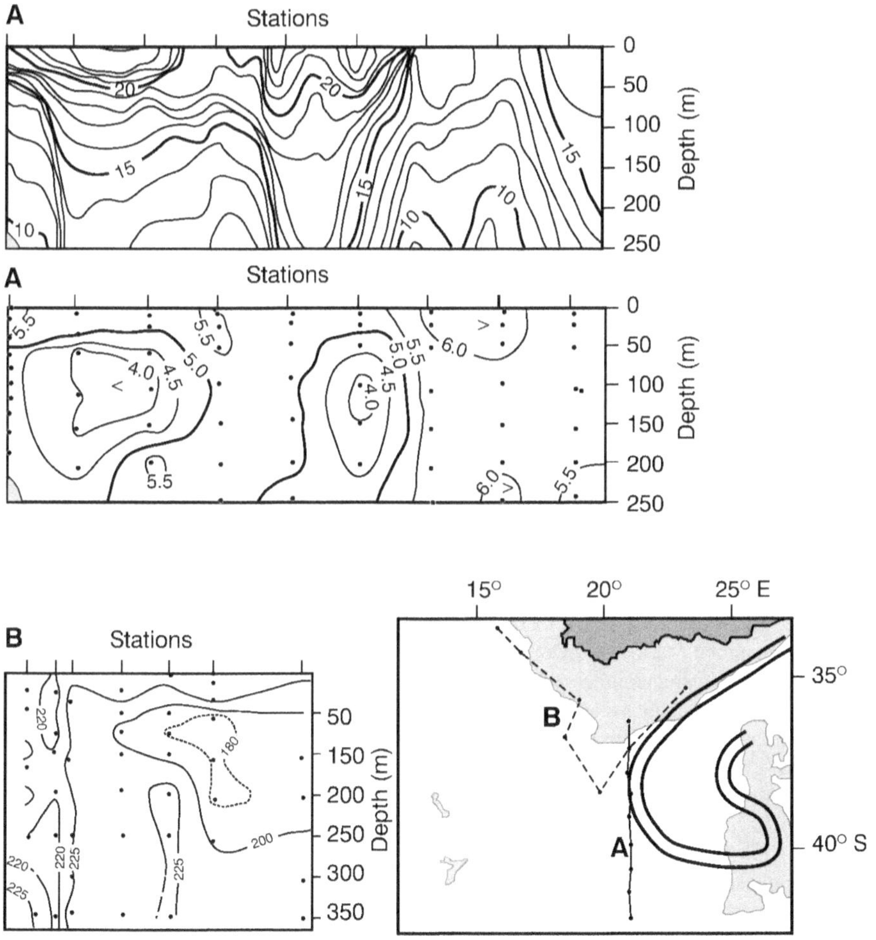

Figure 6.14. A temperature section (uppermost panel) and concomitant dissolved oxygen section (middle panel) across the Agulhas retroflection loop[231] showing the distinctive oxygen minimum layer (ml/l), centred at 100 m, that is associated with the core of the Agulhas Current. The lower panel shows an oxygen section along the southern African shelf break[531] (μmol/m^3) demonstrating the abrupt end of this shallow oxygen minimum at the termination of the Agulhas Current. The location of the sections is shown in the map. See also Figure 6.15.

sistently at the edge of the Agulhas Current, also along its retroflection loop. It is so characteristic of Agulhas Current water that it may be used to trace water from the Agulhas Current retroflection as far north as 32° S and 10° E in the South Atlantic Ocean[231].

Agulhas retroflection nutrients

A hydrographic section across the Agulhas retroflection loop (Figure 6.15) demonstrates that in this region the nutrient concentrations are usually inversely related to temperature. The lowest levels of phosphate and nitrate are thus seen to be associated with the outer rims of the loop, representing the Agulhas Current and the Agulhas Return Current. All kinematic products of the Agulhas retroflection, such as Agulhas rings, eddies and filaments, carry this signature with them. At the Subtropical Convergence the concentrations of these nutrients are much higher (Figure 6.15), Subantarctic Surface Water being characteristically higher in all nutrients except silicate. The outer edges of the Agulhas Bank are also shown to have higher levels of nutrient concentrations, probably as the result of the inshore upwelling between the Agulhas Current and the shelf slope (viz. Figures 5.2, 5.4, 5.9).

Water mass modifications

An inspection of precise temperature and salinity data from the Agulhas Current retroflection (e.g. Figure 6.12) show a number of significant outliers. Outliers in the low-salinity direction are for the greater part due to the effect of water from south of the Subtropical Convergence or from the South East Atlantic Ocean. Subantarctic water may make its presence felt by mixing processes along the lower limb of the Agulhas retroflection loop, i.e. along the confluence of the Agulhas Return Current and the Subtropical Convergence. Its major influence on the temperature–salinity characteristics of the region is probably a result of the spasmodic occurrence of wedges of subantarctic water moving northward into the region when an Agulhas ring is spawned (Figure 6.9). Within the thermocline of the Agulhas retroflection this subantarctic influence increases with depth[230].

Since the high, and seasonally persistent, heat fluxes from the ocean to the atmosphere are well known for this region[121,147], the possibility exists that thermohaline alterations to water above the thermocline would be evident in the water masses found here. This is indeed the case. Gordon et al.[230], using a quasi-synoptic, high-quality hydrographic data set, have found that upper thermocline water in the Agulhas retroflection, upon exposure to the colder overlying atmosphere, forms water that is anomalously salty to that of the Agulhas Current proper. Such modified water is found predominantly as thermostads within the Agulhas retroflection loop, but also in Agulhas rings[230]. Modifications of water masses in this region are particularly important for a number of reasons.

There is evidence that the Agulhas retroflection is a source region for Subtropical Mode Waters in the potential temperature range 17.4 °C to 17.8 °C[382] for the South Indian Ocean and, in a more modified form, for the South Atlantic. This exceptionally cold Subtropical Mode Water is found extensively in the eastern South Atlantic. The convective changes that bring about these modifications have been considered to be highly episodic, while there may be longer periods where the active mixing is restricted to a near-surface, wind-mixed layer[382]. Nevertheless, the water that has been cooled in the Agulhas retroflection is believed to be principally responsible for cooling the near-surface layers in the Indian Ocean and for ventilating thermocline and intermediate waters of the South West Indian Ocean[532]. Although it is considered difficult to use water mass indicators to trace water of South Indian origin in the South Atlantic, it appears possible that water altered at the Agulhas retroflection may play a determining role in the nature of the South Atlantic thermocline.

Fluxes in the Agulhas retroflection

The transport values of the Agulhas Current are known to increase downstream (Figure 6.16), but the rates of increase that have been estimated to date differ markedly, between 2.7×10^6 m^3/s per 100 km[380] to 6×10^6 m^3/s per 100 km[230]. Calculations of what this volume transport increase should be, based on the zonally integrated interior transport of the South Indian Ocean, driven by the known wind-stress curl, lie between 9×10^6 m^3/s per 100 km at 25° S to zero at 37° S[534]. With the adjustment of the wind stress values to more accurate ones, estimates of the volume flux of the Agulhas Current, based on the wind stress, at 37° S have been adjusted downward from 72×10^6 m^3/s[534] to about 55×10^6 m^3/s[187]. This reduces the downstream increase to 2×10^6 m^3/s per 100 km. The observed flux values as well as rates of downstream change are far in excess of those predicted by purely linear, thermohaline and wind-driven dynamics[533]. This has to date not been adequately explained.

The fluxes within the Agulhas retroflection itself have been calculated based on the hydrographic data collected during only a few suitable cruises.

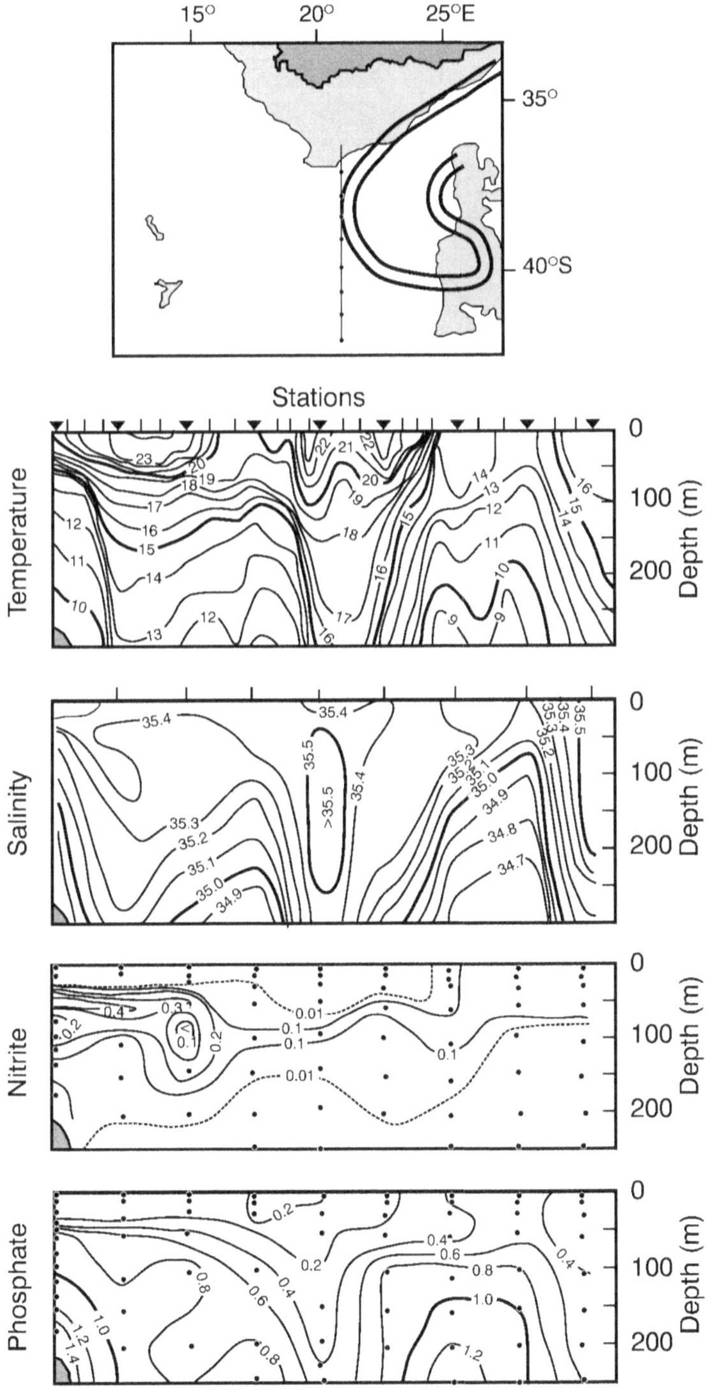

Figure 6.15. A hydrographic section across the Agulhas retroflection loop[531] showing the temperature, salinity, nitrite and phosphate (µmol/m^3) for this feature. The accompanying map shows the location of the section relative to the retroflection at that time. The inverse relationship between temperature and nutrient concentration for these waters is immediately discernible. See also Figure 6.14.

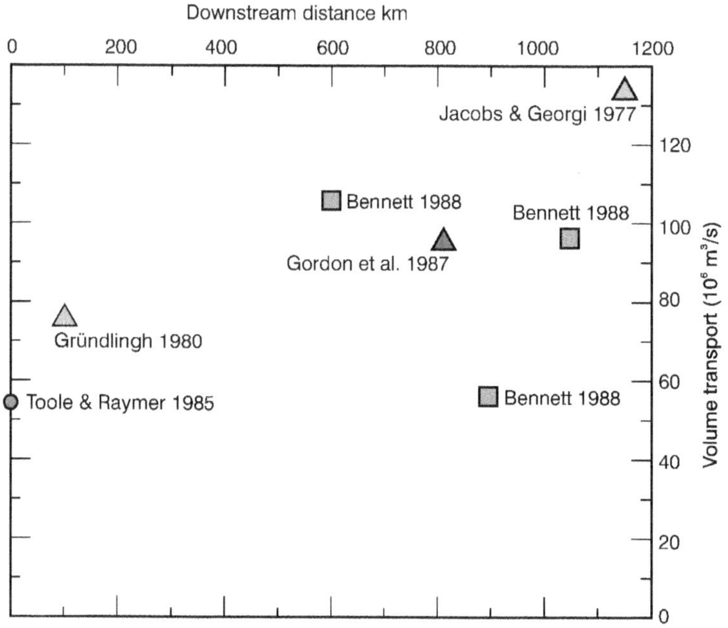

Figure 6.16. Volume transport values for the Agulhas Current downstream of 30° S[533]. The Agulhas retroflection lies downstream of 900 km in this figure. Individual papers referenced here by author(s) and year can be found in the Bibliography.

Gordon et al.[230] have estimated that 70 × 10^6 m^3/s above 1500 m passed into the Agulhas retroflection region through the Agulhas Current (95 × 10^6 m^3/s relative to the sea floor) during one particular cruise. Of this, 10 × 10^6 m^3/s continued to flow west; the rest joined the retroflection proper. There was a recirculation, within the loop of the Agulhas retroflection, of 15 × 10^6 m^3/s. About 55 × 10^6 m^3/s left the retroflection as the Agulhas Return Current. Early, upstream retroflections, local recirculation and time biases in the coverage of the region during one particular cruise make it notoriously difficult to draw up a balanced budget for the water masses entering and leaving the Agulhas retroflection region (e.g. Figure 6.16).

Agulhas rings

The presence of intense vortices near the southern tip of Africa has been surmised[46,59] from or observed[75,511] in hydrographic data for a long time. In Figure 6.5 the dynamic topography of the Agulhas retroflection region and vicinity clearly shows, for instance, the presence of such a substantial eddy south-west of Cape Town. It may be assumed to have been an Agulhas ring and to have had its inception in the Agulhas retroflection. It had a characteristic diameter of about 400 km and a volume transport of 5 to 10 × 10^6 m^3/s to 1100 decibar[92]. On one cruise[230] the spawning of such a feature has actually been observed[61] and the nature of a newly formed ring could be established in detail (Figure 6.17).

Ring characteristics

First, having been recently spawned from the Agulhas Current, the surface expression of an Agulhas ring is that of a warm annulus with Agulhas Current surface water clearly distinguishable as a circular ribbon of high temperatures[61] with a tell-tale, subsurface oxygen minimum[531]. Hence the designation *Agulhas ring*. These characteristics are not seen in other mesoscale eddies cast off from the Agulhas Current[458], particularly along the Subtropical Convergence[63]. These latter features are therefore preferably called *Agulhas eddies*. However, Agulhas rings do not retain this distinctive surface structure for a long time. Convective mixing due to severe heat loss to the atmosphere[98], as well as substantial mixing due to high levels of wind stress, rapidly erases all surface contrasts between rings and their surroundings in the South East Atlantic Ocean. Before this happens to a substantial degree, surface expressions represent the dimensions of Agulhas rings quite well.

Based on more than 600 thermal infrared images

from the METEOSAT satellite, it has been established that, at the sea surface, the average diameter of an Agulhas ring is 324 km, with a standard deviation of 97 km[414]. An investigation on the characteristics of Agulhas rings sufficiently robust to reach the Walvis Ridge[712] has shown that the diameters of these features have no noticeable seasonal variations. Agulhas rings closer to Cape Town that have lost their contrasting surface expression, but that are circumscribed by encircling Agulhas filaments[440], have diameters of 307 (±89) km. This apparent reduction in size may be an artefact of the Agulhas filaments partially overlying Agulhas rings[92], thus making them seem smaller than they are.

Characteristics of a newly spawned ring have been observed at sea a number of times[784–5]. Detailed observations of a ring in March 2000[785] have shown that it had a maximum anomaly of sea surface height of 0.70 m, a radius of 120 km and a volume of 38×10^{12} m^3. A strong azimuthal current of about 1 m/s gave this ring a kinetic energy of 18×10^{15} J. The ring was strongly baroclinic, but also had a significant barotropic component. The hydrographic structure of the ring as well as its velocity extended down to a depth of 4500 m (Figure 6.18). This is the first solid evidence that Agulhas rings extend to such depths. This does not of necessity imply that all individual rings are this deep. The Agulhas Current itself at times extends to the bottom[367], but at other times is much shallower[738]. One would expect this therefore to be true of Agulhas rings as well. The hydrographic properties of the ring so meticulously surveyed in March 2000 differed from those of the ambient waters only at temperatures greater than 12° C. Based on a number of criteria, this specific ring has been considered a very large one[785]. An indication of the range of diameters to be observed is evident from Figure 6.19.

The two rings observed during a particular cruise (Figure 6.17) had different shapes and sizes. The older, more northern ring, was more nearly circular and had a diameter of about 200 km. The southern one was more elliptical with minor and major diameters of 100 km and 250 km respectively[230]. When ring dimensions are defined by their average diameter of maximum radial velocity, they seem smaller. Duncombe Rae[127] has carried out a statistical analysis of 18 rings observed in the general vicinity of Cape Town (Figure

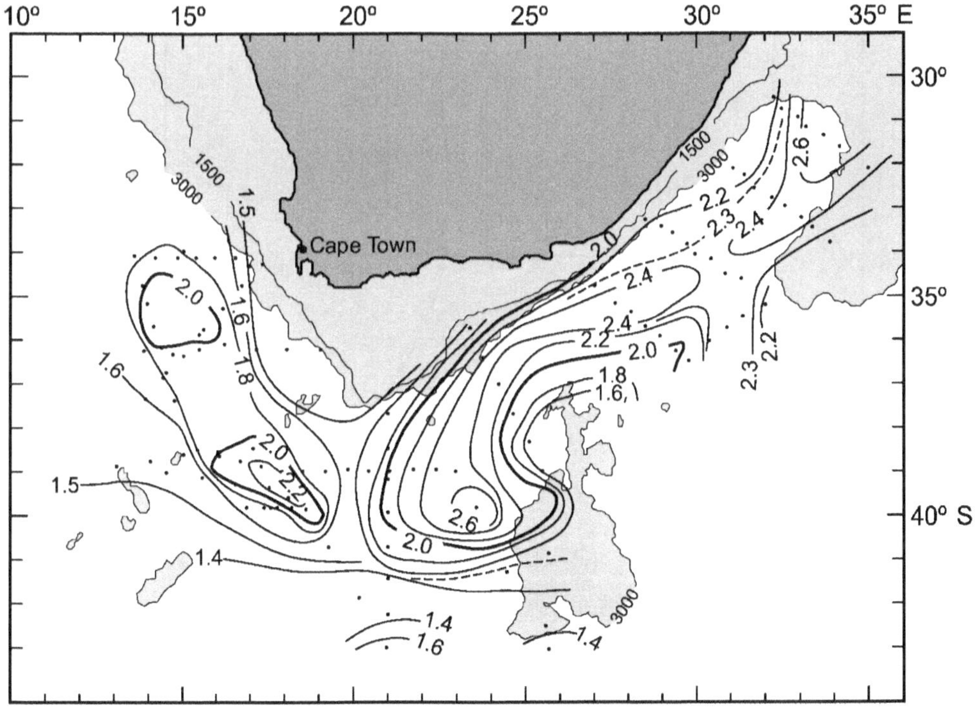

Figure 6.17. The dynamic height anomaly of the sea surface relative to 1500 decibar, given in dynamic metres[230]. The Agulhas retroflection loop is clearly circumscribed, as well as two Agulhas rings to the west of the retroflection. Regions shallower than 3000 m are indicated by shading.

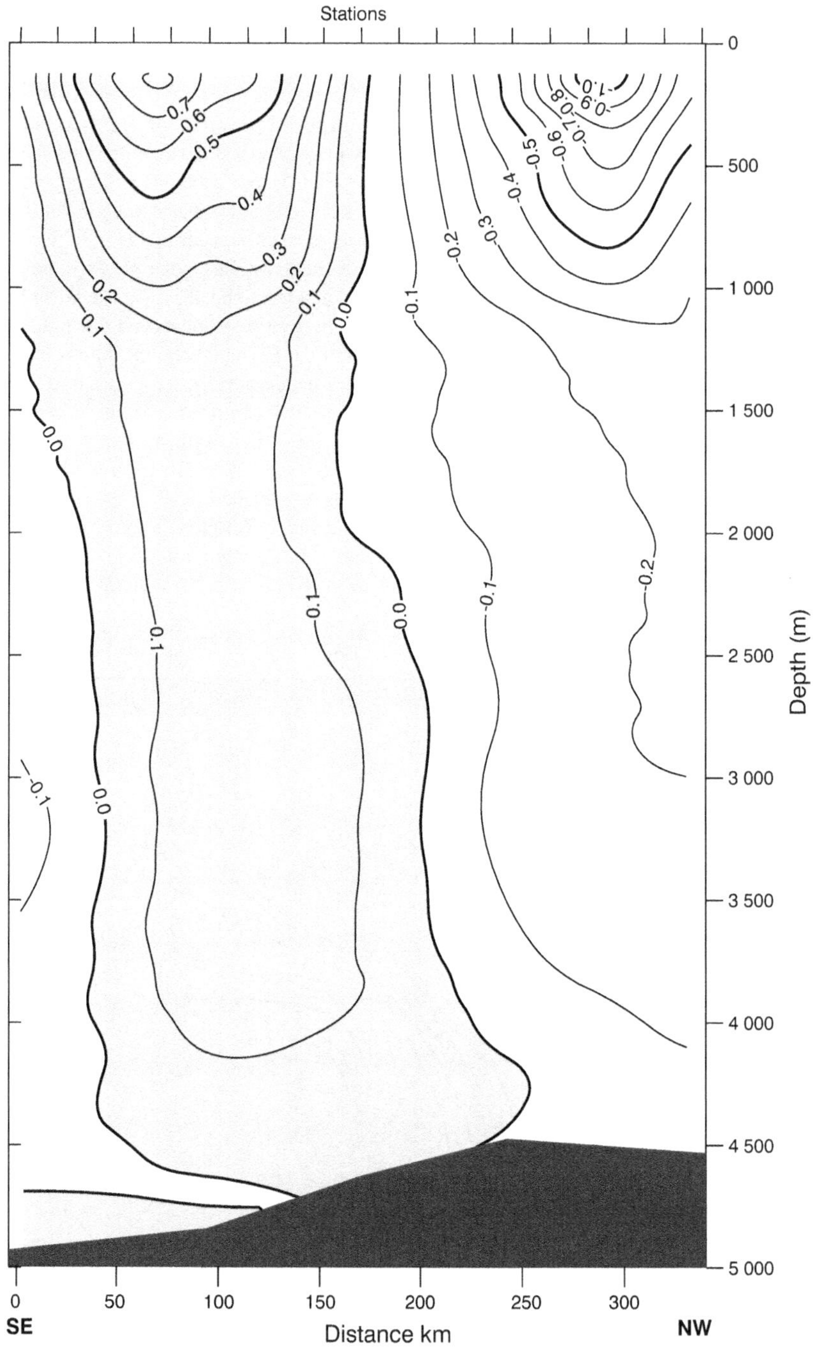

Figure 6.18. The full velocity structure of a relatively young Agulhas ring in the Cape Basin[785]. The speeds are in m/s. The shaded region indicates movement to the north-east. Speeds of slightly less than 0.1 m/s were found right down to the sea floor on this occasion.

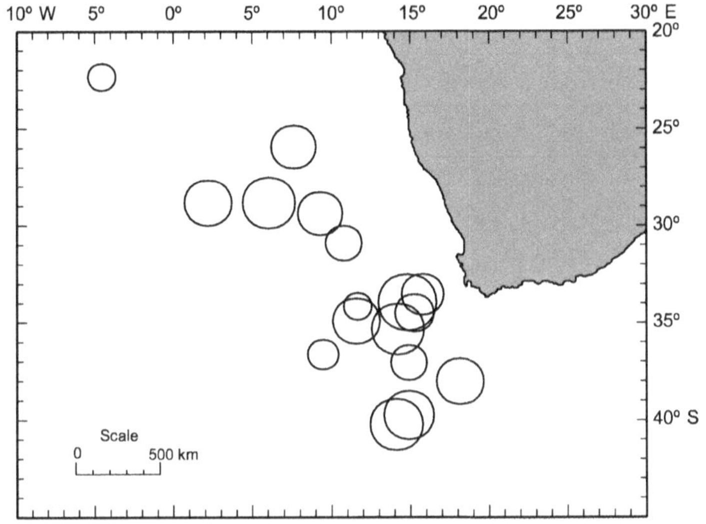

Figure 6.19. A compendium of Agulhas rings, their diameters and their geographical locations from eleven independent sets of hydrographic measurements[127] south-west of South Africa.

6.19) and, according to these hydrographic data, estimates a mean diameter of 240(±40) km. This is probably the most reliable assessment of the dynamic dimensions of the rings to date. The mean depth of the 10 °C isotherm, a good proxy for the dynamic topography of these features[129], is 650 (±130) m for this set (Figure 6.19) of rings. The range of a number of variables connected to Agulhas rings is given in Table 6.3. The geographic distribution of this set of rings suggests a general north-westerly drift.

Rings translating

A subsequent survey of eddies in the Cape Basin[535–7], during three major cruises, located seven eddies in the region[520]. Five of these were positively identified as Agulhas rings. The passage of these rings was associated with depressions in the 10 °C isotherm lasting from 100 to 400 days at a particular spot. After the passage of a ring the thermocline appears to shallow appreciably before relaxing to the local mean depth for that temperature[520], suggesting the passage of an attendant, but smaller, cyclonic eddy.

The depth to which particles are actually trapped in an eddy, and thus move with it, may depend on the ratio between the azimuthal speed and the translational speed of the eddy[538]. Based on the advection rates that have been observed for rings to date[94,464] or calculated[539], it has been estimated[127] that the trapped depths lie between 670 m (for the highest drift speeds) and 110 m. This would imply that, although depressed isotherms may indicate a ring depth of 4000 m or more, only intermediate and shallower waters are carried along, the rest of the signal progressing as a wave in the density field only. It still needs to be established whether this theoretical limitation on the trapped depths of Agulhas rings actually applies. Detailed modelling[271] has suggested that the baroclinic velocity of an Agulhas ring would exhibit an inversion at a depth of about 1250 m (Figure 6.20). There are measurements that seem to show this, but it seems that the deep velocity structure of a ring may very much be a function of its age.

Ring distributions

Using this perturbation of the temperature field to identify mesoscale features in the South Atlantic Ocean during the abovementioned three dedicated cruises has resulted in the overall distribution of eddies for the Cape Basin[520] given in Figure 6.21. These cruises were carried out over a 17 month period[535] in 1992 and 1993 to survey the region, making it a relatively synoptic survey. The distribution is not dissimilar to that portrayed in Figure 6.19. It was shown[517] that two to six rings co-existed in the Cape Basin at any one time. Subsequent altimetric studies[465] as well as investigations with floats[627] have largely substantiated these numbers, but shown that they may vary considerably from year to year. It has for instance been estimated[628] that during the KAPEX endeavour nearly 12 Agulhas rings were to be found in the south-eastern Cape Basin. The abovementioned hydrographic surveys have, how-

Table 6.3. A compendium of ring parameters for a number of published observations of Agulhas rings[785]. V_{max} is the maximum tangential speed, L_{max} is the radius where this maximum tangential speed is found, Vol_{10} is the volume of the ring above the 10° C isotherm, APE is the available potential energy relative to the depth of the thermocline outside the ring, KE is the kinetic energy if the ring is considered as consisting of two layers, AHA is the integrated heat excess above the 10° C isotherm (relative to a temperature profile which is representative for the surrounding water) and ASA is the integrated salinity excess above the 10° C isotherm (relative to a salinity profile which is representative for the surrounding water).

Source	Programme	Ring	V_{max} [m/s]	L_{max} [km]	Vol_{10} [10^{12} m^3]	APE [10^{15} J]	KE [10^{15} J]	AHA [10^{20} J]	ASA [10^{12} kg]
Van Aken et al. (2003)[785]	MARE-1	Astrid	1.0	120	38	20	18	0.8	4.1
Van Ballegooyen et al. (1994)[129]	SCARC	A3		160	34			1.5	8.7
		A4		140	33			2.4	13.1
		A5		95	11			0.7	4.4
		A6		125	17			1.1	4.6
Olson and Evans (1968)[94]	ARC	RE	0.9	130	26	51	9		
		CTE	0.6	115	30	31	6		
Duncombe Rae et al. (1996)[520]	BEST	B1–1	0.4	85		17	2	0.2	1.2
	1992–1994	B2–2	0.5	65	17	11	2	0.6	3.8
		B2–3	0.3	85		7	1	0.6	3.7
		B2–4	0.3	95		23	2	0.4	1.7
McDonagh et al. (1999)[542]	A11	Ring 1	0.6	71	15	5	1	0.4	2.4
		Ring 2	0.8	75	24	8	3	0.2	1.4
Garzoli et al. (1999)[783]	KAPEX	Ring 1	0.4	100	31	43	1	0.9	
	1997	Ring 2	0.2	110	42	52		1.3	
		Ring 3	0.3	100	19	15		0.8	

ever, turned up some novel features that have not been observed here before and that are of considerable importance.

First, a number of cyclonic eddies have been found on the periphery of the retroflection region (viz. Figure 6.21). Modelling of this ocean region with high spatial resolution[273] has suggested the production of dipole pairs of eddies – one anti-cyclonic, the other cyclonic – at the Agulhas retroflection. It is not clear whether the cyclonic features observed during this set of cruises of 1992 and 1993 support the simultaneous shedding of eddies with opposing flow directions. For a proper understanding of the vorticity balance of the ring-spawning process as well as the long-term stability of Agulhas rings it would be of crucial importance to establish the formation mechanism for these cyclonic eddies. There are theoretical results that suggest[654–5] that baroclinic rings, accompanied by weaker cyclones, would inherently have a greater degree of stability.

Subsequent investigations[627–31] have indicated a totally new and important phenomenon in this regard: independent Cape Basin cyclones. It has been demonstrated[627–8] that cyclones are an ubiquitous component of the circulation in this part of the South East Atlantic Ocean. They move in a south-westerly direction[626,628] from the edge of the African continental shelf, crossing the average north-westerly path of Agulhas rings with advection speeds of 3–5 cm/s, very similar to those of Agulhas rings[628]. This phenomenon is most evident in regions surrounding the Agulhas retroflection loop and not at the retroflection itself where Agulhas rings are first formed, indicating that these cyclones do not necessarily form part of the ring-shedding process. The criss-crossing movement of Agulhas rings and cyclones is in agreement with simple theories[538] for vortex propagation on a β-plane[626] through a weak background flow. This crossing of the paths of cyclonic and anti-cyclonic eddies has also been observed[626] in other, similar ocean regions, such as the North East Pacific Ocean and the South East Indian Ocean. What are the currently known characteristics of these Cape Basin cyclones, and what role might they play in the mixing of water from Agulhas rings in this basin?

Cape Basin cyclones

It would seem that three, perhaps related, cyclone types are to be found in the Cape Basin. The first are eddies that were previously imbedded in Natal Pulses[629, 632], the second lee eddies shed from the western side of the Agulhas Bank[630–1] and the third type cyclones shed

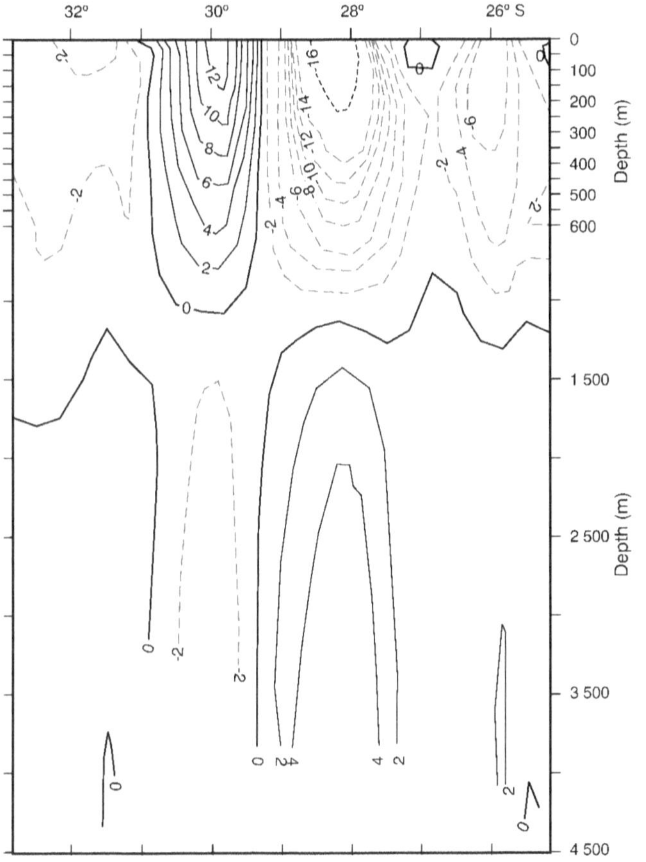

Figure 6.20. The baroclinic speed in a section across a modelled Agulhas ring[271]. Speeds are in cm/s with solid lines representing motion towards the reader. The model suggests that the baroclinic motion will exhibit a velocity inversion at depths greater than 1000 m, but with greatly reduced speeds.

(Below) Figure 6.21. Distribution of mesoscale disturbances to the depth of the 10 °C isotherm in the Cape Basin of the South Atlantic[520]. These locations are based on measurements undertaken during three cruises as part of the BEST series in 1992 and 1993. Thin lines denote the tracks of these cruises. Circulation features are denoted AR (Agulhas Current ring), BR (Brazil Current ring) and C (cyclonic eddy). The temperature–salinity characteristics of the postulated Brazil Current rings, denoted BR1 and BR2, are given in Figure 6.25.

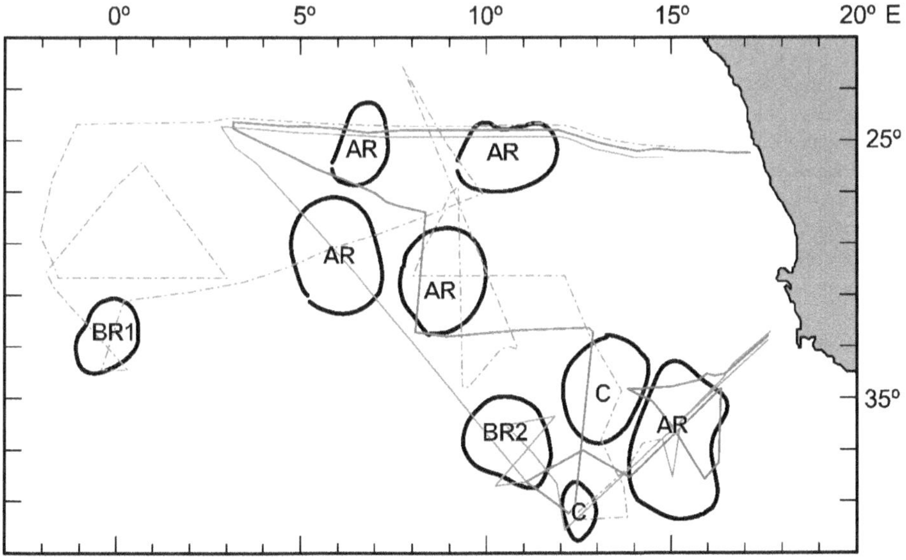

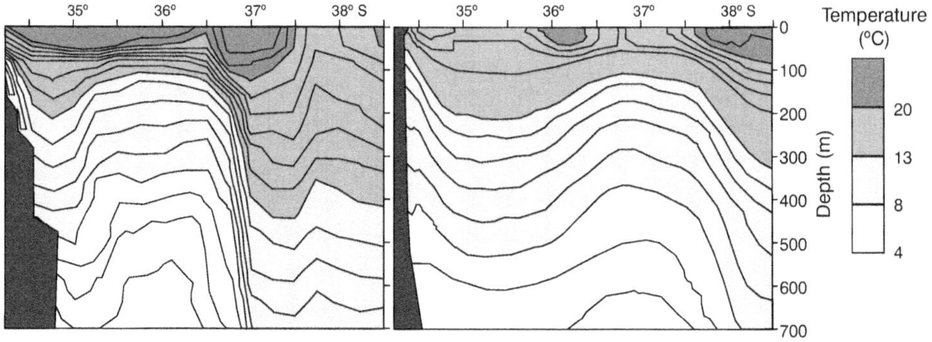

Figure 6.22. Vertical temperature sections across a characteristic eddy generated off the western side of the Agulhas Bank, south-west of Cape Town. The left-hand panel represent actual hydrographic observations; the right-hand panel temperature distributions simulated by a regional model[630].

from the continental shelf edge of south-western Africa[627–8].

The general behaviour of Natal Pulses has always suggested[62,399] that this special meander of the Agulhas Current has an embedded cyclone in its core (viz. Figures 4.24 and 5.15). As mentioned before, it has even been suggested that this particular cyclone originates from the lee eddy often present off Durban[315, 169] (viz. Figures 4.20 and 4.21), just south of the Natal Bight. More recent observations, including current meter records, float trajectories and satellite remote sensing products have demonstrated unequivocally[632] that the cyclone in the loop of a Natal Pulse is persistent along its trajectory and extends to the full depth of the Agulhas Current. What becomes of this well-developed cyclone once a Natal Pulse passes the tip of the Agulhas Bank south of Africa?

Studies combining RAFOS floats and satellite observations have shown[629] that such cyclones are shed into the South Atlantic Ocean and are sometimes instrumental in triggering the occlusion of an Agulhas ring (Figure 6.23). They may even move poleward through the gap between the newly shed Agulhas ring and the Agulhas retroflection loop. In many models[763] the shedding of Agulhas rings is accompanied by the formation of an accompanying cyclone. Once in the Cape Basin cyclones of this category seem to disintegrate rapidly. This behaviour of cyclones in Natal Pulses is mirrored closely by shear edge features of the southern Agulhas Current[629] when they move past the southern tip of the Agulhas Bank. Both these sets of cyclones may have an effect on the circulation on the western side of the Agulhas Bank.

It has been shown[630] that cyclonic motion is a recurrent, but not persistent, feature of the southern part of the shelf edge of the western Agulhas Bank. This motion in the lee of the bank is driven by the passing Agulhas Current and can be quite easily modelled[630–1]. In such a model[633] leakage of vorticity from shear edge – or border – eddies along the eastern edge of the Agulhas Bank (viz. Figures 5.7 and 5.8) feeds into the lee eddy at irregular intervals enhancing it spasmodically. The modelled lee eddy bears a very strong resemblance to those measured at sea (Figure 6.22). Both in the model and in observations this lee eddy is eventually shed into the Cape Basin where it may interact vigorously[628] with Agulhas rings and with other cyclones. In a primitive equation model[761] with a spatial resolution of $1/6° \times 1/6°$, such cyclones are usually paired with Agulhas rings in dipolar or even tripolar structures[760]. The influence of border eddies from the eastern side of the Agulhas Bank on lee eddies as well as on the subsequent behaviour of lee eddies is shown schematically in Figure 6.23.

This figure is based on sea-surface height anomalies and on the simultaneous movement of RAFOS floats placed in two groups; the first at roughly 400 m depth, the others at an average of 800 m depth. The behaviour of both sets of floats was very similar in these cyclones, indicating their coherent depth structure. As can be seen from Figure 6.23 (first panel) floats on this occasion spent some time in border eddies on the eastern side of the Agulhas Bank. At some later stage they moved downstream and were caught up in a vigorous lee eddy (Figure 6.23, third panel). On subsequently breaking away from the shelf edge (Figure 6.23, last panel), this lee eddy crossed the Agulhas retroflection loop, cutting off an Agulhas ring in the process. The south-westward direction of drift of the lee eddy is entirely characteristic for such features; there is insufficient evidence to warrant its influence on ring shedding as equally distinctive. Of importance here is to note that the modelled[631] simulation, that shows vorticity being transferred from border eddies on the eastern side of the

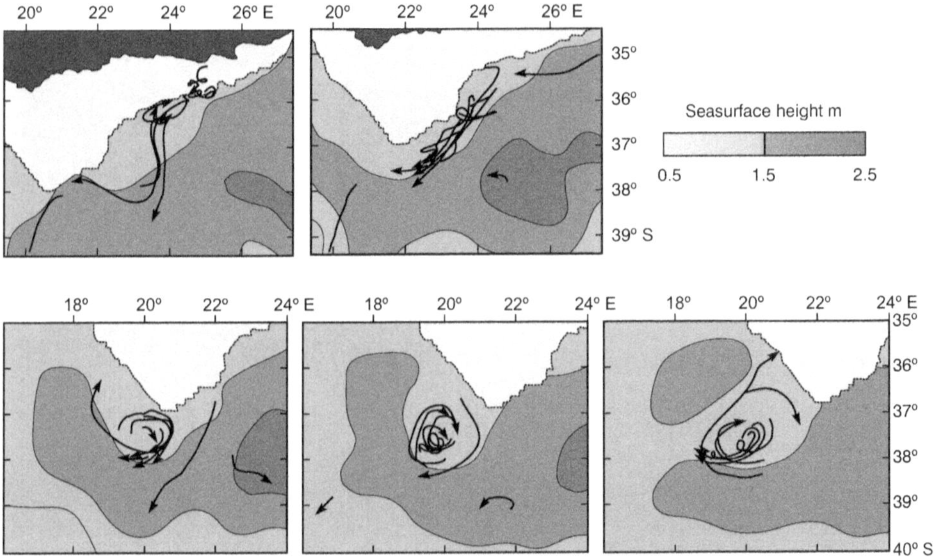

Figure 6.23. Evidence at depth for the role of a shear edge or border eddy in the detachment of an Agulhas ring from the Agulhas Current retroflection[629]. The tracks of RAFOS floats at intermediate depths are shown in these panels relative to concurrent sea height anomalies. Arrows are given at the leading edges of each float track and show the location of the float on the day given. Anti-cyclonic motion is denoted by the grey scale and the land mass deeper than 1000 m is shown as light grey. Panels follow in the conventional sequence.

Agulhas Bank to a lee eddy on its western side, has therefore in fact been observed. However, a study during the period 1997–1999[629] suggested that of six cyclones identified on the landward side of the southern Agulhas Current, two dissipated at their site of formation; four propagated downstream. It therefore seems clear that not all border eddies feed the Agulhas Bank lee eddy. Of especial relevance is the role border eddies may at times play in the triggering of ring shedding events.

The last of the three types of cyclones being discussed: Natal Pulse eddies, Agulhas Bank lee eddies and Cape Basin cyclones, is the least understood. Cape Basin cyclones in most respects behave in a manner very comparable to the lee eddy described above, except that they are formed at the shelf edge equatorward of the Agulhas Bank[628] and therefore do not seem to be directly driven by the Agulhas Current. In general they have a diameter of 120 km, smaller therefore than Agulhas rings that have a typical diameter of 200 km. They are much weaker than Agulhas rings with which they co-exist in the Cape Basin. According to float observations at intermediate depths, Cape Basin cyclones on average exhibit kinetic energy levels 60 per cent less than Agulhas rings[628]. During 1997, 22 such cyclones were observed in the Cape Basin, seemingly a representative figure. The mean zonal velocity of Cape Basin cyclones westward was 3.6(±0.6) cm/s and the mean poleward drift 0.4(±0.5) cm/s (Figure 6.24). Azimuthal speeds measured to date reach 22 cm/s. The lifetime of these cyclones is less than two to three months[628], thus much shorter than that of many Agulhas rings. Because of their short lifetime and the south-westward direction of their drift, few of them are found in the northern part of the Cape Basin[627] and only one has to date been observed[781] to cross the equatorward border of this basin, the Walvis Ridge. The importance of Cape Basin cyclones lies predominantly in their presumed role in mixing of eddy and ring features within this basin.

Notwithstanding their short lifetimes, the number of cyclones in the Cape Basin at any one time would seem to exceed that of Agulhas rings by a ratio[628] of 3 : 2. They therefore are quantitatively an important component part of the circulation here. The periods floats, at intermediate depths, spend in these cyclones (44 days) is similar to that of floats in Agulhas rings (41 days). This short trapping period suggests that there is substantial mixing of water from within both types of features with ambient water masses. It has also been observed[628] that floats are frequently exchanged between Agulhas rings and Cape Basin cyclones, evidence of the entrainment and detrainment of water between these features. This implies that the presence of large numbers of cyclones in the Cape Basin substantially enhances the mixing of

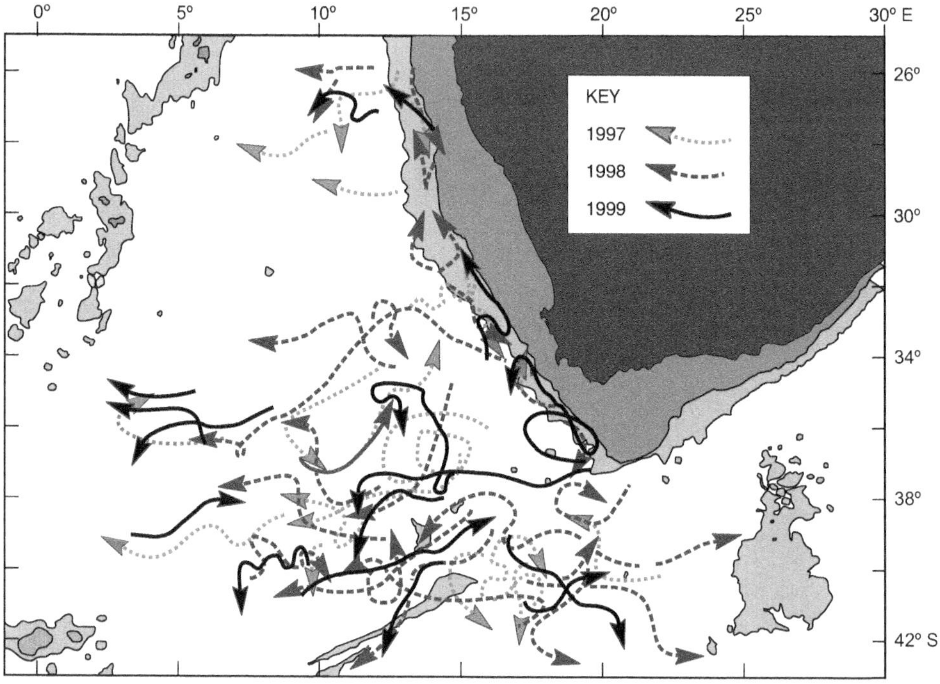

Figure 6.24. The tracks of cyclones in the south-eastern Atlantic Ocean for the period 1997–1999[628] as inferred from sea surface height anomalies. Bottom topography is described by the 1000 m and 2000 m isobaths.

water from Agulhas rings here, thus shortening their lifetime in the south-eastern part of the basin. For this reason it has whimsically been referred to by Boebel et al.[628] as the Cape Cauldron.

Brazil rings?

The second set of unusual, and unexpected, mesoscale features observed during cruises in the Cape Basin are eddies that have hydrographic characteristics somewhat different from those expected of standard Agulhas rings[540]. It has been claimed that these anti-cyclonic features have their origin at the confluence of the Brazil and Falkland Currents in the western Atlantic Ocean and are advected eastward with the high-latitude limb of the South Atlantic gyre until they reach the position shown in Figure 6.21. This somewhat startling conclusion has been based on two sets of data.

First, the potential temperature–salinity characteristics of these two particular eddies below a depth of 600 m, i.e. well away from possible atmospheric influences, is more closely comparable to that of the Brazil Current than that of the Agulhas Current. This is shown in Figure 6.25. Secondly, comparing different CFC tracers, some of which have a declining concentration in the atmosphere, some increasing, an age for water in a feature can be calculated. From such a comparison, the age of ring BR1 (Figure 6.21) has been estimated at three years[540]. With a known eastward drift of 20 to 30 km/day in the southern limb of the subtropical gyre of the South Atlantic, this could place this particular ring at the Brazil Current retroflection when it was formed. Others have come to different conclusions on the origin of these unusual rings.

It has subsequently been shown that the ring that has been claimed to have crossed the South Atlantic from the Brazil Current came from the Agulhas retroflection where it had been spawned 16 months earlier[541]. This conclusion is supported by both altimetric and thermal infrared data for the region for this period. The hydrographic and kinematic nature of this anomalous ring was in all respects akin to that of neighbouring Agulhas rings, except for the higher oxygen and lower nutrient concentrations in its core. Since these are consistent with the characteristics of Subantarctic Mode Water, formed at the Subtropical Convergence in austral winter, it has been suggested[542] that this front is the origin of this water that may occasionally be found within the Agulhas Current. This suggestion has also been put forward by others[783] who have investigated a number

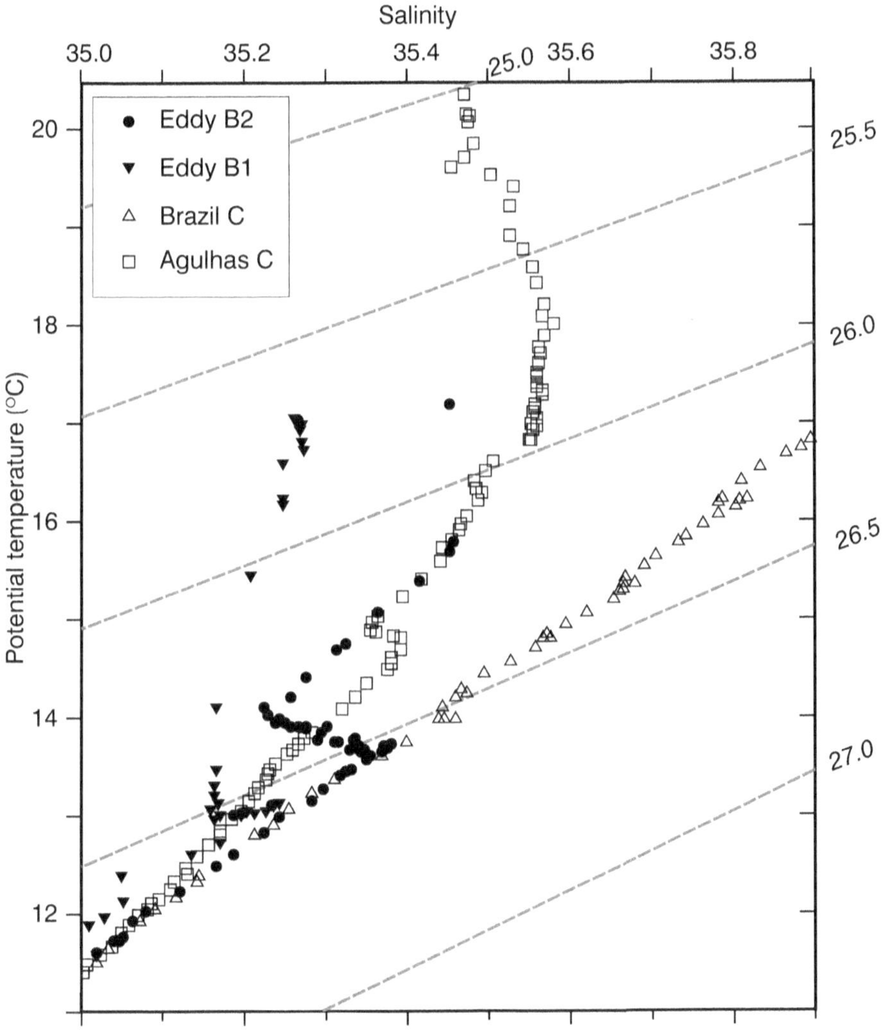

Figure 6.25. Temperature–salinity characteristics of two eddies (B1 and B2) observed in the South East Atlantic Ocean[520] that were believed to have their origin as Brazil Current rings[540]. Their geographic locations are shown in Figure 6.21. These temperature–salinity characteristics are juxtapositional to those distinctive of the Agulhas Current and the Brazil Current respectively.

of rings and found some with high saturations and concentrations of oxygen in a thermostad at a depth between 600 m and 1100 m. Their analysis has led to the tentative conclusion that this particular ring had incorporated a lens of water from the Subtropical Convergence by coalescing with another eddy. An entirely conclusive answer on whether Brazil rings can reach the southeastern Atlantic has therefore not yet been given.

Nonetheless, the possibility of Brazil Current rings remaining essentially intact for what must be a considerable period must be largely dependent on the rate of mixing between the rings and the ambient water masses. This rate of mixing is currently not known, but will most likely be a function of the speed of rotation of these features. This argument will of course also hold for Agulhas rings.

Ring kinematics

Maximum radial speeds of Agulhas rings lie between 0.29 m/s and 0.90 m/s with an average[127] of 0.56 m/s. The azimuthal velocity around a ring, as a function of radius from ring centre, is very variable[94] (see Table 6.3) Agulhas rings do not come with a strongly constrained range of physical characteristics. The available potential energy of Agulhas rings adequately measured to date lies between 26 and 51 × 10^{15} J. Altimetric data suggest[517] up to 70 × 10^{15} J. The kinetic energies fall between 2.3 and 8.7 × 10^{15} J.

Observations of the absolute velocity field of Agulhas rings, using a sophisticated combination of acoustic Doppler current profilers, Global Positioning Systems and high-quality hydrographic data[537] have demonstrated that as much as 50 per cent of the total flow in the core of a mature ring is barotropic. The radius of maximum velocity in this presumed ring had shrunk to only 60 km. On the basis of some of these results Olson and Evans[94] and others[517] have concluded that Agulhas rings are the most energetic in the world and that one ring, by itself, may contribute up to 7 per cent of the annual input of energy by the wind for the entire South Atlantic basin. What does such an intense eddy look like hydrographically? This is shown in Figure 6.26.

Ring hydrography

The general vertical portrayal of this ring is characteristic of others observed in the region[543]. First, two warm lobes of the annulus of warm surface water are still intact. This is not represented in the salinity distributions, but is evident in the subsurface oxygen minimum that is more clearly seen on the inshore side of the ring. Secondly, phosphates are in general low, but in the extensive 16 °C to 17 °C thermostad high nitrite values are to be found (Figure 6.26). This thermostad, fully saturated with oxygen, is believed to be due to cooling and vertical convection of the Indian Ocean water in the centre of the ring[65] over an extended period. Heat losses in early spring of 157 W/m^2 for an Agulhas ring have been estimated[61]; 80 W/m^2 in autumn[543]. Evaporation and convection lead to increased salinities in the surface layers[94]. Observations of Agulhas rings much farther afield[544] accentuate the marked effect on their hydrographic structure of interaction with the atmosphere (Figure 6.27) shortly after they have been spawned.

These rings have been found and properly surveyed[544] by Arhan et al. well beyond the Walvis Ridge in the Angola Basin. One, here called R_2 (Figure 6.27), had a diameter of about 500 km and a core temperature of 17.1 °C. The core of its thermostad lay at a depth of 150 m. Ring R_3 by contrast was only 100 km in cross-section, had a core temperature of only 13.5 °C and its thermostad depth was about 400 m. This is very similar to a ring found even farther west on another cruise[542] and that was assumed to have come from the Brazil Current. From satellite altimetry it could be determined that the former two vortices were the products of one ring, spawned at the Agulhas retroflection about two years earlier. This mother ring had split at the Erica Seamount. Such splitting of Agulhas rings has been inferred for the Vema Seamount as well[465].

On having split at the Erica Seamount, R_2 moved off rapidly into the South Atlantic Ocean, whereas R_3 got stalled in the retroflection region for the full winter. This explains the estimated extra heat loss responsible for a much cooler thermostad in R_3, its lower core salinity and considerably higher dissolved oxygen content (see Figure 6.27). These results all point to the substantial changes in the ring configuration that are driven from the sea surface. Interaction with ambient water masses will also eventually diminish such a feature until it is totally absorbed. This will influence the natural lifetime for an Agulhas ring.

Ring durability

Various estimates of the life-time of Agulhas rings have been made, covering periods from five to ten years[94]. Byrne et al.[95] have calculated the dissipation rates of some Agulhas rings based on both altimetric data and on potential energy estimates of fortuitous measurements of hydrographic anomalies in the South Atlantic that were assumed to be the remnants of Agulhas rings. They have estimated a reduction in surface elevation of rings of 85 per cent over a distance of 5000 km, roughly similar to the potential energy decline over the same distance (Figure 6.28). An e-folding distance of 2600 km[362] to 3000 km[95] seems to apply. These results imply a residence time for Agulhas rings of three to four years[95] in the South Atlantic Ocean.

Using a much more extensive set of sea surface height anomalies as observed by satellite altimetry, other investigators[465] have shown that Agulhas rings dissipate very rapidly in the Cape Basin, losing more than 50 per cent of their sea level expression within four months (Figure 6.29). What is more, more than 40 per cent of all rings thus identified never leave the south-eastern Atlantic Ocean, but seem to disintegrate completely in the Cape Basin. This site-specific diffusion of the anomalous characteristics of such rings will have considerable implications for the nature and circulation of the South Atlantic Ocean. It would mean that nearly 70 per cent of the excess heat, salt and anti-cyclonic vorticity leaked from the South Indian Ocean is absorbed exclusively in this particular corner of the South Atlantic Ocean and subsequently has to make its way equatorwards by a different mechanism than being bodily carried by Agulhas rings.

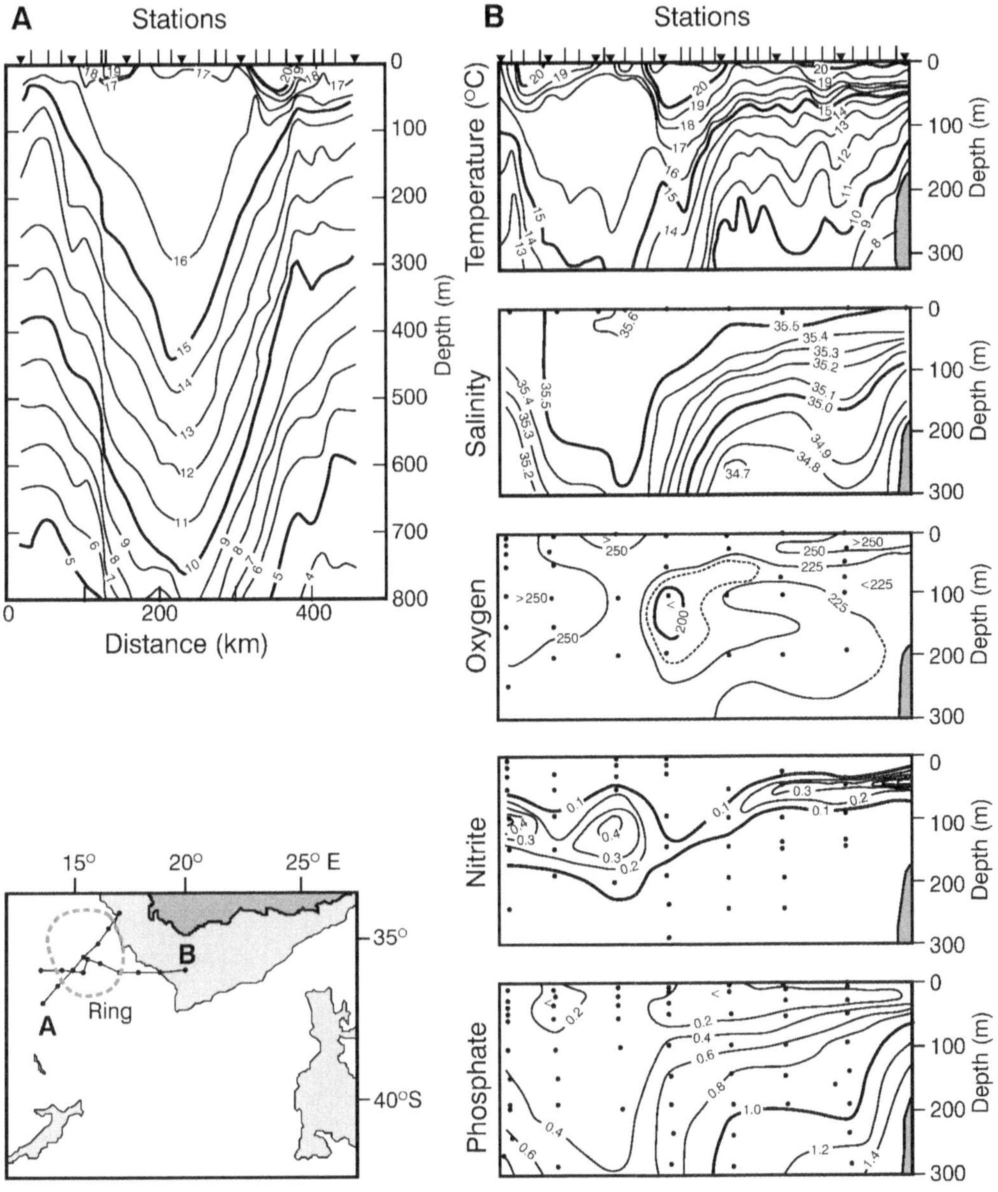

Figure 6.26. Two hydrographic sections across an Agulhas ring off Cape Town[230,531]. The locations of the ring and the sections are shown in the locational map. The hydrographic characteristics of such a ring in the upper layers are portrayed. Both nutrients and dissolved oxygen values are in μmol/m^3. Water depths shallower than 3000 m are shaded.

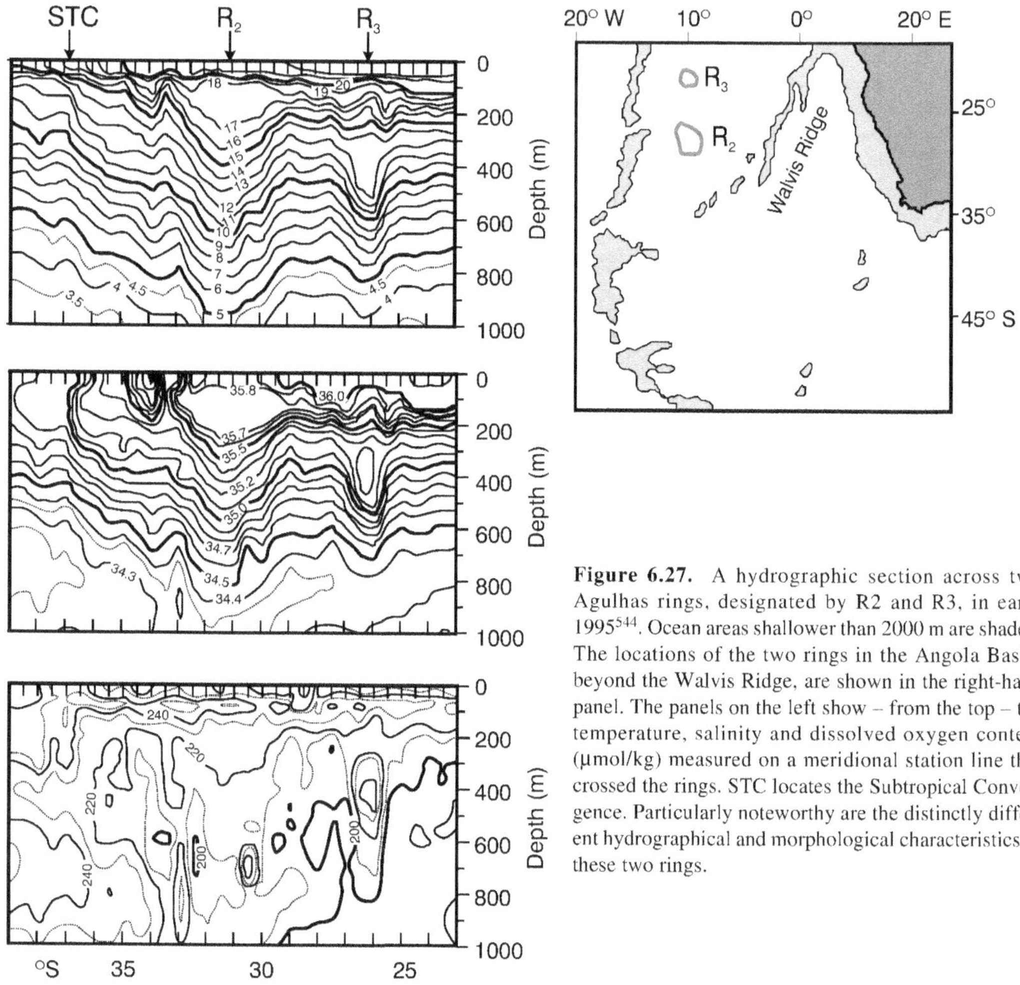

Figure 6.27. A hydrographic section across two Agulhas rings, designated by R2 and R3, in early 1995[544]. Ocean areas shallower than 2000 m are shaded. The locations of the two rings in the Angola Basin, beyond the Walvis Ridge, are shown in the right-hand panel. The panels on the left show – from the top – the temperature, salinity and dissolved oxygen content (μmol/kg) measured on a meridional station line that crossed the rings. STC locates the Subtropical Convergence. Particularly noteworthy are the distinctly different hydrographical and morphological characteristics of these two rings.

Dissipation mechanisms at depth

An analysis focused on low frequency variability in the southern Agulhas Current system[545] has found strong westward radiation of Rossby waves around 32° S to the west of South Africa. The energy for this radiation seems to come from Agulhas rings propagating in a north-westward direction in the south-eastern Atlantic Ocean. It has been claimed that there also is substantial local mixing through Stokes' drift between the water masses of the South Atlantic and the propagating disturbances[545]. These are but some of the mechanism that may be responsible for the local degeneration of Agulhas rings. Modelling Agulhas rings with a realistic two-layer representation suggests[546] that the decay scale for rings that make it across the Walvis Ridge agrees roughly with this numerical simulation.

As will emerge below, rings may also split, although this may not of necessity increase their rate of dissipation[770]. It has been assumed for some time that anticyclones cannot split of their own accord[786]. Using a numerical, multilayer, primitive equation model it has been shown[768] that they can indeed not split by barotropic mechanisms alone. However, barotropic instability is a necessary ingredient for splitting to occur. An extensive analysis of the linear stability of ocean rings[767] has found that they are remarkably robust with respect to changes in ring parameters, like diameter, far field stratification and momentum balance. Nevertheless, realistic rings in theory should be quite unstable based on a linear analysis whereas in reality they do survive for long periods. Investigating the mixing of Agulhas rings using an isopycnals ocean model[770] has shown that the leakage of tracers placed within a simulated

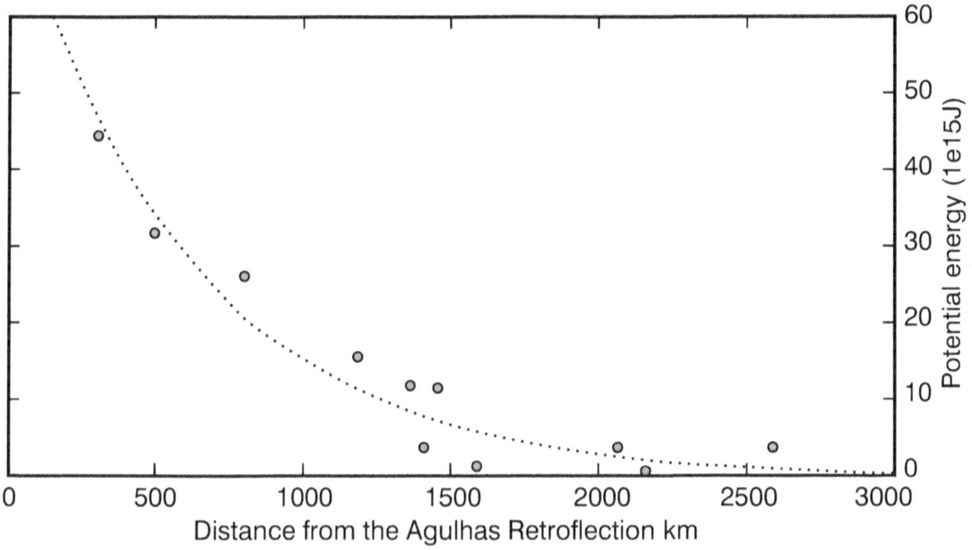

Figure 6.28. The available potential energy of eleven separate Agulhas rings related to their distance from the Agulhas retroflection. Potential energy has been estimated from hydrographic measurements across these features[95].

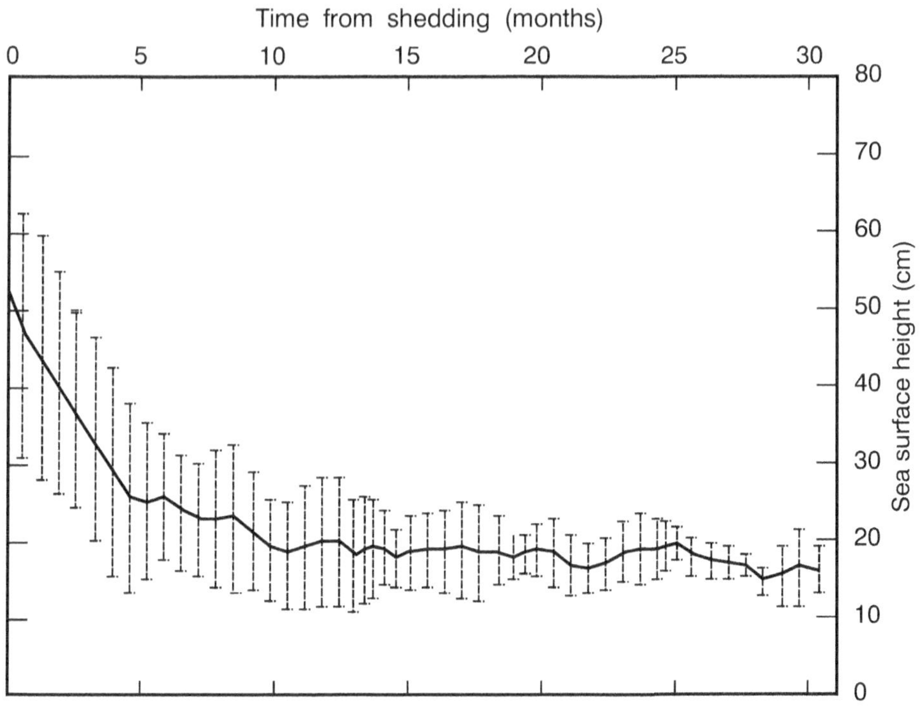

Figure 6.29. The mean sea surface height of all Agulhas rings identified by satellite altimetry between 1993 and 1996 that did not dissipate in the Cape Basin[465]. Error bars denote one standard deviation. More than half the value of surface elevations is lost within the first four months of the rings' existence. After ten months the rings seem to dwindle exceedingly slowly.

ring is associated with the formation of filaments (Figure 6.30). These tracers extend down to the permanent thermocline and therefore contain mostly water in the upper layers. Filaments arise largely because of the elongation of the ring. There are no marked differences between the leakage from one coherent ring and from the combined products of a ring that has split[770]. The main variables contributing to the mixing from this modeled ring are its initial deformation and self-advection. The loss of tracer from a ring shows that in the first months of its existence up to 40 per cent of the water in the ring can be mixed with the environment; in deeper layers up to 90 per cent. These theoretical results agree well with observations[465].

However, as has been mentioned before, Agulhas rings undergo their greatest dissipation while they are in the southern Cape Basin[465], close to their spawning region. This region has therefore been called the Cape Cauldron[628] since on many occasions it is densely populated by a large number of Agulhas rings and cyclones, both with a range of dimensions. This means that few Agulhas rings decay in isolation, at least not before they cross the Walvis Ridge. Comparing theory with actual observations of the decay of an Agulhas ring would therefore be very valuable (Figure 6.31). A ring, called Astrid, was observed hydrographically a number of times[650], showing that its demise corresponded well with previous observations[465] using only sea surface

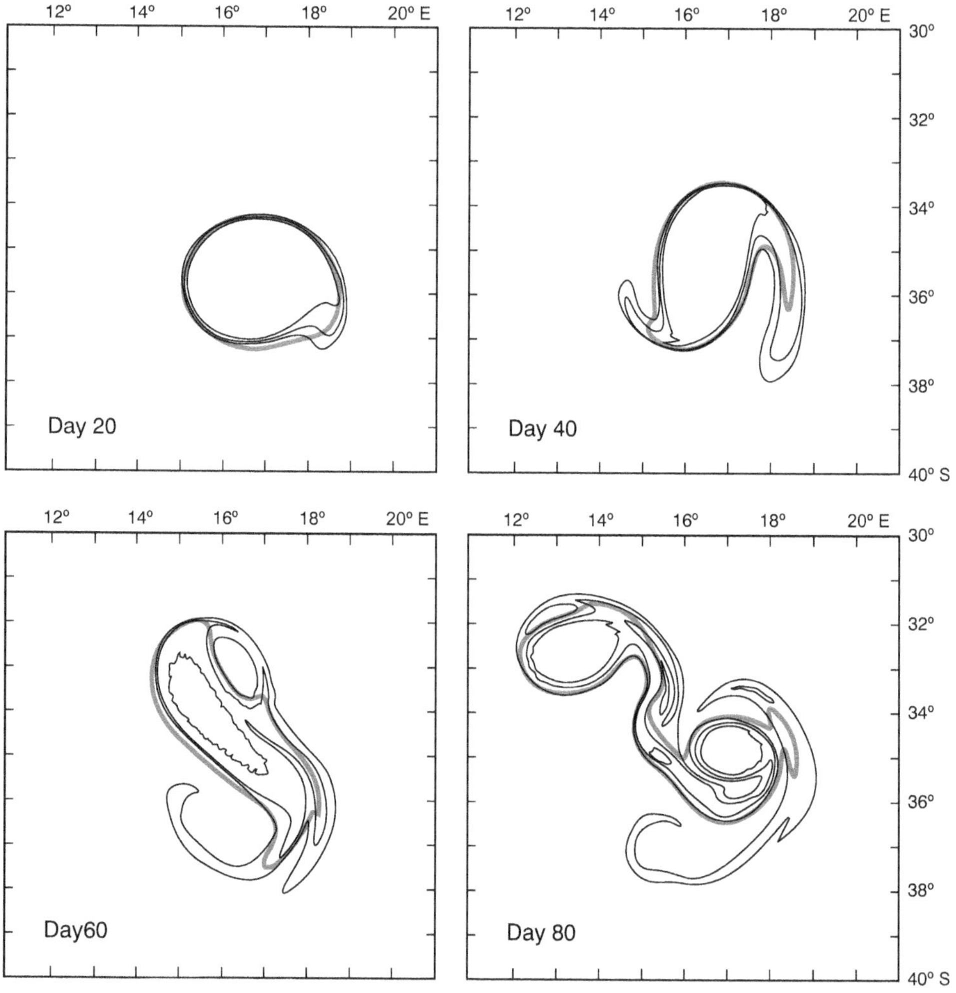

Figure 6.30. The development of the concentration of a tracer in a modelled Agulhas ring as well as the boundary of such a ring[770]. The ring boundary is shown by a thick line.

height anomalies. It decayed most rapidly shortly after being shed. The early evolution of this ring is well predicted by a linear stability analysis. The modelled ring features an energy conversion from its barotropic to its baroclinic components and may eventually split.

Following an individual ring over a period of seven months[784] has shown that its available heat and salt anomaly were reduced by about 30 per cent over this period; its available potential energy by about 70 per cent. It is significant to note that one third of this loss was due to changes at intermediate depth (i.e. between 800 m to 1600 m). This latter process was exemplified by the fact that RAFOS floats placed in this particular ring were detrained after two revolutions in the ring. The vigorous water exchanges at this depth were an underlying cause for the high variability of hydrographic characteristics inside and outside the ring. This is exemplified in Figure 6.32. The temperature and salinity fields at the edge of a well-surveyed Agulhas ring[785] show that inside the ring the distributions of both variables are fairly well-behaved, whereas at the borders there are a variety of disturbances including many boluses of warm water and lenses of saline water. These small scale perturbations to the thermohaline field indicate vigorous mixing at the ring edges. To date these mixing processes at depth have not been quantified. Processes in the upper layers are even more important.

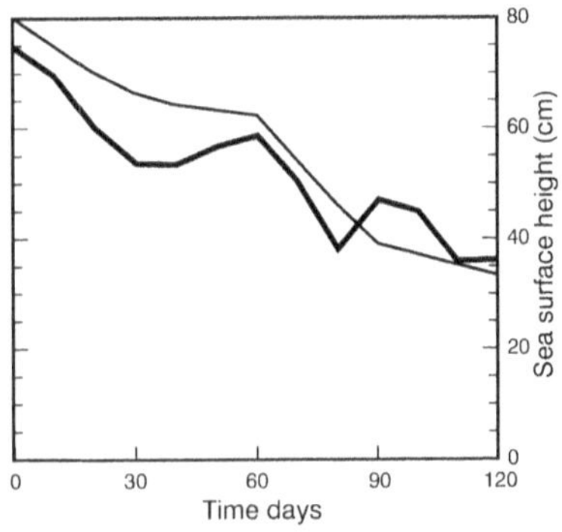

Figure 6.31. Decay of an Agulhas ring as a function of time[766]. The thick line shows the modelled decrease in sea surface height; the thin line the observed decrease.

Dissipation mechanisms at the sea surface

As mentioned elsewhere, the loss of heat from Agulhas rings to the atmosphere is an important consideration in the processes involved in their dissipation, particularly in newly formed rings. In the detailed study[785] of one ring in particular it was shown that the loss of heat to the atmosphere was even severe during summer (54 W/m^2), mainly due to the large mean turbulent flux of latent heat (180 W/m^2). Clearly the heat flux to the atmosphere will show considerable short term variations as different atmospheric systems pass overhead. This has been addressed in a preliminary way[700] by investigating the crossing of weather systems over components of the Agulhas system a number of times. Under an anti-cyclonic atmospheric circulation (i.e. easterly flow) the total turbulent heat flux to the atmosphere over the retroflection was 170 W/m^2 with a maximum at any one time of 360 W/m^2. The latent heat flux made the largest contribution. During the passage of a cold atmospheric front, the total turbulent heat flux remained more or less the same, but the maximum values increased to 500 W/m^2. The last synoptic atmospheric system studied was a cold air outbreak during a post-frontal southerly flow. The mean heat flux increased to 420 W/m^2 with maximum values reaching 630 W/m^2. In this case the effect on the atmospheric boundary layer was substantial[700]. It increased from a convective thermal, internal boundary layer of 500 m height to a well-mixed layer of 900 m. The effect of this heat loss on individual Agulhas rings naturally is substantial.

In the study of two Agulhas rings[544] discussed before, it was found that far from their spawning grounds they showed the disparate structure of a ring that had been formed in summer and that had moved rapidly equatorward, into warmer atmospheric conditions, compared to one that had been formed in winter and that had remained near to the point of formation. Their interaction with the atmosphere in early stages of their development therefore had been crucial. Both had substantial thermostads, but these thermostads had different temperatures and were found at distinctly different depths. (See Figure 6.27.) The effect of winter cooling on one particular ring over a period of seven months[784] has indicated the changes to the upper layers that can be expected. In March the temperatures at the sea surface exceeded 19° C, but after the winter there was no water warmer than 17° C. During the same period the thickness of the mixed layer had increased two-fold. It is clear that this increase in Agulhas rings is due to the convective motion induced by cooling at the sea surface, to increased turbulence induced by strong winter winds and due to increases in the salinity of the surface layers due to evaporation. The latter process has not been quantified in any reliable way. In the case of the ring observed seven months apart[784], and discussed above, the salinity of the upper 300 m had in fact – unexpectedly – decreased. During this period, the depths of all isotherms had decreased, suggesting mixing with the surrounding waters. An attempt to model[757] the impact of cooling on the water mass exchanges of Agulhas rings has produced some interesting results.

A bowl shaped ring was simulated with a diameter of 280 km. Below 800 m depth the Agulhas ring in this model rapidly loses its original water mass. This result agrees substantially with that found by studying the motion of RAFOS floats placed in such rings[625] and with hydrographic observations[785] in rings. In this model strong surface cooling generates a shallow overturning cell with radially outward flow near the surface and a compensating flow at depth. As a result the surface water does not remain trapped in the core of the ring, but exchanges water with the surrounding waters. The overturning cell amplifies this water mass exchange by constantly bringing new water to the edge of the ring where it gets the opportunity to mix with ambient waters.

The question how rapidly rings undergoing all these mixing processes will be absorbed by the ambient water masses may to some extent depend on the trajectories they follow.

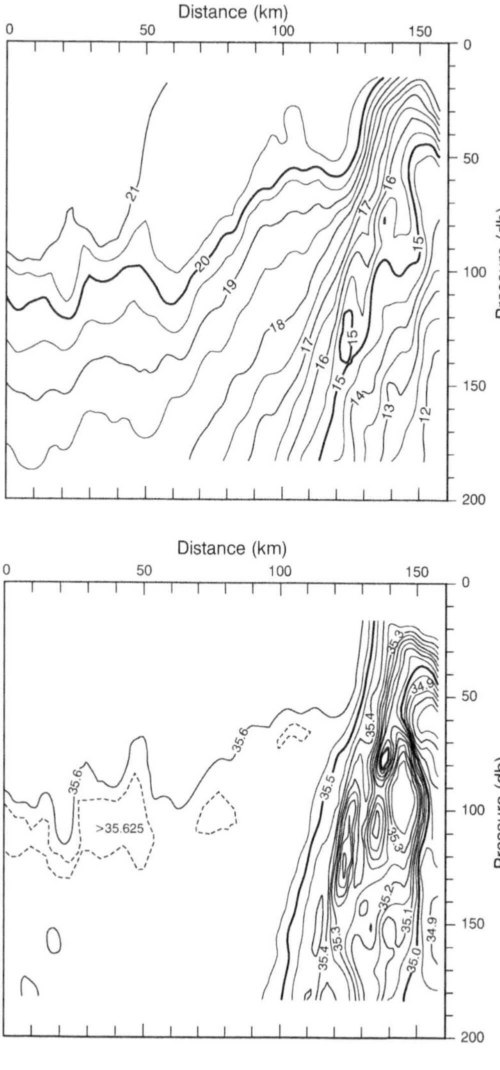

Figure 6.32. Hydrographic disturbances along the edge of a relatively new Agulhas ring[785]. These closely spaced observations were made with a Scanfish undulator resolving the structure to a lateral accuracy of 1.8 km. The upper panel shows the thermal structure; the lower the salinity distribution.

Ring pathways

Drifting buoys have suggested an initial drift rate away from the inception region of 5 to 8 km/day[94]. However, it is now clear that Agulhas rings may stall, change direction, split or speed up during their progression across the South Atlantic Ocean, particularly in the Cape Basin. Whatever the initial drift rate, it has been observed that a number of rings tend to remain close to the retroflection for a considerable time. They have, for instance, been observed to be quite persistent southwest of Cape Town[444] (e.g. Figure 6.26). The presence of Agulhas rings in this particular spot is of added importance because rings situated here will enhance the shelf-edge current as well as the rapid advection of Agulhas filaments into the South Atlantic[92].

In both satellite imagery[61] as well as research cruises[455] up to nine rings and eddies have been observed at the same time clustered around the Agulhas retroflection. Hydrographic data show their further movements (Figure 6.19), although to date only three rings has been visited more than once[543,784–5]. Other types of information therefore have to be employed to evaluate the movement of rings.

Satellite altimetry has proven to be an exceptionally useful tool[70,354,543] to track Agulhas rings, particularly since they have strong signals in sea level elevation[95,464] and drift through a comparatively quiescent region[498]. Comparison between altimetric observations and the measurements from moored current meters and inverted echo sounders has shown that in the southeastern Atlantic Ocean anomalies of the sea surface are significantly related to the thermocline depth and to the dynamic height of the sea surface[547]. Considerable work has been done to refine the analysis[74] and interpretation of altimetric data in the region, including data assimilation into quasi-geostrophic models[548].

Initial investigations[73] using altimetric data have shown rates of translation of 4–8 cm/s, and even[517] up to 16 cm/s, in a general north-westerly direction for the anomalies assumed to be Agulhas rings. Other estimates[362] are 3–7 cm/s. Propagations rates based on RAFOS floats[783] give values of 5.5 to 6.5 cm/s. Modelling studies[546] suggest a rate of about 9 km/day and show that ring trajectories undergo a transition from a turbulent character in the Cape Basin to a much more steady propagation in the rest of the South Atlantic[760], but that this is due to ring decay and not to the topography being crossed. The propagation speeds of those rings that are durable and those that eventually cross the Walvis Ridge show no systematic seasonal variations[712] in their speed of translation. A detailed study of one particular ring[784] for a period of seven months has shown that its north-westerly progression varied between 3 km/day to 20 km/day, possible due to its interaction with other rings and cyclones along its path.

Further studies using altimetry have been able to track anomalies of sea surface height over the greater width of the South Atlantic and were able to positively identify some of these anomalies as Agulhas rings at sea[464,544]. Van Ballegooyen et al.[129] have, for instance, been able to establish a very close correspondence between the hydrographic observations of individual Agulhas rings and their altimetric signatures. During the two-year period of their investigation rings migrated no farther than 1500 km from the Agulhas retroflection.

A similar study[95], for a period of three years (Figure 6.33), has shown inferred Agulhas rings to advect across the South Atlantic Ocean slightly to the left of the mean flow. None crossed the South Atlantic Ocean north of a latitude of 25° S. In a few instances Agulhas rings have been tracked all the way to the coast of Brazil; in one instance[754] there is evidence that a ring was subsequently carried poleward by the Brazil Current over a distance of at least 10° of latitude. Rings probably are advected with the ambient water movement[807], but also exhibit a substantial degree of self-steering due to their own internal dynamics. Calculations have shown that only about 15 per cent of the observed drift of rings is self-induced; advection by the background flow therefore dominating the rate at which they translate. Comparing the movement of floats placed in Agulhas rings, at intermediate depths, with those placed outside rings and therefore in the Benguela Current, Richardson et al.[781] have been able to establish the rate at which rings move through the background waters. The background speeds were about 2 km/day; those of Agulhas rings roughly 6 km/day. This means that rings have a self-induced movement at 750 m of about 4 km/day. The background speed may have its origin in the Benguela Current[807].

Sea height anomalies south of 45° S in general migrate eastwards[362]. Anomalies that migrate eastwards originate at 40° S as far west as 20° W longitude[95]. In all probability these latter ones are not Agulhas rings, but mesoscale eddies shed at the Subtropical Convergence in the South Atlantic[549], or by the South Atlantic Current[83]. The furthest westward Agulhas retroflection observed to date, or Agulhas ring at its inception, has been at 8° E longitude[512]. The other mesoscale features found in the Cape Basin, cyclones, have been shown to migrate roughly at right angles to the mean trajectory of Agulhas rings[628], but since they seem to have a much shorter lifetime have not been followed farther than the confines of the Cape Basin. As mentioned in another context, this cross-traffic of anti-cyclones and cyclones seem to be a characteristic of a number of west coasts of continents, including North America and Australia[780]. In their subsequent journey across the southern Atlantic Ocean Agulhas rings have to cross a number of bottom ridges. Since the rings extend to great depths, it is valid to examine the influence of these bathymetric obstacles on their behaviour.

Interaction with bottom ridges

The influence that distinctive features of the bottom topography may have on the paths taken by Agulhas rings across the South Atlantic is not immediately obvious. In many models[764,758] the effect of the bathymetry is evident, but the magnitude of this effect differs substantially between individual models[763]. One of the major bottom features that rings will have to cross, in order to follow the streamlines of the subtropical gyre, is the Walvis Ridge that lies from about 20° S latitude at the west African coast in a south-westward direction (viz. Figure 6.27). It is interesting to note that of 30 RAFOS floats placed[781] at a depth of roughly 750 m in the Cape Basin, virtually all that crossed the Walvis Ridge did so associated with the passage of Agulhas rings or Cape Basin cyclones. The crossing of the Walvis Ridge by Agulhas rings has been shown to occur irregulary and aperiodically[712].There is evidence that most rings slow down on traversing this feature[95], but there are also some that show signs of drift acceleration[73]. Initial translational speeds of 12 km/day have been observed to decrease to 6 to 7 km/day over the ridge[519].

Schouten et al.[465] have demonstrated that once Agulhas rings have been slowed down on crossing the Walvis Ridge, they never regain their previous propagational speed, but remain sluggish in their subsequent movement. There also is greater directional uniformity amongst rings that have successfully crossed this

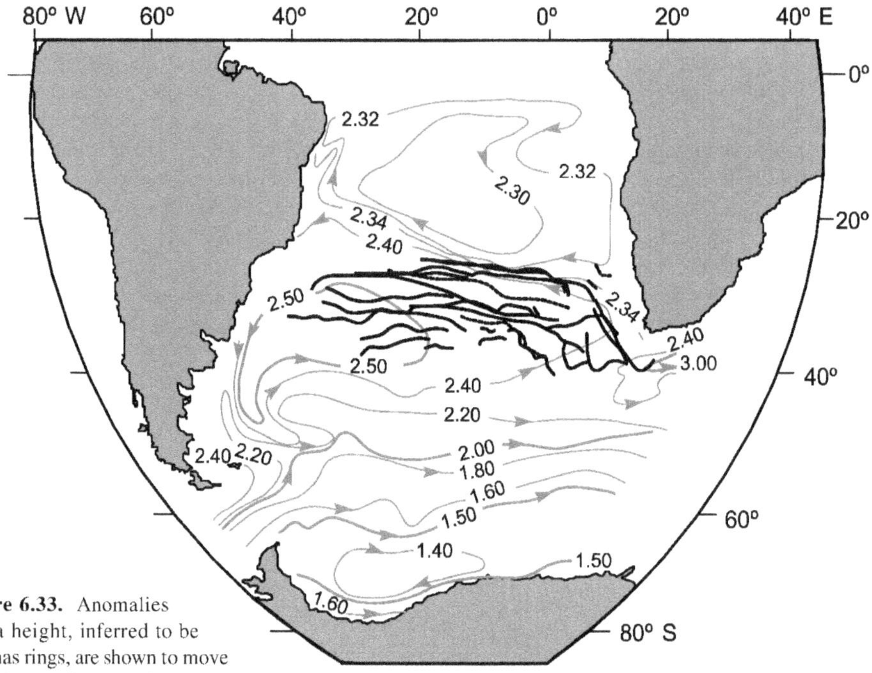

Figure 6.33. Anomalies of sea height, inferred to be Agulhas rings, are shown to move from the Agulhas retroflection region westwards. Data used for this study[95] were altimetric measurements from the Geosat exact repeat mission between 1986 to 1989. Drift tracks are superimposed on the general direction of movement from the steric anomaly at 500 dbar[195]. The rings exhibit a drift tendency slightly to the south of the averaged background flow.

hurdle. Their rate of decay moreover drops markedly on having crossed the Walvis Ridge, but this may be a function of their age (viz. Figure 6.29) and may not only be due to interaction with the ridge.

Modelling Agulhas rings with realistic dimensions and characteristics[546] shows that they may indeed slow down, stall or even be destroyed at the Walvis Ridge, depending on their configurations. Rings with sufficient initial vertical shear can cross the ridge, but ones that are nearly barotropic cannot. In general, baroclinic rings modelled in this manner all slow down to a translational speed of 4 km/day on crossing the ridge, adjusting their vertical structure and intensifying towards the same final, ridge-crossing state[546]. This particular model predicts that the surface elevation of rings will increase measurably on crossing the ridge. This has been observed in some, but not in all, rings in nature. According to this model, the Walvis Ridge may therefore act as a substantial filter, allowing only rings with vertical characteristics under a certain threshold to pass. Other models[760] do not indicate any such function for the Walvis Ridge. A different model[550], using a two-layer ocean at rest and with Gaussian-shaped anomalies, shows that the ridge in stead affects the drift direction of deep-reaching eddies.

On reaching the upslope of the ridge they are forced in a more equatorial direction. This has been seen in some, but again not in all, altimetric trajectories of such rings[465]. Eddy permitting numerical simulations have shown that the trajectories of Agulhas rings that are intensified in their upper layers are changed by the Walvis Ridge to a more westerly direction. The deep compensation generated by the ridge in the model causes an energy loss of about 30 per cent. However, only modelled eddies with a substantial dynamic signal in the lower layer are influenced by the bottom topography. In nature rings have been observed predominantly to cross the ridge where the water is deep[95,465], but this may be the result of the general background advection (Figure 6.33) and not directly due to topographical steering. The waters of the extension of the Benguela Current may move predominantly through these gaps in the Walvis Ridge, carrying Agulhas rings with them. The transits of Agulhas rings cross not only bottom ridges, but seamounts as well.

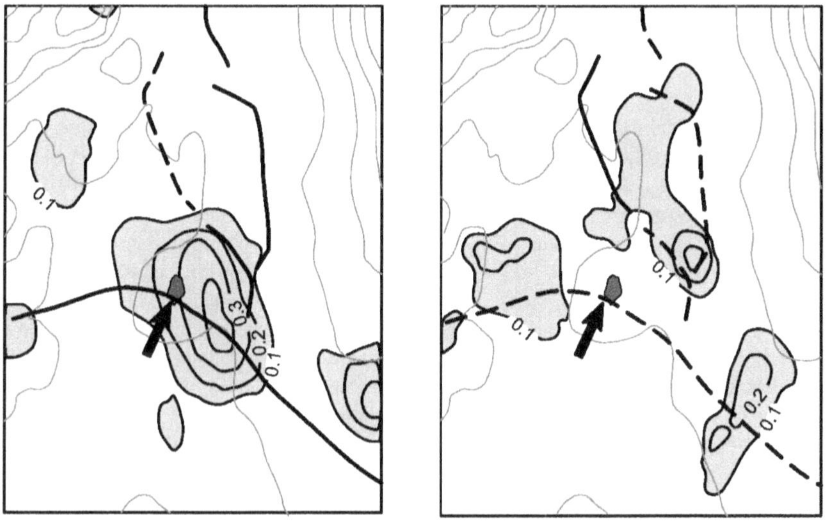

Figure 6.34. The splitting of an Agulhas ring on crossing the Vema Seamount in the Cape Basin of the south-eastern Atlantic Ocean[465] in October 1996. The time difference between the two portrayals is six weeks. The location of the seamount is indicated by an arrow; the sea surface heights are in metre and are derived from altimetric information. Background isobaths slope from left to right. Thick lines denote the trajectory of the single ring on approaching the Vema Seamount and the paths of the collision products subsequently.

Ring interaction with seamounts

A much less geographically prominent feature of the bottom topography in the Cape Basin is the Vema Seamount. This peak rises from an otherwise deep and unremarkable abyssal plain to a reported depth of less than 32 m below the sea surface and would, at first glance, not constitute an insurmountable obstacle to an Agulhas ring. Nonetheless, there is growing evidence from satellite altimetry[465] that rings that come into contact with this oceanic pinnacle have a tendency to split into two or more smaller rings (Figure 6.34). This is similar to the sequence of events that has been detected at the Erica Seamount[544].

Theoretical investigations[769,771] have shown that the advection by a ring of deep fluid parcels generates deep anti-cyclonic and cyclonic circulations near the bathymetry. These circulations exert a strong shear on the upper layers which causes an erosion of the ring by filamentation or, sometimes, the subdivision of the ring. Under certain circumstances[771] an eddy, such as a ring, may be scattered by a topographic obstacle.

The products of a ring-seamount collision subsequently take different routes. Recent high-resolution modelling[325] has suggested a fork in the trajectories for Agulhas rings in the general vicinity of the Vema Seamount, with two distinctly different pathways downstream. This modelling result therefore is consistent with what has been observed from the movement of positive anomalies of sea surface height in this ocean region[465]. It also clearly demarcates the wide-ranging influence that the passage of Agulhas rings may possibly have on the background current of the south-western Atlantic Ocean.

The Benguela Current

To recapitulate briefly what has been dealt with more thoroughly above, the Benguela Current forms the eastern and part of the northern component of the wind-driven, anti-cyclonic gyre of the South Atlantic Ocean[779]. It starts in the south-eastern corner of the Cape Basin and reaches the South American coast at about 18° S. The presence of Agulhas rings in the southern Benguela Current, south of 30° S latitude, has a profound influence on the nature of this current. Whereas the mean flow next to the African continent is more invariant, the western part is dominated by the transient effects of passing rings[519]. Observations of equatorward transport show strong correlations between increases in this transport and Agulhas water influx via rings. The primary inter-annual variability in the transport of the south-eastern part of the Benguela Current therefore derives almost totally from the passage of Agulhas rings and variations in the inflow from the South Atlantic subtropical gyre[547]. In the upper

1000 m the annual contribution to the volume flux of the Benguela by the Agulhas Current varies between 10 per cent and 50 per cent. Over a period of five years[782] the contribution from the South Atlantic Ocean to the Benguela is 58 per cent, that of the Indian Ocean about 30 per cent. However, at intermediate depths the greater part of the waters in the Cape Basin is supplied from the Indian Ocean[628] with minor direct inflow from the Atlantic Ocean. Red Sea Water contributes[775] about 6 per cent. More about this follows below.

An investigation[772] combining sea surface height and sea surface temperature measurements has suggested that a convoluted jet is found in the Benguela Current, probably separating the coastal upwelled water from the deep-sea waters. This jet shows seasonal behaviour, being stronger in summer. This strengthening is due to an increase in coastal upwelling, but also due to an increased injection of Agulhas ring water on the offshore side of the jet. One should at this stage perhaps remind oneself that the movement of these Agulhas rings is of course highly unusual. No other products of a western boundary current are known to move past such a well-developed coastal upwelling regime[551–3].

Agulhas rings and coastal upwelling

The Benguela upwelling system of the South East Atlantic Ocean extends from about 15° S to 35° S latitude[551]. Its northern border is the Angola/Benguela front[554], while the wind-driven upwelling on the Agulhas Bank[473] may be considered to be its southern extremity. This upwelling system has a central point where it is most intense and durable, at Lüderitz, and is otherwise concentrated in a number of relatively distinct upwelling cells[470]. The frequency of upwelling at these cells decreases both north- and southwards with distance from Lüderitz, while the southernmost cells are totally seasonally driven.

Notwithstanding this along-coast variability in upwelling intensity, the instantaneous upwelling expression along this coastline is one of a contiguous strip of cold water at the sea surface that overlies the continental shelf, and a zone about twice as wide that is populated by a range of wisps of cold surface water, upwelling filaments, vortex dipoles[555] and small eddies[556–7]. Some of these mesoscale frontal features seem to occur at random while others[557] seem to be locked to the morphology of the coastline. Some of the upwelling filaments can, probably by a combination of extreme offshore berg winds and entrainment in offshore eddies, be made to extend to distances of 1000 km offshore[558]. This may bring them into the path of passing Agulhas rings.

At least one such interaction between an Agulhas ring and an upwelling filament has been investigated in detail[559–60]. A pioneering investigation of an upwelling filament off the south-western coast of Africa by Shillington et al.[561] has suggested the presence of a warm eddy to the south of this particular filament at the time, the former possibly of Agulhas origin. A subsequent set of cruises, to follow an Agulhas ring in its northward movement along the coastline[543], have established that on one occasion this particular ring was encircled by a filament of cold surface water (e.g. Figure 6.35). Temperature–salinity characteristics of the filament showed it to be nearly pure upwelling water[559]. It was about 50 km wide, 100 m deep and, when entrained around the full circumference of the ring, would have had a length of 1000 km.

It has been suggested[560] that removing this amount of water from the upwelling front could have a profound effect on the biota of the upwelling regime and could, on this particular occasion, have been partially responsible for depressing the anchovy year-class of that year and hence the recruit biomass available for the following year. An eddy-permitting, large-scale model of the whole upwelling system[562] has simulated cases of filament-ring interaction, demonstrating that under the appropriate conditions this may be an inherent part of the system. Subsequent hydrographic studies in the region[563] have produced circumstantial evidence that Agulhas rings draw upwelled water as well as tropical Atlantic Ocean water, found over the upper continental slope, into the ocean interior. It has in fact been suggested that Agulhas rings are the primary removal mechanism for the low oxygen water found on the slope[563]. This process can probably only come about if an Agulhas ring during its equatorward passage reaches the upwelling front itself.

Filaments of water blown offshore by extreme wind events would be lost to the upwelling system irrespective of whether they get entrained in Agulhas rings far offshore[558] or not. The frequency of Agulhas trajectories lying closer inshore would therefore be a prime factor in ascertaining how important Agulhas rings are in this exchange process compared to wind events. Altimetric studies of such trajectories[95,362,465] to date suggest that an Agulhas ring conjunction with the active upwelling front may be a very rare event. These unusual incidents may nonetheless have a significant local effect on a fragile fish recruitment process. The continuous creation of warm Agulhas rings and their steady movement into the South Atlantic, forming the basis of the inter-basin exchange of water south of Africa, has, by contrast, effects on a much larger scale.

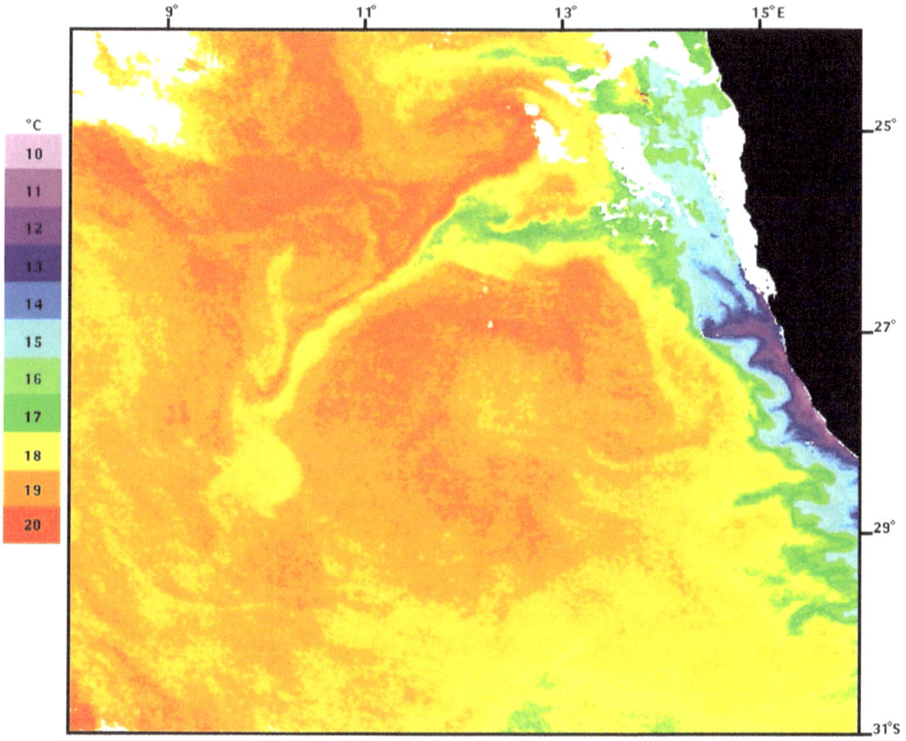

Figure 6.35. A satellite image showing a cold filament, from the coastal upwelling system, being wrapped around a passing Agulhas ring[559]. Note the low temperatures (blue-green) in the upwelling regime off the south-west coast of Africa. These sea surface temperatures are from the radiometer on board the NOAA 17 satellite for 4 December 2005.

Inter-ocean exchange at the Agulhas retroflection

As discussed previously, the Agulhas Current may be a key link in the global thermohaline circulation cell[65,68]. Since leakage of its water into the South Atlantic is the mechanism by which this process is maintained here, it is vitally important to establish how much water is exchanged, how frequently and what the various factors are that may influence, or control, this process[413].

As we have seen, the major process in this inter-ocean exchange south of Africa is the shedding of Agulhas rings[66]. There may also be some direct flux of Indian Ocean water into the Atlantic. Secondary processes are the northward advection of Agulhas filaments[92] (Figure 5.10), and the movement of Agulhas Bank water.

Direct leakage

The geostrophic estimates of the direct volume flux of water past the tip of Africa[516] is summarised in Table 6.4. The estimates vary from 4×10^6 m^3/s to 10×10^6 m^3/s. Data sets on which these estimates are based are from separate cruises and therefore independent. The manner in which they have been calculated – and the reference depths used for the calculations – are not the same, making comparisons difficult. The contribution to the inter-ocean flux by water from the Agulhas Bank has been inadequately studied to make any substantive pronouncements at all. Gordon et al.[230] have estimated it to be about 10×10^6 m^3/s, relative to 1500 decibar, on one occasion. The geographic distribution of observations of water with a shallow oxygen minimum[231], that is usually associated with water from the Agulhas Current core (viz. Figure 6.14), shows such water as extending as far as 32° S along the west coast of South Africa.

A number of studies[773–5] have been undertaken to try to establish how much water and of what water type is exchanged south of Africa. Using compendiums of hydrographic data, these studies do not differentiate between the mechanisms that may have caused the inter-ocean transfers, they only quantify the end results. You[774] has, for instance, investigated the origin of

Antarctic Intermediate Water in the South Atlantic Ocean and come to the conclusion that of this type the water originating from the Drake Passage is dominant. Antarctic Intermediate Water from the Indian Ocean comprises 30–60 per cent of that originating in the South Pacific Ocean in the subtropical latitudes of the South Atlantic Ocean. The meridional volume transport of Antarctic Intermediate Water in these subtropical latitudes consists of 64 per cent water from the Drake Passage, 36 per cent from south of Africa. The former extends to the south-western Indian Ocean in a continuous band[775] whereas the Indian Ocean source waters spread to the southeastern South Atlantic mostly in a patchy distribution, perhaps indicating the intermittency of their generation. The volume transport of Antarctic Intermediate Water south of Africa consists of water from the Drake Passage (63 per cent), from the South Indian Ocean (16 per cent), from the Indonesian Seas (10 per cent) and from the Red Sea (12 per cent). Only a small proportion of Antarctic Intermediate Water from the Drake Passage that moves into the Indian Ocean is eventually returned westward[775].

One known mechanism for interocean exchange south of Africa, albeit it not the major one, is the movement of filaments drawn from the core of the Agulhas Current.

Agulhas filaments

Agulhas filaments are advected past the western edge of the Agulhas Bank (viz. Figure 5.10), carry only surface water from the upper 50 m of the Agulhas Current[92] and are often entrained in the perimeter of passing Agulhas rings. By being captured in the rim of Agulhas rings they may be replenishing the rapidly cooling surface layers of such features, increasing their surface salinity and enhancing convective overturning in these ageing rings. Agulhas filaments presumably carry little net heat into the South Atlantic Ocean, all excess heat being rapidly lost to the colder overlying atmosphere[121]. This may be surmised from the occasional presence of cumulus cloud bands above these filaments[143], suggesting substantial fluxes of heat and moisture to the atmosphere. Agulhas filaments are, nonetheless, estimated as contributing an annual net flux of 3 to 9 × 10^{12} kg salt, or 9 per cent of that due to Agulhas rings. The contribution to inter-ocean exchange by Agulhas filaments is therefore small, but not entirely negligible. Such surface water exchange may occasionally increase substantially[439] through interaction with Agulhas rings and under wind conditions conducive to northward advection[564].

Leakage by Agulhas rings

Nevertheless, the major component of the inter-ocean exchange of heat and salt south of Africa, in the thermocline and surface waters, seems to be due to Agulhas rings[413,516]. Using a box model informed by measurements from an array of inverted echo sounders, Garzoli and Goñi[782] have demonstrated the sources of water crossing the Cape Basin at 30° S. A total of 12 × 10^6 m^3/s water in the upper 1000 m moves across this line. This is an average over five years. Of this 6 × 10^6 m^3/s comes from the South Atlantic, possibly largely from the South Atlantic Current, 2 × 10^6 m^3/s directly from the South Indian Ocean and the rest (3 × 10^6 m^3/s) is a mixture of Agulhas water in filamentous form and tropical Atlantic water originating from the north. The ratios are very variable. During 1995 more than 50 per cent of the volume transport came from the Indian Ocean; in 1996 it was barely 10 per cent. This incorporation of water from the north has also been shown from the drift tracks of floats at intermediate depths[781]. Near 30° S floats placed east of the Walvis Ridge tended to move southward before turning northwestward to join the Benguela Current.

Table 6.4. Geostrophic estimates of the direct flux, i.e. excluding Agulhas rings or filaments, of Indian Ocean water into the South Atlantic[516] in 10^6 m^3/s.

Authors	Flux	Reference	Date
Harris and Van Foreest (1978)[92]	5	1100 db	March 1969
Gordon et al. (1987)[230]	10	1500 db	November–December 1983
Bennett (1988)[533]	6.3	T > 8 °C	November–December 1983
	2.8	T > 8 °C	February–March 1985
Stramma and Peterson (1990)[83]	8	1000 m	November 1983
Gordon et al. (1992)[67]	10	T > 9 °C	December 1989–January 1990
	15	1500 db	December 1989–January 1990
Garzoli et al. (1997)[547]	4	1000 db	September 1992–December 1995

An estimate of the inter-ocean exchange brought about by Agulhas rings that is probably the most accurate to date, being based on the largest number of actual hydrographically measured rings[129], gives a volume flux of 6.2 × 10^6 m^3/s for water warmer than 10 °C and 7.3 × 10^6 m^3/s for water warmer than 8 °C. A heat flux by Agulhas rings of 0.945 PW and a salt flux of 78 × 10^{12} kg/year have thus been calculated. Estimates of the average excess of heat and salt contained in an Agulhas ring relative to the surrounding waters of the South Atlantic, based on the hydrographic surveys of a substantial number[520] of rings (Figure 6.21), give values of 7.1 × 10^{12} kg salt and 2.7 × 10^{20} J heat. This would lead to an inter-ocean heat flux of between 0.034 and 0.051 PW and a salt flux of between 28.4 and 42.6 × 10^{12} kg. This heat flux has been substantially confirmed with independent altimetric estimates[362].

Various other estimates have been made[83,95] of the mean volume flux achieved by these rings. These are all summarised in Table 6.5. The values in this table are for fluxes by individual rings only. They should therefore be viewed in concert with the estimated number of ring-shedding events, given in Table 6.6. By doing this it can be seen that the total fluxes achieved through the process of ring shedding lie between 2 × 10^6 m^3/s[464] and 15 × 10^6 m^3/s[20,65]. Some calculations[517] assign an average of 1 × 10^6 m^3/s to each ring, a value that is in rough agreement with the average for all estimates to date (Table 6.5). It has accordingly been estimated[94] that the replacement time for water above 10 °C in the South Atlantic Ocean by Agulhas rings alone would take only 70 years.

As could be expected, and been suggested above, these volume transports by Agulhas rings are by no means invariant. In fact, they exhibit large interannual variations[782] (Figure 6.36). It is evident that the upper layer transport from the South Indian Ocean to the South Atlantic Ocean varies from 0 × 10^6 m^3/s to nearly 40 × 10^6 m^3/s, that the number of rings shed per year is not constant and that the volume of water in each ring is very different from individual ring to individual ring. For example, during 1997 only four rings were formed, but the volume content of each was much higher than the average for the years 1993 to 1998. As a result the average inter-ocean volume flux for 1997 was much higher than normal. The mean volume transport by rings in 1997 was 2.4 × 10^6 m^3/s whereas it was 0.8 × 10^6 m^3/s in 1993[782]. This high level of variability would also hold for other inter-ocean fluxes such as that of potential and kinetic energy.

An estimate of the mean, available potential energy flux per year due to Agulhas rings of 20 × 10^{16} J has been made[95] with an average, concurrent kinetic energy flux of 22 × 10^{16} J. These results depend directly on the properties of potential energy and kinetic energy found in individual rings. Values for these variables, as calculated from hydrographic measurements, are tabulated in Table 6.7. In these calculations, as for those for those of net salt and heat fluxes, the hydrographic values within rings are compared to those of unsullied South Atlantic water masses. The values accepted as representative for the ambient waters are therefore critical to an accurate estimate. Hydrographic stations so selected are usually chosen to be in the vicinity of the ring in question, but seemingly unaffected by foreign water masses. This selection process remains a hazardous one since all the waters of the south-eastern Atlantic Ocean probably exhibit some influence of Indian Ocean water or other. None would be pristine.

The net energy values calculated to date for transport

Table 6.5. Estimates of inter-ocean volume transports south of Africa by ring translation[516]. Sv is 10^6 m^3/s. Results obtained by Lutjeharms and Cooper (1996)[93] are for Agulhas filaments.

Authors	Flux/ring [Sv]	Reference	Date
Olson and Evans (1986)[94]	0.5–0.6	T > 10 °C	November–December 1983
Duncombe Rae et al. (1989)[543]	1.2	total	April–May 1989
Gordon and Haxby (1990)[464]	1.0–1.5	T > 10 °C	May 1987
	2.0–3.0	total	May 1987
McCartney and Woodgate-Jones (1991)[395]	0.4–1.1	total	February–March 1983
Van Ballegooyen et al. (1994)[129]	1.1	T > 10 °C	February–March 1987
Byrne et al. (1995)[95]	0.8–1.7	1000 db	10 cruises in the 1980s
Clement and Gordon (1995)[537]	0.45–0.90	1500 db	May 1993
Duncombe Rae et al. (1996)[520]	0.65	total	June 1992; May–October 1993
Lutjeharms and Cooper (1996)[93]	0.10	total	November 1983; December 1992
Goni et al. (1997)[517]	1.0	T > 10 °C	September 1992–December 1995

Table 6.6. The number of shedding events per annum for Agulhas rings as estimated by different authors[516]. Results obtained by Lutjeharms and Cooper (1996)[93] are for Agulhas filaments. Numbers in parentheses denote the average number of rings shed per year.

Authors	Number per year	Device	Period
Lutjeharms and Van Ballegooyen (1988)[91]	6–12 (9)	infrared	1978–1983
Gordon and Haxby (1990)[464]	5	altimeter	November 1986–November 1987
Feron et al. (1992)[74]	4–8 (6)	altimeter	November 1986–September 1989
Van Ballegooyen et al. (1994)[129]	6	altimeter	December 1986–December 1988
Byrne et al. (1995)[95]	6	altimeter	November 1986–August 1989
Duncombe Rae et al. (1996)[520]	4–6	echo sounder	June 1992–October 1993
Lutjeharms and Cooper (1996)[93]	6.5	infrared	1987–1991
Goni et al. (1997)[517]	4–7 (6)	altimeter	September 1992–December 1995

by Agulhas rings therefore may differ by more than just the variations in the characteristics of the rings themselves (Table 6.7). The potential energy calculated for Agulhas rings in this way varies between 2.8 to 51.4×10^{15} J; the kinetic energy lay between 2.01 to 8.7×10^{15} J. Notwithstanding the inherent variability to be expected in a collection of Agulhas rings, as well as the variability introduced by the different selection criteria for reference stations used, the level of energy in these features remains enormous. Olson and Evans[94] have consequently judged these rings to be the most energetic in the world ocean. Studies of annual and interannual variability in the South East Atlantic Ocean[572], using sea surface temperatures for the past 80 years, show high correlations for the greater offshore part of this region. Interannual changes in heat input by Agulhas rings can therefore most probably not be resolved in this way. This may be possible using observations of sea surface height.

A study[712] of inter-annual variability[754] of the circulation in the South Atlantic using satellite altimetry for a period of four years has demonstrated that there can be a transition from a state of high mean sea level to a state of lower sea level over a period of months (Figure 6.37) with the commensurate increase and decrease in circulation intensity in the Agulhas retroflection region. This was due to a basin-scale mode consisting of a broad, flat gyre replaced by a more zonally compact gyre, the latter with a stronger western boundary flow. The Agulhas ring corridor in the South Atlantic Ocean also widened when the average sea level was high and shrunk to a narrower one when the average sea level was lower[712], suggesting that the basin-scale mode in the South Atlantic Ocean plays a role in the dispersal of Agulhas rings. The dominant mode of basin-scale, zonal wind forcing in the South Atlantic was in phase with these inter-annual changes in the Agulhas retroflection region[712]. This may well imply that the leakage of heat and salt to the South Atlantic Ocean by Agulhas rings are partially controlled by inter-annual

Table 6.7. Physical properties of Agulhas rings as presented by different authors[516]. Heat flux (F_Q), salt flux (F_S), available potential energy (APE) and kinetic energy (KE) have all been calculated with respect to the properties of the ambient waters in which the rings were found. Results obtained by Lutjeharms and Cooper (1996)[93] are for Agulhas filaments.

Authors	F_Q [10^{-3} PW]	F_S [10^5 kg/s]	APE [10^{15} J]	KE [10^{15} J]
Olson and Evans (1986)[94]			30.5	6.2
			51.4	8.7
Duncombe Rae et al. (1989)[543]	25	6.3		
Duncombe Rae et al. (1992)[559]			38.8	2.3
Van Ballegooyen et al. (1994)[129]	7.5	4.2		
Byrne et al. (1995)[95]			18	4.5
Clement and Gordon (1995)[537]			7.0	7.0
Duncombe Rae et al. (1996)[520]	1.74	1.1	11.3	2.01
Lutjeharms and Cooper (1996)[93]	1.1	0.15–0.46		
Goni et al. (1997)[517]			24	
Garzoli et al. (1996)[519]	1.0–1.6	0.7–1.0	2.8–3.8	

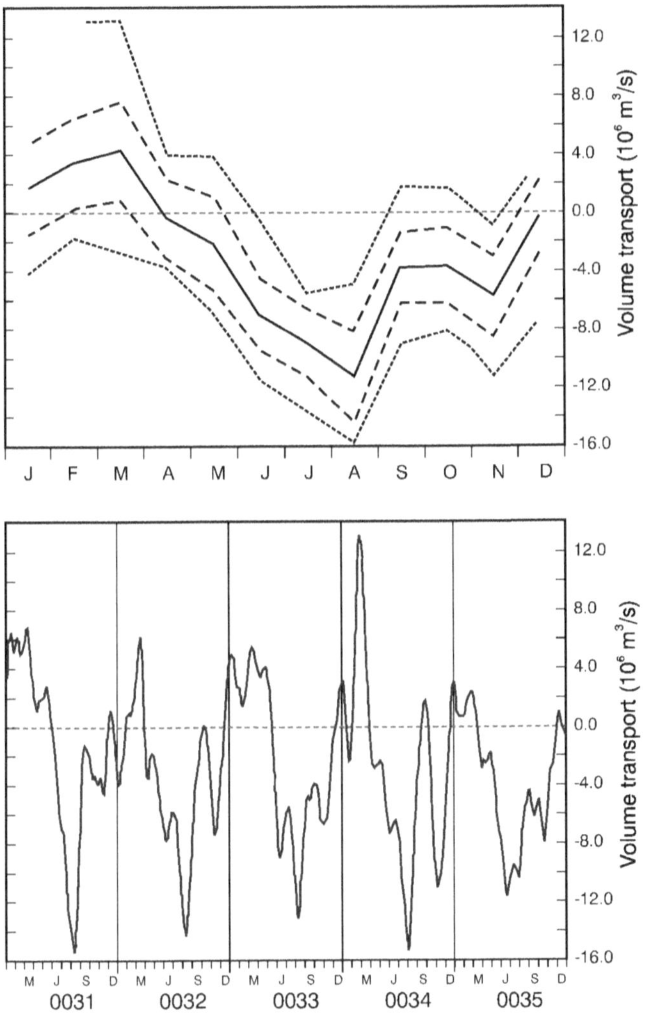

Figure 6.36. The volume transport across the 20° E meridian south of Africa as simulated by a numerical model[777] for a period of five years. The meridional extent from 35 to 45° S covers both the Agulhas Current as well as the Agulhas Return Current. The upper panel shows the mean value as a thick line with one standard deviation bordering the mean and the maximum and minimum values as final borders. The lower panel shows the a time series for these five years highlighting the variability in the flux. Positive values denote eastward transport.

variations of the wind-forced, large-scale circulation. Apart from inter-annual variations in the Agulhas Current retroflection region, there also is some evidence of a seasonal variability[418]. In fact, early results from satellite altimetry[752] suggest that in the South Atlantic Ocean, the strongest seasonality is found at the Agulhas retroflection.

In all this it is crucial to remember that just after spawning Agulhas rings find themselves in an extremely complex and varying environment[628]. Uncomplicated mechanistic visualizations of the subsequent behaviour of Agulhas rings just will not do, as has been experienced in the planning of a number of research expeditions to the region[650]. The waters in the southern Cape Basin constitute a highly energetic field of rings that merge, split, deform and even reconnect to the Agulhas retroflection. To this veritable cauldron may further be added a field of cyclones that interact with the Agulhas rings as well as amongst themselves. An extra complication in estimating the inter-ocean leakage due to Agulhas rings may be the irregular occurrence of upstream retroflection in the Agulhas Current itself.

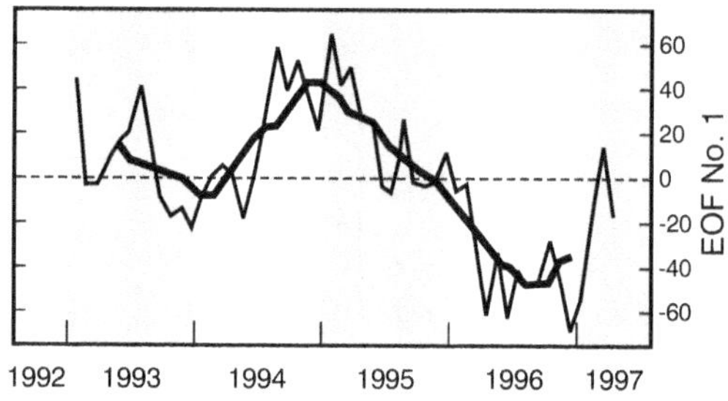

Figure 6.37. The intra-annual variability for the Agulhas retroflection and the Cape Basin for a period of five years[712] as expressed by a regional empirical orthogonal function. This mode 1 explains 45 per cent of the variance in the region. The thick line is a nine-month running mean for the amplitude.

Upstream retroflection

Whereas the normal location of the Agulhas retroflection loop lies west of 20° E[91], there have been instances[64] in which the disposition of sea surface temperatures have suggested that part – if not all – of the Agulhas Current retroflected south of Port Elizabeth (viz. Figure 1.2). These suggestions of an upstream retroflection have been supported by the tracks of drifters. Such early retroflection has been assumed[64] to come about when an exceptionally well-developed Natal Pulse forces the core of the Agulhas Current sufficiently far offshore so that it intersects the shallow bathymetry of the Agulhas Plateau to the south and is thus forced eastward. Few such upstream short circuits in the normal trajectory of the current have been observed to date. On occasion a seemingly incomplete early retroflection has been noticed in thermal infrared imagery[412]. It can be assumed that such short-lived events do not contribute to a major change in the inter-ocean fluxes south of Africa. A long-lasting early retroflection, over the full depth of the current, would naturally have a substantial effect on the interchange south of Africa since Agulhas water would never reach the normal retroflection region. It would therefore not be available for inter-ocean exchange.

It has been assumed[133] that augmentations in the incidence of large Natal Pulses and concurrent increases in early retroflection events would be instrumental in substantial changes in the global thermohaline circulation. For this to be a robust mechanism, early retroflection events would have to be durable. An event lasting a number of months was observed from satellite remote sensing for the first time in 2000–2001[670–1]. This gives an indication of the limited frequency of these events that can be expected. One of the main questions that remain would be the extent to which early retroflections succeed in siphoning off the greater part of the flux of the full Agulhas Current, i.e. how deep do they extend. During the 2000–2001 event fortuitous hydrographic measurements could be made[672] across the path of the current, proving that this particular upstream retroflection involved the greater part of the Agulhas Current. The significance of this finding is substantial.

Global significance

With the ever more accurate estimates of the inter-ocean exchanges by Agulhas rings, the role they play in the global thermohaline circulation cell would seem to be increasingly more reliably quantified. This is not yet the case. Nonetheless, establishing the effect of Agulhas rings remains critical to an understanding of the role in global climate of the one major ocean basin – the Atlantic Ocean – in which there is a substantial net heat flow across the equator in a northward direction. This flow has been estimated[753] to be about 0.29 PW.

Rintoul, using an inverse model, has concluded[565] that no input of warm Indian Ocean is required to account for the net northward heat flux in the South Atlantic Ocean, the flow being totally determined by differences in the water masses entering via the Drake Passage and leaving the South Atlantic sector of the Southern Ocean between Africa and Antarctica. A different model, using as constraint the historical hydrographic data, predicts that an inflow into the South Atlantic Ocean of 4 to 7 × 10^6 m^3/s can be accommodated, but no larger values[247]. A strong correlation is found between the meridional heat transport in this latter model, the strength of the global thermohaline cell and inflow from the Indian Ocean. The model has no Agulhas rings and may therefore be biased.

On the other hand, use of a primitive equation model of the southern hemisphere[566] has suggested that 85 per cent of the northward heat transport into the Atlantic originates in the Indian Ocean, only the remainder coming through the Drake Passage. Others[67] have shown that up to two-thirds of the Benguela Current of the South Atlantic, within and above the thermocline, has its origin in the South Indian Ocean. The total transport has been estimated[759] at 28(±4) × 10^6 m^3/s. Another published value[779] is 25 × 10^6 m^3/s. The results of all inversion studies to date that have estimated the heat flux across 30° S latitude in the South Atlantic are given in Table 6.8; those for modelling studies in Table 6.9. From these tables it is clear that linkages of Atlantic and Indian Ocean waters continue into deep water and, as

Table 6.8. The heat fluxes (F_Q) and the volume transports across 30° S latitude in the South Atlantic Ocean according to a number of inversion studies[516]. Values are given in 10^6 m^3/s; positive values denoting equatorward transport. Values are given for different water masses: SW representing Surface Water; AAIW, Antarctic Intermediate Water; NADW, North Atlantic Deep Water and AABW, Antarctic Bottom Water. *Date* refers to the date on which a hydrographic section was carried out along 30° S latitude.

Authors	SW	AAIW	NADW	AABW	F_Q [PW]	Date
Fu (1981)[567]	15	10	−24	−2	0.85	July–August 1925 METEOR
	9	6	−20	1	0.88	April–June 1959 IGY
Rintoul (1991)[565]	8	5	−17	4	0.25	April–May 1959 IGY
MacDonald (1993)[568]	6.1	7.9	−21.6	7.5	0.3	February 1988–February 1989
Schlitzer (1993)[569]	2.2	10.0	−15.8	3.1	−0.05	historical
Schlitzer (1996)[570]	2.0	11.9	−18.7	4.2	0.3	historical
Boddem and Schlitzer (1995)[247]	−1.9	9.8	−8.9	1.1	0.04	historical
Holfort et al. (1998)[571]					0.26	January 1993 WOCE

could be expected from the relative volumes of the water types[363], are greatest by volume in deep water.

The set of historical hydrographic data in the Agulhas retroflection and in the South Atlantic Ocean show[250] the progression of Antarctic Intermediate Water to mirror that of the Agulhas Current and Agulhas rings, but cannot resolve their influence directly. Agulhas rings do not seem to have an adequately distinct hydrographic signal, or are not present in sufficient abundance, to leave behind a tell-tale record in the hydrographic data at depth to indicate their range of influence or average path[195]. Nevertheless, estimates of potential vorticity on the 27.3 isopycnal[575] shows clear evidence of leakage of Antarctic Intermediate Water from the Indian to the Atlantic Ocean and its penetration in a north-westerly direction across the South Atlantic Ocean.

Modelling of the impact of inter-ocean exchanges on the thermohaline overturning of the Atlantic Ocean[68,668–9] has been very suggestive. Weijer et al.[68] have, as mentioned above, shown that the heat and salt transports by the South Atlantic subtropical gyre play an essential role in the heat and salt budgets of the Atlantic as a whole. It has been shown that in this model the exported North Atlantic Deep Water is fresher than the return flows and that the overturning circulation thus exports freshwater from the Atlantic Ocean. Even small changes in the composition of the return transports of the North Atlantic Deep Water may influence the overturning circulation in this ocean considerably. The model furthermore shows[68] that interocean fluxes of heat and salt are important for the strength and operation of the overturning circulation. Comparing the roles of the inter-ocean exchanges between the Atlantic, the Pacific through the Drake Passage, the Pacific through the Bering Sea and the South Indian Ocean[668] in a global circulation model, it can be shown that it is especially the Indian–Atlantic transfers of heat and salt by leakage from the Agulhas Current that contributes to the strength and the stability of the northern sinking circulation. When the stabilizing effect of the leakage from the Agulhas Current disappears, the destabilizing influence on the overturning circulation by freshwater from the Bering Strait becomes more effective. Of particular importance in these model studies has been an investigation on the influence of water from the Agulhas Current on the Atlantic overturning as a whole[669] (viz. Figure 6.38).

From Figure 6.38 it is clear that the model's overturning circulation is sensitive to changes in the inter-

Table 6.9. The heat fluxes (last column) and volume transports for different water masses across 30° S latitude in the South Atlantic Ocean, according to a few modelling studies[516]. Volume fluxes are given in 10^6 m^3/s. Water masses are: SW: Surface Water; AAIW: Antarctic Intermediate Water; NADW: North Atlantic Deep Water and AABW: Antarctic Bottom Water.

Authors	SW	AAIW	NADW	AABW	F_0 (PW)
FRAM Group (1991)[272]	11	8	−22	3.2	0.56
Semtner and Chervin (1992)[274]	12	4.7	−18	1.3	0.60
Matano and Philander (1993)[573]	6.8	1.6	−10.9	2.5	0.19
Thompson et al. (1997)[574]	12.7	6.8	−20.9	1.4	0.56

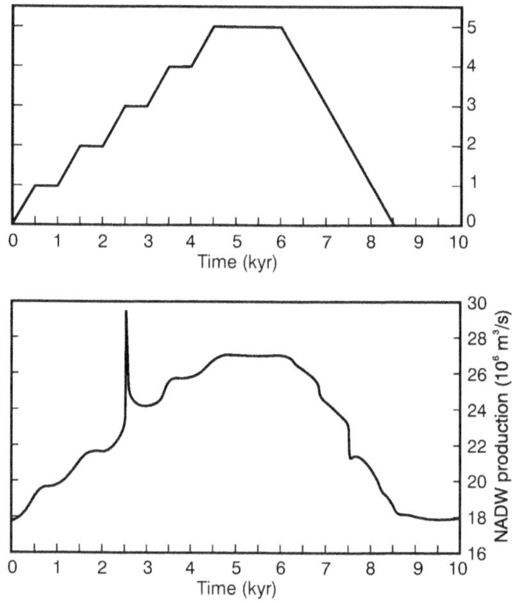

Figure 6.38. Response of the overturning circulation in the Atlantic Ocean to changes in volume flux from the Agulhas Current[669]. The upper panel shows the volume flux applied to a model, where a value of 1 is the value currently estimated by observations at sea[129]; 0.045 PW for heat and 2.52 Gg/s for salt. The lower panel shows the concurrent production of North Atlantic Deep Water. It closely reflects the source function in the upper panel, following each step in the increased volume flux as applied to the model. Stopping the throughflow entirely results in a remaining flux of about 18 × 10^6 m^3/s in the production of North Atlantic Deep Water.

ocean leakage of water from the Indian Ocean. The response of the overturning strength to changes in the inter-ocean transfers is mainly linear. Changes in the transfers of buoyancy affect the strength of the Atlantic Ocean's overturning by the modification of the basin-scale meridional density and pressure gradients. This response takes place within a few years, being the time it takes for barotropic and baroclinic Kelvin waves to reach the northern Atlantic Ocean. The heat and salt anomalies inserted into the model's South Atlantic by contrast take three decades to be advected all the way to the northern North Atlantic. These model studies suggest the decisive influence alterations in the inter-ocean exchanges of heat and salt south of Africa may have on the global overturning circulation. The importance of this influence has been confirmed by palaeoceanographic results[667] discussed elsewhere. But how to quantify this importance?

Using Lagrangian path following techniques it has been shown[756] that 90 per cent of the upper branch of the overturning circulation in the Atlantic Ocean is derived from inflow of Indian Ocean water. One may wonder about this, since the inter-ocean leakage from the Agulhas Current takes place in the upper 2000 m of the water column, but it has been demonstrated that 95 per cent of all the volume transport that contributes to the upper branch of the thermohaline overturning circulation is found in the upper 1000 m. In contrast to other studies[774–5] this analysis indicates that almost all water from the Drake Passage moves eastward, past the tip of Africa.

Loss of water from the Indian Ocean by the formation of mesoscale eddies at the Agulhas retroflection is not, however, restricted to Agulhas rings.

Agulhas Current eddies

The portrayal of temperature fronts at the Agulhas Current retroflection presented in Figure 6.2 shows a number of circular features to the south of the retroflection. The first vortex of this region that has been hydrographically observed[511] may well have been one of these. This seems clear from the fact that it was well-imbedded in surface water colder than 14 °C, the mean temperature for the front of the Subtropical Convergence in this region[97]. Eddies of this kind are continually being formed, carrying substantial amounts of heat poleward across the Subtropical Convergence[458] thus contributing to the global, meridional heat flux of the ocean. However, they may lose up to 800 W/m^2 of heat to the atmosphere[496] under the cold and stormy conditions found in this region and thus exhibit considerable effects of convective overturning in their upper layers.

Those that have been observed hydrographically extend into deep water[262,455] and have azimuthal velocities similar to that of the parent current. How many are shed per year is not known. Very persistent cloud coverage over the region has limited the use of satellite observations in the thermal infrared, while altimetric measurements have not been able to resolve such features well[362]. This could possibly be explained by the features remaining virtually stationary. With a strong eastward current and substantial meridional shear in the flow, this is not what one would expect. Analyses of these features[63] have suggested that they populate only a restricted region, but this is perhaps best included in the discussion of the Agulhas Return Current and the South Indian Ocean Current that follows.

The important concept that needs to be kept in mind here is that the loss of water that occurs between the southern Agulhas Current and the Agulhas Return

Current – in other words, while the current passes through the retroflection – is not only lost to the South Atlantic Ocean, but also to the Southern Ocean. The processes in both cases might be much better quantified if the dynamics were better understood. To this end a fair degree of modelling, both analytical and numerical, has been carried out.

The dynamics of the Agulhas retroflection

The forcing of the processes that occur at the Agulhas retroflection has been investigated by modelling using a wide range of approaches. These may, for ease of description, be grouped into four broad categories. First, there have been attempts to simulate the flow path of the Agulhas Current by using an inertial jet[130,506]. Secondly, a series of wind-driven models have been constructed specifically for the Agulhas System employing increasingly realistic coastal outlines and bottom topographies[268,576–7]. Thirdly, modelling the system incorporating data assimilation[265,548,578–9] – mostly satellite altimeter observations – has considerably increased the verisimilitude of modelling results and, fourthly, a number of global circulation models have simulated certain aspects of the Agulhas retroflection[273,276] fairly well, and therefore are instructive about the forcing involved. Modelling[469,580] of a more analytic nature, to address certain fundamental problems concerning the reasons for a retroflection[741,778], have been discussed above.

Inertial jet models

Hydrographic investigations as well as satellite remote sensing in the thermal infrared have all shown that the Agulhas Current follows the edge of the continental shelf quite religiously and that downstream of the Agulhas retroflection the path of the Agulhas Return Current is noticeably affected by the presence of shallower regions. Early modelling efforts have therefore concentrated on the sensitivity of the current trajectory to the bottom topography and thus its role on the retroflection, and shown that in some model configurations the Agulhas Current is very sensitive to small changes in current speeds at the bottom[506].

Using a more realistic polygonal velocity profile[130], it has been demonstrated that penetration of the Agulhas Current into the Atlantic Ocean is a function of high current shear and high bottom velocities. Penetrations of the Agulhas retroflection loop are a function of volume transport in the current; the more westerly retroflections occurring, according to this model, with lower volume transports (Figure 6.39). Although the simulated path resembles a retroflection loop, these loops are unstable, the jet trajectories crossing themselves further upstream. This problem has subsequently been addressed[469] by using a model in which the boundary current is confined to the upper layer.

The point at which the simulated boundary current in this specific model leaves the coastline is a function of the volume flux, upstream separation occurring with increased volume transport. This more sophisticated model furthermore suggests that both the inertial and the beta (β) effect play an important role in the retroflection of the Agulhas Current. Using a transport model of the large-scale wind-driven ocean circulation in the subtropical region of the Atlantic and Indian Ocean[580] De Ruijter has demonstrated that inertia must be incorporated in model configurations in order to achieve a retroflection for the Agulhas Current.

Wind-driven models

In such a model the meridional gradient in the wind-stress curl over the Indian Ocean domain is a controlling factor for the Agulhas retroflection. If the wind-stress curl decreases substantially southward, most Agulhas Current water ends up in the South Indian Ocean Current. If not, a larger proportion of its transport will bend westward[580]. Increasing the spatial resolution of such wind-driven, barotropic models[576] has shown that the retroflection of the simulated Agulhas Current is largely due to the net accumulation of β-generated, anti-cyclonic, relative vorticity as the current follows an inertially driven southward path after having separated from the tip of Africa. The degree to which Agulhas water from the Agulhas retroflection penetrates into the South Atlantic has been shown primarily to be determined by the latitude of the zone of zero wind-stress curl[576]. Since this zone may wander with season, as well as interannually, the degree of isolation of the two anti-cyclonic gyres east and west of Africa may change commensurably.

By increasing the complexity of this particular model through the inclusion of baroclinicity[581], Boudra and De Ruijter have successfully increased the intensity of the simulated retroflection loop, but decreased the inter-basin leakage of surface water (Figure 6.40). The process responsible for the Agulhas retroflection in this model remains an adjustment to the change in the vorticity balance as the current overshoots the continent. Along the coast the current gains relative vorticity by the β-effect, but loses this as friction to the shelf edge. After separation, however, gain in relative vorticity is accommodated by an eastward turn. Agulhas ring formation in the model may thus require inter-

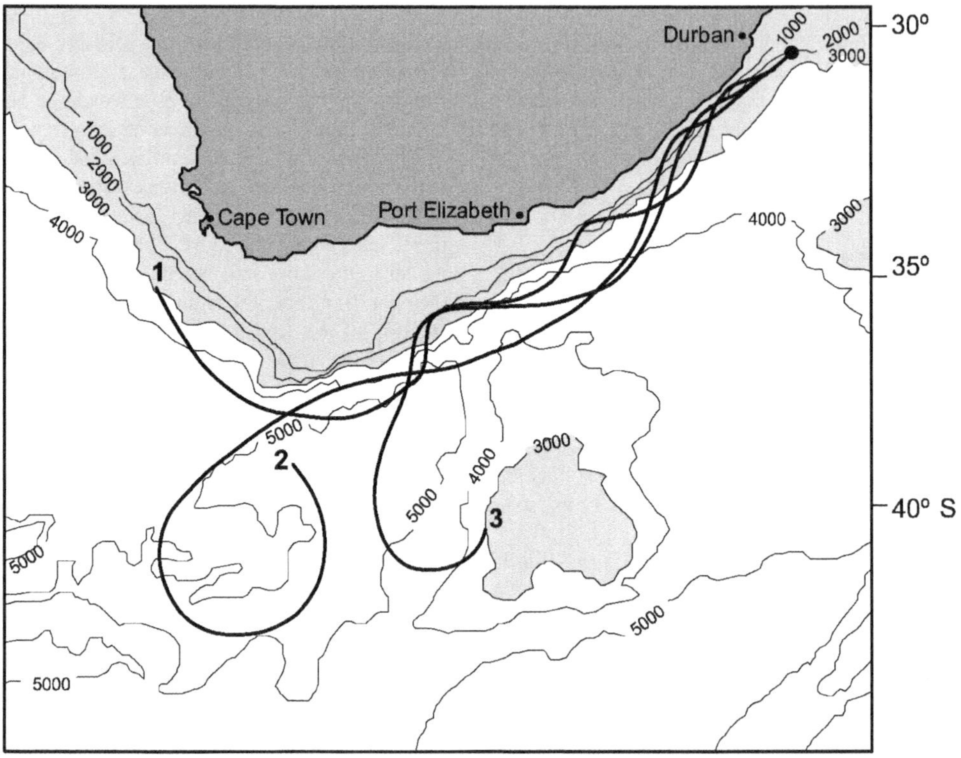

Figure 6.39. Trajectories of a free inertial jet used to simulate the flow path of the Agulhas Current in the retroflection region[130]. Farther westward penetration of the retroflection loop is achieved by decreasing the volume flux of the jet.

action with the eastward flow to the south[581].

The importance of a substantial viscous stress curl along the coast of Africa in determining the nature of the Agulhas retroflection has become evident in more advanced model simulations[582]. Whether rings will form in this model configuration has been shown to be primarily a factor of the southward inertia and the baroclinicity of the overshooting Agulhas Current[132]. Baroclinic–barotropic instabilities have been suggested by this model version as being associated with ring formation. No attempt has been made in these models to include thermodynamic forcing[529], but allowing isopycnal outcropping has made this series of wind-driven models more realistic. The model results from these latter numerical experiments, shown in Figure 6.41, show the type of realism that has been achieved. These simulated rings have a coherent structure all the way to the ocean floor[577]. The model shows that rings that have drifted into the South Atlantic move westwards predominantly due to the large-scale ambient water movement in the gyre.

Pichevin et al.[583] have attempted to understand, from an analytical viewpoint, why rings are shed from the Agulhas retroflection at all. They come to some unconventional conclusions. Using a reduced gravity, one-and-a-half layer, primitive equation model they show that the generation of rings from a retroflecting current is inevitable. They conclude that the triggering of ring spawning is not necessarily due to instabilities but, rather, is due to the zonal momentum flux of an Agulhas jet that curves back on itself. To compensate for this momentum flux, rings have to be produced. Spawned rings exert a compensating momentum effect analogous to the backward push when a rifle is fired. The fact that the observed rings are considerably larger than what the local Rossby radius of deformation would suggest that they should be, is explained in a novel way. Vortices at the Rossby radius would come about due to normal flow instability; here the rings need to balance the momentum flux of a large retroflecting current, hence their size. In this model[583] a simulated Natal Pulse has no obvious relationship to ring occlusion.

The causes of the Agulhas retroflection have also been investigated using the Princeton Ocean Model, a primitive equation model in sigma co-ordinates[268]. A series of process-oriented studies, using different wind-

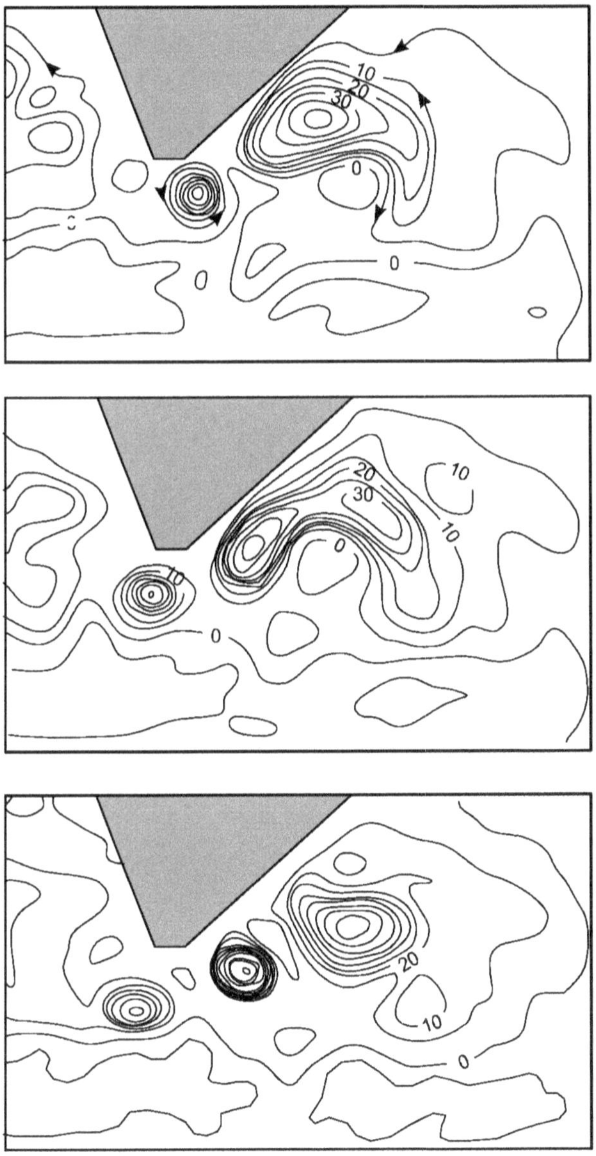

Figure 6.40. Modelling results that simulate the shedding of an Agulhas ring at the Agulhas retroflection and the movement of a previously shed ring into the South Atlantic Ocean[132]. The mass transport stream function is shown for the upper layer of the model. Contour intervals are at 7×10^6 m^3/s. The results are given, from top to bottom, for days 2950, 2990 and 3010.

stress distributions and different degrees of smoothing of the bathymetry, has produced some interesting results. It has been shown that in this model the simulated Agulhas retroflection is more strongly affected by the torques exerted by the bottom topography than by the effect of β-accumulated vorticity or the effect of coastline curvature. An adaptation of the Princeton Ocean Model, NORWECOM, has been used[765] to study biological aspects of the southern part of the Agulhas Current system as well as the Benguela upwelling regime. This includes primary production. The near oligotrophic nature of the south Agulhas Current is successfully simulated.

An eddy-permitting model that focuses on the Cape

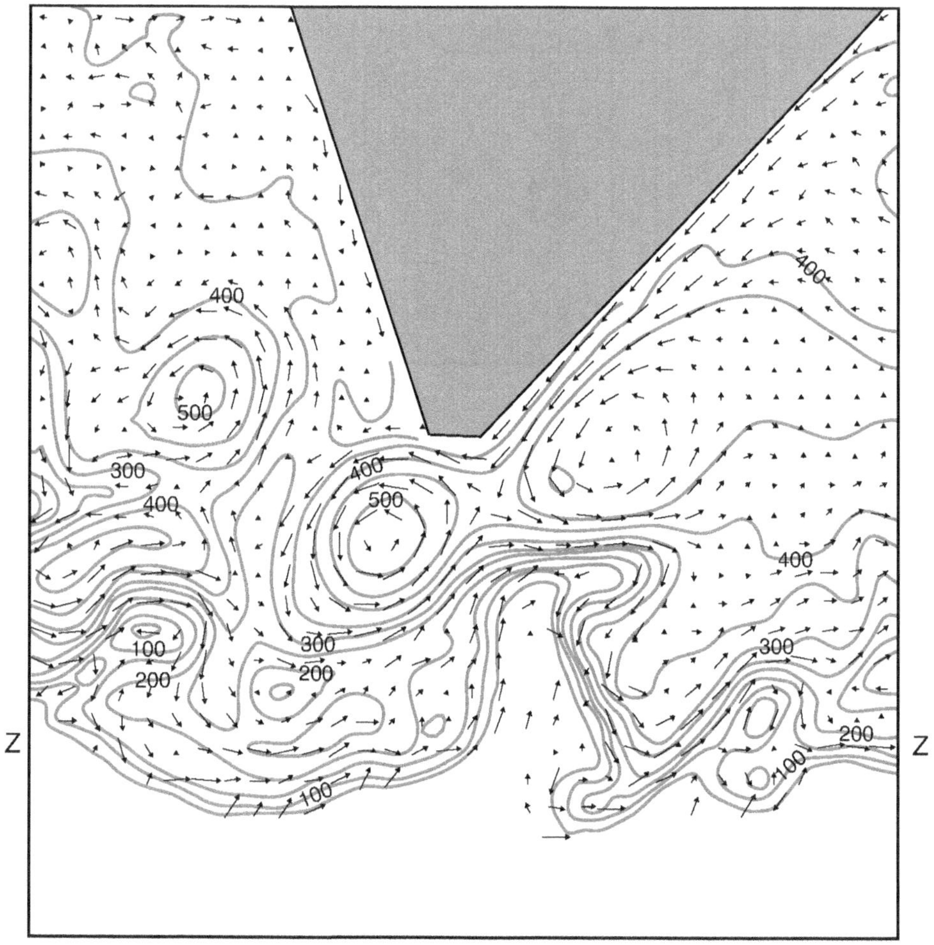

Figure 6.41. Detail of the Agulhas retroflection loop, a ring-spawning event, progression of a previously spawned ring and meanders in the Subtropical Convergence and Agulhas Return Current, all as simulated by a pure-isopycnic, numerical model with three layers[529]. Shown is the velocity and thickness field of the intermediate layer. Flow vectors are given for every other grid point. The latitude where the wind stress curl is nil is shown by the letter Z along the borders of the figure. Note the substantial meander in the Agulhas Return Current over the Agulhas Plateau.

Basin[776], but that includes a large part of the South West Indian Ocean and the South Atlantic Ocean, has successfully simulated the role that Agulhas rings play not only in the transients of the region, but also fluxes associated with the mean circulation. Modelled rings, correctly, indicate that most of the energy in the Benguela Current is supplied by themselves. This model shows the co-existence of anti-cyclonic rings and cyclones in firm dipole structures. This modeled configuration should not be confused with the freely moving cyclones derived from the west African coastline that were described in detail above[628]. The modelled cyclones[776] are bottom intensified vortices with barotropic structures. Their passage is blocked by the Walvis Ridge and the Vema Seamount. Using such a eddy-permitting model to evaluate the variability in the inter-ocean fluxes south of Africa[777] has shown a seasonal variation of about 10 per cent across a section at 35° S in the South Atlantic Ocean and around 20 per cent through a section at 20° E. Simulated volume transports of the Agulhas Current through a section at 35° S are about 58 × 10^6 m^3/s in summer/autumn and about 64 × 10^6 m^3/s in winter/spring (viz. Figure 6.42). Short term variability in this model simulation is large and seems realistic.

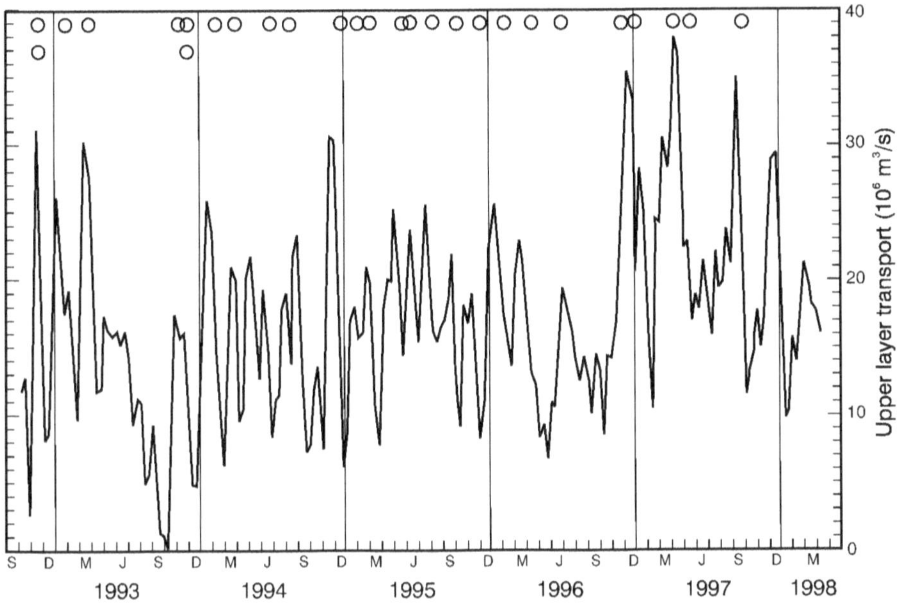

Figure 6.42. A time series of the upper layer volume transport of the Agulhas Current into the South Atlantic across the 19° E meridian[782]. The circles indicate the times when each Agulhas ring that contributed to the flux was first detected. Particularly noteworthy is the high level of temporal variability.

Modelling with data assimilation

One of the major limitations of most of these models of the Agulhas retroflection and of the ring shedding process is the inability of these models to predict events with sufficient accuracy that they may reliably be used prognostically. This could conceivably be achieved if real-time data of sufficient spatial resolution were to be assimilated on a regular basis. A first attempt to do this for the region around southern Africa has been carried out[265] with promising results. A next attempt[548] has used an ensemble Kalman filter to assimilate Geosat altimeter data into a two-layer quasi-geostrophic ocean model.

This method has increased the frequency of ring shedding which in most other quasi-geostrophic models is too low. It has therefore been concluded that this type of data assimilation system accommodates ageostrophic effects that cannot be accounted for in other models of this kind. This procedure has been further developed[578] by combining the time-varying part of altimetric data with a two-layer, quasi-geostrophic model, imposing the time-mean circulation as an unknown. This data assimilation experiment has been successful in reducing errors in the time-mean, sea surface topography from about 10 cm to 3 cm. Van Leeuwen has subsequently shown[584] that in such data assimilation models designed to study Agulhas Current processes a smoother will give superior results to a filter.

The most successful of the data assimilation models for the Agulhas retroflection region to date has a $1/6°$ grid, with four layers, is quasi-geostrophic and incorporates altimetric data from both the TOPEX/Poseidon and the ERS 1 satellites[579]. Not only the large-scale, time-mean circulation is simulated well by this model; the meso-scale processes also are very realistic (Figure 6.43). In the Agulhas Current proper a surface speed of 1.3 m/s is simulated; the volume flux above 1200 m is 75×10^6 m^3/s and the general disposition of the retroflection in the model agrees closely with what has been observed. The model suggests an interesting decrease in the core speed of the Agulhas Current from 130 cm/s to 80 cm/s along the Agulhas Bank. This needs to be confirmed by appropriate measurements in this region. The ring-shedding process is simulated well, as are the subsequent drifts of Agulhas rings across the South Atlantic, even including the dissipation of three rings in the Cape Basin[579]. This is not particularly remarkable, however, since information on these rings is assimilated from altimetric anomalies. All the above-mentioned models either have a relatively coarse horizontal resolution or a small number of layers in the vertical. In most of the global circulation models this is very different.

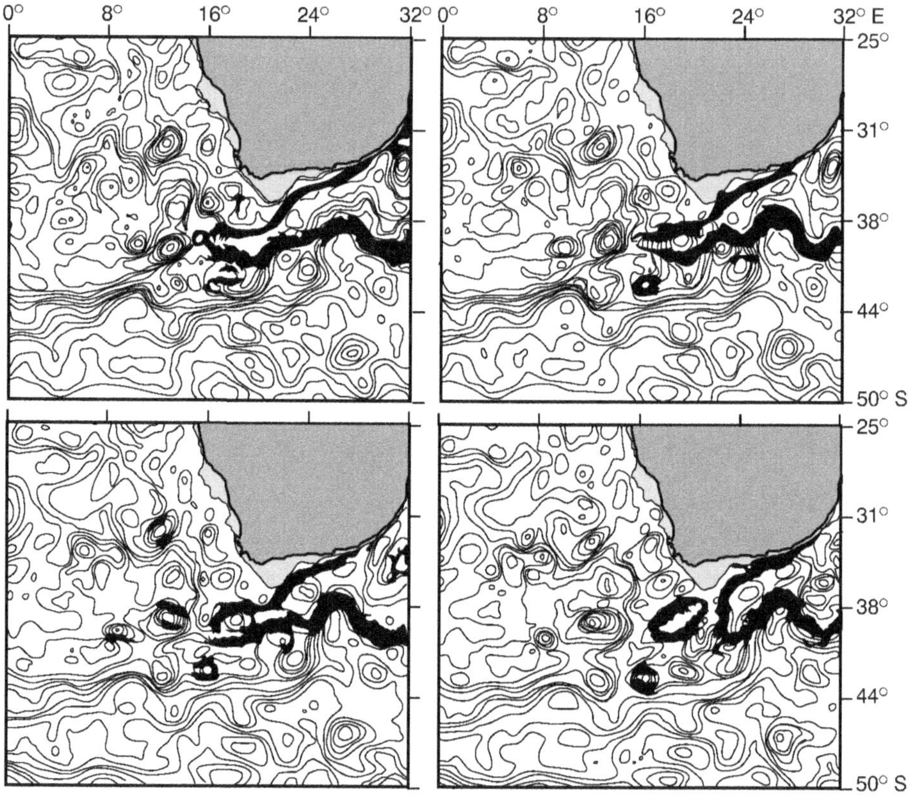

Figure 6.43. Ring shedding at the Agulhas Current retroflection as simulated by a four-layer, quasi-geostrophic numerical model with data assimilation[776]. The panels are representations of the instantaneous stream function at ten-day intervals, starting on 4 August 1994. The northward penetration of a meander in the Agulhas Return Current and its role in the occlusion of an Agulhas ring is evident. This corresponds well with what has been observed for this process using satellite thermal infrared observations[91].

Global circulation models

With a ½° latitude by ½° longitude spatial grid, 20 vertical levels, a realistic geometry and annual mean wind forcing, the eddy resolving model by Semtner and Chervin[273] has simulated the spawning of warm-core rings which enter the South Atlantic Ocean and move off in a northwest direction. The spawning of both anticyclonic and cyclonic disturbances has been produced by this model. These dipoles have not been unambiguously observed in most of the hydrographic data to date, but there are some suggestions of their presence in satellite altimetry[362]. An important result of this particular model has been its simulation of the global thermohaline circulation cell and, particularly, the warm water path due to Agulhas rings.

Changing the wind forcing to climatological monthly forcing[274] in this global model does not change this result substantially. The geographic distribution of eddy-variability produced in the model resembles closely that for the Agulhas retroflection region found from altimetric data[73]. This is also reflected in further studies[275] that have compared the eddy kinetic energy in a model simulation of the southern Agulhas Current, Agulhas Retroflection and Agulhas Return Current with altimeter, drifter and current-meter data. More advanced forms of the Semtner model[325], driven by very realistic atmospheric forcing, have simulated the shedding of Agulhas rings even more realistically. In this global, eddy-resolving model 3.6 rings are shed per year and they take about three years to cross the South Atlantic Ocean. It is of interest to note that the rings take two preferred paths in this version of the model[325] and that the split seems to coincide with the location of the Vema Seamount.

The FRAM (*Fine Resolution Antarctic Model*)[272] is,

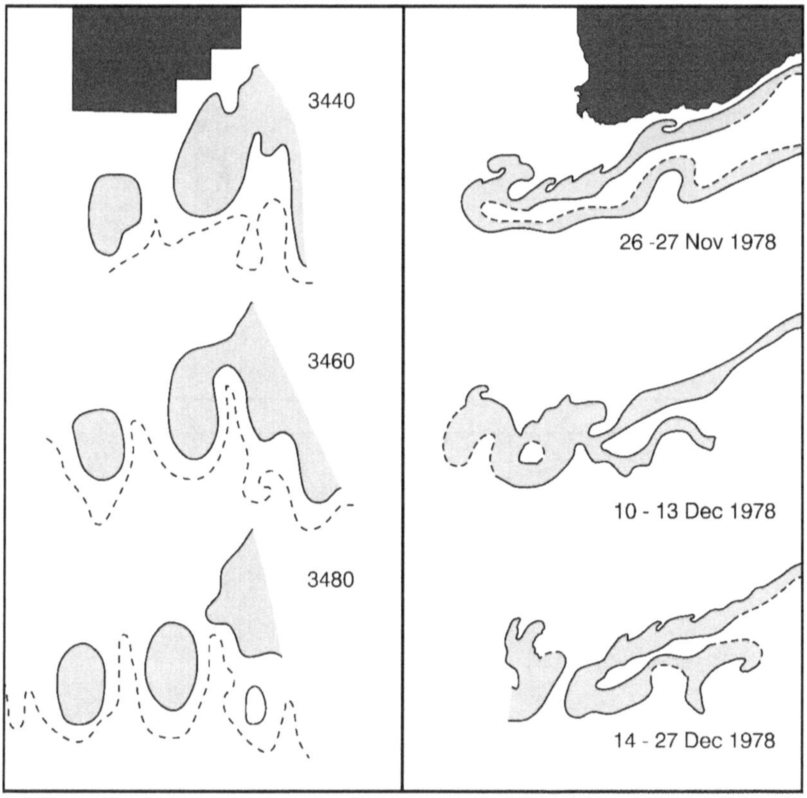

Figure 6.44. The ring shedding process at the Agulhas Current retroflection as observed from satellite infrared imagery[60] (right-hand panel) and simulated by the FRAM (Fine Resolution Antarctic Model; left-hand panel)[277]. Broken lines denote the Subtropical Convergence; dotted lines regions where temperatures were poorly resolved. A comparison between actual dates (right) and model days (left) shows that the process occurs too slowly in this model.

by contrast, restricted to the region south of 24° S latitude, with a ¼° north–south spatial resolution; ½° in the east–west direction. It has 32 horizontal levels, spaced at increasing intervals with depth. It was run with a weak relaxation to the mean values produced by Levitus[348] and allowed to run freely after that. It models the creation of Agulhas rings very convincingly[276] (Figure 6.44). They start with an internal volume transport of $140 \times 10^6\ m^3/s$, and are shed at 160 day intervals. Both these values are too large when compared to observation[74,91]. Fortuitously these values partially compensate each other so that the net heat flux is realistic[585].

The rings modelled by the FRAM drift off into the South Atlantic slowly losing their kinematic and hydrographic characteristics, but do not stray from a singular track followed by all rings[277] (Figure 6.45). This is clearly at odds with observation (e.g. Figures 6.19, 6.33). Rings in the model are also shed too far upstream. The regular cycles of wind-stress and the simplified bottom topography may be the respective culprits. The FRAM shows no direct interaction of Agulhas rings with the coastal upwelling system of the South East Atlantic[562], possibly due to the invariant, offshore tracks of all simulated rings. The model does indicate that the thermal structure of the South Atlantic Ocean, and in consequence the meridional heat transport, depends heavily on the input of heat via Agulhas rings[574]. Models that do not include this inter-ocean exchange south of Africa[544,565] exhibit a much lower equatorward transfer of heat.

Clearly, even the most sophisticated numerical models are at present incapable of simulating the Agulhas retroflection and the ring-shedding process with a verisimilitude that makes them reliable tools for prediction or even experimentation. Nevertheless, the realism with which these processes are already represented by models suggests that most of the underlying physics of

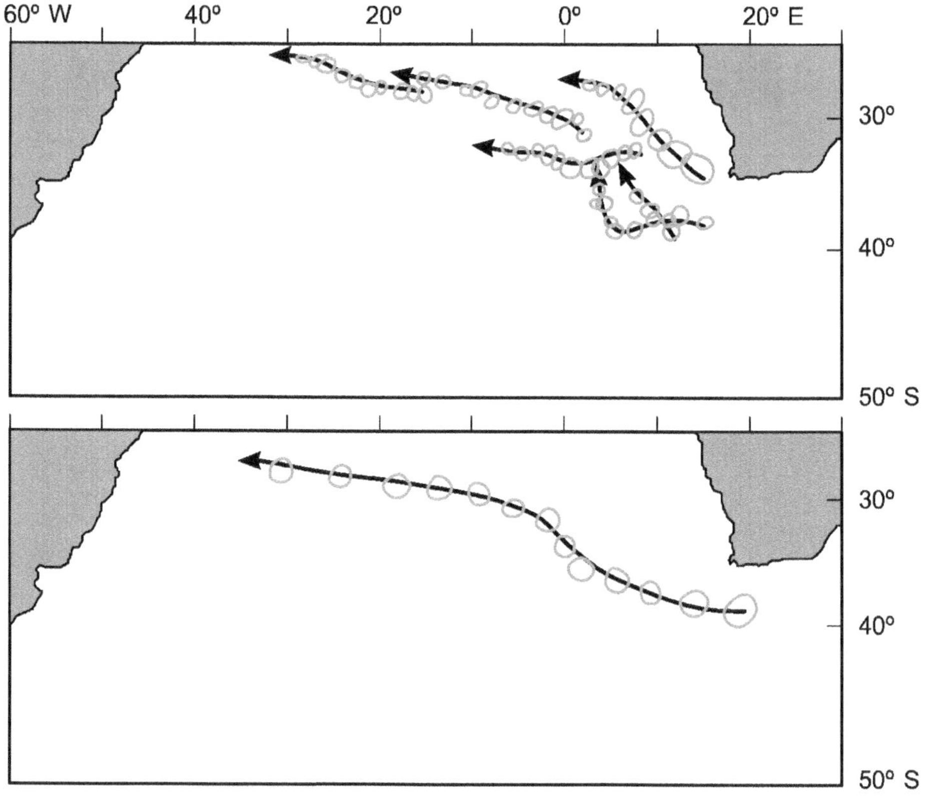

Figure 6.45. Trajectories of Agulhas rings in the South Atlantic Ocean from satellite altimetry[464] and as represented by a streamline field for model-year 8 in the FRAM (Fine Resolution Antarctic Model)[277]. Rings in this model follow the path shown without exception, whereas rings in nature move over a wide range.

these processes is adequately understood. With an increase in spatial and temporal resolution for the models, with more realistic bottom configurations and wind-stress forcing, an Agulhas retroflection simulation closer to that available to date can be expected from numerical models in the near future.

One of the smaller-scale processes that may have a decided effect on the rate and timing of ring shedding at the Agulhas retroflection is the downstream movement of the Natal Pulse[62]. As discussed previously, it has been shown that a well-developed Natal Pulse may cause upstream retroflection between Port Elizabeth and the Agulhas Plateau[64] (viz. Figure 1.2). The logical question would then be[131] whether the downstream progress by an average Natal Pulse all the way to the retroflection would precipitate ring shedding in an already far-prograded retroflection loop. Some numerical models suggest this, and recent results from satellite altimetry[401] have largely substantiated this process (viz. Figure 6.8).

Overview

The development, spatial scales and temporal behaviour of the Agulhas Current retroflection are now fairly well known. Forced mainly by a balance between inertia, planetary vorticity and bottom topography, the retroflective behaviour is thus increasingly well-modelled by a range of numerical models. This suggests that the underlying physics may be adequately understood. However, accurate predictive capability has not been reached yet. This might occur sooner for the ring shedding events at the retroflection.

The process of Agulhas ring spawning seems to be partially, but not totally, due to an imbalance in momentum. The timing of shedding events seems to be a result of the arrival of Natal Pulses. The downstream translation of these triggering features seems highly predictable so that there may be a great deal of prognostic potential in monitoring the onset of Natal Pulses at the Natal Bight.

The behaviour of Agulhas rings subsequent to spawning has been intensively studied. Their observed drift behaviour across the South Atlantic Ocean has nonetheless raised a number of key questions. A large number seem totally to disintegrate in the Cape Basin. This raises questions about the mixing processes involved in their decay. A large observational programme has been undertaken to investigate these. A substantial number split into smaller eddies that may, conceivably, mix out faster. This splitting process may be strongly influenced by the presence of seamounts. The physics of such processes needs to be investigated. There is no clear-cut indication of how rings are affected by the Walvis Ridge. All disparate model predictions are accommodated by the behaviour of at least some observed rings on crossing this ridge. On having passed this obstacle, the behaviour of the remaining rings seems more uniform, suggesting the filtering behaviour of the Walvis Ridge. From the diverse behaviours of Agulhas rings in the Cape Basin it is clear that there is a wide spectrum of natural histories for Agulhas rings once shed, substantially affecting the interbasin exchanges south of Africa.

Models currently in use for studying the global thermohaline cell show the decisive influence of the leakage of water from the South Indian Ocean into the South Atlantic Ocean on the overturning behaviour. The fluxes at depth are still poorly understood, but are being investigated. The fluxes in the upper layers are dominated by the shedding of the huge Agulhas rings. Direct exchanges and Agulhas filaments play minor roles. In this region of extreme mesoscale flow variability there is also a substantial exchange between the subtropics and the subantarctic, mostly by the shedding of Agulhas eddies from the Agulhas Return Current.

7

The Agulhas return flow

Once the water of the Agulhas Current has successfully negotiated the Agulhas retroflection, it flows eastwards as the southern limb of the South Indian anti-cyclonic-gyre. At this stage of its life cycle the current has lost a lot of heat from its upper layers to the colder ambient atmosphere. It has also lost parts of its volume flux to the South Atlantic Ocean, both by the shedding of Agulhas rings and by Agulhas filaments. It has passed through a region of extreme variability and mesoscale turbulence and has therefore also undergone a substantial degree of mixing. Nevertheless, it emerges sufficiently intact that its Agulhas characteristics are still clearly evident.

Since the very earliest days[14] of scientific interest in this region it has been recognised that some of the water that leaves the retroflection moves, more or less zonally, in an eastward direction, notwithstanding the fact that early investigations[44] had very few data farther to the south. During 1899[586], for example, a disabled steamer, the *Waikato*, drifted with this return flow. Its drift route adumbrated the results of drifting weather buoys a century later. It was also clear, from an early stage in hydrographic investigations[40] of the region, that this Agulhas return flow was part of, or moved parallel to, the Subtropical Convergence in the South Indian Ocean[92]. This front is, in itself, a very remarkable ocean feature.

Not only does the Subtropical Convergence of the South West Indian Ocean exhibit some of the highest horizontal temperature gradients in the world ocean[96]; it also forms part of the most highly variable part of the world ocean[183,587] and produces a very characteristic and unusual biological habitat. It has, for instance, been shown[441, 588] that the levels of chlorophyll-*a* at this front are considerably enhanced and that this is a result not of mechanical accumulation of immotile organisms, but is due to increased potential primary productivity[589]. This is evident in the distribution and density of diatoms[590] between Africa and the subantarctic waters to the south and may also be reflected in the increased number of birds[591] and cetaceans along the Subtropical Convergence. The count of birds and whales in the Agulhas Current retroflection[592] and in its rings is, by contrast, very low.

The juxtaposition of the narrow flow of Agulhas Current water eastward and the Subtropical Convergence is complicated by a very variable bottom topography that affects the flow path, the dynamic meandering and shedding of mesoscale eddies to either side of the flow as well as the leakage of water northward to join the recirculation of various components of the subtropical gyre of the South Indian Ocean. As was clear from the previous discussions, this anti-cyclonic gyre is by no means a simple system but consists of a number of subcells. The sub-gyre of the South West Indian Ocean (viz. Figures 3.15 and 3.30) is particularly strong and well demarcated. The near-closure of the subtropical gyre at about 70° E is suggested by some recent results[252]. What effects these sub-gyres have on the flow of Agulhas water eastwards still needs to be determined with precision[67]. Much of what happens to this flow farther downstream is determined by what happens to this water just beyond the Agulhas retroflection in the Agulhas Return Current proper.

The Agulhas Return Current

The eastward flow of Agulhas water is usually called the Agulhas Return Current up to that longitude where the South West Indian Ocean sub-gyre is believed to terminate, i.e. more or less at the Kerguelen Plateau, at 70° E longitude (viz. Figure 3.30). Beyond this it is known as the South Indian Ocean Current[100] (see inset on page 220). Most meridional hydrographic sections across the Agulhas Current that extend poleward beyond 40° S show evidence for the presence of the Agulhas Return Current, as do most analyses of historical hydrographic data for the region (e.g. Figure 7.1). One of the major problems in tracing this water is apparent from Figure 7.1, constructed in the late 1960s. There were then very few hydrographic data east of about the longitude 30° E. This has altered somewhat during the past 25 years, helping to give a clearer indication of the course Agulhas water takes across the South Indian Ocean.

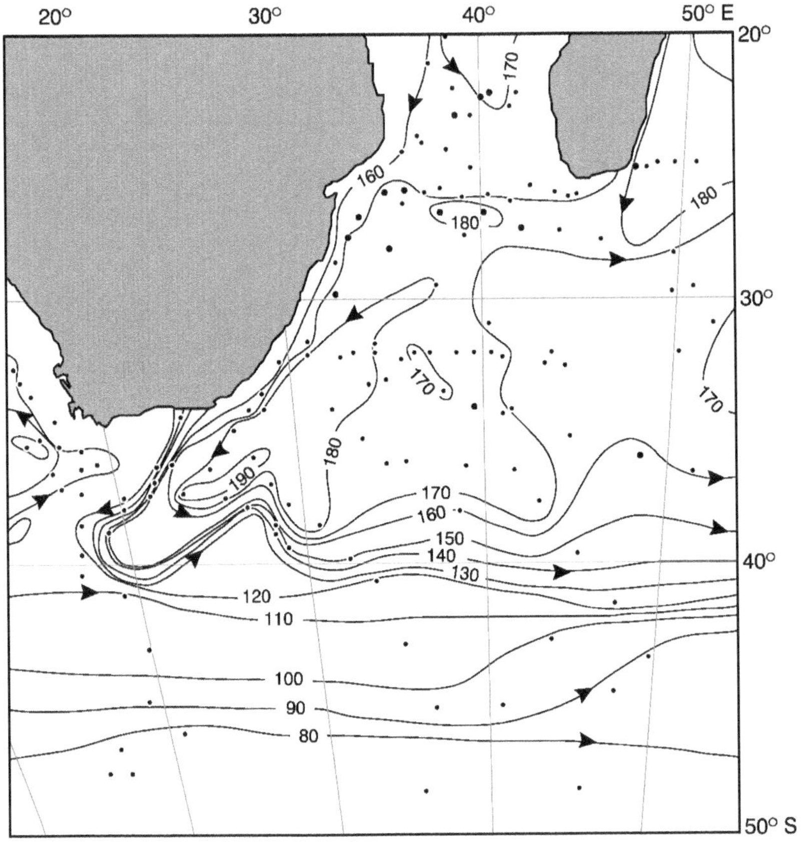

Figure 7.1. The Agulhas Return Current is expressed by the dynamic topography of the sea surface (in dynamic centimetres) as a zonal bunching of isolines relative to the 1000 decibar surface roughly at the 40° S parallel. Historical hydrographic data collected during the calendar months June, July and August were used[75] for this portrayal.

Path of the current

From the abovementioned Figure 7.1 two aspects of the location of the path of the Agulhas Return Current may nonetheless be seen. First, on average it lies at about 40° S; secondly, it meanders strongly at certain specific locations. These meanders have been apparent in the very first studies of the system using historical hydrographic data[40] and in most subsequent ones as well. Dedicated research cruises identified this flow (Figure 5.2) on numerous occasions[215] and the meridional meanders have been recognised to be a characteristic part of the current. The first detailed, quasi-synoptic, hydrographic data set for the system, collected in 1969, circumscribed the first downstream meander with great clarity[60,92] (Figure 6.5). It was established that the geographic location of this meander reflected the bottom topography very closely (Figure 7.2). In order to maintain its potential vorticity, a column of water moving eastwards in the southern hemisphere will have to move equatorward when vertically compressed, i.e. when the depth decreases. The behaviour of the Agulhas Return Current therefore implies that it extends to depths exceeding 4000 m.

Satellite thermal infrared imagery is not particularly useful in following the Agulhas Return Current itself, since it has lost its contrasting surface temperature. On the assumption that this current follows the Subtropical Convergence, that has a distinct thermal signature at the sea surface, it was soon seen[59,60] that not only the Agulhas Plateau, but also ridges farther downstream, had an effect on the trajectory of this water. It has subsequently been shown that the Agulhas Return Current, on a significant number of occasions, does not follow the Subtropical Convergence[97] closely, but may lie slightly to the north forming a secondary front in the process, the Agulhas Front.

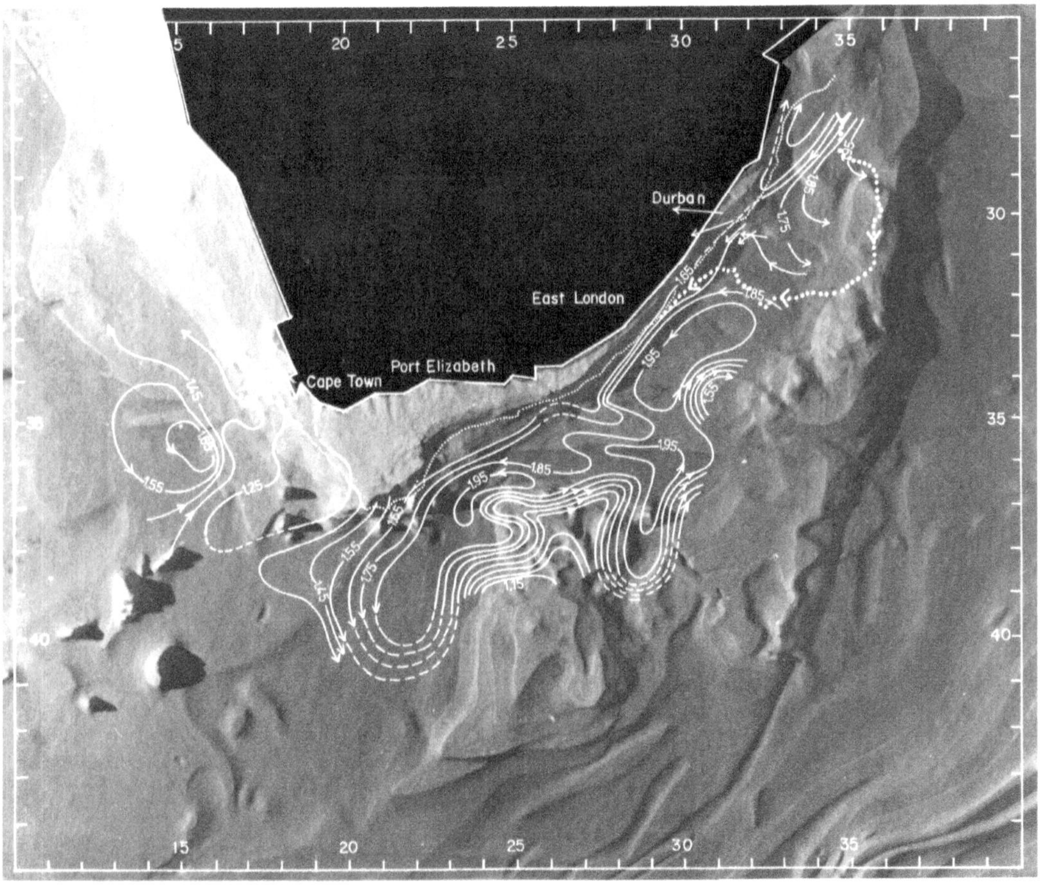

Figure 7.2. The dynamic height of the sea surface over the southern Agulhas Current system relative to the 1000 decibar level, in dynamic metres, superimposed on a model of the bottom topography[92]. The core of the Agulhas Return Current follows the bottom topography. The dotted line shows the movement of a drifter.

Meanders

Nonetheless, the evidence for regular meanders in the eastward flow of the Agulhas Return Current is strong, both from hydrographic data, thermal infrared imagery where appropriate (see Figure 6.2) and even ocean colour[170]. The drifts of satellite-reporting buoys that have been placed in the current have demonstrated these excursions very vividly (Figure 7.3). Boebel et al.[750] have shown the same from the tracks of floats at intermediate depths (Figure 7.4). Eddies of various sizes, both over ridges and between them, are also suggested by the trajectory given in Figure 7.3. Comparing the flow path of a drifter with simultaneous hydrographic measurements to establish the current axis[513], it has been established that they give valid Lagrangian measurements for the Agulhas Return Current.

These meanders, due to the interaction of current and bottom, depend largely on the transport, velocity and vertical profile of the current. In such a turbulent region neither of these parameters can be expected to remain very stable and temporal shifts in these trajectories should therefore come as no surprise. Observations from satellite microwave[670] indicate that the first meanders, over the Agulhas Plateau, may be relatively stable over periods of six months, but that the further, downstream meanders may move more freely. Nevertheless, northward excursions of the Agulhas Return Current at intermediate depths (viz. Figure 7.4) are found[750] predominantly at 26.8° E, 32.6° E and 38.9° E, resulting in a typical wavelength of about 500 km. There have been suggestions[261] of seasonal behaviour in the location of the meander, but subsequent analyses[673] have found this signal only for a short period.

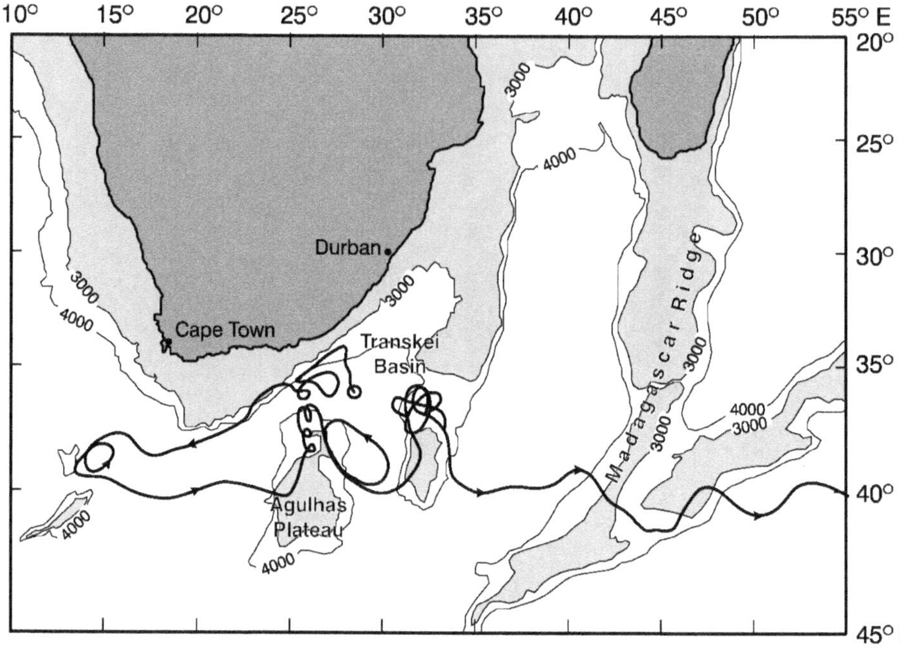

Figure 7.3. Drift track of a surface buoy that moved with the waters of the southern Agulhas Current system and the Agulhas Return Current from September 1975 to June 1976[350]. The bottom topography (in metres) is shown; that shallower than 3000 m has been shaded. The buoy's zonal trajectory is strongly deflected northward by the Agulhas Plateau and the extension of the Mozambique Plateau; less so by the Madagascar Ridge and the Indian Mid-ocean ridge.

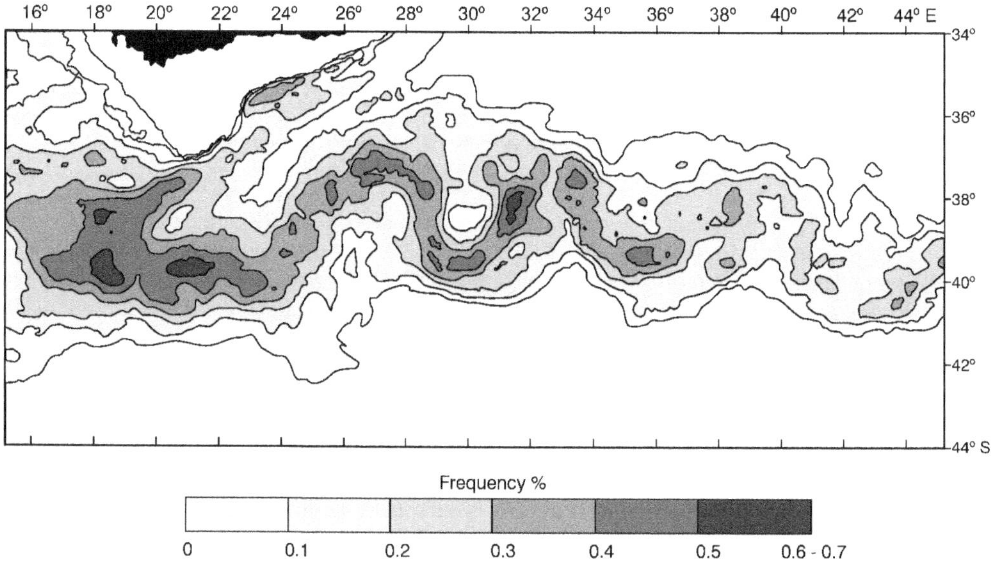

Figure 7.4. The geographic distribution of speeds greater than 0.5 m/s for the Agulhas Return Current at intermediate depths. The land mass in demarcated by the 1000 m isobath in the upper left-hand corner of the diagram. This frequency diagram is based on the drift patterns of RAFOS floats in the region[750]. This may be compared to the variability in Figures 3.14 and 6.3 for the same region.

Hydrographic data have on one occasion even shown the Agulhas Return Current crossing the Agulhas Plateau[533]. Although considerable variations in location are therefore evident over even short periods[130], time averages over longer periods show up the dominant effect of the bottom topography very clearly (viz. Figure 6.2). This suggests strongly that these meanders are Rossby waves[99,354]. A simple calculation of the wavelength at this latitude gives a result of 470 km, similar to that shown by float tracks at intermediate depths. Using satellite images of ocean colour, that clearly demarcates the Subtropical Convergence[681–3], as well as sea level height anomalies, it has been shown[683] that the wavelength of the meanders may vary between at least 500 and 620 km.

Vertical hydrographic sections across this current in the vicinity of the Agulhas Plateau show little significant difference to those of the Agulhas Current proper[230] (Figure 7.5). Farther downstream[302], this current has been observed to extend to the sea floor. Would the speeds in the Agulhas Return Current also be closely similar to that of the mother Agulhas Current?

Current speeds

Speeds at the sea surface, as established from the drift speed of buoys[350] (Figure 7.3), decreased from 1.3 m/s between 15° and 25° E to 0.4 m/s between 55° and 63° E. This agrees well with speeds of 1.2 m/s estimated from hydrographic data. Average speeds, calculated for large numbers of drifters[310], have shown no unidirectional decline in current speeds, but a very complex distribution. Current speeds of 1.5 m/s have been measured as far east as 47° E. Geostrophic speeds calculated from quasi-synoptic cruises[220] have shown a drop of about 10 per cent in the surface speeds relative to 1500 decibar of the Agulhas Return Current compared to the southern Agulhas Current. Current speeds, averaged between stations, of 1.0 m/s are not uncommon.

Combining a large set of *in situ* current observations made with Acoustic Doppler Current Profiler over the top 400 m of the Agulhas Return Current with the drift rates of RAFOS floats at intermediate depths through the region[750] the most representative velocity section to

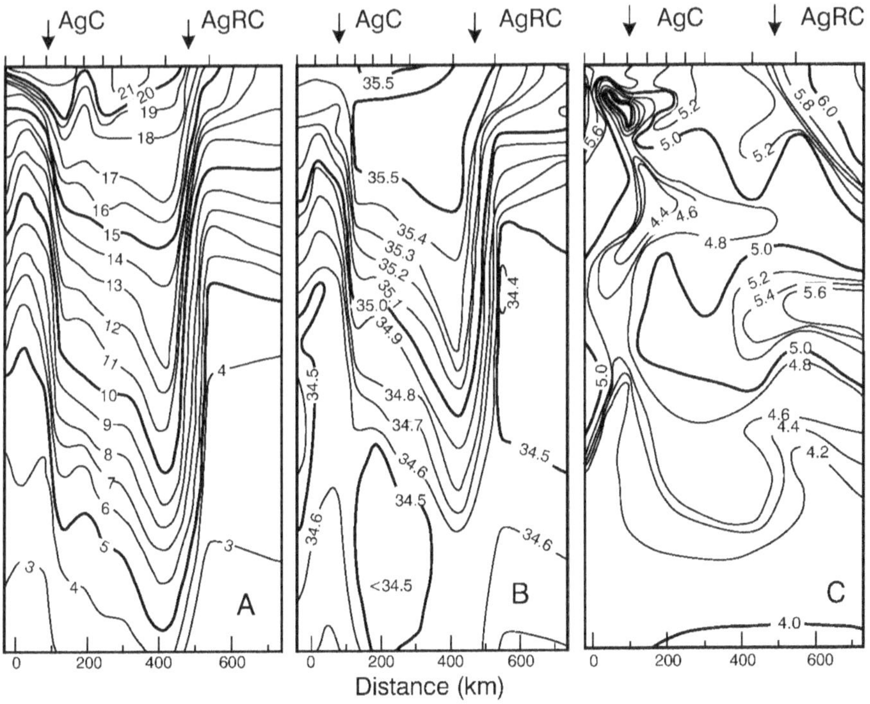

Figure 7.5. Meridional hydrographic sections southward from Africa across the Agulhas Current (AgC) and the Agulhas Return Current (AgRC) near the Agulhas Plateau: **A**: temperature, **B**: salinity[128], **C**: oxygen.

date has been produced (Figure 7.6). It shows that, on average, the core of the Agulhas Current between 20° and 44° E has surface velocities exceeding 1.4 m/s. Even at 1200 m speeds of 0.3 m/s are still evident. The core width of the current has been estimated[750] at 70 km, but the velocity section in Figure 7.6 shows that at the surface speeds in excess of 0.6 m/s extend over a width in excess of 100 km. It should be stressed that these are averages over an extensive zonal strip. The current intensity, as reflected in current speeds at the sea surface, in fact diminished rapidly with distance from the Agulhas retroflection (Figure 7.6). Whereas speeds in excess of 0.5 m/s were found up to 60 per cent of the time in the retroflection region, at 40° E the likelihood of finding such speeds had been reduced to 20 per cent. Peak surface speeds decreased from 2.1 m/s near the retroflection to 1.1 m/s at 32° E.

In a number of instances[215] the Agulhas Return Current was in fact more intense, with higher velocities, than the Agulhas Current itself at the same time. Since the Agulhas Return Current may merge with the flow along the Subtropical Convergence, this does not seem altogether unexpected. What kind of volume transports are then to be expected in the Agulhas Return Current?

Current transport

Based on an analysis of historical hydrographic data, Duncan[75] has estimated values of 39 × 10^6 m^3/s to 40 × 10^6 m^3/s relative to a 1000 m reference surface, decreasing to between 19 × 10^6 m^3/s to 32 × 10^6 m^3/s at the longitude of Madagascar. This flux he has considered to be separate from the flow in the Subtropical Convergence itself which should lie between 16 × 10^6 m^3/s and 20 × 10^6 m^3/sa and which should be added to the previous values for a total eastward transport. This result does not agree with more precise lines across the Agulhas and Agulhas Return Currents that have been carried out subsequently[278].

Close to the retroflection (Figure 7.7) the baroclinic transport of the Agulhas Current as well as of the Agulhas Return Current was about 140 × 10^6 m^3/s to

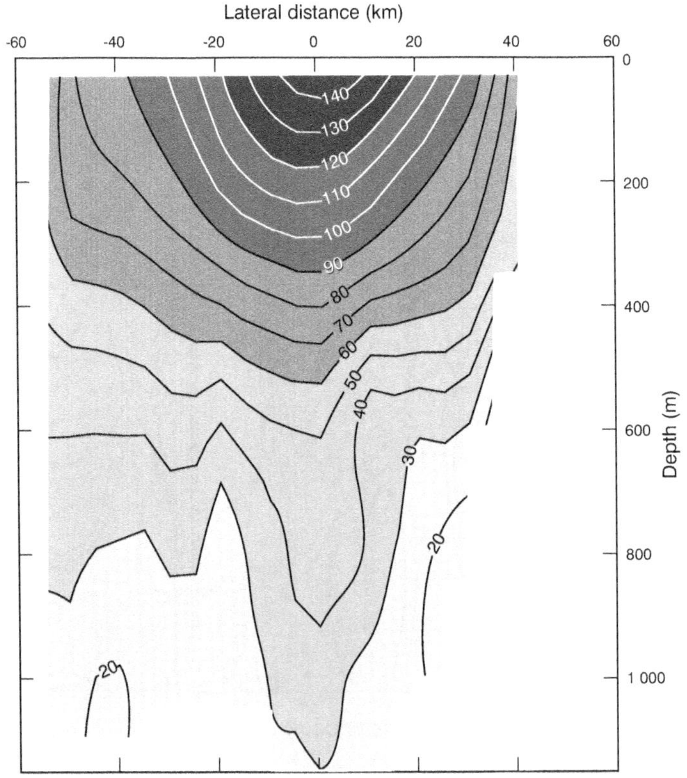

Figure 7.6. A velocity section across the Agulhas Return Current[750], based on speeds of RAFOS floats at intermediate depths in combination with acoustic Doppler current meter readings in the upper 400 m. Speeds are in the downstream direction and are given in cm/s.

the sea floor. Both were confined to a narrow ribbon and more or less cancelled each other out for a net eastwards flux of zero. Between Durban and the Crozet Islands, i.e. crossing the northern Agulhas Current and the Agulhas Return Current south of Madagascar (Figure 7.7), the flows are more or less in balance once more, but are more diffuse. Wider station spacing may have played a role in this particular estimate, but is not the only factor involved. Others have variously estimated 50×10^6 m^3/s to 75×10^6 m^3/s for the Agulhas Return Current, relative to 1500 decibar. The only set of reliable measurements at the longitude of Madagascar[302] shows a combined volume transport at the Agulhas Front and the Subtropical Convergence of 104×10^6 m^3/s relative to the deepest level measured. Based on accurate current observations in the upper layers, combined with float drift rates at intermediate depths (viz. Figure 7.6), an average transport of $44(\pm 5) \times 10^6$ m^3/s for the upper 1000 m has been calculated[750]. Using a large data set of hydrographic sections across this current, Stramma[100] has shown high variability in the volume flux of the Agulhas Return Current but no statistically reliable decrease between its inception and 45° E. Such an expected flux decrease might come about largely as a result of mixing with adjacent water masses.

Mixing

Apart from a surface cooling, the Agulhas Return Current will mix with the ambient waters and therefore will exhibit a downstream change in its general temperature–salinity characteristics. Tracing the very characteristic subsurface oxygen minimum of Agulhas Current surface water shows that it may pass through the Agulhas retroflection fairly unscathed[231], but that in most cases it is no longer distinguishable far down the trajectory of the Agulhas Return Current. This is particularly clear in Figure 7.5 (panel C) where the subsurface oxygen minimum is clearly seen at 100 m depth in the Agulhas Current, but is entirely absent in the Agulhas Return Current.

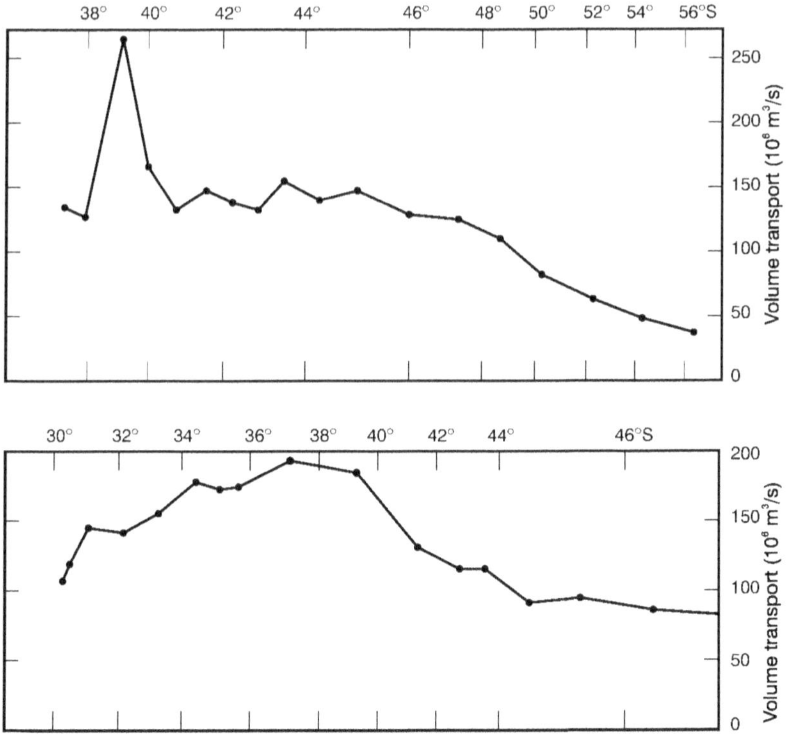

Figure 7.7. Baroclinic transport observed on two meridional sections from Cape Agulhas (upper panel) and from Durban, both in a south-eastward direction[278]. Values are cumulative from the south and calculated down to the sea floor. The zonal transport values for the Cape Agulhas section show the increased intensity of the Agulhas as well as the Agulhas Return Current here compared to further east in the basin gyre (see also Figure 5.1).

A comparison between the hydrographic characteristics of the upstream part of the Agulhas Return Current (e.g. Figure 7.5) and at 40° E longitude[302] (Figure 7.8) demonstrates the degree of mixing that occurs over this distance. South of Africa the 10 °C isotherm rises from a depth of 1060 m to 260 m in the Agulhas Return Current. At 40° E, at the longitude of the Mozambique Channel, it rises from only 800 m to 300 m, giving an indication of the erosion this current undergoes over this distance. South of Africa salinities of 35.50 are found to depths of 400 m in the Agulhas Return Current; at 40° E only to 280 m. Although the intensity of the Agulhas Return Current is much reduced at 40° E the combination of this current and the Subtropical Convergence remains formidable. For the section shown in Figure 7.6 a total volume transport of 103×10^6 m^3/s was calculated for the two features combined. They were 84×10^6 m^3/s and 19×10^6 m^3/s respectively. Maximum surface speeds at the core of the Agulhas Return Current were 0.81 m/s, while in the Subtropical Convergence itself it was 0.29 m/s[302].

Up till this point the discussion has dealt almost exclusively with the upstream, or initial, part of the Agulhas Return Current. The question now arises whether there may not be some inherent characteristics of the flow in the South West Indian Ocean that would define a natural point for the termination of the current. There is substantial circumstantial evidence that suggests that in fact there is such a natural discontinuation brought about largely by continuous downstream emissions from the current.

Leakage from the current

First, the leakage from the Agulhas Return Current, partially feeding the South West Indian Ocean sub-gyre (Figure 3.30), is very strong from the inception of the current up to about 60° to 70° E longitude[218,242], but downstream from here very little of this type of leakage remains[241]. This is evident from the volume transport of the Agulhas Return Current portrayed in Figure 7.9. Except for the geographic region directly south of South Africa, this analysis is regrettably supported by only two cruises east of 30° E[96]. Nonetheless, it exhibits a consistent downstream decrease in volume flux that may conceivably be adjusted with the advent of more data in the region, but more than likely not reversed.

The major leakage seems to occur between 40° and 50° E[242], with a drop from 54×10^6 m^3/s (relative to 1500 m) to 40×10^6 m^3/s. The location of this largest downstream leak corresponds to the site of the South West Indian mid-ocean ridge. At 70° E longitude only 15×10^6 m^3/s, or 28 per cent, of the original volume flux of the Agulhas Return Current remains. At this point the current may well be considered to have petered out altogether. A number of investigations have been carried out to study the frontal systems of the Southern Ocean in this region using a variety of

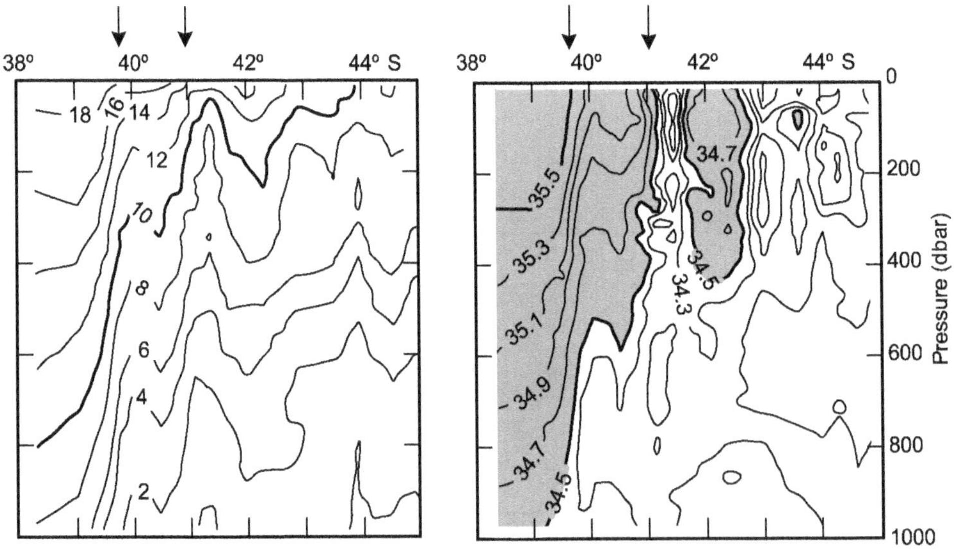

Figure 7.8. Hydrographic sections across the Agulhas Return Current at 40° E; potential temperature (left-hand panel), salinity (right-hand panel)[302]. Arrows denote the double front representing the current and the Subtropical Convergence. A warmer, more saline body south of the arrows is probably a warm eddy.

hydrographic data. In most of these studies[97,594–6] a distinct and separate front, north of the Subtropical Convergence, is found, associated with the Agulhas Current or the Agulhas Return Current[597]. Whether based on full hydrographic stations, XBT (expendable bathythermograph) observations or thermograph readings, all these show the separate front associated with the Agulhas Return Current as disappearing at about 70° E or west of here.

A number of studies have also been carried out on the fronts in this region using satellite observations of sea surface temperature[644] and sea surface height[643]. Using these data, Kostianoy et al. have demonstrated that this separate front associated with the Agulhas Return Current, the Agulhas Front, extends eastward to 80° E, remaining roughly between the latitudes 40° S and 42° S along its zonal range[644]. Its core temperature decreases from about 17° C south of Africa to about 13° C at 80° E (Figure 7.10); its meridional temperature gradient from 0.06 °C/km at 20° E (the highest of any front here, even the Subtropical Convergence) to nearly 0.02 °C/km at 80° E (the lowest of any front at this longitude). These data support the concept of a steady leakage from the Agulhas Return Current along its full length and its termination at about 70° E.

Cessation of intense variability

The second major collection of data that may be suggestive for the termination of the Agulhas Return Current is that which describes the mesoscale variability of currents in the Southern Hemisphere. A number of investigations have, for instance, demonstrated[130,418,643] that the very high levels of variability in sea surface temperatures found in the Agulhas retroflection region extend as a tongue of high values in an easterly direction. These high values of temperature variability are mirrored by the high measures of sea level variability[218,260,643] for roughly the same region[183,418], as derived from satellite altimetry[73,587] (viz. Figure 6.3). Of particular interest is the longitudinal extent eastward of these tongues of high variability.

There are some slight differences from data set[183] to data set[260] and from year to year but, as a rule, levels of variability decline on proceeding eastward and reach the background level of variability at about 70° E to 80° E[643]. This result is also found in some global circulation models[60]. This would suggest that the turbulent effect of a zonal jet, in this case the Agulhas Return Current, will have run its course at this distance from its origin.

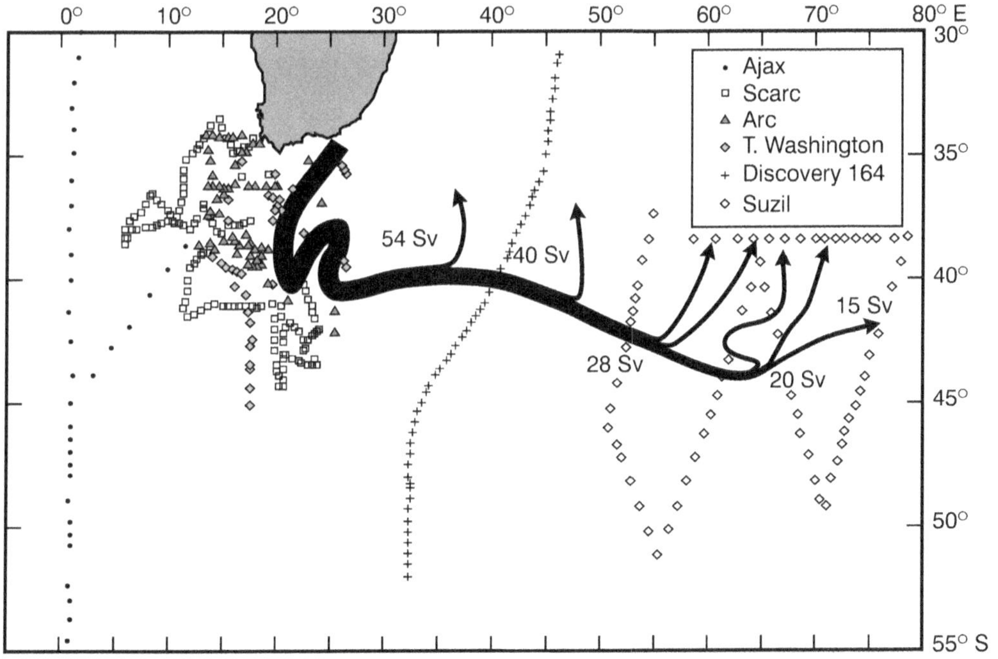

Figure 7.9. The volume flux of the Agulhas Return Current in 10^6 m^3/s (Sv) based on a set of high-quality, modern hydrographic data[96]. Cruises used for this analysis are given in the inset opposite to the symbols representing hydrographic stations from each individual cruise. By 65° E the current has all but petered out.

All these data sets suggest that little of the Agulhas Return Current remains at 70° E. Is this sufficient to consider it as having been terminated? The station distribution shown in Figure 7.9 includes modern data upstream of the Agulhas Return Current, i.e. where the flow is totally outside its region of influence. In going zonally eastward, one may reasonably consider the Agulhas Return Current to have terminated where both the volume transport and the geostrophic velocities along the Subtropical Convergence have returned to the values observed west of where the influence of the Agulhas Return Current was first encountered and where – in addition – the temperature/salinity values show no remains of Agulhas Current water. The results presented in Figures 7.11 and 7.12 demonstrate that this does indeed occur at about 70° E[96].

Decreases in transport and speed

East of 20° E longitude the geostrophic volume transport to 1500 decibar (Figure 7.11) does not exceed 22×10^6 m^3/s. At the Agulhas retroflection there is, as might be expected, considerably enhanced transport, but also much greater variability. Values from 35 to 75×10^6 m^3/s have been observed here. West of the retroflection the volume transport decreases rapidly. At a longitude of about 65° E it has returned to the values observed in the South East Atlantic Ocean. Geostrophic velocities referenced to 1500 decibar (Figure 7.12) mirror these volume transports[96].

Maximum speed values of 1.25 m/s have been observed in the Agulhas retroflection, but downstream, at 70° E, this diminishes to about 20 cm/s, the speeds normally occurring along the Subtropical Convergence in the South East Atlantic Ocean[83]. Once again one may argue that therefore the Agulhas Return Current may be considered not to extend beyond the 70° E meridian.

Park et al. have analysed historic hydrographic data, supported by a numerical model, and have shown that an Agulhas Return Front, or Agulhas Front[598–9] and its associated jet, terminates at 75° E[242]. This may amount to much the same thing. Intensified eastward flow along the Subtropical Convergence east of here is therefore entirely due to the South Indian Ocean Current[100]. This latter flow can most probably be considered the longitudinally consistent background flow associated with the Subtropical Convergence whereas the Agulhas Return Current has a geographically limited range downstream of the termination of the Agulhas Current.

In all the hydrographic sections across the Agulhas Return Current[302] the current axis has always been considered to lie north of the Subtropical Convergence. Realistically, this feature can rather be considered a frontal zone than a single, sharply defined front. Furthermore, its expression is very often complicated by multiple fronts and eddies.

The Subtropical Convergence south of Africa

Classically[351–2] the front located circumpolarly, except in the South East Pacific Ocean, at about 40° S latitude has been considered to constitute the northern boundary to the Southern Ocean. Here Subantarctic Surface Water with a slight northward flow component converges with Subtropical Surface Water and subducts below it. South of Africa this front has also been considered as forming the limit to the influence of the Agulhas Current with the Agulhas Return Current merging with the Antarctic Circumpolar Current[333] in one, gradual, eastward movement of water. Others,

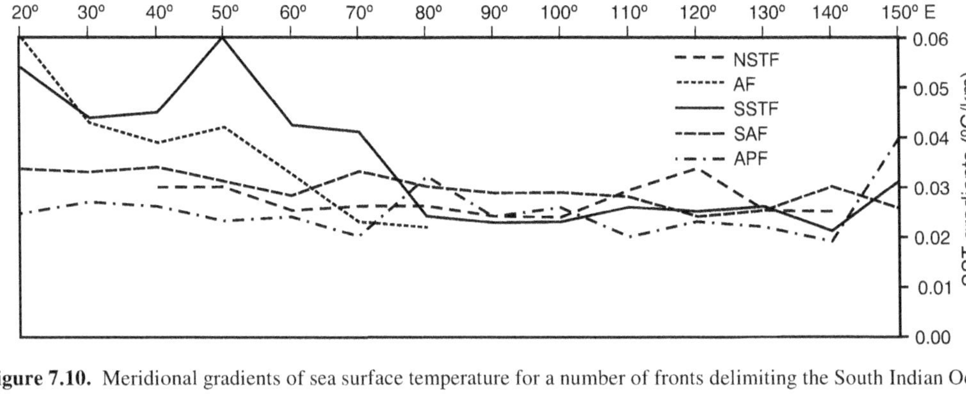

Figure 7.10. Meridional gradients of sea surface temperature for a number of fronts delimiting the South Indian Ocean[644] and the Southern Ocean. These values have been inferred from satellite remote sensing. The fronts are: the North Subtropical Convergence (NSTF), the Agulhas Front (AF), the South Subtropical Convergence (SSTF), the Subantarctic Front (SAF) and the Antarctic Polar Front (APF).

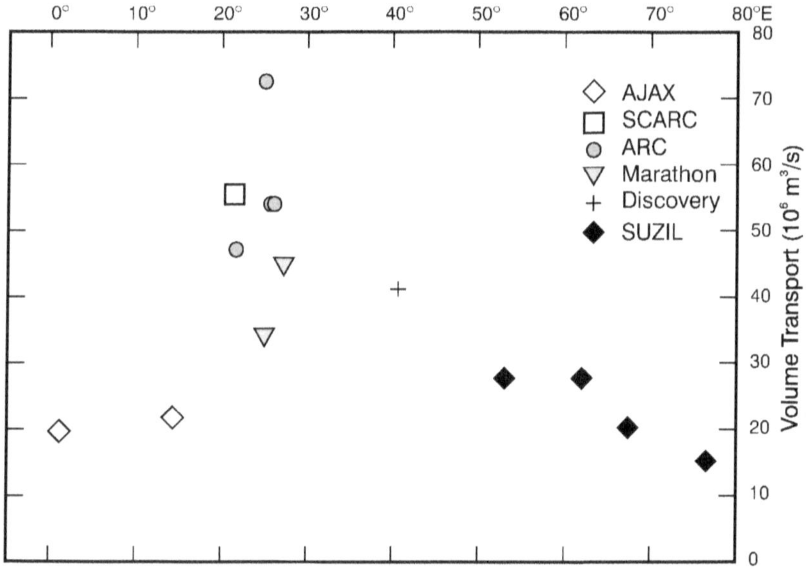

Figure 7.11. The volume transport to 1500 decibar along the Subtropical Convergence in the South Atlantic Ocean (up to about 20° E) and in the South Indian Ocean (20° to 80° E longitude)[96]. Volume transport is given in Sv, i.e. $10^6\ m^3/s$. Geographic station positions for the cruises identified in the inset are shown in Figure 7.9. The influence of the Agulhas Return Current is evident from about 20° E and has terminated at about 70° E longitude.

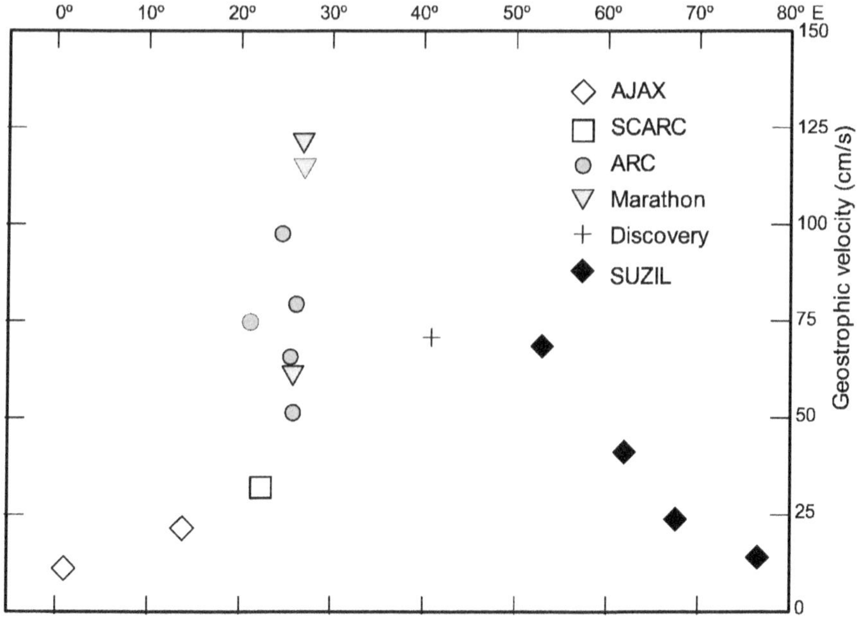

Figure 7.12. The geostrophic velocity of the flow adjacent to the Subtropical Convergence, referenced to 1500 db, in the South Atlantic Ocean (up to about 20° E) and in the South Indian Ocean farther east[96]. Cruises from which these results are derived are given in the inset; the station distributions in Figure 7.9. The influence of the Agulhas Return Current starts at about 20° E and is seen up to about 70° E.

probably more accurately[75], have considered the flow eastwards of subantarctic water and subtropical water to be separate and their common boundary to be demarcated by the location of the Subtropical Convergence. The Subtropical Convergence is also a border region between waters with high concentrations of nutrients (except silica) to the south[751] and low concentrations to the north.

There have been well-substantiated proposals[643] (viz. Figure 7.6) that the Agulhas Front, also called Agulhas Return Front, and the Subtropical Convergence are two independent, separately identifiable fronts[242,819]. Each would support a zonal jet, the Agulhas Return Front's being the Agulhas Return Current. In practice it is not always possible to show unambiguously where the Agulhas Return Current ends and where the Subtropical Convergence, and the geostrophic flow associated with it, begins (viz. Figure 7.5, 7.18). Furthermore, in this very highly variable region, fronts may shift meridionally, amalgamate and separate over relatively short periods. By making regular hydrographic sections across the Agulhas Front and the Subtropical Convergence south of Africa[656] it has been demonstrated that the former may move more than 40 km over a period of a month, the latter front at least 110 km. The difficulty in demarcating each front definitively becomes particularly clear when considering the effect of the Agulhas Current on the Subtropical Convergence (Figure 7.13).

The Subtropical Convergence in the South Atlantic Ocean

West of the Agulhas Current retroflection the nature of the Subtropical Convergence is considerably different to that east of this feature. Although relatively distinct in subsurface gradients of temperature and salinity to a depth of about 800 m[600] in the South Atlantic Ocean, the surface expression here is weak and very ephemeral, with strong horizontal temperature gradients persisting for periods of days to weeks before reassembling at a slightly different location[549]. This leads to a much

Why the Agulhas *Return* Current?

That part of the Agulhas Current system just downstream of the Agulhas Current retroflection is usually called the Agulhas Return Current. Its trajectory is strongly influenced by the bottom topography it traverses, and somewhere along its eastward path it becomes known as the South Indian Ocean Current[100]. The designation of this latter part of the flow is analogous to the nomenclature for the South Atlantic Ocean where the southern component of the anti-cyclonic gyre is called the South Atlantic Current[83] along the full breadth of that ocean. Why should the terminology for the South West Indian Ocean be different and, for that matter, where does the Agulhas Return Current end?

T. F. W. "Sandy" Harris

The name Agulhas Return Current was coined in about 1970 by Dr Sandy Harris of the Department of Oceanography at the University of Cape Town. He had noticed[81], based on the hydrographic data available at the time[92,354], that a sub-gyre existed in the South West Indian Ocean, west of the Madagascar Ridge. In the contemporary current portrayals of the sea surface currents, based on ships' drift[286,291], there was also some supporting evidence of northward setting currents at the Madagascar Ridge. The alongcurrent distance between the Agulhas Plateau, the Mozambique Ridge extension and the Madagascar Ridge is such that it might create an instability in a planetary-type wave at the Mozambique Ridge, forcing the eastward-flowing current northward[99], thus completing a closed circulation in the South West Indian Ocean. Hence the name Agulhas *Return* Current.

Subsequent drifter studies[311,593] and altimetric investigations have shown no evidence for such a continuous northward current, although the northward meanders over these ridges may be meridionally quite extensive[350]. Based on present data, it seems likely that very little Agulhas surface water is directly involved in this sub-gyre. The first part of the Agulhas Return Current therefore *returns* hardly anything at the surface, equatorward leakage from the current taking place along its full length. Nonetheless, the name is now widely used and attempts to retract or revoke it will most probably be futile. The name should consequently most profitably be restricted[96] to only that part of the South Indian Ocean Current west of the Kerguelen Plateau at about 70° E (viz. Figure 2.4).

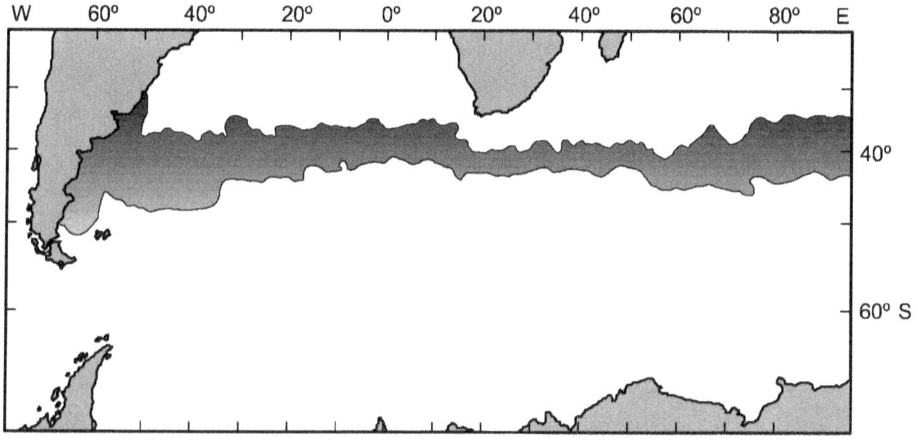

Figure 7.13. Limits of the meridional range of surface expressions of the Subtropical Convergence in the South Atlantic and South Indian Oceans[596]. This portrayal is based on the location of the 13 °C isotherm, a good proxy for the Subtropical Convergence[97], as observed from satellite thermal infrared imagery[826].

wider meridional dispersion of its location (Figure 7.13). This weakness in frontal gradients has even led to a proposal to consider the existence of a separate northern and southern Subtropical Convergence[594,601] in the South Atlantic. It has been suggested[601] that the South Atlantic Current[83] flows between these two expressions of the Subtropical Convergence. These fronts are believed to migrate between 1.5° and 2.5° of latitude on a seasonal basis.

The different hydrodynamic forcing of the Subtropical Convergence in the central South Atlantic Ocean and south of Africa is also evident in the pelagic community structure[602]. Higher biological diversity and weaker zonation patterns were observed in the mid-Atlantic sector. Some degree of cross-frontal mixing is prevalent everywhere. The mean position of this front in the south-eastern Atlantic is about 37° S. Close to the African continent both position and characteristics of the Subtropical Convergence change dramatically.

Impact of the Agulhas Current

The effect of the Agulhas retroflection may on occasion be felt as far west as 5° E longitude[512]. Altimetry results suggest[362] that some cyclonic disturbances may originate even further west. The main effect of the Agulhas retroflection seems to be restricted to a zone between 10° to 20° E longitude and here the Subtropical Convergence is forced to lie at least 5° of latitude southward of its general location in the South Atlantic, at about 42° S (viz. Figure 6.2). Its latitudinal range is also considerably compressed[530] (Figure 7.13). Along the Subtropical Convergence in the South Atlantic the South Atlantic Current forms part of the southernmost component of the subtropical gyre[83]. What part of this current flows northwards to feed the Benguela Current and what part continues along the, now more poleward, Subtropical Convergence into the South Indian Ocean is not known. It has been surmised[67,597] that the latter part may be substantial.

The horizontal temperature and salinity gradients across the Subtropical Convergence south of Africa are extremely high (Figure 7.14). Meridional changes of more than 5 °C over distances of less than 35 km are not uncommon here[97,597]. Average values based on satellite observations[643–4] give meridional gradient values in excess of 0.05 °C/km south of Africa (Figure 7.10). Very high salinity gradients are also observed in the upper 500 m[595] and in many instances this haline gradient is a better indicator of the true location of the Subtropical Convergence (Figure 7.14, panel A) than the temperature gradient. A careful analysis of 89 crossings of the front using continuous surface observations[97] has shown that on many occasions a multiple front may be observed[597], one along the Agulhas Return Current, the Agulhas Front, and the other the Subtropical Convergence. This double front is particularly well defined whenever the Agulhas Return Current does not follow the meandering Subtropical Convergence closely.

As will become apparent when the time and space variability of these fronts is discussed, not only are there considerable inter- and intra-annual changes in all characteristics of these fronts, but there is also a very high level of mesoscale turbulence brought about by eddy shedding at the fronts. This causes the rather simplified conceptual version of two major fronts used here

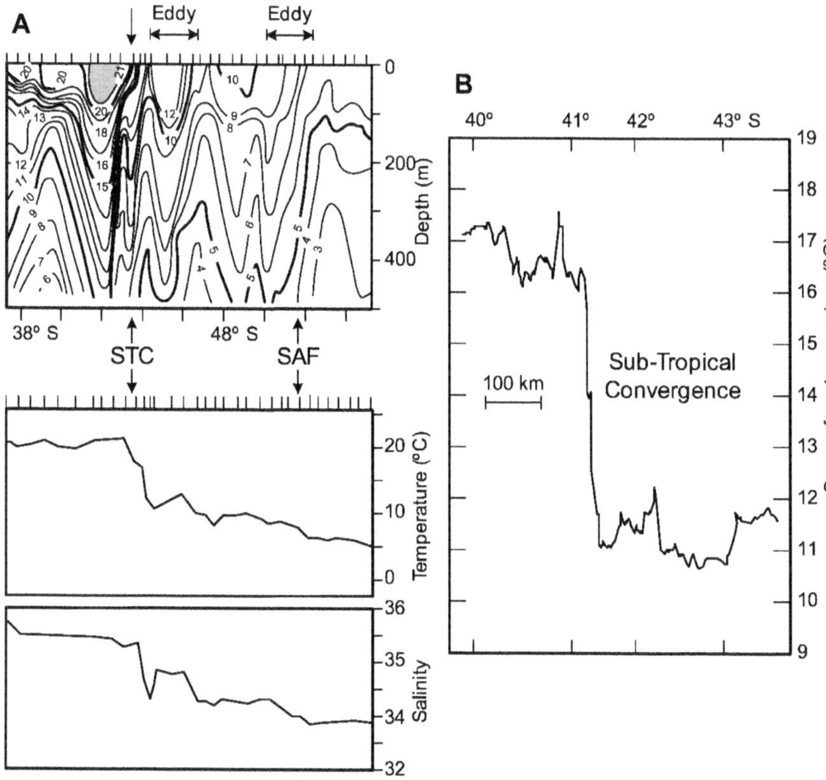

Figure 7.14. The surface expression of the Subtropical Convergence and the adjacent Agulhas Return Current south of Africa[97]. Panel **A** shows the Agulhas Return Current lying next to the Subtropical Convergence, based on a closely spaced line of temperature measurements. Two warm eddies are also evident, the one closer to the front still retaining much of its original high surface salinity, visible in the salinity trace for the sea surface (lower panels). The high surface gradients in temperature at the Subtropical Convergence are demonstrated by the continuous thermograph trace of panel **B**, extending from Cape Town to a latitude of 44° S.

to be complicated by a host of minor and sub-fronts[595]. This has led to a proliferation of names[597] for what are considered distinct fronts in their own right, including the Agulhas Front, Agulhas Return Front, Northern Subtropical Front, Southern Subtropical Front and others. The temptation to get involved in this nomenclatural musical chairs will be strenuously resisted by this author. An analysis of the surface thermal expressions of both major fronts is given in Table 7.1.

The Agulhas Front

The Agulhas Front has been considered, on average, to be the weaker one because whenever the two fronts are merged this resultant stronger front is generally considered to be the Subtropical Convergence. Others have shown[644] that the meridional gradient of the Agulhas Front is stronger south of Africa (Figure 7.15), but becomes weaker than that of the Subtropical Convergence east of 30° E. There is some overlap in the geographic position of these two fronts but, whether defined only by their surface expression[97] or by their subsurface expression as well[594–5], the meridional range of both fronts is considerable. There is even some overlap in meridional range between the Subtropical Convergence and the Subantarctic Front[603] particularly in the region at the longitude of Madagascar[594] (Figure 7.10). Since, in general, the location of the surface expression of both these fronts is representative of the major part of the subsurface expression[595], these more plentiful data may be used to establish the location of this front with some reliability (Table 7.1). Studies using all available hydrographic data for the region[594] have not substantially extended this meridional range. As mentioned above, identifying the Agulhas Front by its surface expression only has shown it to be limited zonally to a range from 15° to 56° E[597]. Satellite data[644] suggest to 80° E. Considering the Agulhas Front instead

Table 7.1. Calculated locations and thermal characteristics for the surface expressions of four oceanic fronts south of Africa[97].

		Latitudinal position			Width [km]	Temperature [°C]				
		From	To	Middle		From	To	Middle	Range	Gradient [°C km⁻¹]
Agulhas Front (24 crossings)	Mean	39°09'	40°01'	39°37'	96.3	21.0	15.7	18.4	5.4	0.102
	s.d.	01°16'	01°16'	01°14'	69.1	1.6	1.5	1.2	1.6	0.106
Subtropical Convergence (70 crossings)	Mean	40°35'	42°36'	41°40'	225.1	17.9	10.6	14.2	7.3	0.047
	s.d.	01°23'	01°32'	01°19'	140.6	2.1	1.8	1.7	1.9	0.043
Subantarctic Front (61 crossings)	Mean	45°15'	47°25'	46°23'	241.4	9.0	5.1	7.0	3.9	0.018
	s.d.	01°12'	01°06'	01°04'	100.4	1.6	1.6	1.2	1.3	0.009
Antarctic Polar Front (65 crossings)	Mean	49°39'	50°47'	50°18'	126.0	4.1	2.5	3.4	1.8	0.018
	s.d.	01°16'	01°35'	01°20'	63.3	1.1	1.1	1.0	0.6	0.012

The greater Agulhas Current system and southern African climate

Consistent with what one would expect from the strong temperature contrasts which exist between the surface of the Agulhas Current, particularly in its southern reaches, and the overlying atmosphere, a very high average heat flux is found over this current. In the Agulhas retroflection region its mean lies between 75 and 175 W/m^2, depending on season, but instantaneous values of 380 W/m^2 have been measured[98]. The dramatic effect these upward heat fluxes have on the atmospheric boundary layer has been measured as part of a number of investigations[13,148,449,494].

It has now been established that this atmospheric modification causes enhanced precipitation along the coastline bordering the northern Agulhas Current[149] (Figure 1.4), but what measurable effect the warm Agulhas Current might have in intensifying storms approaching from the west has not as yet been unequivocally established, notwithstanding attempts from 1858[11] to the present[12]. Nonetheless, studies of ocean temperature conditions preceding severe storms over land[604] suggest a notable influence.

Since the Agulhas retroflection region exhibits an unusual year-long flux to the atmosphere[147], it is quite possible that this region may influence long-term rainfall trends over the adjacent land mass. A careful analysis by Walker[12,134] of the possible links between South African summer rainfall and temperature variability[605] of the Agulhas and Benguela Current systems has indeed shown, unequivocally, that positive anomalies of sea surface temperature in the Agulhas retroflection region are positively correlated with increased precipitation over the highveld of southern Africa. This correlation also holds for temperatures of the coastal upwelling off the west coast of southern Africa, but since the latter are directly due to local winds this correlation may be considered to be more complex. A positive, but weaker, teleconnection exists as well between sea surface temperatures over the greater part of the South West Indian Ocean[134] and summer rainfall over southern Africa[606]. The mechanisms proposed for these teleconnections include the intensification, south of Africa, of mid-latitude frontal systems as well as the enhancement of both tropical and temperature synoptic systems that bring rainfall to the subcontinent[606]. This is being further investigated[262].

Investigating possible cycles in ocean temperature anomalies, Mason[607,608] and Tyson have shown a clear 10 to 12-year oscillation in parts of the southern Agulhas Current region. This is important since there is substantial evidence that annual rainfall over southern Africa has varied systematically in both time and space[609]. Rainfall oscillation with periods of 18 and 2 to 3 years are prominent. It has furthermore been shown that the association between the quasi-biennial atmospheric oscillation (2–3 years) and rainfall is modulated by sea surface temperatures[608] and it has even been suggested that the temporal variability of sea surface temperatures around southern Africa may be a forcing mechanism for the 18-year rainfall oscillation[607]. Sea surface temperatures in the greater Agulhas system have also been shown[497,606] to dampen the rainfall effect of the Southern Oscillation for southern Africa, thus causing the occasional faltering of this otherwise relatively strong association.

Much of this research is still in its infancy[497], but already there are intriguing indications that the greater Agulhas Current system may have a significant role in controlling the inter-annual climate variations over southern Africa and an even more prominent role over longer periods, as has been shown from the geological records[341].

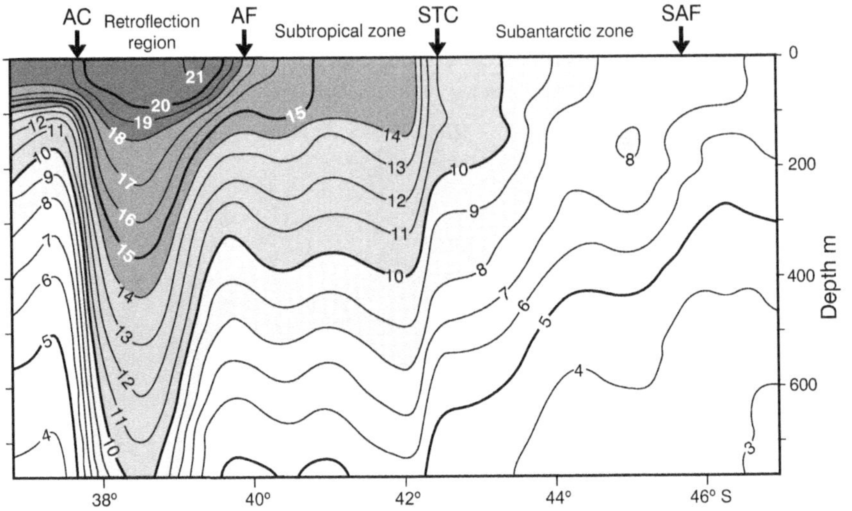

Figure 7.15. A vertical temperature section south-east of South Africa[656]. The characterisitic zones are given as well as the thermal fronts that separate them. These fronts are: the Agulhas Current (AC), the Agulhas Front (AF), the Subtropical Convergence (STC) and the Subantarctic Front (SAF).

as a subsurface front at intermediate depths, it can be considered identical to the Agulhas Return Current and its location extended to 75° E[594].

Dynamics of the front

The averaged circulation inherent to the Subtropical Convergence is not well understood. In general, because of the aforementioned large-scale subduction, it has been accepted that a confluence of water at the upper layers is generally present. This supposition is roughly supported by the observed drift of buoys in the region[593] that shows a tendency for these drifters to accumulate at the frontal region. It is of particular interest that no drifters in the Agulhas Return Current or along the Agulhas Return Current, west of 60° E longitude, drifted either north or south. In fact, a portrayal of the drift tracks of about 300 buoys during the FGGE experiment, from 1978 to 1981, shows conspicuous empty zones, both north and south of here[593], in which no buoys moved at all. This suggests a high degree of convergence, by which most drifters in this zone get trapped and remain in the along-front flow.

The flow pattern along the Subtropical Convergence is also of considerable importance to a proper understanding of the observed enhanced primary productivity here[441,588]. May this be due to the accumulation of material caused by the convergent motion only?

It has been demonstrated[462] that the intermittently enhanced pigment levels at the Subtropical Convergence in the South West Indian Ocean are a function of increased frontal intensity of this front. Observations of the detail of the Subtropical Convergence south of Africa have shown[610] that the surface expression of this front moves southward with the advancing summer season while the subsurface expression does not. As a result a relatively thin surface layer of increased vertical stability is formed in which higher concentrations of nutrients, usually found south of this front only, may be found. This configuration of a band of higher stability and increased nutrients will inevitably lead to increased biological productivity. This frontal mechanism may not be restricted to seasonal departures, but may also occur as short-lived events, with accompanying blooms of phytoplankton[170]. Numerical models[689] have successfully simulated the blooms at the Subtropical Convergence and have shown that the determining factors are indeed vertical stratification in the upper ocean layers as well as the availability of nutrients. Cyclonic eddies in the southern hemisphere by their very presence enhance vertical stratification[690] and the high levels of mesoscale turbulence at the Subtropical Convergence south of Africa should be no exception. Detailed observations of the pigment concentrations at the Subtropical Convergence, using satellite observations, have confirmed[462] how extremely variable this front is in both time and space. This is particularly true downstream of the Agulhas Retroflection.

These event-scale occurrences are particularly likely since the region is known for its very high levels of mesoscale variability. This variability can take a number of different forms.

Variability and eddy shedding

Any line of closely spaced hydrographic measurements across the Agulhas Return Current and the Subtropical Convergence invariably shows, if not a series of multiple fronts[595,643–4], at least a number of distinct disturbances in the temperature and salinity field[611]. These can usually be interpreted as eddies that have spun off the Agulhas Return Current or the Subtropical Convergence. First documented in the 1960s[511], all subsequent detailed hydrographic investigations in the region[90,444] have demonstrated the presence of eddies in the region (Figure 6.2).

General circulation models for this region[420] have indicated that the Agulhas Return Current is baroclinically unstable and will therefore exhibit meridional meanders, but that it is barotropic instability that is the main mechanism by which eddies are generated, predominantly on the northern edge of the Agulhas Return Current. A comprehensive study of the variability of the Agulhas Return Current has shown[418] some persistent patterns.

Meridional meandering

In general the band of high variability follows the meanders of the Subtropical Convergence over the bottom topography. There are variations in the geographic distribution of this variability from month to month, but on the whole these patterns are quite persistent[418]. The large-scale meandering of the Agulhas Return Current or Subtropical Convergence is probably responsible for the observed low-frequency variability, since this variability has been shown to be clustered around these features[545].

An analysis of the detail of such imagery[414] has shown that the width of the meander over the Agulhas Plateau is 290 km with a standard deviation in this width of 65 km only, thus being relatively stable. The wavelength between this and the next meander is

Maury and the Agulhas Return Current

The popular scientific press has posthumously elevated Lieutenant M. Maury of the United States Navy to the position of "Father of Oceanography" and "Pathfinder of the Seas".

The near-mythical aura with which he is currently being surrounded stems largely from the popular success of his book of the 1850s, *The Physical Geography of the Sea*[613], that went through numerous editions and was translated into several European languages. In addition he played a substantial role in furthering the establishment of the KNMI (*Koninklijk Nederlandsch Meteorologisch Instituut*: Royal Dutch Meteorological Institute)[614] and was subsequently awarded a special medal by His Majesty King Willem II of the Netherlands in 1856[14] for his efforts. Nevertheless, the best oceanographic minds of the time did not approve of his outdated theses on ocean circulation.

His biographer[613] has therefore stated quite bluntly that Maury ". . . could not be looked upon as representative of the best scientific thought of the eighteen-fifties". His treatment of information on the Agulhas Return Current is an enlightening case study of the cavalier fashion in which this famous individual dealt with information conflicting with his dearly held ideas on ocean circulation.

Maury's concept of water movement in the surface layers of the ocean was simple: to balance the distribution of heat in the ocean, warm water has to flow poleward and cold water equatorward. This dogmatically held notion led him into some serious conceptual mistakes about the currents around southern Africa. He, for instance, claimed[28] that ". . . the most unexpected discovery of all is that of the warm [southward] flow along the west [sic] coast of South Africa, its junction with the Lagullas current, called higher up, the Mozambique, and then starting off as one stream to the southward".

He was advised by a close friend and collaborator, Marin Jansen of the Dutch Navy, that this portrayal was incorrect[28]. This constructive criticism was most probably based on the excellent work by Andrau[10,615] and Van Gogh[505] of the KNMI that gave a truly prescient portrayal of the Agulhas retroflection as well as of the Agulhas Return Current. These important new results on the current patterns south of Africa were in fact rapidly taken over by a number of other geographers[29,284,616], but not by Maury. Even more solid information on the Agulhas Return Current came to hand.

A ship's captain, one N. B. Grant, sailing from New York to Australia, reported unexpectedly high water temperatures south of Africa at 39° S. Temperatures of between 13 °C and 23 °C were found to extend to at least 41° E at this latitude. Maury correctly concluded "Here therefore, was a stream – a mighty 'river in the ocean' – one thousand six hundred miles across from east to west, having water in the middle of it 23° [Fahrenheit; 13 °C] higher than at the sides." Having this new information adequately portrayed in the main figure of his book (Plate IX)[28] would conceivably have contributed to the wide dissemination of this first proper understanding of the retroflection of the Agulhas Current and of the existence of the Agulhas Return Current. This did not happen.

Maury was not to be disabused of his main thesis. He successfully swept all the offending new information under the proverbial carpet by assigning it to ". . . an illustration of the sort of *spasmodic* efforts – the heaves and throes – which the sea, in the performance of its ceaseless task, has sometimes to make". An exception therefore, that could legitimately be ignored in the greater scheme of things.

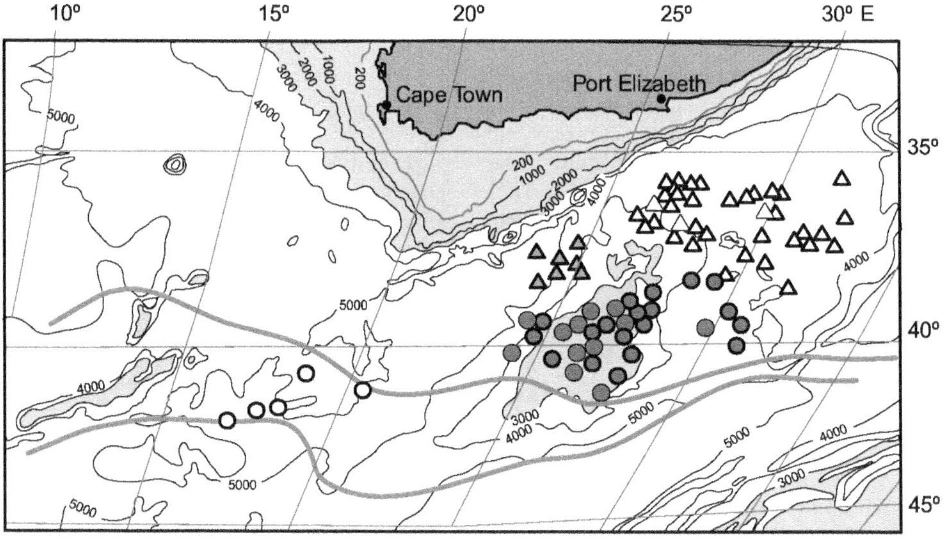

Figure 7.16. The geographic distribution of different types of mesoscale eddies associated with the Agulhas Return Current[63]. Circles represent warm, anti-cyclonic Agulhas eddies shed at the Agulhas retroflection; dots, large warm eddies at the Agulhas Plateau vicinity; solid triangles, small cold eddies forming part of the processes at the Agulhas Plateau and open triangles, larger cold eddies shed at the Subtropical Convergence. Portrayals of these eddy types are given in Figure 7.17. The heavy lines give the envelope of the surface expression of the Subtropical Convergence in this region[97]. Bottom topography isobaths are given in kilometres; regions shallower than 3000 m being shaded.

450(±110) km, therefore being less persistent. At intermediate depths[750] this was – as mentioned above – about 500 km. The amplitude of the first meander was a very stable 323(±60) km. An analysis of three years of sea surface temperatures[261] has demonstrated that the planetary wave downstream of the Agulhas Plateau shifts[545] seasonally by half a wavelength. It has also been claimed that upstream of the Agulhas Plateau there exist westward propagating baroclinic Rossby waves in the Agulhas Return Current[545] and there is evidence from temperature patterns[261] to support this contention. The Agulhas Plateau clearly affects the trajectory of the Agulhas Return Current, but there does not seem to be a close correspondence between this and other major bottom topographic features and the spatial structure of the low-frequency variability[545]. There is substantial evidence from satellite altimetry that there exists an intra-annual variability with a period of about eight months in the low-frequency variability. All these portrayals are based on time-averages for a very turbulent region, with not only meanders, but also eddies with which to contend.

Mesoscale eddies

The high-frequency variability in the Agulhas Return Current is due to general turbulence, most likely associated with the shedding of mesoscale eddies owing to baroclinic instability. Studies using satellite thermal infrared imagery[60] have shown this to be the case. Cold eddies shed northwards from the first meander downstream of the Agulhas Plateau are present[414] very frequently and with a very predictably north–south diameter (276±52 km) as well as east–west diameter (228±24 km). This suggests that these, and other eddies in the region, are regularly formed due to processes that are geographically stable. The results of further studies of this kind are in substantial agreement with such a hypothesis.

Using both satellite thermal infrared imagery and many detailed thermal sections across the Agulhas Return Current and the Subtropical Convergence, it has been demonstrated (Figure 7.16) that eddies with quite clearly definable characteristics are found in clearly circumscribed geographical regions[63,205]. These features rapidly lose their distinguishable surface temperatures. Monitoring of one particular cold eddy[361] has shown that it disappears from infrared imagery in a month. The geographic distribution presented in Figure 7.16 can therefore be questioned on these grounds. Circumstantial evidence from altimetric measurements suggests[362] that these eddies, once shed, do not move very far, thus by implication supporting the tentative portrayal of Figure 7.16. The movement of floats has also

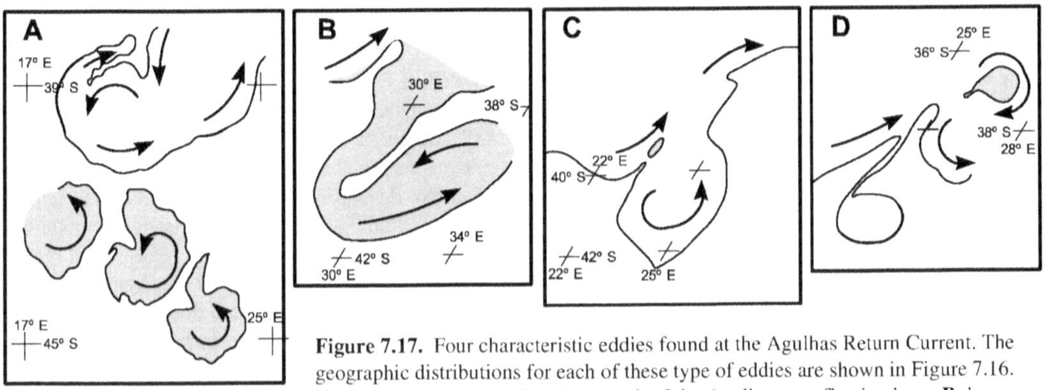

Figure 7.17. Four characteristic eddies found at the Agulhas Return Current. The geographic distributions for each of these type of eddies are shown in Figure 7.16. They are, **A**: Agulhas eddies just south of the Agulhas retroflection loop, **B**: large, warm eddies over the Agulhas Plateau and beyond, **C**: small cold eddies formed at the shear zone where large warm eddies are formed, and **D**: large cold eddies shed from the Subtropical Convergence. They are here roughly drawn to scale from the thermal gradients observed in satellite thermal infrared images[63].

demonstrated[750] the formation of cold eddies, proving that they extend to at least intermediate depths. The cold cyclonic eddies so formed propagate – unexpectedly – westward at an average phase speed of 5 cm/s. They are subsequently absorbed by the first meander located in their westward paths. On one occasion a recently shed Agulhas eddy, that had crossed the Subtropical Convergence, has been hydrographically measured and seeded with a drifter[458].

Its thermohaline characteristics were identical to that of the Agulhas Return Current. It remained in one location for about two months after which it seems to have been sheared apart by the strong neighbouring currents. It revolved with a period of 3.1 days and average azimuthal speeds of 65 cm/s. The volume transport around the eddy, calculated relative to 1500 m, was 32×10^6 m^3/s, or about half that of the Agulhas Return Current at the time[458]. This particular eddy seems to have been representative of those shed across the Subtropical Convergence close to the Agulhas Retroflection (Figures 7.16, 7.17), but not of those found further downstream.

As mentioned above, the domain of intense mesoscale activity along the axis of the Agulhas Return Current has distinct borders. It is widest at the Agulhas retroflection where the variations in sea surface height – implying mesoscale eddy activity – are also greatest[643]. It shrinks in meridional width towards the east and dies out completely between 70° to 80° E.

Classes of eddies

The characteristic shapes of the eddies found along the Subtropical Convergence make it possible to assign each one to a rather definite category. This also suggests that the spawning dynamics are consistent for each and that it may be site specific[63,361]. Eddies at the Agulhas retroflection are usually quasi-circular with diameters of about 210 km and seem to be formed by the strong shearing motion in the region. Over the Agulhas Plateau, and in its lee, large warm pools occur with east–west diameters of up to 600 km and north–south diameters of 200 km. These elliptical eddies are formed in an unusual way.

Southward meanders of the Subtropical Convergence create pools of warmer water that are rapidly closed by cold tongues of subantarctic water moving eastwards (Figure 7.17). The anti-cyclonic eddies in these pools draw cold subantarctic water around themselves at a higher latitude, thus leading to a retort-shaped eddy (viz. Figure 7.17B and D). As part of this oft-recurring process, small cold eddies with diameters between 40 and 80 km are shed as well (Figure 7.17C). Larger cold eddies with diameters of 100 to 150 km are also shed at the Agulhas Plateau (Figure 7.17D) and these have been observed to populate an extensive part of the Transkei Basin[63] (Figure 7.16).

A hydrographic section across such a cold eddy (Figure 7.18) shows that it extended to a depth of at least 1500 m, had temperature and salinity characteristics of subantarctic waters and estimated azimuthal speeds exceeding 2 m/s[444]. The Agulhas Return Current flowed partly around this cold eddy.

A number of important questions arise as a result of these results. How far do all these different eddies drift and what are their contributions to meridional heat transport and thermocline ventilation? Are the eddies with these very clearly circumscribed morphologies characteristic of the interaction of the Agulhas Return Current with the Agulhas Plateau only, or are they also formed farther downstream? Are the high horizontal current shears the main forcing functions for the forming of these eddies? There are suggestive indications that this mesoscale turbulence plays a key role in the functioning of the whole Agulhas Current system[131] and that there is substantial merit in studying and understanding the processes involved. Some of this has been accomplished by numerical modelling.

Dynamics of the Agulhas Return Current

The major finding of modelling efforts to understand the behaviour of the Agulhas Return Current has been – not unexpectedly – that the bottom topography of the region steers the movement of the eastward flow.

Inertial jet models

Modelling the Agulhas Return Current as an inertial jet has been shown[350,506] to give results that correspond well with those in nature (Figure 7.19). The meander on passing the Agulhas Plateau simulates the drift patterns of surface drifters in the region well. A two-layer, quasi-geostrophic model that uses an ensemble smoother[578] has shown that a cyclonic recirculation cell may be imbedded in this meander over the Agulhas Plateau. Further meanders downstream do not seem to be well correlated with the bottom topography, but instead to be a function of the latitude at which the current reaches this subsurface obstruction as well as other variables such as bottom velocity. Elongations and thus shedding

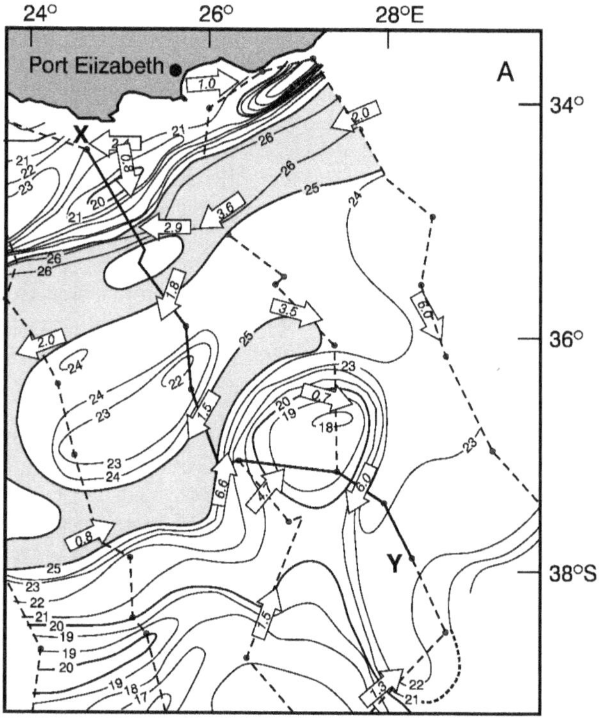

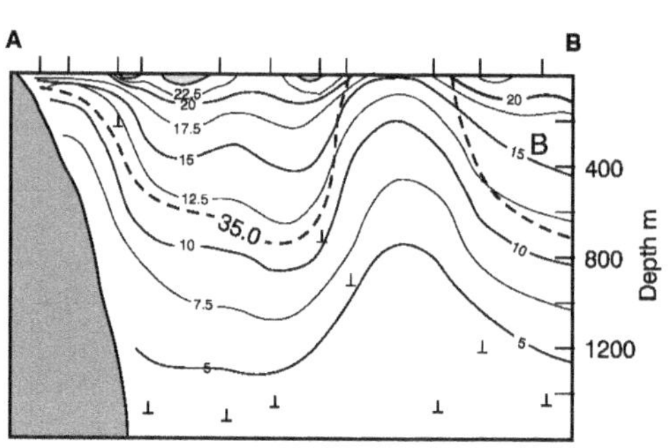

Figure 7.18. A cold eddy from the Subtropical Convergence at the Agulhas Return Current[444]; the warm surface water of the latter is shown as hatched in panel **A**. Panel **A** also shows the cruise track of the research vessel *Thomas B. Davie*, station positions and ship's set and drift (in open arrows) in knots. Panel **B** gives a temperature section across the cold eddy along station line AB (**X**, **Y** in upper panel), with the isohaline 35.0 superimposed as a broken line.

of cold eddies from these meanders are also affected by changing bottom velocities[130]. Measurements of bottom velocities at 4800 m depths here have shown[612] variations of 1 to 20 cm/s, therefore not ruling out this particular mechanism for changing the shape of the meanders in the Agulhas Return Current with time.

The growth of such meanders at the Subtropical Convergence in this region seems not to be totally dependent on the bottom topography. A numerical model with four or five layers through the water column, but with a flat bottom topography[529], shows the development of meanders, their elongation (viz. Figure 6.41) and eventually the shedding of a cold eddy northwards as observed with satellite thermography (e.g. Figure 7.17D). This result suggests that the presence of the Agulhas Plateau may initiate downstream meanders

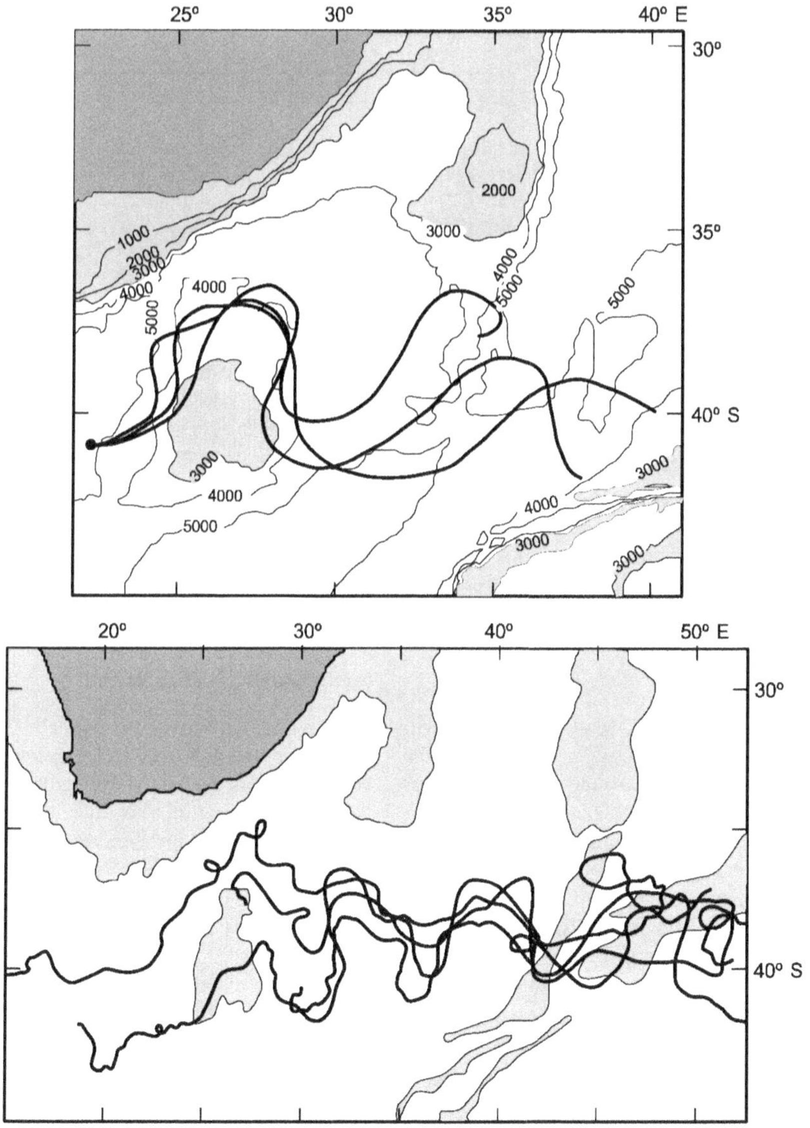

Figure 7.19. Trajectories of the Agulhas Return Current as simulated by an inertial jet model (upper panel) and by the drift tracks of free-drifting buoys[130] (lower panel). Bottom topography shallower than 3000 m has been shaded. For the jet trajectories the bottom current has been modified between 2 to 6 cm/s, leading to slightly different trajectories.

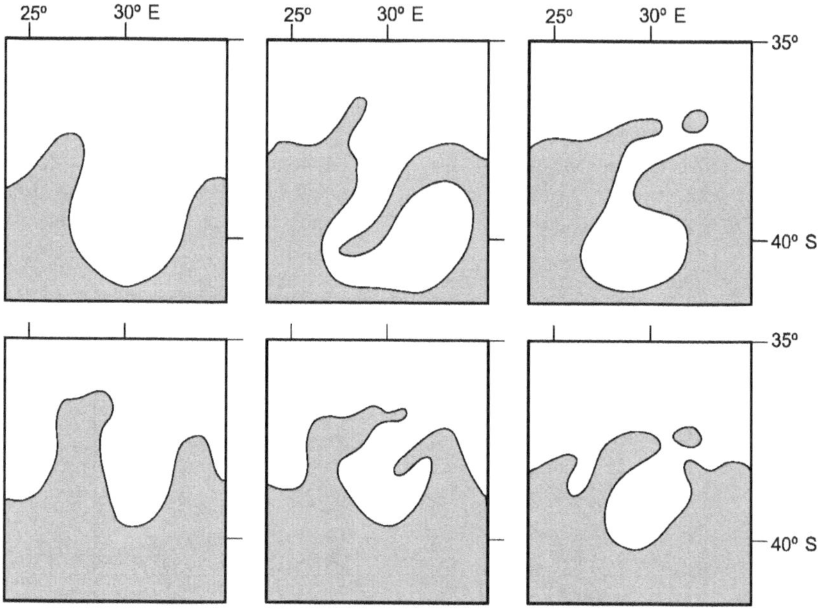

Figure 7.20. The formation of a warm eddy by a process of enclosure at the Subtropical Convergence[277] in the vicinity of the Agulhas Plateau according to satellite thermal infrared observations (upper panels) and as simulated by the FRAM (Fine Resolution Antarctic Model; lower panels). Simulations in the lower row of panels are ten model days apart.

in the Agulhas Return Current and in the Subtropical Convergence, but may anchor some of the eddy-creating processes at the plateau. General meandering and eddy-shedding may be expected farther downstream, but probably are not so geographically bound.

Large-scale, eddy-allowing models

The Fine Resolution Antarctic Model[272] (FRAM) with 32 layers in the vertical and a spatial resolution in this region of about 32 km x 5 km would seem an appropriate numerical tool with which to investigate the dynamics of the Agulhas Return Current[276]. It has a detailed representation of the bottom topography. The trajectory of the Agulhas Return Current, although variable, in general terms agrees well with that which has been observed to date[277]. Of even greater value are the coherent simulations the FRAM gives of the mesoscale processes inherent in the variability of the eastwards flow (Figure 7.20). The shedding of cold eddies equatorward and the enclosure of bodies of warm water to the south are all modelled with what would seem to be great realism. The results of these and similar global circulation models[274] hold out hope for a greater growth in understanding of these processes in the future.

It is of interest also to note that a number of the models[59,130] show regular meanders in the eastward current much farther east than the range of what is known as the Agulhas Return Current. This is particularly important since so few hydrographic measurements of this current have been made here.

South Indian Ocean Current

Almost all global circulation models[274] show a zone of high speed, but with very variable flow, along a tongue coincident with the path of the Agulhas Return Current. As mentioned before, this zone of high variability is also evident in sea surface temperature changes[130] and in sea level variations[183,335,587]. It has usually vanished at about 70° E longitude, from where the South Indian Ocean Current can therefore be expected to be the main flow along the Subtropical Convergence. This current can in some respects be considered to lie outside the greater Agulhas Current system.

Careful analyses of the few historic hydrographic sections[100] across the South Indian Ocean Current have shown that the core of this current lies just north of the Subtropical Convergence in the region of interest at about 40° S. The maximum geostrophic velocities at the sea surface are about 30 cm/s at 35° E longitude, in the Agulhas Return Current. In the South Indian Ocean

Current, beyond 60° E, they are less than 10 cm/s. Because of large station spacing in the oceanographic sections across this current these values are most probably all underestimates. Strong flows extend to 1000 m and deeper. South of the Subtropical Convergence the flow is much weaker, indicating that one has left the southern limb of the anti-cyclonic gyre and entered the Antarctic Circumpolar Current.

Hydrographic sections that have been carried out between 40° and 70° E longitude[603], i.e. in the run-up to the South Indian Ocean Current proper, show frontal systems with strong gradients from north to south with salinities dropping from 35.4 to 34.5 and temperatures from 16 °C to 14 °C. Multiple fronts, eddies and meandering, as observed in the Agulhas Return Current, are evident in the South Indian Ocean Current as well, but with considerably reduced incidence[643–4].

In general the South Indian Ocean Current, east of 70° E, is one of the poorest known currents of the whole Indian Ocean. Few hydrographic data are available for this remote region and those that have been collected here have been gathered for different reasons than for a better understanding of the South Indian Ocean Current. Global-scale, eddy-resolving modelling may give valuable indications, but without a chance of verifying these, the substantial degree of ignorance remains.

Summation

In most subtropical, anti-cyclonic circulations the poleward limb of the gyre has a much greater intensity close to the western edge than further along this flow. This is true for the South Atlantic Ocean, but it is especially true for the southern Indian Ocean. In addition, the gyral flow in the South Indian Ocean is complicated by a strong intensification towards the west. As a result the circulation in the South West Indian Ocean can be considered as a somewhat closed system; the greater Agulhas Current system. This allows one to distinguish between a southern limb for the Agulhas Current system and one for the South Indian Ocean as a whole. The closing current of the Agulhas Current system is the Agulhas Return Current.

This current in general follows the flow along the Subtropical Current, considerably strengthening it, enhancing its meridional shear and causing extraordinary high levels of mesoscale variability. The eastward flow is affected by a number of bathymetric obstructions along its path that are made evident in the current paths by large equatorward meanders. These meanders are locked to the bottom topography, but the disturbances to the flow created in this manner may conceivably radiate outward, thus adding to the general variability of the region. The possible influence of the ridges on the intensification of the wind-driven gyre is not known. The effect of eddies that have been shed to both sides of the Agulhas Return Current and the Subtropical Convergence may be different.

A previous concept that these eddies would diminish the meridional gradients of current and front and would radiate outward, carrying their characteristic water masses with them, seems to be wrong. Observed eddies in general tend to stay put, or to be re-incorporated into the current and front at a later stage. They thus strengthen and intensify the jet-like flow, creating and maintaining a special biological habitat at the Subtropical Convergence.

At about 70° E the last remnants of Agulhas water seem to have been completely absorbed into the subtropical gyre. The remaining flow along the Subtropical Convergence from here eastward – the South Indian Ocean Current – is much slower, has a lower volume flux and exhibits far less variability. The lack of solid observational data on which the conclusions pertaining to this current are based points to some of the clearly identifiable problems that remain if the circulation of the South Indian Ocean in general, and the Agulhas Return Current in particular, is to be properly understood.

8

The greater Agulhas Current: some reflections

Conceptual thinking on the role of the greater Agulhas Current system in oceanic circulation has progressed through a number of distinct phases. However, in looking back over this historical development one is invariably tempted to read the past by the agendas of the present. The prevailing interest in inter-ocean exchanges and their effects on global ocean circulation is fundamentally motivated by contemporary concerns about climate change and climate variability. This was not the case when the main momentum in Agulhas Current studies was initiated.

Nonetheless, this modern incentive has helped in lifting investigations on the Agulhas Current from a sphere of largely localised, parochial interests to one where the Agulhas Current system is instead seen in its more comprehensive role as a conduit for inter-ocean exchanges of heat and salt. This wider perspective has had some distinct advantages, but may ultimately also have a few drawbacks.

A global perspective

In thus tracing the development of research on the Agulhas Current, one also cannot but be struck by the increasing geographic scope that is considered to be required for a proper understanding of the system.

Far-ranging controls

Some numerical models, for instance, indicate that the Indonesian throughflow at the Timor Sea has a statistically significant effect on the flow behaviour south of Africa. The variability of this quasi-Pacific Ocean discharge has therefore, in principle, to be included in a proper study of the Agulhas Current system. Somewhat closer to home, more information on the current dynamics in the Mozambique Channel and to the east of Madagascar is generally considered to be of cardinal importance to a better understanding of the sources of the Agulhas Current proper. This at first glance does not seem a highly debatable thesis.

However, the influence of these sources may derive from much further afield. Recent investigations have indicated that the behaviour of the southern branch of the East Madagascar Current may be directly influenced by specific flow perturbations moving across the full width of the South Indian Ocean. These disturbances might well be of considerable importance to an understanding of Agulhas Current dynamics as a whole. Pursuing them to their source would, however, excessively increase the oceanic expanse to be investigated as part of the Agulhas Current system.

Similarly, as concepts on the influence of the Agulhas Current on the South Atlantic Ocean develop, its geographical range is continually extended. Satellite altimetry has allowed investigators to follow Agulhas rings for distances and periods previously unimaginable. Studies on the interactions of rings in the Cape Basin, effects on their crossing of the Walvis Ridge and the South Atlantic Mid-ocean Ridge as well as their further movement across the South Atlantic Ocean are all getting to be standard. This has even led to speculations – perhaps too precipitously – on the effects of Agulhas rings being absorbed by the Brazil Current. Numerical models allow one to go even farther afield. Adjusting the number and content of Agulhas rings in such models has demonstrated their effect on the large-scale, thermohaline, overturning cell of the Atlantic Ocean as a whole.

The obvious danger for those closely involved in these exciting investigations is to lose a sense of proportion. The Agulhas Current system may well play an important and even central role in some aspects of global circulation, but it remains a geographically limited system and its influence may be closely circumscribed in the greater scheme of global ocean circulation. What then, from a global perspective, are the exceptional elements of the Agulhas Current system that make it especially noteworthy?

Agulhas Current peculiarities

The first and, on further consideration, most obvious factor in the whole circulation system is the limited poleward extent of the African continent. This allows for a direct inter-ocean exchange of subtropical water masses and even for the penetration of a western boundary current from one basin to an adjacent one. The role of the Agulhas Current in the global thermohaline circulation hinges entirely on this fact of continental geography. This factor also allows for the interaction of circulation elements of this western boundary current with an extensive coastal upwelling system and with an eastern boundary current.

Furthermore, by the process of its retroflection enormous amounts of warm surface water brought southward by the Agulhas Current are retained for lengthy periods in geographical regions critical to the atmospheric systems that affect the southern African subcontinent. This unusual peculiarity of the Agulhas Current system has been shown to have a marked effect on rainfall over the adjacent land mass.

In addition, little is as yet known about the distribution of pelagic organisms and their larvae in these regions, but a great deal of ingenuity is not required to recognise that a freer exchange of water masses between adjacent basins would be conducive to a freer exchange of biota as well. This could conceivably extend to tropical species from the Indo-Pacific system as well. Organisms retained in Agulhas rings could remain sheltered from the impacts of life-threatening environmental conditions at higher latitudes as they pass through two subtropical systems. The passage from one tropical marine ecosystem to another would thus be made considerably easier. It would constitute one more special aspect of the Agulhas Current system. Although this particular biological impact of the current system has to remain largely speculative at present, a glance back over Agulhas Current research does accentuate how much has been learnt over the past two decades in other spheres.

An inventory of things learnt

A listing of scientific progress on the Agulhas Current system since about 1979 could relatively easily turn into a synopsis of this book – so much has been gained during this specific period alone. Nonetheless, a number of scientific achievements during this and preceding periods are conspicuous by being particularly fundamental to our understanding of the system. A selection of these foremost achievements may be subjective, but nevertheless instructive.

Agulhas Current retroflection

The discovery of the retroflective behaviour of the southern termination of the Agulhas Current was a significant departure from previous thinking. It had been adumbrated by some earlier work and had even been incorporated in hypotheses based on hydrographic cruises. With the regular acquisition of the products of satellite remote sensing it became an established oceanographic fact for the first time, as did the ring-shedding process that was shown to be an inherent part of this behaviour of the current. The results of some directed research cruises in the region then demonstrated these processes at sea and allowed a reliable estimate of the total fluxes of heat and salt from the South Indian to the South Atlantic Ocean to be made for the first time. Most of the subsequent scientific interest in the Agulhas Current has been based on this ring-shedding process and its wider consequences.

These observational and modelling investigations have been focused especially on the behaviour of Agulhas rings once they have been spawned. With hydrographic observations, bottom moorings, satellite observations and subsurface floats the movement of these features through the South Atlantic has been described and monitored. It has become clear that rings are subject to a host of influences and that each individual ring may behave in radically different ways from the next. This has important consequences for a proper understanding of the inter-ocean fluxes and the eventual dispersion of the contents of Agulhas rings. The timing of ring spawning events themselves has remained somewhat of a mystery until the impact of the Natal Pulse was investigated.

Natal Pulses

As in almost all aspects of the growth in knowledge on the Agulhas Current, it has been the access to satellite remote sensing that has illuminated the role of this solitary meander on the behaviour of the current. Observed in hydrographic data previously, it was thermal infrared imagery that brought its recurrent presence, its invariant downstream speed and other unusual characteristics to light. This meander feature is unique to this particular western boundary current. Analytical studies have suggested why this should be so and these investigations are continuing. Since it has now been demonstrated that the Natal Pulse may trigger ring shedding events at the Agulhas retroflection and may occasionally even cause upstream retroflections, its place in studies on inter-basin exchange seems assured.

The triggering of these unusual meanders in the Natal Bight has also received substantial theoretical and observational attention and this continues. The presence of intense, mid-ocean eddies had previously been thought possibly to play a role in this process, but the existence of these intense vortices in the centre of a subtropical gyre – far removed from any intense currents that might generate them – remained an enigma. The judicious use of altimetric and other observations has now explained the origin of these features and indicated their role on the initiation of Natal Pulses. More research on these unusual eddies is continuing.

By contrast, not as much has perhaps been learnt about the characteristics of the northern Agulhas Current itself. Its trajectorial stability has been admirably established and its speed, volume flux and temporal behaviour closely monitored with a set of deep-sea moorings for a period of more than a year. Most important, the existence of a compensating undercurrent has been established unambiguously.

The impact of the Agulhas Current on the adjacent shelf circulation has been shown, not unexpectedly, to be pervasive. However, the manner of this impact is unusual. The site-specific upwelling of deeper water onto the shelf and the moulding of the bottom layers and the seasonal thermocline over the shelf by this process was totally unanticipated. On the broader canvas of the circulation of the South West Indian Ocean as a whole the level of ignorance only increases.

Agulhas Current sources

It has become clear that the inflow into the Agulhas Current proper comes about mainly through recirculation in a wind-driven subgyre. Lack of appropriately placed hydrographic data limits our current concept of the exact dimensions of this subgyre, its possible seasonality as well as the time-varying nature of the Agulhas Return Current that forms the poleward limb of this gyre. The generation of sea surface temperature anomalies in this general region and their influence on continental rainfall also continues to be a subject of intense research. Observational programmes over the last two decades have presented very good information on the direct influence of the Agulhas Current on the overlying atmosphere.

The reasons why most of the scientific advances have been made in the southern parts of the greater Agulhas Current system rather than farther north are instructive. The horizontal temperature contrast in the south – facilitating the use of thermal infrared imagery – are amongst the highest in the world ocean. The gradients in sea surface heights in the southern Agulhas Current and its rings are also unsurpassed. Even atmospherical lightning strikes, unusual over the ocean, are more prevalent over this part of the Agulhas Current than anywhere else. All these, and other, extreme values for ocean and atmospheric parameters have made this a favourable region for observational studies. Theoretical and modelling studies have not been far behind. By contrast the gradients for all these parameters in the northern parts of the Agulhas Current system are much lower and growth in knowledge has been commensurably less with perhaps one prominent exception.

The flow in the Mozambique Channel had remained shrouded in mystery for some time. Few hydrographic data were available. The altimetric results and the modelled current behaviour conflicted dramatically; many models suggesting an unperturbed poleward flow, the altimetry indicating very high levels of variability. One dedicated research cruise has shattered any concept of a continuous Mozambique Current and firmly established that the flow in the Mozambique Channel consists of a train of eddies. A particularly important finding has been that some of these eddies may trigger Natal Pulses. By contrast, knowledge on the southern limb of the East Madagascar Current remains rudimentary.

This current has also been conceived as a tributary to the Agulhas Current. Results gathered during the past decades have, instead, shown it to be a miniature western boundary current that may retroflect completely south of Madagascar. Hydrographic data to pin this concept down have stayed in abeyance. Altimetric data have indicated that the southern termination of this current is a source of considerable mesoscale turbulence. Such turbulence has also been shown to come across the South Indian Ocean in two distinct zonal bands. The exact role of the perturbations arriving from these different sources on the behaviour of the Agulhas Current proper is being investigated.

Temporal variations

Whether farther to the south, where the use of some remote sensing products seems more efficacious, or farther to the north, where it is less so, satellite remote sensing has not been available for a sufficient duration to allow clear-cut investigations into the seasonality of the system. Some tentative and enticing studies have nonetheless attempted this. Numerical models have to date presented a range of mutually conflicting results in this regard. Once again, hard data to resolve these model disparities are lacking. A number of other serious limitations to our current knowledge can also be identified.

Knowledge gaps remaining

As mentioned, the major scientific problems – as currently perceived – are largely concentrated in the peripheral regions of the Agulhas Current system. These include the inflow and the outflow regions.

Input into the Agulhas Current

The source regions of the Agulhas Current, i.e. the South West Indian Ocean subgyre, the flow through the Mozambique Channel and the flow from east of Madagascar all remain relatively poorly known. The circulation in the Mozambique Channel as a unit has last been studied by research cruises in the 1960s. This was done in a very limited and preliminary manner and was never followed up properly by subsequent cruises, although some high-quality data have been gathered in the channel as part of WOCE (World Ocean Circulation Experiment) and directed cruises have changed thinking on the poleward flow. Long-lasting current meter observations in the narrows of the channel have been of particular value in ascertaining the formation of eddies and its variability. Some models for the region suggest that a seasonal surface flow through the channel would have a significant influence on the Agulhas Current, others that there is little or no such net meridional flow.

The geographic disposition and the retroflective behaviour of the southern limb of the East Madagascar Current have as yet not been studied adequately at sea. This gives this current the dubious distinction of probably being the least studied western boundary current on the globe. If our understanding at present is fairly accurate, the return flow of this current would be in the opposite zonal direction to that of the general wind-driven flow of the region. What effect would this have? This truly is a problem crying out for a solution.

Most of these problematic aspects of the circulation in this wind-driven, subtropical gyre are a product of the unique existence of a major obstruction to the expected anti-cyclonic flow pattern, i.e. the land mass of Madagascar. Its presence does, naturally, make itself felt in global circulation models, but a true analytic understanding of its effect on gyral flow dynamics is still lacking.

Output from the Agulhas Current

The major component of the outflow from the Agulhas Current system is the Agulhas Return Current. Although never specifically the focus of a dedicated research programme or a cruise, a number of modern cruise lines have fortuitously crossed this current and have given some general concept of its speed, water masses and volume flux. How it is influenced by each of the ridges it crosses, how it varies with time and how and where its waters are again taken up into the South West Indian Ocean subgyre all still need to be better understood. The lamentable ignorance on the basic hydrography of these parts of the inflow and outflow of the Agulhas Current is due, for the greater part, to a lack of appropriate data on these systems. This is not so much the case with that other component of the outflow, the Agulhas rings.

The spawning of rings, their dynamic and hydrographic characteristics and their subsequent drift patterns have all received considerable international attention over the past two decades. This, as could be expected, has led to an even larger number of questions on their mixing rates, their tendency to split, the influence of the bottom topography on specific types of rings, etc. How do Agulhas rings distribute their excess heat, salt and vorticity over the South Atlantic Ocean and what effect does this have on the behaviour of this ocean? Do rings of a specific size and vertical penetration all act in the same way? What role may convective overturning in rings, brought about by atmospheric influences, have on their subsequent behaviour? And could the frequency of ring shedding, their sizes and the inter-ocean exchanges brought about as a result be modified in some way by, for instance, changing wind patterns over the South Indian Ocean? A number of these questions have recently begun to be addressed and initial results have engendered tantalising further questions. Only intense observational programmes will be able to give authoritative answers to many of these questions.

Two of the most aggravating problems in a better understanding of the Agulhas Current system concern the effect of smaller scales on the circulation and the temporal behaviour of the circulation. The lack of understanding of the temporal behaviour is particularly vexing.

Transient changes

To address the problem of temporal behaviour one naturally requires observations taken at regular intervals and at intervals short enough to prevent aliasing. For many questions one may have on the dynamics of the Agulhas Current system such a research project would be prohibitively expensive. It would, for instance, be essential to know how the Agulhas Current system reacts to temporal changes in the wind stress over the Indian Ocean. It would be important to know this at inter-annual, seasonal and even shorter time scales. Numerical modelling could perhaps give very good indications, but without proper monitoring the reliability and verisimilitude of model results would remain unknown.

One would want to know how fast the subtropical gyre of the South Indian Ocean reacts to changes in the winds. How does the Agulhas Current itself react to such changes? Does its volume flux change? Does the ratio between its component water masses change? There is no reason to believe that the amounts of Red Sea Water – or even Tropical Surface Water – it carries stay constant. Is there some seasonality in the presence of these water masses or is this input totally random? The excellent results of a current meter mooring in the Agulhas Current proper have begun to answer some of these questions, but only regular monitoring in many other parts of the system could start giving some definitive answers. However both the temporal and spatial scales at which this needs to be done might be problematic.

Small-scale effects

In recent years it has become clear that there may be a number of mesoscale processes acting in the greater Agulhas Current system that have an important effect on the large-scale circulation. This effect has been shown to be totally out of proportion to their size. Perhaps this is also true of other western boundary currents, but in the Agulhas Current system it is particularly conspicuous. The most obvious of these mesoscale features is the Natal Pulse. This meander may be triggered by interaction between the Agulhas Current and other mesoscale flow elements. The Natal Pulse starts off as a feature of very modest dimensions, but may cause early retroflection of the entire Agulhas Current or ring shedding from the Agulhas retroflection far downstream. A role for the mid-ocean eddies, yet another mesoscale feature of the system, has been shown for triggering Natal Pulses, but this needs to be established in greater detail.

The evidence that a Mozambique Current does not exist, but that instead the entire poleward flow along the east coast of Mozambique consists of mesoscale eddies, forces further studies on this phenomenon to be focussed on the scale of eddy sizes. The behaviour of the southern limb of the East Madagascar Current also seems to fall entirely within the spectrum of the mesoscale.

The problem with numerical modelling of the entire Agulhas Current system is that these mesoscale features all need to be adequately resolved before the dynamics of the system can be understood to the level where it can be predicted. To date this has been too demanding of computer time and has therefore not been done adequately. It is possible – even likely – that those characteristics and that behaviour of the Agulhas Current that makes it unique amongst western boundary currents is due to the spatial dynamics at the mesoscale that pertain here. The upshot of all these, and other, questions is that a number of new research programmes, new data acquisition methods and new modelling foci are required in the immediate future. What can realistically be expected?

A look behind the curtain

It is a foregone conclusion that the spatial and temporal resolutions of numerical models for the hydrodynamics in the ocean are to increase. This is obvious from any extrapolation from modelling trends over the past decade. The problem of resolving mesoscale features in the Agulhas Current system may thus succumb to this onslaught in the near future. A spurt would be given if more attention could immediately be given to nested models for the Agulhas Current, preferably including sophisticated data assimilation. A number of individual projects of this kind are underway. It therefore seems highly likely that this specific problem is going to receive the required attention.

An international, multi-disciplinary project has studied mixing in Agulhas rings. This project has been a good start to a better understanding of these processes in rings and may suggest some even more focused questions on the role of these processes in future. With the steady growth in the length of the record of altimetric information for the South Atlantic Ocean the behaviour of Agulhas rings in this ocean is going to be analysed with greater statistical reliability in the next decade.

A number of dedicated cruises are planned to study the flows in the Mozambique Channel and east of Madagascar. These efforts will in many respects be covering *mare incognitum*. Their results will no doubt confirm some currently held concepts, but will more than likely also present a number of unexpected and disturbing results that will require further investigations at sea. Of particular importance will be a better understanding of the temporal variability of many aspects of the flow.

A detailed description of the geoid has recently become available. This will allow absolute surface topography of the ocean, and not just anomalies as has been the case up till now, to be derived from satellite altimetry. As a result, it will become possible to describe temporal changes to all components of the greater Agulhas Current system. The enduring question of a seasonality in the Agulhas Current will then probably be laid to rest. If satellite altimetry becomes available with higher spatial resolution, many of the smaller mesoscale features could also be resolved and their interaction with the larger-scale flow elements then pinned down. However, the flow at depth and the water masses involved can not be observed in this way.

It would not be a very debatable thesis that the likelihood of a preponderance of hydrographic cruises in the South West Indian Ocean over the next decades is low. However, some plans are already afoot for current meter moorings in different parts of the system and for the maintenance of those that are there. The important results of a large, international float programme are starting to become available. This constantly growing hydrographic data set is bound to revolutionise knowledge on the water masses of the region, but will not replace dedicated, well-focused research cruises. So, also in this respect there is a high likelihood that substantial progress is going to be made during the next decade. This will hopefully build on the firm foundation of knowledge on the greater Agulhas Current system that has been summarised in these pages.

References

1. Tchernia, P. (1980). *Descriptive Regional Oceanography*. Pergamon Press, Oxford, 253 pp.
2. Lutjeharms, J.R.E. (1972). *A Guide to Research Done Concerning Ocean Currents and Water Masses in the South West Indian Ocean*. University of Cape Town, Cape Town, 577 pp.
3. Lutjeharms, J.R.E. (1980). Fisiese Oseanologie van die Suidwes Indiese Oseaan: 'n bronnelys en oorsig van 1970 tot 1980. *WNNR Verslag*, **T/SEA 8016**, 82 pp.
4. Lutjeharms, J.R.E. (1987). The Physical Oceanology of the South West Indian Ocean; a bibliography for 1980 to 1985. *WNNR-verslag*, **T/SEA 8712**, 131 pp.
5. Lutjeharms, J.R.E. (1992). The Physical Oceanology of the South West Indian Ocean; a bibliography from 1986 to 1990. *CSIR Research Report*, **707**, 95 pp.
6. Basto, A. de M. (1945). *Diario da viagem de Vasco da Gama*. Vol. 1. Editora, Porto, 150 pp.
7. Rennell, J. (1832). *An investigation of the currents of the Atlantic Ocean and of those which prevail between the Indian Ocean and the Atlantic*. J.G. and F. Rivington, London, 3xx pp.
8. Deacon, M. (1971). *Scientists and the Sea 1650–1900; a study of marine science*. Academic Press, London, 445 pp.
9. Kortum, G. (1990). An unpublished manuscript of Alexander von Humboldt on the Gulf Stream. In *Ocean Sciences. Their history and relation to man,* editors W. Lenz and M. Deacon. *Deutsche Hydrographische Zeitschrift*, **B**(22): 122–129.
10. Koninklijk Nederlandsch Meteorologisch Instituut (1857). De Agulhas stroom, afgeleid uit de temperatuur van het zeewater aan de oppervlakte; en de invloed dien deze op de atmosfeer uitoefent. In *Uitkomsten van Wetenschap en Ervaring, aangaande winden en zeestromingen in sommige gedeelten van den oceaan,* Koninklijk Nederlandsch Meteorologisch Instituut, L.E. Bosch en Zoon, Utrecht, pp. 40–50.
11. Van Gogh, J. (1858). De stormen nabij de Kaap de Goede Hoop in verband beschouwd met de temperatuur der zee. *Verhandelingen en Mededelingen der Koninklijke Akademie, afdeling Natuurkunde*, **8**, 23 pp.
12. Brundrit, G.B. and L.V. Shannon (1989). Cape storms and the Agulhas Current: a glimpse of the future? *South African Journal of Science*, **85**(10): 619–620.
13. Jury, M.R. (1994). A thermal front within the marine atmospheric boundary layer over the Agulhas Current south of Africa: Composite aircraft observations. *Journal of Geophysical Research*, **99**(C2): 3297–3304.
14. Lutjeharms, J.R.E., W.P.M. de Ruijter and R.G. Peterson (1992). Interbasin exchange and the Agulhas retroflection; the development of some oceanographic concepts. *Deep-Sea Research*, **39**(10): 1791–1807.
15. Thomson, C.W. and J. Murray (editors) (1881–1895). *Report on the scientific results of the voyage of H.M.S. Challenger during the years 1873–1876.* His Majesty's Stationery Office, London, 50 volumes.
16. Von Schleinitz, B. (1889–1890). *Die Forschungsreise S.M.S. "Gazelle" in den Jahren 1874 bis 1876 unter Kommando des Kapitän zur Zee Freiherrn von Schleinitz*. Hydrographische Amt, Reichs-Marine-Ampts, Berlin, Bände I–V.
17. Hydrographic Department (1896). *List of oceanic depths and serial temperature observations during the years 1895*. Great Britain, Hydrographic Department, London.
18. Chun, C. (*et al.*) (1902–1940). *Wissenschaftliche Ergebnisse der Deutschen Tiefsee-Expedition auf dem Dampfer Valdivia, 1898–1899.* Gustaf Fischer, Jena, 24 volumes.
19. Chun, C. (1900). *Aus den Tiefen des Weltmeeres. Schilderungen von der Deutschen Tiefsee-Expedition.* Gustav Fisher, Jena, 549 pp.
20. Von Drygalski, E. (editor) (1908–1952). *Ergebnisse der Deutschen Südpolar-Expedition 1901–1903.* Reichsamtes des Innern, Berlin, Bände I–XX.
21. Reichs-marine Ampt (1909). *Forschungsreise S.M.S. "Planet" 1906–1907.* Reichs-marine Ampt, Berlin, Bände I–V.
22. Brennecke, W. (1915). Oceanographische Arbeiten S.M.S. Möwe im westlichen Indischen Ozean 1913. *Annalen der Hydrographie und Maritimen Meteorologie*, **43**(8): 337–343.
23. Anonymous (1970). Agulhas. In *Standard Encyclopaedia of Southern Africa*, editors D.J. Potgieter and P.C. du Plessis, Vol.1, Nasou, Cape Town, p. 254.
24. Kerhallet, M.C.P. de (1852). Considérations générales sur l'océan Atlantique. *Annales Hydrographiques*, **1er Sem**. pp. 33–143.
25. Zimmermann, W.F.A. (1865). *Der Erdball und Seine Naturwunder*. Gustav Hempel, Berlin, 544 pp. (as shown in Peterson *et al.*, 1996[26]).
26. Peterson, R.G., L. Stramma and G. Kortum (1996). Early concepts and charts of ocean circulation. *Progress in Oceanography,* **37**(1): 1–115.
27. Rennell, J. (1778). *Chart of the Bank of Lagullus and southern Coast of Africa,* London.

28. Maury, M.F. (1855). *The Physical Geography of the Sea*. Harper & Brothers, New York, 274 pp., 12 plates.
29. Findlay, A.G. (1866). *A directory for the navigation of the Indian Ocean*. Richard Holmes Laurie, London, 113 pp.
30. Krümmel, O. (1882). Bemerkungen über die Meereströmungen und Temperaturen in der Falklandsee. *Aus den Archiv der Deutschen Seewarte*, **V**, 2, 25 pp. (as shown by Peterson *et al.*, 199626).
31. Krümmel, O. (1911). *Handbuch der Ozeanographie. Band II. Die Bewegungsformen des Meeres*. von J. Engelhorns Nachf., Stuttgart, 766 pp.
32. Merz, A. (1925). Die Deutsche Atlantische Expedition auf dem Vermessungs- und Forschungsschiff "Meteor". *Sitzungsberichten der Physikalish-Matematischen Klasse der Preussischen Akademie der Wissenschaften*, **31**, pp. 562–586.
33. Defant, A. (1932–1960). *Wissenschaftlichen Ergebnisse der Deutschen Atlantischen Expedition auf dem Forschung- und Vermessungschiff "Meteor" 1925–1927*. Bände I–XVI, Berlin.
34. Discovery Committee (1929). Discovery Investigations Station List, 1925–1927. *Discovery Reports*, **1**: 3–138.
35. B[arlow], E.W. (1932). Major James Rennell. *Marine Observer*, **9**(101): 94–98. 496.
36. Pollard, R. and G. Griffiths (1993). James Rennell, the father of oceanography. *Ocean Challenge*, **4**(1/2): 24–25.
37. Bravo, M. (1993). James Rennell: antiquarian of ocean currents. *Ocean Challenge*, **4**(1/2): 41–50.
38. Gould, J. (1993). James Rennell's view of the Atlantic circulation: a comparison with our present knowledge. *Ocean Challenge*, **4**(1/2): 26–32.
39. Markham, C.R. (1901). *Major James Rennell and the rise of modern English geography*. Cassell and Company, London, 232 pp.
40. Dietrich, G. (1935). Aufbau und Dynamik des südlichen Agulhasstromgebietes. *Veröffentlichungen des Institut für Meereskunde an der Universität Berlin*, n.f. **A**(27): 79 pp.
41. Dietrich, G. (1936). Aufbau und Bewegung von Golfstrom und Agulhasstrom, eine vergleichende Betrachtung. *Naturwissenschaften*, **24**(15): 225–230.
42. Dietrich, G. (1935). Zur Dynamik des Atlantischen Zweiges des Agulhasstromes. *Annalen der Hydrographie und Maritimen Meteorologie*, **63**: 383–387.
43. Möller, L. (1929). Die Zirkulation des Indischen Ozeans; auf Grund von Temperatur- und Salzgehaltstiefenmessungen und Oberflächenstrombeobachtungen. *Veröffentlichungen des Instituts für Meereskunde an der Universität Berlin*, n.f. **A21**: 48 pp.
44. Michaelis, G. (1923). Die Wasserbewegung an der Oberfläche des Indischen Ozeans im Januar und Juli. *Veröffentlichungen des Instituts für Meereskunde an der Universität Berlin*, n.f. **A8**(16): 32 pp.
45. Paech, H. (1926). Die Oberflächenströmungen um Madagascar in ihrem Jährlichen Gang. *Veröffentlichungen des Instituts für Meereskunde an der Universität Berlin*, n.f. **A**(16): 39 pp.
46. Möller, L. (1933). Zur Frage der Tiefenzirkulation im Indischen Ozean. *Annalen des Hydrographie Berlin*, **61**(8,9): 233–236.
47. Thomsen, H. (1935). Entstehung und Verbreitung einiger charakteristischer Wassermassen in dem Indischen und südlichen Pazifischen Ozean. *Annalen der Hydrographie und Maritimen Meteorologie*, **63**(8): 293–305.
48. Thomsen, H. (1933). The circulation in the depths of the Indian Ocean. *Journal Conseil*, **8**(1): 73–79.
49. Schott, G. (1935). *Geographie des Indischen und Stillen Ozeans*. C. Boysen, 413 pp., 37 plates.
50. Sverdrup, H.U., H.W. Johnson and R.H. Flemming (1942). *The oceans; their physics, chemistry and general biology*. Prentice-Hall, Englewood Cliffs, 1087 pp.
51. Brown, A.C. (1997). John D.F. Gilchrist and the early years of marine science in South Africa. In *A Century of Marine Science in South Africa*, editors A.I.L. Payne and J.R.E. Lutjeharms, Sea Fisheries Research Institute, Royal Society of South Africa, pp. 2–16 and *Transactions of the Royal Society of South Africa*, **52**(1): 2–16.
52. Summers, R.F.H. (1975). *A History of the South African Museum, 1825–1975*. A. A. Balkema, Cape Town, 245 pp.
53. Gilchrist, J.D.F. (1923). The South African seas. *South African Journal of Science*, **20**(1): 1–27.
54. Gilchrist, J.D.F. (1902). Observations on the temperature and salinity of the sea around the Cape Peninsula. *Marine Investigations in South Africa*, **1**: 181–216 + 10 plates + p. 238.
55. Gilchrist, J.D.F. (1904). Currents on the South African coast, as indicated by the course of drift bottles. *Marine Investigations in South Africa*, **2**: 101–113.
56. Lutjeharms, J.R.E. and L.V. Shannon (1997). A century of physical oceanography in South Africa; in search of the legacy of John D. Gilchrist. In *A Century of Marine Science in South Africa*, editors A.I.L. Payne and J.R.E. Lutjeharms, Sea Fisheries Research Institute, Royal Society of South Africa, pp. 17–30 and *Transactions of the Royal Society of South Africa*, **52**(1): 17–30.
57. Gilchrist, J.D.F. (1925). II. Report of Director of Survey. In *Report of the Fisheries and Marine Biological Survey, Union of South Africa, June, 1923–June, 1925*. **4**. Cape Town; Government Printers: xix–lx.
58. Gilchrist, J.D.F. (1905). The South African marine fauna and its environment. In *Science in South Africa*, editors W. Flint and J.D.F. Gilchrist, Maskew Miller, Cape Town, pp. 182–197.
59. Harris, T.F.W., R. Legeckis and D. van Forest [sic] (1978). Satellite infra-red images in the Agulhas Current system. *Deep-Sea Research*, **25**(6): 543–548.
60. Lutjeharms, J.R.E. (1981). Features of the southern Agulhas Current circulation from satellite remote sensing. *South African Journal of Science*, **77**(5): 231–236.
61. Lutjeharms, J.R.E. and A.L. Gordon (1987). Shedding

of an Agulhas Ring observed at sea. *Nature*, **325**(7000): 138–140.
62. Lutjeharms, J.R.E. and H.R. Roberts (1988). The Natal Pulse; an extreme transient on the Agulhas Current. *Journal of Geophysical Research*, **93**(C1): 631–645.
63. Lutjeharms, J.R.E. and H.R. Valentine (1988). Eddies at the Sub-Tropical Convergence south of Africa. *Journal of Physical Oceanography*, **18**(5): 761–774.
64. Lutjeharms, J.R.E. and R.C. van Ballegooyen (1988). Anomalous upstream retroflection in the Agulhas Current. *Science*, **240**(4860): 1770–1772.
65. Gordon, A.L. (1985). Indian–Atlantic transfer of thermocline water at the Agulhas retroflection. *Science*, **227** (4690): 1030–1033.
66. Gordon, A.L. (1986). Inter-ocean exchange of thermocline water. *Journal of Geophysical Research*, **91**(C4): 5037–5046.
67. Gordon, A.L., R.F. Weiss, W.M. Smethie and M.J. Warner (1992). Thermocline and intermediate water communication between the South Atlantic and Indian Oceans. *Journal of Geophysical Research*, **97**(C5): 7223–7240.
68. Weijer, W., W.P.M. de Ruijter, H.A. Dijkstra and P.J. van Leeuwen (1999). Impact of interbasin exchange on the Atlantic overturning circulation. *Journal of Physical Oceanography*, **29**(9): 2266–2284.
69. Zlotnicki, V. (1991). Sea level differences across the Gulf Stream and the Kuroshio. *Journal of Physical Oceanography*, **21**(4): 599–609.
70. Gründlingh, M.L. (1988). Altimetry in the southwest Indian Ocean. *South African Journal of Science*, **84**(7): 568–573.
71. Gründlingh, M.L. (1993). On the satellite-derived wave climate off southern Africa. *South African Journal of Marine Science*, **13**: 223–235.
72. Wakker, K.F., M.C. Naeije, R. Scharroo and B.A.C. Ambrosius (1991). Extraction of mesoscale ocean currents information from Geosat altimeter data. *Proceedings of the Space & Sea Colloquium*, Paris, ESA SP-312, pp. 221–226.
73. Wakker, K.F., R.C.A. Zandbergen, M.C. Naeije and B.A.C. Ambrosius (1990). Geosat altimeter data analysis for the oceans around South Africa. *Journal of Geophysical Research*, **95**(C3): 2991–3006.
74. Feron, R.C.V., W.P.M. de Ruijter and D. Oskam (1992). Ring-shedding process in the Agulhas Current system. *Journal of Geophysical Research*, **97**(C6): 9467–9477.
75. Duncan, C.P. (1970). *The Agulhas Current*. Ph.D. dissertation, University of Hawaii, 76 pp.
76. Pearce, A.F., E.H. Schumann and G.S.H. Lundie (1978). Features of the shelf circulation off the Natal coast. *South African Journal of Science*, **74**(9): 328–331.
77. Christensen, M.S. (1980). Sea-surface temperature charts for Southern Africa, south of 26°S. *South African Journal of Science*, **76**(2): 541–546.
78. Engelmann, G. (1969). Christian Gottfried Ehrenberg, Ein Wegbereiter der deutschen Tiefseeforschung. *Deutsche Hydrographische Zeitschrift*, **22**(2): 145–157.
79. Sætre, R. and A. Jorge da Silva (1984). The circulation of the Mozambique Channel. *Deep-Sea Research*, **31**(5): 485–508.
80. Lutjeharms, J.R.E., N.D. Bang and C.P. Duncan (1981). Characteristics of the currents east and south of Madagascar. *Deep-Sea Research*, **28**(9): 879–899.
81. Harris, T.F.W. (1972). Sources of the Agulhas Current in the spring of 1964. *Deep-Sea Research*, **19**(9): 633–650.
82. Lutjeharms, J.R.E. (1976). The Agulhas Current system during the Northeast Monsoon Season. *Journal of Physical Oceanography*, **6**(5): 665–670.
83. Stramma, L. and R.G. Peterson (1990). The South Atlantic Current. *Journal of Physical Oceanography*, **20**(6): 846–859.
84. Wyrtki, K. (1971). *Oceanographic Atlas of the International Indian Ocean Expedition*. National Science Foundation, Washington, D.C. 531 pp.
85. Peterson, R.G. and L. Stramma (1991). Upper-level circulation in the South Atlantic Ocean. *Progress in Oceanography*, **26**: 1–75.
86. Gründlingh, M.L. (1983). On the course of the Agulhas Current. *South African Geographical Journal*, **65**(1): 49–57.
87. De Ruijter, W.P.M., P.J. van Leeuwen and J.R.E. Lutjeharms (1999). Generation and evolution of Natal Pulses: solitary meanders in the Agulhas Current. *Journal of Physical Oceanography*, **29**(12): 3043–3055.
88. Lutjeharms, J.R.E., R. Catzel and H.R. Valentine (1989). Eddies and other border phenomena of the Agulhas Current. *Continental Shelf Research*, **9**(7): 597–616.
89. Schumann, E.H. and I. Li [sic] van Heerden (1988). Observations of Agulhas Current frontal features south of Africa, October 1983. *Deep-Sea Research*, **35**(8): 1355–1362.
90. Bang, N.D. (1970). Dynamic interpretations of a detailed surface temperature chart of the Agulhas Current retroflexion [sic] and fragmentation area. *South African Geographical Journal*, **52**(12): 67–76.
91. Lutjeharms, J.R.E. and R.C. van Ballegooyen (1988). The retroflection of the Agulhas Current. *Journal of Physical Oceanography*, **18**(11): 1570–1583.
92. Harris, T.F.W. and D. van Foreest (1978). The Agulhas Current in March 1969. *Deep-Sea Research*, **25**(6): 549–561.
93. Lutjeharms, J.R.E. and J. Cooper (1996): Interbasin leakage through Agulhas Current filaments. *Deep-Sea Research I*, **43**(2): 213–238.
94. Olson, D.B. and R.H. Evans (1986). Rings of the Agulhas Current. *Deep-Sea Research*, **33**(1): 27–42.
95. Byrne, D.A., A.L. Gordon and W.F. Haxby (1995). Agulhas Eddies: a synoptic view using Geosat ERM data. *Journal of Physical Oceanography*, **25**(5): 902–917.
96. Lutjeharms, J.R.E. and I. Ansorge (2001). The Agulhas

Return Current. *Journal of Marine Systems*, **30**(1/2): 115–138.

97. Lutjeharms, J.R E. and H.R. Valentine (1984). Southern Ocean thermal fronts south of Africa. *Deep-Sea Research*, **31**(12): 1461–1476.
98. Mey, R.D., N.D. Walker and M.R. Jury (1990). Surface heat fluxes and marine boundary layer modification in the Agulhas retroflection region. *Journal of Geophysical Research*, **95**(C9): 15 997–16 015.
99. Harris, T.F.W. (1970). Planetary-type waves in the South West Indian Ocean. *Nature*, **227**(5262): 1043–1044.
100. Stramma, L. (1992). The South Indian Ocean Current. *Journal of Physical Oceanography*, **22**(4): 421–430.
101. Stramma, L. and J.R.E. Lutjeharms (1997). The flow field of the subtropical gyre of the South Indian Ocean. *Journal of Geophysical Research*, **102**(C3): 5513–5530.
102. Richardson, P.L. (1983). Gulf Stream rings. In *Eddies in Marine Science*, editor A.R. Robinson, Springer-Verlag, Berlin, pp. 19–45.
103. Watts, D.R. (1983). Gulf Stream variability. In *Eddies in Marine Science*, editor A.R. Robinson, Springer-Verlag, Berlin, pp. 114–144.
104. Gutshabash, Ye. Sh. and I.V. Lavrenov (1986). Swell transformation in the Cape [sic] Agulhas Current. *Izvestiya, Atmospheric and Oceanic Physics*, **22**(6): 494–497.
105. Fuglister, F.C. (1972). Cyclonic rings formed by the Gulf Stream 1965–1966. In: *Studies in Physical Oceanography 1,* editor A.L. Gordon, Gordon and Breach, New York, pp. 137–168.
106. Joyce, T.M. (1985). Gulf Stream warm-core ring collection: an introduction. *Journal of Geophysical Research*, **90**(C5): 8801–8802.
107. Lai, D.Y. and P.L. Richardson (1977). Distribution and movement of Gulf Stream rings. *Journal of Physical Oceanography*, **7**(5): 67–683.
108. The Ring Group (1981). Gulf Stream cold-core rings: their physics, chemistry and biology. *Science*, **212** (4499): 1091–1100.
109. Bennett, A.F. (1983). The South Pacific including the East Australian Current. In *Eddies in Marine Science*, editor A. R. Robinson, Springer-Verlag, Berlin, pp. 219–244.
110. Boland, F.M. and J.A. Church (1981). The East Australian Current 1978. *Deep-Sea Research*, **28**(9): 937–957.
111. Godfrey, J.S., G.R. Cresswell, T.J. Golding, A.F. Pearce and R. Boyd (1980). The separation of the East Australian Current. *Journal of Physical Oceanography*, **10**(3): 430–440.
112. Nilsson, C.S. and G.R. Cresswell (1981). The formation and evolution of East Australia Current warm-core eddies. *Progress in Oceanography*, **9**: 133–183.
113. Boland, F.M. and B.V. Hamon (1970). The East Australian Current. *Deep-Sea Research*, **17**(4): 777–794.
114. Lutjeharms, J.R.E. (1988). Remote sensing corroboration of retroflection of the East Madagascar Current. *Deep-Sea Research*, **35**(12): 2045–2050.
115. Peterson, R.G. (1992). The boundary currents in the western Argentine Basin. *Deep-Sea Research*, **39**(3/4): 623–644.
116. Kawaii, H. (1972). Hydrography of the Kuroshio Extension. In *Kuroshio; physical aspects of the Japan Current*, editors H. Stommel and K. Yoshida, University of Washington Press, London, pp. 235–352.
117. Taft, B.A. (1978). Structure of Kuroshio south of Japan. *Journal of Marine Research*, **36**(1): 77–117.
118. Shoji, D. and H. Nitani (1966). On the variability of the velocity of the Kuroshio. *Journal of the Oceanographic Society of Japan*, **22**: 192–196.
119. Taft, B.A. (1972). Characteristics of the flow of the Kuroshio south of Japan. In *Kuroshio; physical aspects of the Japan Current*, editors H. Stommel and K. Yoshida, University of Washington Press, London, pp. 165–214.
120. Leetma, A., D.R. Quadfasel and D. Wilson (1982). Development of the flow field during the onset of the Somali Current, 1979. *Journal of Physical Oceanography*, **12** (12): 1325–1342.
121. Bunker, A.F. (1988). Surface energy fluxes of the South Atlantic Ocean. *Monthly Weather Review*, **116**(4): 809–823.
122. Bunker, A.F. (1980). Trends of variables and energy fluxes over the Atlantic Ocean from 1948 to 1972. *Monthly Weather Review*, **108**(6): 720–732.
123. Houghton, J.T., G.J. Jenkins and J.J. Ephraums (editors) (1990). *Climate Change. The IPCC Scientific Assessment.* Cambridge University Press, Cambridge, 365 pp.
124. Broecker, W.S. (1987). The biggest chill. *Natural History Magazine*, **97**: 74–82.
125. Broecker, W.S., D. Peteet and D. Rind (1985). Does the ocean–atmosphere have more than one stable mode of operation? *Nature*, **315**(6014): 21–25.
126. Broecker, W.S. (1991). The great ocean conveyor. *Oceanography*, **4**(2): 79–89.
127. Duncombe Rae, C.M. (1991). Agulhas retroflection rings in the South Atlantic Ocean; an overview. *South African Journal of Marine Science*, **11**: 327–344.
128. Gordon, A.L., J.R.E. Lutjeharms and M.L. Gründlingh (1987). Select hydrographic sections from the Agulhas research cruises of the research vessels *Knorr* and *Meiring Naude* - 1983. Lamont-Doherty Geological Observatory of Columbia University, *Technical Report* **LDGO-87-1**.
129. Van Ballegooyen, R.C., M.L. Gründlingh and J.R.E. Lutjeharms (1994). Eddy fluxes of heat and salt from the southwest Indian Ocean into the southeast Atlantic Ocean: a case study. *Journal of Geophysical Research*, **99**(C7): 14 053–14 070.
130. Lutjeharms, J.R.E. and R.C. van Ballegooyen (1984). Topographic control in the Agulhas Current system.

Deep-Sea Research, **31**(11): 1321–1337.

131. Lutjeharms, J.R.E. (1989). The role of mesoscale turbulence in the Agulhas Current system. In *Mesoscale/synoptic Coherent Structures in Geophysical Turbulence,* editors J.C.J. Nihoul and B.M. Jamart, Elsevier, Amsterdam, pp. 357–372.
132. Chassignet, E.P. and D.B. Boudra (1988). Dynamics of Agulhas retroflection and ring formation in a numerical model. II. Energetics and ring formation. *Journal of Physical Oceanography*, **18**(2): 304–319.
133. Lutjeharms, J.R.E. and W.P.M. de Ruijter (1996). The influence of the Agulhas Current on the adjacent coastal zone: possible impacts of climate change. *Journal of Marine Systems*, **7**(2–4): 321–336.
134. Walker, N.D. (1990). Links between South African summer rainfall and temperature variability of the Agulhas and Benguela Current Systems. *Journal of Geophysical Research*, **95**(C3): 3297–3319.
135. Jury, M.R. and J.R.E. Lutjeharms (1993). Die struktuur en die aandrywingskragte van die 1991–1992 droogte in suidelike Afrika. *Suid-Afrikaanse Tydskrif vir Natuurwetenskap en Tegnologie*, **12**(1): 8–16.
136. Jury, M.R., B. Pathack, C.J. de W [sic] Rautenbach and J. vanHeerden [sic] (1996). Drought over South Africa and Indian Ocean SST: statistical and GCM results. *The Global Atmosphere and Ocean System*, **4**(1):47–63.
137. Jury, M.R. and M. Majodina (1997). Preliminary climatology of southern Africa extreme weather 1973–1992. *Theoretical and Applied Climatology*, **56**(1–2): 103–112.
138. Tennant, W.J. (1996). Influence of Indian Ocean sea-surface temperature anomalies on the general circulation of southern Africa. *South African Journal of Science*, **92**(6): 289–295.
139. Crimp, S.J., J.R.E. Lutjeharms and S.J. Mason (1998). Sensitivity of a tropical-temperate trough to sea-surface temperature anomalies in the Agulhas retroflection region. *Water SA*, **24**(2): 93–101.
140. Reason, C.J.C. and J.R.E. Lutjeharms (1998). Variability of the South Indian Ocean and implications for southern African rainfall. *South African Journal of Science*, **94**(3): 115–123.
141. Reason, C.J. C., C.R. Godfred-Spenning, R.J. Allan and J.A. Lindesay (1998). Air–sea interaction mechanisms and low frequency climate variability in the South Indian Ocean region. *International Journal of Climatology*, **18**(4) : 391–405.
142. Lutjeharms, J.R.E., R.D. Mey and I.T. Hunter (1986). Cloud lines over the Agulhas Current. *South African Journal of Science*, **82**(11/12): 635–640.
143. Lutjeharms, J.R.E., and M. Rouault (2000). Observations of cloud formation above Agulhas Current intrusions in the South-east Atlantic. *South African Journal of Science*, **96**(11/12): 577–580.
144. Rouault, M. and J.R.E. Lutjeharms (1994). Air–sea interaction in the marine atmosphere boundary layer: a new South African research venture. *South African Journal of Science*, **90**(1): 11–12.
145. Lee-Thorp, A.M., M. Rouault and J.R.E. Lutjeharms (1998). Cumulus cloud formation above the Agulhas Current. *South African Journal of Science*, **94**(7): 351–354.
146. Lee-Thorp, A.M., M. Rouault and J.R.E. Lutjeharms (1999). Moisture uptake in the boundary layer above the Agulhas Current: a case study. *Journal of Geophysical Research*, **104**(C1): 1423–1430.
147. Walker, N.D. and R.D. Mey (1988). Ocean/atmosphere heat fluxes within the Agulhas Retroflection region. *Journal of Geophysical Research*, **93**(C12): 15473–15483.
148. Jury, M.(R.) and N. Walker (1988). Marine boundary layer modification across the edge of the Agulhas Current. *Journal of Geophysical Research*, **93**(C1): 647–654.
149. Jury, M.R., H.R. Valentine and J.R.E. Lutjeharms (1993). Influence of the Agulhas Current on summer rainfall on the southeast coast of South Africa. *Journal of Applied Meteorology*, **32**(7): 1282–1287.
150. Ryan, P.G. (1988). The characteristics and distribution of plastic particles at the sea-surface off the southwestern Cape Province, South Africa. *Marine Environmental Research*, **25**(4): 249–273.
151. Heydorn, A.E.F., N.D. Bang, A.F. Pearce, B.W. Flemming, R.A. Carter, M.H. Schleyer, P.F. Berry, G.R. Hughes, A.J. Bass, J.H. Wallace, R.P. van der Elst, R.J.M. Crawford and P.A. Shelton (1978). Ecology of the Agulhas Current region: an assessment of biological responses to environmental parameters in the South-West Indian Ocean. *Transactions of the Royal Society of South Africa*, **43**(2): 151–190.
152. Hughes, G.R. (1975). The marine turtles of Tongaland. *Lammergeyer*, **22**: 9–18.
153. Randall, R.M., B.M. Randall and D. Baird (1981). Speed of movement of Jackass penguins over long distances and their possible use of ocean currents. *South African Journal of Science*, **77**(9): 420–421.
154. Findlay, K.P., P.B. Best, G.J.B. Ross and V.G. Cockcroft (1992). The distribution of small odontocete cetaceans off the coasts of South Africa and Namibia. *South African Journal of Marine Science*, **12**: 237–270.
155. Prell, W.L. and W.H. Hutson (1979). Zonal temperature-anomaly maps of Indian Ocean surface waters – modern and ice-age patterns. *Science*, **206**(4417): 454–456.
156. De Decker, A.H.B. (1984). Near-surface copepod distribution in the south-western Indian and south-eastern Atlantic Ocean. *Annals of the South African Museum*, **93**(5): 303–370.
157. Lazarus, B.I. and D. Dowler (1979). Pelagic tunicata off the west and south-west coasts of South Africa, 1964–1965. *Fisheries Bulletin*, **12**: 93–119.
158. Schleyer, M.H. (1985). Chaetognaths as indicators of water masses in the Agulhas Current system. *Investigations Reports, Oceanographic Research Institute,*

Durban, **61**: 1–20.

159. De Decker, A.H.B. (1973). Agulhas Bank plankton. In *The Biology of the Indian Ocean*, editor B. Zeitzschel, Springer-Verlag, Berlin, pp. 189–219.
160. Prell, W., W. Hutson, D. Williams, A. Bé, K. Geitzenauer and B. Molfine (1980). Surface circulation of the Indian Ocean during the last glacial maximum, approximately 18 000 yr. B.P. *Quarternary Research*, **14**(3): 309–336.
161. Hulley, P.A. and J.R.E. Lutjeharms (1995). The south-western limit for the warm water, mesopelagic ichthyofauna of the Indo-West Pacific: Lanternfish (*Myctophidae*), a case study. *South African Journal of Marine Science*, **15**: 185–205.
162. Nafpaktitis, B.G. (1978). Systematics and distribution of laternfishes of the genera Lobianchia – and Diaphus (*Myctophidae*) in the Indian Ocean, *Scientific Bulletin, Natural History Museum, Los Angeles*, **30**: 1–91.
163. Lutjeharms, J.R.E., M.L. Gründlingh and R.A. Carter (1989). Topographically induced upwelling in the Natal Bight. *South African Journal of Science*, **85**(5): 310–316.
164. Lutjeharms, J.R.E., H.R. Valentine and R.C. van Ballegooyen (2000). The hydrography and water masses of the Natal Bight, South Africa. *Continental Shelf Research*, **20**(14): 1907–1939.
165. Lutjeharms, J.R.E., A.A. Meyer, I.J. Ansorge, G.A. Eagle and M.J. Orren (1997). The nutrient characteristics of the Agulhas Bank. *South African Journal of Marine Science*, **17**: 253–274.
166. Lutjeharms, J.R.E., J. Cooper and M. Roberts (2000). Upwelling at the inshore edge of the Agulhas Current. *Continental Shelf Research*, **20**(7): 737–761.
167. Gill, A.E. and E.H. Schumann (1979). Topographically induced changes in the structure of an inertial coastal jet: application to the Agulhas Current. *Journal of Physical Oceanography*, **9**(5): 975–991.
168. Lutjeharms, J.R.E. and A.A. Meyer (2006). On the origin and circulation of bottom water on the Agulhas Bank, South Africa. *Continental Shelf Research*, in preparation.
169. Meyer, A.A., J.R.E. Lutjeharms and S. de Villiers (2002). The nutrient characteristics of the Natal Bight, South Africa. *Journal of Marine Systems*, 35(1–2): 11–37.
170. Weeks, S.J. and F.A. Shillington (1994). Interannual scales of variation of pigment concentrations from coastal zone colour scanner data in the Benguela upwelling system and the Subtropical Convergence zone south of Africa. *Journal of Geophysical Research*, **99**(C4): 7385–7399.
171. Mallory, J.K. (1984). Abnormal waves off the south-east coast of South Africa. *Marine Observer*, **54**(283): 29–38.
172. Schumann, E.H. (1976). Changes in energy of surface gravity waves in the Agulhas Current. *Deep-Sea Research*, **23**(6): 509–518.
173. Irvine, D.E. and D.G. Tilley (1988). Ocean wave directional spectra and wave–current interaction in the Agulhas from the Shuttle Imaging Radar–B synthetic aperture radar. *Journal of Geophysical Research*, **93**(C12): 15 389–15 401.
174. Gründlingh, M.L. (1994). Some characteristics of satellite-derived wave heights in the South Atlantic Ocean. *Deep-Sea Research I*, **41**(2): 413–428.
175. Gründlingh, M.L. (1994). Evidence of surface wave enhancement in the southwest Indian Ocean. *Journal of Geophysical Research*, **99**(C4): 7917–7927.
176. Hirst, A.C. and J.S. Godfrey (1993). The role of Indonesian throughflow in a global ocean GCM. *Journal of Physical Oceanography*, **23**(6): 1057–1086.
177. Deutsches Hydrographisches Institut (1960). *Monatskarten für den Indischen Ozean*. Deutsches Hydrographisches Institut, Hamburg, Public. 2422.
178. Wyrtki, K. (1973). Physical oceanography of the Indian Ocean. In *Biology of the Indian Ocean*, editor B. Zeitzschel, Springer-Verlag, Berlin, pp. 18–36.
179. Cresswell, G.R. and T.J. Golding (1980). Observations of a south-flowing current in the south-eastern Indian Ocean. *Deep-Sea Research*, **27**(6): 449–466.
180. Smith, R.L., A. Huyer, J.S. Godfrey and J.A. Church (1991). The Leeuwin Current off western Australia. *Journal of Physical Oceanography*, **21**(2): 323–345.
181. McCartney, M.S. (1977). Subantarctic Mode Water. *Deep-Sea Research*, **24** (Supplement): 103–119.
182. You, Y. (1996). Dianeutral mixing in the thermocline of the Indian Ocean. *Deep-Sea Research I*, **43**(3): 291–320.
183. Cheney, R.A., J.G. Marsh and B.D. Beckley (1983). Global mesoscale variability from collinear tracks of SEASAT altimeter data. *Journal of Geophysical Research*, **88**(C7): 4343–4354.
184. Lutjeharms, J.R.E. (1977). The need for oceanological research in the South-west Indian Ocean. *South African Journal of Science*, **73**(2): 40–43.
185. Ramage, C.S. (1984). Climate of the Indian Ocean north of 35°S. In *Climates of the Oceans*, editor H. van Loon, World Survey of Climatology, Volume 15, Elsevier, Amsterdam, pp. 693–697.
186. Taljaard, J.J. and H. van Loon (1984). Climate of the Indian Ocean south of 35°S. In *Climates of the Oceans*, editor H. van Loon, World Survey of Climatology Volume 15, Elsevier, Amsterdam, pp. 505–601.
187. Hellerman, S. and M. Rosenstein (1993). Normal monthly wind stress over the world ocean with error estimates. *Journal of Physical Oceanography*, **13**(7): 1093–1104.
188. Gill, A.E. (1977). Coastally trapped waves in the atmosphere. *Quarterly Journal of the Royal Meteorological Society*, **103**(437): 431–440.
189. Schumann, E.H. (1989). The propagation of air pressure and wind systems along the South African coast. *South African Journal of Science*, **85**(6): 382–385.
190. Schumann, E.H. and J.A. Martin (1991). Climato-

logical aspects of the coastal wind field at Cape Town, Port Elizabeth and Durban. *South African Geographical Journal*, **73**(1): 48–51.
191. Naval Oceanographic Office (1977). *Surface Currents. Southeast South Atlantic Ocean*. U.S. Naval Oceanographic Office, Special Publication 1401-SA4.
192. Naval Oceanographic Office (1977). *Surface Currents. West Central Indian Ocean Including Mozambique Channel*. U.S. Naval Oceanographic Office, Special Publication 1404-IN3.
193. Naval Oceanographic Office (1977). *Surface Currents. Southwest Indian Ocean*. U.S. Naval Oceanographic Office, Special Publication. 1404-IN6.
194. United Kingdom Hydrographic Office (1970). *Routing Charts – Indian Ocean*. British Admiralty, Hydrographic Office, London, No.5126.
195. Reid, J.L. (1989). On the total geostrophic circulation of the South Atlantic Ocean: flow patterns, tracers and transports. *Progress in Oceanography*, **23**(3): 149–244.
196. Lutjeharms, J.R.E. (1982). Die geskiedkundige ontwikkeling van fisiese navorsing oor die oseane grensend aan Suider-Afrika. *Suid-Afrikaanse Tydskrif vir Natuurwetenskap en Tegnologie*, **1**(1): 42–52.
197. Buchan, A. (1895). Report on oceanic circulations based on observations made on board H.M.S. Challenger and other observations. *Challenger Reports*, II (Physics and Chemistry). Part VIII, appendix to Summary, 38 pp.
198. Lutjeharms, J.R.E. and G. Kortum (2005). German research on the Agulhas Current system between the World Wars; a lost scientific achievement. *Historisch-meereskundliches Jahrbuch*, in preparation.
199. Harries, H.D. (1932). Über die Veränderlichkeit von Monatswerten meteorologischer und hydrologischer Elementen der Äquatorialsee. *Annalen der Hydrographie, Berlin*, **60**(12): 496–499.
200. Becker, R. (1938). Über den jährlichen Temperaturgang auf dem Indischen und Stillen Ozean. *Annalen der Hydrographie, Berlin*, **66**(7): 338–340.
201. Lütgens, R. (1905). Oberflächentemperaturen im südlichen Indischen Ozean 1901 bis 1903. *Annalen der Hydrographie Berlin*, **33**(11): 498–513.
202. Schott, G. (1928). Die Verteilung des Salzgehaltes im Oberflachenwasser der Ozeane. *Annalen der Hydrographie, Berlin*, **5**: 145–166.
203. Van Riel, P.M. (1932). Einige ozeanographische Beobachtungen im Roten Meer, Golf van Aden und Indischen Ozean. *Annalen der Hydrographie, Berlin*, **60**(10): 401–407.
204. Willimzik, M. (1929). Die Strömungen im Subtropischen Konvergenzgebiet des Indischen Ozeans. *Veröffentlichungen des Instituts für Meereskunde an der Universität Berlin, n.f.* **A**(14): 1–27.
205. Defant, A. (1936). Ausbreitung- und Vermischungsvorgänge im antarktischen Bodenstrom und im subantarktischen Zwischenwasser. *'Meteor' Forschungsergebnisse*, **6**(2): 53–96.
206. Defant, A. (1941). Die absolute Topografie des physikalischen Meeresniveaus und der Druckflächen sowie die Wasserbewegung im Atlantischen Ozean. *'Meteor' Forschungsergebnisse*, **6**(2): 191–260.
207. Defant, A. (1941). Die relative Topografie eizelner Druckflächen im Atlantischen Ozean. *'Meteor' Forschungsergebnisse*, **6**(2/4): 183–190.
208. Schott, G. (1944). *Geographie des Atlantischen Ozeans*. C. Boysen, 438 pp., 38 plates.
209. Schott, G. (1926). Die Tiefenwasserbewegung des Indischen Ozeans. *Annalen der Hydrographie, Berlin*, **54**(2): 417–431.
210. Wüst, G. (1935). Zur Frage des Indischen Tiefenstroms. *Naturwissenschaften*, **23**(9): 137–139.
211. Wüst, G. (1934). Anzeichen von Beziehungen zwischen Bodenstrom und Relief in der Tiefsee des Indischen Ozeans. *Naturwissenschaften*, **22**(16): 241–244.
212. Wüst, G. (1935). Die Ausbreitung des antarktischen Bodenwassers im Atlantischen und Indischen Ozean. *Zeitschrift zur Geophysik*, **11**: 40–49.
213. Lutjeharms, J.R.E. and G. Kortum (2003). The influence of German marine research on South African physical oceanography since 1890. *Historisch-meereskundliches Jahrbuch*, **9**: 81–100.
214. Clowes, A.J. (1950). An introduction to the hydrology of South African waters. Fisheries and Marine, Biological Survey Division, Union of South Africa, *Investigational Reports*, **12**: 42 pp.
215. Darbyshire, J. (1964). A hydrological investigation of the Agulhas Current area. *Deep-Sea Research*, **11**(5): 781–815.
216. Shannon, L.V. (1966). Hydrology of the South and West coasts of South Africa. *Investigational Report, Division of Sea Fisheries*, **58**: 1–52.
217. Darbyshire, M. (1966). Agulhas Current (and Mozambique Current). In *The Encyclopaedia of Oceanography*, editor R. W. Fairbridge, Reinhold Publishing Corporation, New York, pp. 23–28.
218. Park, Y.-H. and L. Gambéroni (1995). Large-scale circulation and its variability in the South Indian Ocean from TOPEX/POSEIDON altimetry. *Journal of Geophysical Research*, **100**(C12): 24 911–24 929.
219. Ffield, A., J.M. Toole and W.D. Wilson (1997). Seasonal circulation in the South Indian Ocean. *Geophysical Research Letters*, **24**(22): 2773–2776.
220. Emery, W.J. and J. Meincke (1986). Global water masses: summary and review. *Oceanologica Acta*, **9**(4): 383–391.
221. Pollak, M.J. (1958). Frequency distribution of potential temperatures and salinities in the Indian Ocean. *Deep-Sea Research*, **5**(2): 128–133.
222. Anonymous (1965). Development of the International Indian Ocean Expedition. *International Indian Ocean Expedition, Collected Reprints*, **I**: ix–xi.
223. Wüst, G. (1960). Proposed International Indian Ocean Expedition, 1962–1863. *Deep-Sea Research*, 6(3): 245–249.

224. Deacon, G.E.R. (1960). The Indian Ocean Expedition. *Nature*, **187**(4737): 561–562.
225. Knauss, J.A. (1961). The International Indian Ocean Expedition. *Science*, **134**(3491): 1674–1676.
226. Anonymous (1970). Symposium 'Oceanography in South Africa 1970', South African National Committee for Oceanographic Research, Durban.
227. Lutjeharms, J.R.E. (1991). Twintig jaar fisiese oseanologie in Suid-Afrika: 'n persoonlike beskouing. *South African Journal for Marine Research*, **10**: 305–320.
228. Schumann, E.H., J.R.E. Lutjeharms, A.J. Boyd, M.L. Gründlingh and G.B. Brundrit (1991). Physical oceanography in South Africa: 1987 to 1990. *South African Journal of Science*, **87**(10): 486–492.
229. Lutjeharms, J.R.E. and F.P. Anderson (1987). The World Ocean Circulation Experiment (WOCE): a South African perspective. *South African Journal of Science*, **83**(1): 10–14.
230. Gordon, A.L., J.R.E. Lutjeharms and M.L. Gründlingh (1987). Stratification and circulation at the Agulhas Retroflection. *Deep-Sea Research*, **34**(4): 565–599.
231. Chapman, P. (1988). On the occurrence of oxygen-depleted water south of Africa and its implications for Agulhas–Atlantic mixing. *South African Journal of Marine Science*, **7**: 267–294.
232. Warren, B.A. (1981). The shallow oxygen minimum of the South Indian Ocean. *Deep-Sea Research*, **28**(8): 859–864.
233. Rochford, D. (1964). Salinity maximum in the upper 1000 waters of the north Indian Ocean. *Australian Journal of Marine and Freshwater Research*, **15**(1): 1–24.
234. Clowes, A.J. and G.E.R. Deacon (1935). The deep-water circulation of the Indian Ocean. *Nature*, **136** (3450): 936–938.
235. Gründlingh, M.L. (1985). Occurrence of Red Sea Water in the Southwestern Indian Ocean, 1981. *Journal of Physical Oceanography*, **15**(2): 207–212.
236. Valentine, H.R., J.R.E. Lutjeharms and G.B. Brundrit (1993). The water masses and volumetry of the southern Agulhas Current region. *Deep-Sea Research*, **40**(6): 1285–1305.
237. Lutjeharms, J.R.E. (1991). The temperature/salinity relationships of the South West Indian Ocean. *South African Geographer*, **18**(1/2): 15–31.
238. Mantyla, A.W. and J.L. Reid (1995). On the origins of deep and bottom waters of the Indian Ocean. *Journal of Geophysical Research*, **100**(C2): 2417–2439.
239. Warren, B.A. (1981). Transindian hydrographic section at Lat. 18°S: Property distributions and circulation in the South Indian Ocean. *Deep-Sea Research*, **28**(8): 759–788.
240. Kolla, V., L. Sullivan, S.S. Streeter and M.G. Langseth (1976). Spreading of Antarctic Bottom Water and its effects on the floor of the Indian Ocean inferred from bottom-water potential temperature, turbidity, and sea-floor photography. *Marine Geology*, **21**: 171–189.
241. Park, Y.-H. and L. Gambéroni (1997). Cross-frontal exchanges of Antarctic Intermediate Water and Antarctic Bottom Water in the Crozet Basin. *Deep-Sea Research I*, **44**(5): 963–986.
242. Sparrow, M.D., K.J. Heywood, J. Brown and D.P. Stevens (1996). Current structure of the south Indian Ocean. *Journal of Geophysical Research*, **101**(C3): 6377–6391.
243. Read, J.F. and R.T. Pollard (1999). Deep inflow into the Mozambique Basin. *Journal of Geophysical Research*, **104**(C2): 3075–3090.
244. Warren, B.A. (1974). Deep flow in the Madagascar and Mascarene basins. *Deep-Sea Research*, **21**(1): 1–21.
245. Fieux, M., F. Schott and J.C. Swallow (1986). Deep boundary currents in the western Indian Ocean revisited. *Deep-Sea Research*, **33**(4): 415–526.
246. Swallow, J.C. and R.T. Pollard (1988). Flow of bottom water through the Madagascar Basin. *Deep-Sea Research*, **35**(8): 1437–1440.
247. Boddem, J. and R. Schlitzer (1995). Interocean exchange and meridional mass and heat fluxes in the South Atlantic. *Journal of Geophysical Research*, **100**(C8): 15 821–15 834.
248. Metzl, N., B. Moore and A. Poisson (1990). Resolving the intermediate and deep convective flows in the Indian Ocean by using temperature, salinity, oxygen and phosphate data: the interplay of biogeochemical and geophysical tracers. *Palaeogeography, Palaeoclimatology, Palaeoecology*, **89**(1–2): 81–111.
249. Toole, J.M. and B.A. Warren (1993). A hydrographic section across the subtropical South Indian Ocean. *Deep-Sea Research*, **40**(10): 1973–2019.
250. Shannon, L.V. and D. Hunter (1988). Notes on Antarctic Intermediate Water around Southern Africa. *South African Journal of Marine Science*, **6**: 107–117.
251. Colborn, J.G. (1975). *The thermal structure of the Indian Ocean.* International Indian Ocean Expedition Oceanographic Monographs, Number 2, An East-west Center Book, The University Press of Hawaü, Honolulu, 173 pp.
252. Fine, R.A. (1993). Circulation of Antarctic Intermediate Water in the South Indian Ocean. *Deep-Sea Research I*, **40**(10): 2021–2042.
253. Boebel, O., C. Duncombe Rae, S. Garzoli, J. Lutjeharms, P. Richardson, T. Rossby, C. Schmid and W. Zenk (1997). Float experiment studies interocean exchanges at the tip of Africa. *Eos, Transactions of the American Geophysical Union*, **79**(1): 1, 7–8.
254. Lutjeharms, J.R.E., O. Boebel and T. Rossby (1997). KAPEX: an international experiment to study deep water movement around southern Africa. *South African Journal of Science*, **93**(9): 377–381.
255. Tomczak, M. (1984). Ausbreitung und Vermischung der Zentralwassermassen in den Tropengebieten der Ozeane. 2. Indischer und Pazifischer Ozean. *Oceanologica Acta*, **7**(3): 271–288.
256. You, Y. and M. Tomczak (1993). Thermocline circu-

lation and ventilation in the Indian Ocean derived from water mass analysis. *Deep-Sea Research I*, **40**(1): 13–56.

257. You, Y. (1998). Dianeutral mixing and transformation of Antarctic Intermediate Water in the Indian Ocean. *Journal of Geophysical Research*, **103**(C13): 30 941–30 971.
258. Swallow, J.C., F. Schott and M. Fieux (1991). Structure and transport of the East African Coastal Current *Journal of Geophysical Research*, **96**(C12): 22 245–22 257.
259. Lutjeharms, J.R.E. (1972). A quantitative assessment of year-to-year variability in water movement in the Southwest Indian Ocean. *Nature*, **239**(91): 59–69.
260. Villacastin-Herrero, C.A., L.G. Underhill, R.J.M. Crawford and L.V. Shannon (1996). Sea surface temperature of oceans surrounding subequatorial Africa: seasonal patterns, spatial coherence and long-term trends. *South African Journal of Science*, **92**(4): 189–197.
261. Weeks, S.J., F.A. Shillington and G.B. Brundrit (1998). Seasonal and spatial SST variability in the Agulhas retroflection and Agulhas return current. *Deep-Sea Research I*, **45**(10): 1611–1625.
262. Lutjeharms, J.R.E., M.I. Lucas, R. Perissinotto, R.C. van Ballegooyen and M. Rouault (1994). Oceanic processes at the Subtropical Convergence: report of research cruise SAAMES III. *South African Journal of Science*, **90**(7): 367–370.
263. Gründlingh, M.L. (1987). On the seasonal temperature variation in the southwestern Indian Ocean. *South African Geographical Journal*, **69**(2): 129–138.
264. Holland, W.R. (1978). The role of mesoscale eddies in the general circulation in the ocean – numerical experiments using a wind-driven quasigeostrophic model. *Journal of Physical Oceanography*, **8**(3): 363–392.
265. Holland, W.R., V. Zlotnicki and L.-L. Fu (1991). Modelled time-dependent flow in the Agulhas retroflection region as deduced from altimeter data assimilation. *South African Journal of Marine Science*, **10**: 407–427.
266. Luther, M.E. and J.J. O'Brien (1989). Modelling the variability in the Somali Current. In *Mesoscale/Synoptic Coherent Structures in Geophysical Turbulence*, editors J.C.J. Nihoul and B.M. Jamart, Elsevier, Amsterdam, pp. 373–386.
267. Woodberry, K.E., M.E. Luther and J.J. O'Brien (1989). The wind-driven seasonal circulation in the southern tropical Indian Ocean. *Journal of Geophysical Research*, **94**(C12): 17 985–18 002.
268. Matano, R.P. (1996). A numerical study of the Agulhas retroflection: the role of bottom topography. *Journal of Physical Oceanography*, **26**(10): 2267–2279.
269. Matano, R.P., C.G. Simionato, W.P.(M.) de Ruijter, P.J. van Leeuween [sic], P.T. Strub, D.B. Chelton and M.G. Schlax (1998). Seasonal variability in the Agulhas Retroflection region. *Geophysical Research Letters*, **25**(23): 4361–4364.
270. Matano, R.P., C.G. Simionato and P.T. Strub (1999). Modelling the wind driven variability of the South Indian Ocean. *Journal of Physical Oceanography*, 29(2); 217–230.
271. Biastoch, A. (1998). Zirkulation und Dynamik in der Agulhasregion anhand eines numerischen Modells. *Berichte aus dem Institut für Meereskunde an der Christian-Albrechts-Universität, Kiel*, **301**, 118 pp.
272. FRAM Group (1991). An eddy-resolving model of the Southern Ocean. *Eos, Transactions of the American Geophysical Union*, **72**(15): 169, 174–175.
273. Semtner, A.J. and R.M. Chervin (1988). A simulation of the global ocean circulation with resolved eddies. *Journal of Geophysical Research*, **93**(C12): 15 502–15 522. 420.
274. Semtner, A.J. and R.M. Chervin (1992). Ocean general circulation from a global eddy-resolving model. *Journal of Geophysical Research*, **97**(C4): 5493–5550.
275. Wilkin, J.L. and R.A. Morrow (1994). Eddy kinetic energy and momentum flux in the Southern Ocean: comparison of a global eddy-resolving model with altimeter, drifter, and current-meter data. *Journal of Geophysical Research*, **99**(C4): 7903–7916.
276. Lutjeharms, J.R.E., D.J. Webb and B.A. de Cuevas (1991). Applying the Fine Resolution Antarctic Model (FRAM) to the ocean circulation around southern Africa. *South African Journal of Science*, **87**(8): 346–349.
277. Lutjeharms, J.R.E. and D.J. Webb (1995). Modelling the Agulhas Current system with FRAM (*Fine Resolution Antarctic Model*). *Deep-Sea Research I*, **42**(4): 523–551.
278. Jacobs, S.S. and D.T. Georgi (1977). Observations on the Southwest Indian/Antarctic Ocean. *Deep-Sea Research*, **24** (supplement): 43–84.
279. Reason, C.J.C., R.J. Allan and J.A. Lindesay (1996). Evidence for the influence of remote forcing on interdecadal variability in the southern Indian Ocean. *Journal of Geophysical Research*, **101**(C5): 11 867–11 882.
280. Reason, C.J.C., R.J. Allan and J.A. Lindesay (1996). Dynamical response of the oceanic circulation and temperature to interdecadal variability in the surface winds over the Indian Ocean. *Journal of Climate*, **9**(1): 97–114.
281. Berghaus, H. (1845). *Physikalischer Atlas*. Erster Band. Verlag J. Perthes, Gotha, 234 pp.
282. Kerhallet, M.C.P. de (1851). Considérations générales sur l'océan Indien. *Annales Hydrographiques*, **2e Sem**, pp. 219–264.
283. Petermann, A. (1850). *Atlas of Physical Geography*. Wm. S. Orr and Co., London, 124 pp., 15 plates.
284. Petermann, A. (1865). Karte der Arktischen und Antarktischen Regimen zur Übersicht des geographischen Standpundtes im Jahre 1865. *Petermanns Mitteilungen*, Tafel 5.
285. Pearce, A.F. (1980). Early observations and historical notes on the Agulhas Current circulation. *Transactions of the Royal Society of South Africa*, **44**(2): 205–212.

286. Barlow, E.W. (1931). Currents in the western portion of the Indian Ocean. I – Introduction. *Marine Observer*, **8**(90): 130–132.

287. Barlow, E.W. (1931). Currents in the western portion of the Indian Ocean. II – Currents during the S.W. Monsoon period. *Marine Observer*, **8**(93): 193–194.

288. Barlow, E.W. (1931). Currents in the western portion of the Indian Ocean. III – Currents during the N.E. Monsoon period (northern winter and southern summer) and general summary. *Marine Observer*, **8**(96): 254–259.

289. Barlow, E.W. (1933). Currents in the southern Indian Ocean, winter season. *Marine Observer*, **10**(111): 99.

290. Barlow, E.W. (1933). Currents in the southern Indian Ocean, summer season and general summary. *Marine Observer*, **10**(112): 132–135.

291. Barlow, E.W. (1935). The 1910 to 1935 survey of the currents of the Indian Ocean and China Seas. *Marine Observer*, **12**(120): 153–163.

292. Pearce, A.F. and M.L. Gründlingh (1982). Is there a seasonal variation in the Agulhas Current? *Journal of Marine Research*, **40**(1): 177–184.

293. Lutjeharms, J.R.E., P.M. Wedepohl and J.M. Meeuwis (2000). On the surface drift of the East Madagascar and the Mozambique Currents. *South African Journal of Science*, **96**(3): 141–147.

294. Bang, N.D. (1973). Oceanography: Oceanographic Environment of Southern Africa. In *Standard Encyclopaedia of Southern Africa,* editors J. J. Spies and P.C. du Plessis, Vol.8, Nasou, Cape Town, pp. 282–286.

295. Swallow, J.C., M. Fieux and F. Schott (1988). The boundary currents east and north of Madagascar. Part I. Geostrophic currents and transports. *Journal of Geophysical Research*, **93**(C5): 4951–4962.

296. Duncan, C.P. and S.G. Schladow (1981). World surface currents from ship's drift observations. *International Hydrographic Review*, **58**(2): 101–112.

297. Schott, F., M. Fieux, J. Kindle, J. Swallow and R. Zantopp (1988). The boundary currents east and north of Madagascar. Part II. Direct measurements and model comparisons. *Journal of Geophysical Research*, **93** (C5): 4963–4974.

298. Piton, B. and J.F.Poulain (1974). Resultat des mesures de courant superficiels au GEK effectuees avec le N.O. "Vauban" dans le sud-ouest de l'ocean Indien, 1973–74. *ORSTOM Centre de Nosy, Bé, Document Scientifique*, **47**, 77 pp.

299. Biastoch, A., C.J.C. Reason, J.R.E. Lutjeharms and O. Boebel (1999). The importance of flow in the Mozambique Channel to the seasonality in greater Agulhas Current system. *Geophysical Research Letters*, **26**(21): 3321–3324.

300. Fu, L.-L. (1986). Mass, heat and freshwater fluxes in the South Indian Ocean. *Journal of Physical Oceanography*, **16**(10): 1683–1693.

301. Biastoch, A. and W. Krauss (1999). The role of mesoscale eddies in the source regions of the Agulhas Current. *Journal of Physical Oceanography*, **29**(9): 2303–2317.

302. Read, J.F. and R.T. Pollard (1993). Structure and transport of the Antarctic Circumpolar Current and Agulhas Return Current at 40°E. *Journal of Geophysical Research*, **98** (C7): 12 281–12 295.

303. Lutjeharms, J.R.E., C.-T. Liu, W.-S. Chuan and C.-Z. Shyu (1993). On some similarities between the oceanic circulations off Southern Africa and off Taiwan. *South African Journal of Science*, **89**(8): 367–371.

304. Lutjeharms, J.R.E., and E. Machu (2000). An upwelling cell inshore of the East Madagascar Current. *Deep-Sea Research I*, **47**(12): 2405–2411.

305. Gordon, A.L., K.-I. Horai and M. Donn (1983). Southern hemisphere western boundary current variability revealed by GEOS 3 altimeter. *Journal of Geophysical Research*, **88**(1): 755–762.

306. Haines, M., M. Luther, Z. Ji and R. Fine (1995). Particle trajectories in an Indian Ocean Model. In *1995 U. S. WOCE Report*, U. S. WOCE Implementation Report Number 10, pp. 34–35.

307. Gründlingh, M.L. (1989). Drift of the SAA *Helderberg* wreckage in the Indian Ocean. *South African Journal of Science*, **85**(4): 209–211.

308. Lutjeharms, J.R.E. (1988). On the role of the East Madagascar Current as a source of the Agulhas Current. *South African Journal of Science*, **84**(4): 236–238.

309. Gründlingh, M.L., R.A. Carter and R.C. Stanton (1991). Circulation and water properties of the Southwest Indian Ocean, Spring 1987. *Progress in Oceanography*, **28**(4): 305–342.

310. Patterson, S.L. (1985). Surface circulation and kinetic energy distribution in the southern hemisphere oceans from FGGE drifting buoys. *Journal of Physical Oceanography*, **15**(7): 865–884.

311. Daniault, N. and Y. Menard (1985). Eddy kinetic energy distribution in the Southern Ocean from altimetry and FGGE drifting buoys. *Journal of Geophyscial Research*, **90**(C6): 18 877–11 889.

312. Kingwill, D.G. (1990). *The CSIR, the first 40 years.* CSIR, Pretoria, 352 pp.

313. Carte, A.E. (1989). An informal historical review of the National Physical Research Laboratory. *South African Journal of Science*, **85**(2): 76–79.

314. Natal Regional Research Committee (1961). *Marine studies off the Natal coast.* Council for Scientific and Industrial Research, CSIR Symposium **S2**, 134 pp.

315. Anderson, F.P., M.L. Gründlingh and C.C. Stavropoulos (1988). Kinematics of the southern Natal coastal circulation: some historic measurements 1962–63. *South African Journal of Science*, **84**(10): 857–860.

316. Anderson, F.P. (1972). Circulation patterns along the east coast (coastal zone and open ocean) in relation to pollution. *South African Journal of Science*, **68**(5): 121–123.

317. Gründlingh, M.L., C.C. Stavropoulos and L.J. Watt (1988). The R.V. Meiring Naudé: Two decades of sup-

port to physical oceanography. *South African Journal of Science*, **84**(9): 746–748.
318. CSIR (1985). *A decade in oceanology. Activities of the National Research Institute for Oceanology, 1974–1984*. Council for Scientific and Industrial Research, Stellenbosch, CSIR Report T/SEA 8514, 28 pp.
319. CSIR (1988). *CSIR. Review of activities in oceanology, 1985–1987*. Council for Scientific and Industrial Research, Stellenbosch, 40 pp.
320. Lutjeharms, J.R.E. (1992). Hoe gaan dit met die fisiese oseanologie in Suid-Afrika? *South African Journal of Science*, **88**(2): 73.
321. Schumann, E.H. (1988). Physical oceanography off Natal. In *Coastal Ocean Studies off Natal, South Africa*, editor E. H. Schumann, Lecture Notes on Coastal and Estuarine Studies 26, Springer-Verlag, Berlin, pp. 101–130.
322. Lutjeharms, J.R.E. and J.A. Thomson (1993). Commercializing the CSIR and the death of science. *South African Journal of Science*, **89**(1): 8–14.
323. Gründlingh, M.L., G.W. Bailey, F.A. Shillington, E.H. Schumann and J.J. Agenbag (1995). Physical oceanographic activities in South Africa. 1991–1994. *South African Journal of Science*, **91**(5): 247–254.
324. Warren, B.A. (1978). Bottom water transport through the Southwest Indian Ridge. *Deep-Sea Research*, **25**(3): 315–321.
325. Maltrud, M.E., R.D. Smith, A.J. Semtner and R.C. Malone (1998). Global eddy-resolving ocean simulations driven by 1985–1995 atmospheric winds. *Journal of Geophysical Research*, **103**(C13): 30 825–30 853.
326. Piton, B. (1989). Quelques aspects nouveaux sur la circulation superficielle dans le canal de Mozambique (Ocean Indien). *Document Scientifique, Centre Brest ORSTOM*, **54**: 31 pp.
327. Piton, B., J.-H. Pointeau and J.-S. Nigoumbi (1981). Atlas hydrologique du Canal de Mozambique (Océan Indien). *Travaux et Documents de l'ORSTOM*, **132**: 41 pp.
328. Menache, M. (1963). Premiere Campagne oceanographique du "Commandant Robert Giraud" dans le Canal de Mozambique 11 Octobre–28 Novembre 1957. *Cahiers Oceanographiques*, **15**(4): 224–235.
329. Donguy, J.R. and B. Piton (1969). Aperçu des conditions hydrologiques de la partie nord du Canal de Mozambique. *Cahiers O.R.S.T.O.M., série Océanographie*, **7**(2): 3–26.
330. Donguy, J.R. (1975). Les eaux superficielles tropicales de la partie accidentale de l'Ocean Indien en 1966–1967. *Cahiers O.R.S.T.O.M.*, **13**(1): 31–47.
331. Zahn, W. (1984). Eine Abschätzung des Volumentransportes im Kanal von Moçambique während des Zeitraumes Oktober–November 1957. *Beiträge zur Meereskunde*, **51**: 67–74.
332. Sætre, R. (1985). Surface currents in the Mozambique Channel. *Deep-Sea Research*, **32**(12): 1457–1467.
333. Zahn, W. (1984). Influence of bottom topography on currents in the Mozambique Channel. *Tropical Ocean–Atmosphere Newsletter*, **26**: 22–23.
334. DiMarco, S.F., W.D. Nowlin and P. Chapman (1998). Properties and transport of the Mozambique Channel. In *1998 U. S. WOCE Report*, U. S. WOCE Implementation Report Number 10, pp. 34–35.
335. Naeije, M.C., K.F. Wakker, R. Scharoo and B.C.C. Ambrosius (1992). Observation of mesoscale ocean currents from GEOSAT altimeter data. *ISPRS Journal of Photogrammetry and Remote Sensing*, **47**: 347–368.
336. Nehring, D. (editor) (1984). The oceanological conditions in the western part of the Mozambique Channel in February–March 1980. *Geodätische und geophysikalische Veröffentlichungen*, **4**(9): 163 pp.
337. Schemainda, R. and E. Hagen (1983). On steady state intermediate vertical currents induced by the Mozambique Current. *Océanographie Tropicale*, **18**(1): 81–88.
338. Lutjeharms, J.R.E. and A. Jorge da Silva (1988). The Delagoa Bight eddy. *Deep-Sea Research*, **35**(4): 619–634.
339. Gründlingh, M.L. (1992). Agulhas Current meanders: review and a case study. *South African Geographical Journal*, **74**(1): 19–28.
340. Gründlingh, M.L. (1993). On the winter flow in the southern Mozambique Channel. *Deep-Sea Research I*, **40**(2): 409–418.
341. Martin, A.K. (1981). Evolution of the Agulhas Current and its palaeo-ecological implications. *South African Journal of Science*, **77**(12): 547–554.
342. Martin, A.K. (1981). The influence of the Agulhas Current on the physiographic development of the northernmost Natal Valley (S.W. Indian Ocean). *Marine Geology*, **39**: 259–276.
343. Gründlingh, M.L. (1991). Draaikolke suid van die Mosambiekkanaal. *Suid-Afrikaanse Tydskrif vir Natuurwetenskap en Tegnologie*, **10**(3): 137–141.
344. Gründlingh, M.L. (1984). An eddy over the northern Mozambique Ridge. *South African Journal of Science*, **80**(7): 324–329.
345. Magnier, Y. and B. Piton (1973). Les masses d'eau de l'Ocean Indien a l'ouest et au nord de Madagascar au début de l'été austral (Novembre–Décembre). *Cahiers ORSTOM.*, **11**: 97–113.
346. Piola, A.R., H.A. Figueroa and A.A. Bianchi (1987). Some aspects of the surface circulation south of 20°S revealed by First GARP Global Experiment drifters. *Journal of Geophysical Research*, **92**(C5): 5101–5114.
347. Feron, R.C.V., W.P.M. de Ruijter and P.J. van Leeuwen (1998). A new method to determine the mean sea surface dynamic topography from satellite altimeter observations. *Journal of Geophysical Research*, **103** (C1): 1343–1362.
348. Levitus, S. (1982). Climatological atlas of the world ocean. *NOAA Professional Papers*, **13**: 1–173.
349. Gründlingh, M.L. and J.R.E. Lutjeharms (1979). Large scale flow patterns of the Agulhas Current system.

South African Journal of Science, **75**(6): 269–270.
350. Gründlingh, M.L. (1978). Drift of a satellite-tracked buoy in the southern Agulhas Current and Agulhas Return Current. *Deep-Sea Research*, **25**(12): 1209–1224.
351. Deacon, G.E.R. (1933). A general account of the hydrology of the South Atlantic Ocean. *Discovery Report*, **7**: 171–238.
352. Deacon, G.E.R. (1937). The hydrology of the Southern Ocean. *Discovery Report*, **15**: 124 pp.
353. Deacon, G.E.R. (1982). Physical and biological zonation in the Southern Ocean. *Deep-Sea Research*, **29** (1):1–15.
354. Harris, T.F.W. and N.D. Bang (1974). Topographic Rossby waves in the Agulhas Current. *South African Journal of Science*, **70**(7): 212–214.
355. Gründlingh, M.L. (1988). Review of cyclonic eddies in the Mozambique Ridge Current. *South African Journal of Marine Science*, **6**: 193–206.
356. Gründlingh, M.L. (1985). An intense cyclonic eddy east of the Mozambique Ridge. *Journal of Geophysical Research*, **90**(C4): 7163–7167.
357. Gründlingh, M.L. (1987). Cyclogenesis in the Mozambique Ridge Current. *Deep-Sea Research*, **34**(1): 89–103.
358. Gründlingh, M.L. (1987). Anatomy of a cyclonic eddy of the Mozambique Ridge Current. *Deep-Sea Research*, **34**(2): 237–251.
359. Gründlingh, M.L. (1989). Two contra-rotating eddies of the Mozambique Ridge Current. *Deep-Sea Research*, **36**(1): 149–153.
360. Gründlingh, M.L. (1985). Features of the circulation in the Mozambique Basin in 1981. *Journal of Marine Research*, **43**(4): 779–792.
361. Lutjeharms, J.R.E. and H.R. Valentine (1988). On mesoscale ocean eddies at the Agulhas Plateau. *South African Journal of Science*, **84**(3): 194–200.
362. Gründlingh, M.L. (1995) Tracking eddies in the southeast Atlantic and southwest Indian oceans with TOPEX/POSEIDON. *Journal of Geophysical Research*, **100**(C12): 24 977–24 986.
363. Wyrtki, K., L. Magaard and J. Hagar (1976). Eddy energy in the oceans. *Journal of Geophysical Research*, **81**(15): 2641–2646.
364. Schumann, E.H. (1982). Inshore circulation of the Agulhas Current off Natal. *Journal of Marine Research*, **40**(1): 43–55.
365. Pearce, A.F. (1977). Some features of the upper 500m of the Agulhas Current. *Journal of Marine Research*, **35**(4): 731–753.
366. Beal, L. and H. Bryden (1996). The water masses, velocity structure and volume transport of the Agulhas Current at 32°S. *International WOCE Newsletter*, **22**: 20, 23, 25–27.
367. Beal, L.M. and H.L. Bryden (1999). The velocity and vorticity structure of the Agulhas Current at 32° S. *Journal of Geophysical Research*, **104**(C3): 5151–5176.
368. Beal, L.M. and H L. Bryden (1997). Observations of an Agulhas Undercurrent. *Deep-Sea Research I*, **44**(9–10): 1715–1724.
369. Swallow, J.C. and L.V. Worthington (1961). An observation of a deep countercurrent in the Western North Atlantic. *Deep-Sea Research*, **8**(1): 1–19.
370. Volkmann, G. (1962). Deep current observations in the western North Atlantic. *Deep-Sea Research*, **9**(5): 493–500.
371. Worthington, L.V. and H. Kawai (1972). Comparison between deep sections across the Kuroshio and the Florida Current and Gulf Stream. In *Kuroshio: Its Physical Aspects*, editors H. Stommel and K. Yoshida, University of Tokyo Press, pp. 371–385.
372. Flemming, B.W. (1978). Underwater sand dunes along the southeast African continental margin – observations and implications. *Marine Geology*, **26**(1/2): 177–198.
373. Flemming, B.W. (1980). Sand transport and bedform patterns on the continental shelf between Durban and Port Elizabeth (Southeast African Continental Margin). *Sedimentary Geology*, **26**(1/3): 179–205.
374. Gründlingh, M.L. (1977). Drift observations from Nimbus VI satellite-tracked buoys in the southwestern Indian Ocean. *Deep-Sea Research*, **24**(10): 903–913.
375. Gründlingh, M.L. (1974). A description of inshore current reversals off Richards Bay based on airborne radiation thermometry. *Deep-Sea Research*, **21**(1): 47–55.
376. Gründlingh, M.L. and A.F. Pearce (1990). Frontal features of the Agulhas Current in the Natal Bight. *South African Geographical Journal*, **72**(1): 11–14.
377. Malan, O.G. and E.H. Schumann (1979). Natal shelf circulation revealed by Landsat imagery. *South African Journal of Science*, **75**(3): 136–137.
378. Lee, T.N. and D.A. Mayer (1977). Low-frequency variability and spin-off eddies along the shelf off Southeast Florida. *Journal of Marine Research*, **35**(1): 193–220.
379. Hiray, M. (1985). Satellite observation of characteristic features of frontal eddies in the vicinity of the Kuroshio front. *Bulletin of the Tohoku Regional Fisheries Research Laboratory*, **47**: 69–78.
380. Gründlingh, M.L. (1980). On the volume transport of the Agulhas Current. *Deep-Sea Research*, **27**(7): 557–563.
381. Pearce, A.F. (1978). Seasonal variations of temperature and salinity on the northern Natal continental shelf. *South African Geographical Journal*, **60**(2): 135–143.
382. Olson, D.B., R.A. Fine and A.L. Gordon (1992). Convective modifications of water masses in the Agulhas. *Deep-Sea Research*, **39** (Supplement 1): s163–s181.
383. Schumann, E.H. (1981). Low frequency fluctuations off the Natal coast. *Journal of Geophysical Research*, **86** (C7): 6499–6508.
384. Schumann, E.H. and L.-A. Perrins (1982). Tidal and inertial currents around South Africa. In *Proceedings of the Eighteenth International Coastal Engineering Conference*, American Society of Civil Engineers, Cape Town, South Africa, Nov. 14–19, 1982, pp. 2562–2580.

385. Anderson, F.P. (1967). Time-variations in the Agulhas Current near Durban. National Physical Research Laboratory, Council for Scientific and Industrial Research, *internal publication* **IG 67/8**, 13 pp.
386. Hunter, I.(T.) (1981). On the land breeze circulation off the Natal coast. *South African Journal of Science*, **77**(8): 376–378.
387. Preston-Whyte, R.A. (1974). Land breezes and mountain–plain winds over the Natal coast. *South African Geographical Journal*, **56**(1): 27–35.
388. Flemming, B. and R. Hay (1988). Sediment distribution and dynamics of the Natal continental shelf. In *Coastal Ocean Studies off Natal, South Africa*, editor E.H. Schumann, Lecture Notes on Coastal and Estuarine Studies 26, Springer-Verlag, Berlin, pp. 47–80.
389. Gründlingh, M.L. (1983). Eddies in the southern Indian Ocean and Agulhas Current. In *Eddies in Marine Science*, editor A. R. Robinson, Springer-Verlag, Berlin, pp. 245–264.
390. Pearce, A.F. (1977). The shelf circulation off the east coast of South Africa. *Council for Scientific and Industrial Research, CSIR Research Report*, **361**, 220 pp.
391. Carter, R. and J. d'Aubrey (1988). Inorganic nutrients in Natal continental shelf waters. In *Coastal Ocean Studies off Natal, South Africa*, editor E. H. Schumann, Lecture Notes on Coastal and Estuarine Studies 26, Springer-verlag, Berlin, pp. 131–151.
392. Carter, R.A. and M.H. Schleyer (1988). Plankton distributions in Natal coastal waters. In *Coastal Ocean Studies off Natal, South Africa*, editor E. H. Schumann, Lecture Notes on Coastal and Estuarine Studies 26, Springer-Verlag, Berlin, pp. 152–177.
393. Hseuh, Y. and J.J. O'Brien (1971). Steady coastal upwelling induced by an along-shore current. *Journal of Physical Oceanography*, **1**(3): 180–186.
394. Schumann, E.H. (1986). The bottom boundary layer inshore of the Agulhas Current off Natal in August 1975. *South African Journal of Marine Science*, **4**: 93–102.
395. Van der Elst, R.P. (1988). Shelf ichthyofauna off Natal. In *Coastal Ocean Studies off Natal, South Africa*, editor E. H. Schumann, Lecture Notes of Coastal and Estuarine Studies, 26, Springer-Verlag, Berlin, pp. 209–225.
396. Eagle, G.A. and M.J. Orren (1985). A seasonal investigation of the nutrients and dissolved oxygen in the water column along two lines of stations south and west of South Africa. *National Research Institute for Oceanology, CSIR. Research Report*, **567**: 52 pp.
397. Schumann, E.H. (1987). The coastal ocean off the east coast of South Africa. *Transactions of the Royal Society of South Africa*, **46**(3): 215–229.
398. Gründlingh, M.L. and A.F. Pearce (1984). Large vortices in the northern Agulhas Current. *Deep-Sea Research*, **31**(9): 1149–1156.
399. Gründlingh, M.L. (1979). Observation of a large meander in the Agulhas Current. *Journal of Geophysical Research*, **84**(C7): 3776–3778.
400. Gründlingh, M.L. (1986). Features of the northern Agulhas Current in spring, 1983. *South African Journal of Science*, **82**(1): 18–20.
401. Van Leeuwen, P.J., W.P.M. de Ruijter and J.R.E. Lutjeharms (2000). Natal Pulses and the formation of Agulhas rings. *Journal of Geophysical Research*, **105**(C3): 6425–6436.
402. Römer, E. (1939). The counter current off the south and southeast African coast. *Hydrographic Review*, **61**: 88–91.
403. Mallory, J.K. (1961). Bathymetric and hydrographic aspects of marine studies off the Natal Coast. In *Marine Studies off the Natal Coast*, Natal Regional Research Committee, *CSIR Symposium*, **S2**, pp. 31–39.
404. Lutjeharms, J.R.E. and A.D. Connell (1989). The Natal Pulse and inshore counter currents off the South African east coast. *South African Journal of Science*, **85**(8): 533–535.
405. Nof, D. (1986). The collision between the Gulf Stream and warm-core rings. *Deep-Sea Research*, **33**(3): 359–378.
406. Watts, D.R. and D.B. Olson (1978). Gulf Stream ring coalescence with the Gulf Stream off Cape Hatteras. *Science*, **202**(4371): 971–972.
407. Smith, D.C. and G.P. Davis (1989). A numerical study of eddy interaction with an ocean jet. *Journal of Physical Oceanography*, **19**(7): 975–986.
408. Rossouw, J. (1984). Review of existing wave data, wave climate and design waves for South African and South West African (Namibian) coastal waters. *National Research Institute for Oceanology, CSIR Report*, **T/SEA 8401**, 66 pp. + tables + figures.
409. Gerber, M. (1993). The interaction of deep-water gravity waves and an annular current: Linear theory. *Journal of Fluid Mechanics*, **248**: 153–172.
410. Gründlingh, M.(L.) and M. Rossouw (1995). Wave attenuation in the Agulhas Current. *South African Journal of Science*, **91**(7): 357–359.
411. Stavropoulos, C.C. and C.P. Duncan (1974). A satellite-tracked buoy in the Agulhas Current. *Journal of Geophysical Research*, **79**(C18): 2744–2746.
412. Lutjeharms, J.R.E., S.J. Weeks, R.D. [sic] van Ballegooyen and F.A. Shillington (1992). Shedding of an eddy from the seaward front of the Agulhas Current. *South African Journal of Science*, **88**(8): 430–433.
413. Lutjeharms, J.R.E. (1996). The exchange of water between the South Indian and the South Atlantic. In *The South Atlantic: Present and Past Circulation*, editors G. Wefer, W.H. Berger, G. Siedler and D. Webb, Springer-Verlag, Berlin, pp. 125–162.
414. Lutjeharms, J.R.E. (1981). Spatial scales and intensities of circulation in the ocean areas adjacent to South Africa. *Deep-Sea Research*, **28**(11): 1289–1302.
415. Blanton, J.O., L.P. Atkinson, L.J. Pietrafesa and T.N. Lee (1981). The intrusion of Gulf Stream Water across the continental shelf due to topographically-induced

upwelling. *Deep-Sea Research*, **28**(4): 393–405.
416. Blanton, J.O. (1971). Exchange of Gulf Stream Water with North Carolina Shelf Water in Onslow Bay during stratified conditions. *Deep-Sea Research*, **18**(2): 167–178.
417. Lee, T.N., L.P. Atkinson and R. Legeckis (1981). Observations of a Gulf Stream frontal eddy on the Georgia continental shelf, April 1977. *Deep-Sea Research*, **29**(4): 347–378.
418. Quartly, G.D. and M.A. Srokosz (1993). Seasonal variations in the region of the Agulhas retroflection: studies with Geosat and FRAM. *Journal of Physical Oceanography*, **23**(9): 2107–2124.
419. Bang, N.D. (1972). Logic intensive versus data intensive methods in oceanographic planning, analysis and prediction. In *Proceedings of 'Conference 72'*, South African Geophysical Society, Johannesburg.
420. Wells, N.C., V. Ivchenko and S.E. Best (2000). Instabilities in the Agulhas retroflection current system – A comparative model study. *Journal of Geophysical Research*, 105(C2): 3233–3241.
421. Goschen, W.S. and E.H. Schumann (1990). Agulhas Current variability and inshore structures off the Cape Province, South Africa. *Journal of Geophysical Research*, **95**(1): 667–678.
422. Goschen, W.S. and E.H. Schumann (1994). An Agulhas Current intrusion into Algoa Bay during August 1988. *South African Journal of Marine Science*, **14**: 47–57.
423. Martin, A.K. and B.W. Fleming (1988). Physiography, structure and geological evolution of the Natal continental shelf. In *Coastal Ocean Studies off Natal, South Africa*, editor E.H. Schumann, Lecture Notes on Coastal and Estuarine Studies 26, Springer-Verlag, Berlin, pp. 11–46.
424. Hutson, W.H. (1980). The Agulhas Current during the late Pleistocene: analysis of modern faunal analogs. *Science*, **207**(4426): 64–66.
425. Winter, A. and K. Martin (1990). Late quaternary history of the Agulhas Current. *Paleoceanography*, **5**(4): 479–486.
426. Vincent, E. and N.J. Shackleton (1980). Agulhas Current temperature distribution delineated by oxygen isotope analysis of foraminifer in surface sediments. *Cushman Foundation Special Publication*, **19**: 84–95.
427. Bé, A.W.H. and J.C. Duplessy (1976). Subtropical convergence fluctuations and quaternary climates in the middle latitudes of the Indian ocean. *Science*, **194** (4263): 419–422.
428. Prell, W.L., W.H. Hutson and D.F. Williams (1979). The subtropical convergence and Late Quaternary circulation in the southern Indian Ocean. *Marine Macropaleontology*, **4**: 225–234.
429. Schumann, E.H., G.J.B. Ross and W.S. Goschen (1988). Cold water events in Algoa Bay and along the Cape south coast, South Africa, in March/April 1987. *South African Journal of Science*, **84**(7): 579–584.
430. Swart, V.P. and J.L. Largier (1987). Thermal structure of Agulhas Bank water. *South African Journal of Marine Science*, **5**: 243–253.
431. Goschen, W.S. and E.H. Schumann (1988). Ocean current and temperature structures in Algoa Bay and beyond in November 1986. *South African Journal of Marine Science*, **7**: 101–116.
432. Olson, D.B., G.B. Brown and S.R. Emmerson (1983). Gulf Stream frontal statistics from Florida Straits to Cape Hatteras derived from satellite and historical data. *Journal of Geophysical Research*, **88**(C8): 4569–4577.
433. Watts, D.R. and W.E. Johns (1982). Gulf Stream meanders: observations on propagation and growth. *Journal of Geophysical Research*, **87**(C12): 9467–9475.
434. Vukovich, F.M. and M.W. Crissman (1980). Some aspects of Gulf Stream western boundary eddies from satellite and *in situ* data. *Journal of Physical Oceanography*, **10**(11): 1792–1813.
435. Chew, F. (1981). Shingles, spin-off eddies and an hypothesis. *Deep-Sea Research*, **28**(4): 379–391.
436. Churchill, J.H., P.C. Cornillon and G.W. Milkowski (1986). A cyclonic eddy and shelf-slope water exchange associated with Gulf Stream warm-core rings. *Journal of Geophysical Research*, **91**(C8): 9615–9623.
437. Bang, N.D. and W.R.H. Andrews (1974). Direct current measurements of a shelf-edge frontal jet in the southern Benguela system. *Journal of Marine Research*, **32**(3): 405–417.
438. Lutjeharms, J.R.E. (1981). Satellite studies of the South Atlantic upwelling system. In *Oceanography from Space*, editor J.F.R. Gower, Plenum Press, New York, pp. 195–199.
439. Shannon, L.V., J.J. Agenbag, N.D. Walker and J.R.E. Lutjeharms (1990). A major perturbation in the Agulhas retroflection area in 1986. *Deep-Sea Research*, **37**(3): 493–512.
440. Lutjeharms, J.R.E. and H.R. Valentine (1988). Evidence for persistent Agulhas rings southwest of Cape Town. *South African Journal of Science*, **84**(9): 781–783.
441. Lutjeharms, J.R.E. and N.M. Walters (1985). Ocean colour and thermal fronts south of Africa. In *South African Ocean Colour and Upwelling Experiment*, editor L.V. Shannon, Sea Fisheries Research Institute, Cape Town, pp. 227–237.
442. McClean-Padman, J. and L. Padman (1991). Summer upwelling on the Sydney inner continental shelf. The relative roles of local wind forcing and mesoscale eddy encroachment. *Continental Shelf Research*, **11**(4): 321–345.
443. Tranter, D.J., D.J. Carpenter and G.S. Leech (1986). The coastal enrichment of the East Australian Current eddy field. *Deep-Sea Research*, **33**(11/12): 1703–1728.
444. Bang, N.D. (1970). Major eddies and frontal structures in the Agulhas Current retroflexion [sic] area in March, 1969. In *Symposium "Oceanography South Africa 1970"*, South African National Committee for Oceano-

graphic Research, 16 pp.
445. Walker, N.D. (1986). Satellite observations of the Agulhas Current and episodic upwelling south of Africa. *Deep-Sea Research*, **33**(8): 1083–1106.
446. Beckley, L.E. and R.C. van Ballegooyen (1992). Oceanographic conditions during three ichthyoplankton surveys of the Agulhas Current in 1990/91. *South African Journal of Marine Science*, **12**: 3–17.
447. Rouault, M., A.M. Lee-Thorp and J.R.E. Lutjeharms (1999). An atmospheric moisture and thermal front in the boundary layer above the Agulhas Current. *Journal of Geophysical Research*, submitted.
448. Rouault, M., A.M. Lee-Thorp and J.R.E. Lutjeharms (2000). The atmospheric boundary layer above the Agulhas Current during alongcurrent winds. *Journal of Physical Oceanography*, **30**(1): 40–50.
449. Rouault, M., A. Lee-Thorp, I. Ansorge and J. Lutjeharms (1995). Agulhas Current Air–Sea Interaction Experiment. *South African Journal of Science*, **91**(10): 493–496.
450. Beckley, L.E. (1988). Spatial and temporal variability in sea temperature in Algoa Bay, South Africa. *South African Journal of Science*, **84**(1): 67–69.
451. Beckley, L.E. (1983). Sea-surface temperature variability around Cape Recife, South Africa. *South African Journal of Science*, **79**(11): 436–438.
452. Harris, T.F.W. and C.C. Stavropoulos (1967). Some experience with a radiation thermometer over the Agulhas Current. *South African Journal of Science*, **63**(4): 132–136.
453. Schumann, E.H., L.-A. Perrins and I.T. Hunter (1982). Upwelling along the south coast of the Cape Province, South Africa. *South African Journal of Science*, **78**(6): 238–242.
454. Goschen, W.S. and E.H. Schumann (1995). Upwelling and the occurrence of cold water around Cape Recive, Algoa Bay, South Africa. *South African Journal of Marine Science*, **16**: 57–67.
455. Lutjeharms, J.R.E. (1987). Die Subtropiese Konvergensie en Agulhasretrofleksievaart (SCARC). *South African Journal of Science*, **83**(8): 454–456.
456. Lutjeharms, J.R.E. and I.J. Marais (1986). Real-time use of satellite remote sensing and telemetry on oceanologic research cruises. *South African Journal of Photogrammetry, Remote Sensing and Cartography*, **14**(6): 313–319.
457. Lutjeharms, J.R.E. (1978). Shipping disasters: a role for satellite oceanology. *South African Journal of Science*, **74**(10): 385–387.
458. Lutjeharms, J.R.E. (1987). Meridional heat transport across the Sub-Tropical Convergence by a warm eddy. *Nature*, **331**(6153): 251–253.
459. Shannon, L.V. (1979). The Cape Nimbus-7 CZCS programme: An overview and preliminary results. *South African Journal of Science*, **75**(12): 564.
460. Shannon, L.V., S.A. Mostert, N.M. Walters and F.P. Anderson (1983). Chlorophyll concentrations in the southern Benguela Current region as determined by satellite (Nimbus-7 Coastal Zone Colour Scanner). *Journal of Plankton Research*, **5**(4): 565–583.
461. Shannon, L.V., P. Schlittenhardt and S.A. Mostert (1984). The NIMBUS 7 CZCS experiment in the Benguela Current region off southern Africa, February 1980. 2. Interpretation of imagery and oceanographic implications. *Journal of Geophysical Research*, **89** (D4): 4968–4976.
462. Weeks, S.J. and F.A. Shillington (1996). Phytoplankton pigment distribution and frontal structure in the subtropical convergence region south of Africa. *Deep-Sea ResearchI*, **43**(5): 739–768.
463. Shannon, L.V. (editor) (1985). *South African Ocean Colour and Upwelling Experiment*. Sea Fisheries Research Institute, Cape Town, 270 pp.
464. Gordon, A.L. and W.F. Haxby (1990). Agulhas eddies invade the South Atlantic – evidence from GEOSAT altimeter and shipboard conductivity–temperature–depth survey. *Journal of Geophysical Research*, **95** (C3): 3117–3125.
465. Schouten, M.W., W.P.M. de Ruijter, P.J. van Leeuwen and J.R.E. Lutjeharms (2000). Translation, decay and splitting of Agulhas rings in the south-eastern Atlantic Ocean. *Journal of Geophysical Research*, **105**(C9): 21,913–21,925.
466. Dingle, R.V., G.V. Birch, J.M. Bremner, R.H. de Decker, A. du Plessis, J.A. Engelbrecht, M.J. Fincham, T. Fitton, B.W. Flemming, R.I. Gentle, S.H. Goodlad, A.K. Martin, E.G. Mills, G.J. Moir, R.J. Parker, S.H. Robson, J. Rogers, D.A. Salmon, W.G. Sieser, E.S. W. Simpson, C.P. Summerhayes, F. Westall, A. Winter and M.W. Woodborne (1987). Deep-sea sedimentary environments around southern Africa (South-East Atlantic and South-West Indian Oceans). *Annals of the South African Museum*, **98**(1): 1–27.
467. Boyd, A.J. and F.A. Shillington (1994). Physical forcing and circulation patterns on the Agulhas Bank. *South African Journal of Science*, **90**(3): 114–122.
468. Probyn, T.A., B.A. Mitchell-Innes, P.C. Brown, L. Hutchings and R.A. Carter (1994). A review of primary productivity and related processes on the Agulhas Bank. *South African Journal of Science*, **90**(3): 166–173.
469. Ou, H.W. and W.P.M. de Ruijter (1986). Separation of an inertial boundary current from a curved coastline. *Journal of Physical Oceanography*, **16**(2): 280–289.
470. Lutjeharms, J.R.E. and J.M. Meeuwis (1987). The extent and variability of the South East Atlantic upwelling. *South African Journal of Marine Science*, **5**: 51–62.
471. Lutjeharms, J.R.E. and P.L. Stockton (1991). Aspects of the upwelling regime between Cape Point and Cape Agulhas. *South African Journal of Marine Science*, **10**: 91–102.
472. Lutjeharms, J.R.E., J. Olivier and E. Lourens (1991). Surface fronts of False Bay and vicinity. *Transactions of the Royal Society of South Africa*, **47**(415): 433–445.

473. Boyd, A.J., B.B.S. Tromp and D.A. Horstman (1985). The hydrology of the South African south-western coast between Cape Point and Danger Point in 1975. *South African Journal of Marine Science*, **3**: 145–168.
474. Chapman, P. and J.L. Largier (1989). On the origin of Agulhas Bank bottom water. *South African Journal of Science*, **85**(8): 515–519.
475. Largier, J.L., P. Chapman, W.T. Peterson and V.P. Swart (1992). The western Agulhas Bank: circulation, stratification and ecology. *South African Journal of Marine Science*, **12**: 319–339.
476. Schumann, E.H. (1992). Interannual wind variability on the south and east coasts of South Africa. *Journal of Geophysical Research*, **97**(D18): 20 397–20 403.
477. Schumann, E.H., A.L. Cohen and M.R. Jury (1995). Coastal sea surface temperature variability along the south coast of South Africa and the relationship to regional and global climate. *Journal of Marine Research*, **53**(2): 231–248.
478. Lutjeharms, J.R.E. (1998). Coastal hydrography. In *A Field Guide to the Eastern and Southern Cape Coast*, editors R. Lubke and I. de Moor, The Wildlife and Environment Society of Southern Africa, Grahamstown, University of Cape Town Press, Rondebosch, pp. 50–61.
479. Lutjeharms, J.R.E., N.D. Bang and H.R. Valentine (1981). Die fisiese oseanologie van die Agulhasbank. 1. Vaart 170 van die N.S. *Thomas B. Davie*. South African Council for Scientific and Industrial Research *CSIR Research Report*, **386**, 38 pp.
480. Largier, J.L. and V.P. Swart (1987). East–west variation in thermocline breakdown on the Agulhas Bank. *South African Journal of Marine Science*, **5**: 263–272.
481. Schumann, E.H. and L.J. Beekman (1984). Ocean temperature structures on the Agulhas Bank. *Transactions of the Royal Society of South Africa*, **34**(2): 191–203.
482. Shannon, L.V. and P. Chapman (1983). Suggested mechanisms for the chronic pollution by oil of beaches east of Cape Agulhas, South Africa. *South African Journal of Marine Science*, **1**: 231–244.
483. Lutjeharms, J.R.E., D. Baird and I.T. Hunter (1986). Seeoppervlak dryfgedrag aan die Suid-Afrikaanse suidkus in 1979. *South African Journal of Science*, **82**(6): 324–326.
484. Orren, M.J. (1963). Hydrological observations in the South West Indian Ocean. *Investigational Reports of the Division of Sea Fisheries* (of South Africa), **45**: 61 pp.
485. Blanton, J.O. and L.J. Pietrafesa (1978). Flushing of the continental shelf south of Cape Hateras by the Gulf Stream. *Geophysical Research Letters*, **5**(6): 495–498.
486. Boyd, A.J., J. Taunton-Clark and G.P.J. Oberholster (1992). Spatial features of the near-surface and midwater circulation patterns off western and southern South Africa and their role in the life histories of various commercially fished species. *South African Journal of Marine Science*, **12**: 189–206.
487. Preston-Whyte, R.A. and P.D. Tyson (1988). *The Atmosphere and Weather of Southern Africa*. Oxford University Press, Cape Town, 374 pp.
488. Hunter, I.T. (1987). The weather of the Agulhas Bank and Cape south coast. (South African) Council for Scientific and Industrial Research. *CSIR Research Report*, **634**, 184 pp.
489. Jury, M.R. (1994). A review of the meteorology of the eastern Agulhas Bank. *South African Journal of Science*, **90**(3): 109–113.
490. Schumann, E.H. (1983). Long period coastal trapped waves off the southeast coast of Southern Africa. *Continental Shelf Research*, **2**(2/3): 97–107.
491. Schumann, E.H. and K.H. Brink (1990). Coastal-trapped waves off the coast of South Africa: generation, propagation and current structures. *Journal of Physical Oceanography*, **20**(8): 1206–1218.
492. De Cuevas, B.A., G.B. Brundrit and A.M. Shipley (1986). Low-frequency sea-level fluctuations along the coasts of Namibia and South Africa. *Geophysical Journal of the Royal Astronomical Society*, **87**(1): 1–15.
493. Shillington, F.A. (1984). Long period edge waves off southern Africa. *Continental Shelf Research*, **3**(4): 343–357.
494. Jury, M.R. and S. Courtney (1991). A transition in weather over the Agulhas Current. *South African Journal of Marine Science*, **19**: 159–171.
495. Wai, M.M. and S.A. Stage (1989). Dynamical analyses of marine atmospheric boundary layer structure near the Gulf Stream oceanic front. *Quarterly Journal of the Royal Meteorological Society*, **115**(485): 29–44.
496. Rouault, M. and J.R.E. Lutjeharms (2000). Air–sea exchange over an Agulhas eddy at the Subtropical Convergence. *The Global Atmosphere and Ocean System*, **7**(2): 125–150.
497. Jury, M.R. (1995). A review of research on ocean–atmosphere interactions and South African climate variability. *South African Journal of Science*, **91**(6): 289–294.
498. Wedepohl, P.M., J.R.E. Lutjeharms and J.M. Meeuwis (2000). The surface drift in the South East Atlantic Ocean. *South African Journal of Marine Science*, 22: 71–79.
499. Lutjeharms, J.R.E. and D.J. Baker (1980). A statistical analysis of the meso-scale dynamics of the Southern Ocean. *Deep-Sea Research*, **27**(2): 145–159.
500. Schmitz, W.J. and J.R. Luyten (1991). Spectral time scales for mid-latitude eddies. *Journal of Marine Research*, **49**(1): 75–107.
501. Colton, M.T. and R.R.P. Chase (1983). Interaction of the Antarctic Circumpolar Current with bottom topography: an investigation using satellite altimetry. *Journal of Geophysical Research*, **88**(C3): 1825–1843.
502. Zlotnicki, V., L.L. Fu and W. Patzert (1989). Seasonal variability in global sea level observed with Geosat altimetry. *Journal of Geophysical Research*, **94**(C): 17 959–17 969.

503. Garraffo, Z., S.L. Garzoli, W. Haxby and D. Olson (1992). Analysis of a general circulation model. 2. Distribution of kinetic energy in the South Atlantic and Kuroshio/Oyashio systems. *Journal of Geophysical Research*, **97**(C12): 20 139–20 153.
504. Mühry, A. (1864). Die Meeresströmungen an der Südspitze Afrikas. *Petermanns Geographische Mitteilungen*, **10**: 34–35.
505. Van Gogh, J. (1857). *Uitkomsten van Wetenskap en Ervaring aangaande winden en zeestromingen in sommige gedeelten van den oseaan*. Koninklijk Nederlandsch Meteorologisch Instituut, Bosch en Zoon, 74 pp.
506. Darbyshire, J. (1972). The effect of bottom topography on the Agulhas Current. *Reviews of Pure and Applied Geophysics*, **101**(9): 208–220.
507. Johns, W.E., T.N. Lee, F.A. Schott, R.J. Zantrop and R.H. Evans (1990). The North Brazil Current retroflection: seasonal structure and eddy variability. *Journal of Geophysical Research*, **95**(C12): 22,103–22,120.
508. Luyten, J.(R.), A. Spencer, S. Tarbell, K. Luetkemeyer, P. Flament, J. Toole, M. Francis and S. Bennett (1990). Moored current meter, AVHRR, CTD, and drifter data from the Agulhas Current and Retroflexion [sic] region (1985–1987), Vol, XLII. Woods Hole Oceanographic Institution, Technical Report, WHOI-90-30, 100 pp.
509. Schmitz, W.J. (1996). On the eddy field in the Agulhas Retroflection, with some global considerations *Journal of Geophysical Research*, **101**(C7): 16 259–16 271.
510. Schmitz, W.J. and W.R. Holland (1986). Observed and modeled mesoscale variability near the Gulf Stream and Kuroshio Extension. *Journal of Geophysical Research*, **91**(C8): 9624–9638.
511. Duncan, C.P. (1968). An eddy in the Subtropical Convergence southwest of South Africa. *Journal of Geophysical Research*, **73**(2): 531–534.
512. Lutjeharms, J.R.E. (1988). Examples of extreme circulation events of the Agulhas Retroflection. *South African Journal of Science*, **84**(7): 584–586.
513. Luyten, J.R. and P.F. Smith (1985). Agulhas Current trajectory from new Argos drifter compared with simultaneous shipboard measurements. In *Oceans '85 Ocean Engineering and the Environment*, San Diego, 12–24 November 1985, 2, pp. 1165–1167.
514. Fuglister, F.C. and A.D. Voorhis (1965). A new method of tracking the Gulf Stream. *Limnology and Oceanography*, **10 suppl.**: R115–R124.
515. Fuglister, F.C. and L.V. Worthington (1951). Some results of a multiple ship survey of the Gulf Stream. *Tellus*, **3**(1): 1–14.
516. De Ruijter, W.P.M., A. Biastoch, S.S. Drijfhout, J.R.E. Lutjeharms, R.P. Matano, T. Pichevin, P.J. van Leeuwen and W. Weijer (1999). Indian–Atlantic inter-ocean exchange: dynamics, estimation and impact. *Journal of Geophysical Research*, **104**(C9): 20 885–20 911.
517. Goni [sic], G.J., S.L. Garzoli, A.J. Roubicek, D.B. Olson and O.T. Brown (1997). Agulhas ring dynamics from TOPEX/POSEIDON satellite altimeter data. *Journal of Marine Research*, **55**(5): 861–883.
518. Chassignet, E.P. and D.B. Boudra (1987). Dynamics of Agulhas retroflection and ring formation in a quasi-isopycnic coordinate numerical model. In *Three Dimensional Models of Marine and Estuarine Dynamics*, editors J.C.J. Nihoul and B.M. Jamart, Elsevier, Amsterdam, pp. 169–194.
519. Garzoli, S.L., A.L. Gordon, V. Kamenkovich, D. Pillsbury and C. Duncombe-Rae [sic] (1996). Variability and sources of the southeastern Atlantic circulation. *Journal of Marine Research*, **54**(6): 1039–1071.
520. Duncombe Rae, C.M., S.L. Garzoli and A.L. Gordon (1996). The eddy field of the southeast Atlantic Ocean: a statistical census from the Benguela Sources and Transports Project. *Journal of Geophysical Research*, **101**(C5): 11 949–11 964.
521. Gründlingh, M.L. (1977). A Nimbus VII satellite-tracked buoy moored on the Mozambique Ridge. *South African Journal of Science*, **73**(12): 384–385.
522. Harris, T.F.W. and C.C. Stavropoulos (1978). Satellite tracked drifters between Africa and Antarctica. *Bulletin of the American Meterological Society*, **59**(1): 51–59.
523. Garrett, J. (1981). The performance of the FGGE drifting buoy system. *Advances in Space Research*, **1**(4): 87–94.
524. Garrett, J. (1981). Oceanographic features revealed by the FGGE drifting buoy array. In *Oceanography from Space*, editor J.F.R. Gower, Plenum Press, New York, pp. 61–69.
525. Taljaard, J.J. (1977). South Africa's contribution to the FGGE drifting buoy programme for the southern hemisphere. *News Letter, South African Weather Bureau*, **339**: 198–207.
526. Lutjeharms, J.R.E. and H.R. Valentine (1981). Ocean circulation studies in the vicinity of southern Africa: preliminary results using FGGE drifters and remote sensing. *Advances in Space Research*, **1**(4): 211–223.
527. Taljaard, J.J., J. van Heerden, R.D. Sewell and W. Jordaan (1983). South African drifting buoy programmes. *South African Weather Bureau, Technical Paper*, **12**: 1–57.
528. Shannon, L.V., J.R.E. Lutjeharms and J.J. Agenbag (1989). Episodic input of Subantarctic water into the Benguela region. *South African Journal of Science*, **85**(5): 317–332.
529. Boudra, D.B., K.A. Maillet and E.P. Chassignet (1989). Numerical modeling of Agulhas retroflection and ring formation with isopycnal outcropping. In *Meso-scale/Synoptic Coherent Structures in Geophysical Turbulence*, editors J.C.J. Nihoul and B.M. Jamart, Elsevier, Amsterdam, pp. 315–335.
530. Shannon, L.V., J.R.E. Lutjeharms and G. Nelson (1990). Causative mechanisms for intra-annual and inter-annual variability in the marine environment around Southern Africa. *South African Journal of Science*, **86**(7/8/9/10): 356–373.

531. Chapman, P., C.M. Duncombe Rae and B.R. Allanson (1987). Nutrients, chlorophyll and oxygen relationships in the surface layers at the Agulhas retroflection. *Deep-Sea Research*, **34**(8): 1399–1416.
532. Fine, R.A., M.J. Warner and R.F. Weiss (1988). Water mass modification of the Agulhas retroflection: chlorofluoromethane studies. *Deep-Sea Research*, **35**(3): 311–332.
533. Bennett, S.L. (1988). *Where Three Oceans Meet: the Agulhas retroflection region*. Doctoral dissertation, Woods Hole Oceanographic Institution and Massachusetts Institute of Technology, WHOI-88-51, 367 pp.
534. Veronis, G. (1973). Model of world ocean circulation. I. Wind-driven, two layer. *Journal of Marine Research*, **31** (3): 228–288 (see also p. 289).
535. Garzoli, S.L., A.L. Gordon and D. Pillsbury (1994). Initial results in from BEST cruises. *WOCE Notes*, **6**(10): 10–11, 15.
536. Garzoli, S.L. and A.L. Gordon (1996). Origin and variability of the Benguela Current. *Journal of Geophysical Research*, **101**(C11): 22 591–22 601.
537. Clement, A.C. and A.L. Gordon (1995). The absolute velocity field of Agulhas eddies and the Benguela Current. *Journal of Geophysical Research*, **100**(C11): 22 591–22 601.
538. Flierl, G.R. (1981). Particle motions in large-amplitude wave field. *Geophysical and Astrophysical Fluid Dynamics*, **18**(1): 311–332.
539. McCartney, M.S. and M.E. Woodgate-Jones (1991). A deep-reaching anticyclonic eddy in the subtropical gyre of the eastern South Atlantic. *Deep-Sea Research*, **38** (supplement 1): s411–s443.
540. Smythe-Wright, D., A.L. Gordon, P. Chapman and M.S. Jones (1996). CFC–113 shows Brazil eddy crossing the South Atlantic to the Agulhas retroflection region. *Journal of Geophysical Research*, **101**(C1): 885–895.
541. McDonagh, E.L. and K.J. Heywood (1999). The origin of an anomalous ring in the Southeast Atlantic. *Journal of Physical Oceanography*, **29**(8): 2025–2064.
542. McDonagh, E.L., K.J. Heywood and M.P. Meredith (1999). On the structure, paths and fluxes associated with Agulhas rings. *Journal of Geophysical Research*, **104**(C9): 21,007–21,020.
543. Duncombe Rae, C.M., L.V. Shannon and F.A. Shillington (1989). An Agulhas ring in the South Atlantic Ocean. *South African Journal of Science*, **85**(11): 747–748.
544. Arhan, M., H. Mercier and J.R.E. Lutjeharms (1999). The disparate evolution of three Agulhas rings in the South Atlantic Ocean. *Journal of Geophysical Research*, **104** (C9): 20 987–21 005.
545. Wang, L. and C.J. Koblinsky (1996). Low-frequency variability in the region of the Agulhas Retroflection. *Journal of Geophysical Research*, **101**(C2): 3597–3614.
546. Kamenkovich, V.M., Y.P. Leonov, D.A. Nechaev, D.A. Byrne and A.L. Gordon (1996). On the influence of bottom topography on the Agulhas eddy. *Journal of Physical Oceanography*, **26**(6): 892–192.
547. Garzoli, S.L., G.J. Goñi, A.J. Mariano and D.B. Olson (1997). Monitoring the upper southeasterrn Atlantic transports using altimeter data. *Journal of Marine Research*, **55**(3): 453–481.
548. Evenson, G. and P.J. van Leeuwen (1996). Assimilation of Geosat altimeter data for the Agulhas Current using the ensemble Kalman filter with a quasigeostrophic model. *Monthly Weather Review*, **124**(1):85–96.
549. Lutjeharms, J.R.E., H.R. Valentine and R.C. van Ballegooyen (1993). The Subtropical Convergence in the South Atlantic Ocean. *South African Journal of Science*, **89**(11/12): 552–559.
550. Beismann, J.-O., R.H. Käse and J.R.E. Lutjeharms (1999). Numerical experiments on the influence of submarine ridges on translation and stability of Agulhas rings. *Journal of Geophysical Research*, **104**(C4): 7897–7906.
551. Shannon, L.V. (1985). The Benguela Ecosystem. 1. Evolution of the Benguela, physical features and processes. *Oceanography and Marine Biology, An Annual Review*, **23**: 105–182.
552. Shannon, L.V. and G. Nelson (1996). The Benguela: large scale features and processes and system variability. In *The South Atlantic: Present and Past Circulation*, editors G. Wefer, W. H. Berger, G. Siedler and D. Webb, Springer-Verlag, Berlin, pp. 163–210.
553. Shillington, F.A. (1998). The Benguela upwelling system off southwestern Africa. In *The Sea,* Volume 11, Chapter 20, editors A. R. Robinson and K. H. Brink, John Wiley & Sons, pp. 583–604.
554. Meeuwis, J.M. and J.R.E. Lutjeharms (1990). Surface thermal characteristics of the Angola-Benguela front. *South African Journal of Marine Science*, **9**: 261–279.
555. Stockton, P.L. and J.R.E. Lutjeharms (1988). Observations of vortex dipoles on the Benguela upwelling front. *South African Geographer*, **15**(1/2): 27–35.
556. Lutjeharms, J.R.E. and P.L. Stockton (1987). Kinematics of the upwelling front off Southern Africa. *South African Journal of Marine Science*, **5**: 35–49.
557. Lutjeharms, J.R.E. and C.P. Matthysen (1995). A recurrent eddy in the upwelling front off Cape Town. *South African Journal of Science*, **91**(7): 355–357.
558. Lutjeharms, J.R.E., F.A. Shillington and C.M. Duncombe Rae (1991). Observations of extreme upwelling filaments in the South East Atlantic Ocean. *Science*, **253**(5021): 774–776.
559. Duncombe Rae, C.M., F.A. Shillington, J.J. Agenbag, J. Taunton-Clark and M.L. Gründlingh (1992). An Agulhas Ring in the South East Atlantic Ocean and its interaction with the Benguela upwelling frontal system. *Deep-Sea Research*, **39**(11/12): 2009–2027.
560. Duncombe Rae, C.M., A.J. Boyd and R.J.M. Crawford (1992). "Predation" of anchovy by an Agulhas ring: a possible contributory cause of the very poor year class of 1989. *South African Journal of Marine Science*, **12**:

167–173.
561. Shillington, F.A., W.T. Peterson, L. Hutchings, T.A. Probyn, H.N. Waldron and J.J. Agenbag (1990). A cool upwelling filament off Namibia, southwest Africa: preliminary measurements of physical and biological features. *Deep-Sea Research*, **37**(11): 1753–1772.
562. Lutjeharms, J.R.E., D.J. Webb, B.A. de Cuevas and S.R. Thompson (1996). Large-scale modelling of the South-East Atlantic upwelling system. *South African Journal of Marine Science*, **16**: 205–225.
563. Gordon, A.L., K.T. Bosley and F. Aikman (1995). Tropical Atlantic water within the Benguela upwelling system at 27°S. *Deep-Sea Research*, **24**(1): 1–12.
564. Schell, I.I. (1968). On the relation between the winds off southwest Africa and the Benguela Current and the Agulhas Current penetration in the South Atlantic. *Deutsche Hydrographische Zeitschrift*, **21**(3): 109–117.
565. Rintoul, S.R. (1991). South Atlantic interbasin exchange. *Journal of Geophysical Research*, **96**(C2): 2675–2692.
566. Döös, K. (1995). Interocean exchange of water masses. *Journal of Geophysical Research*, **100**(C7): 13 499–13 514.
567. Fu, L.-L. (1981). The general circulation and meridional heat transport of the subtropical South Atlantic determined by inverse methods. *Journal of Physical Oceanography*, **11**( ): 1171–1193.
568. MacDonald, A.M. (1993). Property fluxes at 30° S and their implications for the Pacific–Indian throughflow and the global heat budget. *Journal of Geophysical Research*, **98**(C4): 6851–6868.
569. Schlitzer, R. (1993). Determining the mean, large-scale circulation of the Atlantic with the Adjoint Method. *Journal of Physical Oceanography*, **23**(9): 1935–1952.
570. Schlitzer, R. (1996). Mass and heat transports in the South Atlantic derived from historical hydrographic data. In *The South Atlantic: Present and Past Circulations*, editors G. Wefer, W.H. Berger, G. Siedler and D.J. Webb, Springer-Verlag, Berlin, pp. 305–323.
571. Holfort, J., G. Siedler and K. Speer (1998). The oceanic transport of heat and nutrients in the South Atlantic. *Journal of Geophysical Research*, submitted.
572. Taunton-Clark, J. and L.V. Shannon (1988). Annual and interannual variability in the South-East Atlantic during the 20th century. *South African Journal of Marine Science*, **6**: 97–196.
573. Matano, R.P. and G.H. Philander (1993). Heat and mass balances of the South Atlantic Ocean calculated from a numerical model. *Journal of Geophysical Research*, **98**(C): 977–984.
574. Thompson, S.R., D.P. Stevens and K. Döös (1997). The importance of interocean exchange south of Africa in a numerical model. *Journal of Geophysical Research*, **102**(C2): 3303–3315.
575. Suga, T. and L.D. Talley (1995). Antarctic Intermediate Water circulation in the tropical and subtropical South Atlantic. *Journal of Geophysical Research*, **100**(C7): 13 441–13 453.
576. De Ruijter, W.P.M. and D.B. Boudra (1985). The wind-driven circulation in the South Atlantic–Indian Ocean – I. Numerical experiments in a one-layer model. *Deep-Sea Research*, **32**(5): 557–574.
577. Chassignet, E.P., D.B. Olson and D.B. Boudra (1989). Evolution of rings in numerical models and observations. In *Mesoscale/Synoptic Coherent Structures in Geophysical Turbulence*, editors J.C.J. Nihoul and B.M. Jamart, Elsevier, Amsterdam, pp. 337–356.
578. Van Leeuwen, P.J. (1999). The time-mean circulation in the Agulhas region determined with the ensemble smoother. *Journal of Geophysical Research*, **104**(C1): 1393–1404.
579. Florenchie, P. and J. Veron (1998). South Atlantic Ocean circulation: Simulation experiments with a quasi-geostrophic model and assimilation of TOPEX/POSEIDON and ERS 1 altimeter data. *Journal of Geophysical Research*, **103**(C11): 24 737–24 758.
580. De Ruijter, W. (1982). Asymptotic analysis of the Agulhas and Brazil Current systems. *Journal of Physical Oceanography*, **12**(4): 361–373.
581. Boudra, D.B. and W.P.M. de Ruijter (1986). The wind-driven circulation in the South Atlantic–Indian Ocean – II. Experiments using a multi-layer numerical model. *Deep-Sea Research*, **33**(4): 447–482.
582. Boudra, D.B. and E.P. Chassignet (1988). Dynamics of Agulhas retroflection and ring formation in a numerical model. I. The vorticity balance. *Journal of Physical Oceanography*, **18**(2): 280–303.
583. Pichevin, T., D. Nof and J. Lutjeharms (1999). Why are there Agulhas rings? *Journal of Physical Oceanography*, **29**(4): 693–707.
584. Van Leeuwen, P.J. (2001) An ensemble smoother with error estimates. *Monthly Weather Review*, 129(4): 709–728.
585. Stevens, D.P. and S.R. Thompson (1994). The South Atlantic in the Fine-Resolution Antarctic Model. *Annales Geophysicae*, **12**: 826–839.
586. Pearce, A.F. (1983). Adrift in the Agulhas Return Current. *Deep-Sea Research*, **30**(3): 343–347.
587. Minster, J.-F. and M.C. Gennero (1995). High-frequency variability of western boundary currents using ERS 1 three-day repeat altimeter data. *Journal of Geophysical Research*, **100**(C11): 22603–22612.
588. Lutjeharms, J.R.E., B.R. Allanson and L. Parker (1986). Frontal zones, chlorophyll and primary production patterns in the surface waters of the Southern Ocean south of Africa. In *Marine Interfaces Ecohydrodynamics*, editor J.C.J. Nihoul, Elsevier, Amsterdam, pp. 105–117.
589. Allanson, B.R., R.C. Hart and J.R.E. Lutjeharms (1981). Observations on the nutrients, chlorophyll and primary production of the Southern Ocean south of Africa. *South African Journal of Antarctic Research*, **10/11**: 3–14.
590. Boden, B.P., C.M. Duncombe Rae and J.R.E.

Lutjeharms (1988). The distribution of the diatoms of the S. W. Indian Ocean waters between Cape Town and the Prince Edward Island archipelago. *South African Journal of Science*, **84**(10): 811–818.

591. Abrams, R.W. and J.R.E. Lutjeharms (1986). Relationship between seabirds and meso-scale hydrographic features in the Agulhas Current retroflection region. In *Acta XIX Congressus Internationalis Ornithologici* Vol. I, Ottawa, Canada, pp. 991–996.
592. Cockroft, V.G., V.M. Peddemors, P.G. Ryan and J.R.E. Lutjeharms (1990). Cetacean sightings in the Agulhas Retroflection, Agulhas Rings and Subtropical Convergence. *South African Journal of Antarctic Research*, **20**(2): 64–67.
593. Hofmann, E.E. (1985). The large-scale horizontal structure of the Antarctic Circumpolar Current from FGGE drifters. *Journal of Geophysical Research*, **90**(C7): 7087–7097.
594. Belkin, E.M. and A.L. Gordon (1996). Southern Ocean fronts from the Greenwich meridian to Tasmania. *Journal of Geophysical Research*, **101**(C2): 3675–3696.
595. Lutjeharms, J.R.E. (1985). Location of frontal systems between Africa and Antarctica: some preliminary results. *Deep-Sea Research*, **32**(12): 1499–1509.
596. Lutjeharms, J.R.E. and J.M. Meeuwis (1999). The location of the surface expression of oceanic fronts in the South Atlantic and South Indian Oceans. In preparation.
597. Holliday, N.P. and J.F. Read (1998). Surface oceanic fronts between Africa and Antarctica. *Deep-Sea Research I*, **45**(2–3): 217–238.
598. Park, Y.-H., L. Gambéroni and E. Charriaud (1993). Frontal structure, water masses, and circulation in the Crozet Basin. *Journal of Geophysical Research*, **98** (C7): 12 361–12 385.
599. Park, Y.-H., L. Gambéroni and E. Charriaud (1991). Frontal structure and transport of the Antarctic Circumpolar Current in the South Indian Ocean sector, 40–80°D. *Marine Chemistry*, **35**(1): 45–62.
600. Whitworth, T. and W.D. Nowlin (1987). Water masses and currents of the Southern Ocean at the Greenwich meridian. *Journal of Geophysical Research*, **92**(C6): 6462–6476.
601. Smythe-Wright, D., P. Chapman, C. Duncombe Rae, L.V. Shannon and S.M. Boswell (1998). Characteristics of the South Atlantic subtropical frontal zone between 15°W and 5°E. *Deep-Sea Research*, **45**(1): 167–192.
602. Barange, M., E.A. Pakhomov, R. Perissinotto, P.W. Froneman, H.M. Verheye, J. Taunton-Clark and M.I. Lucas (1998). Pelagic community structure of the subtropical convergence region south of Africa and in the mid-Atlantic Ocean. *Deep-Sea Research I*, **45**(10): 1663–1687.
603. Bryantsev, V.A., S.V. Pavlukhin and A.S. Pelevin (1979). Characteristics of geostrophic currents and the Subtropical Convergence zone in the Southwestern Indian Ocean. *Oceanology*, **19**(4): 395–399.
604. Walker N.D. and J.A. Lindesay (1989). Preliminary observations of oceanic influences on the February–March 1988 floods in central South Africa. *South African Journal of Science*, **85**(3): 164–169. 367.
605. Walker, N.D. and F.A. Shillington (1990). The effect of oceanographic variability on South African weather and climate. *South African Journal of Science*, **86**(7/8/9/10): 382–386.
606. Mason, S.J. (1995). Sea-surface temperature – South African rainfall associations, 1910–1989. *International Journal of Climatology*, **15**(2): 119–135.
607. Mason, S.J. (1990). Temporal variability of sea surface temperatures around Southern Africa: a possible forcing mechanism for the 18-year rainfall oscillation. *South African Journal of Science*, **86**(5/6): 243–252.
608. Mason, S.J. and P.D. Tyson (1992). The modulation of sea surface temperature and rainfall associations over southern Africa with solar activity and the quasi-biennial oscillation. *Journal of Geophysical Research*, **97**(D5): 5847–5856.
609. Tyson, P.D., T.G.J. Dyer and M.N. Manetse (1975). Secular changes in South African rainfall: 1880 to 1972. *Quarterly Journal of the Royal Meteorological Society*, **101**(430): 817–833.
610. Lutjeharms, J.R.E. and L.H. McQuaid (1986). Changes in the structure of thermal ocean fronts south of Africa over a three-month period. *South African Journal of Science*, **82**(9): 470–476.
611. Lutjeharms, J.R.E. (1990). Temperatuurstruktuur van die oseaanbolaag tussen Kaapstad en Marion-eiland. *South African Journal of Antarctic Research*, **20**(1): 21–32.
612. Camden-Smith, F., L.-A. Perrins, R.V. Dingle and G.B. Brundrit (1981). A preliminary report on long-term bottom-current measurements and sediment transport/erosion in the Agulhas Passage, Southwest Indian Ocean. *Marine Geology*, **39**: M81–M88.
613. Maury, M.F. (edited J. Leighly, 1963). *The physical geography of the sea; and its meteorology.* The Belknap Press of Harvard University Press, Cambridge, MA, 432 pp., 10 plates.
614. Anonymous (1954). *1854 Koninklijk Nederlands Meteorologisch Instituut 1954.* Staatsdrukkerij en Uitgeversbedrijf, 's Gravenhage, 469 pp.
615. Koninklijk Nederlandsch Meteorologisch Instituut (1856). *Maury's passatkaart van den Atlantischen Oceaan, vermeerderd met Hollandsche gegevens door den lt. t. z. K.F.R. Andrau.* Koninklijk Nederlandsch Meteorologisch Instituut.
616. Berghaus, H. (1867). Welt-karte zur Übersicht der Meeres-Stömungen und das Schnellverkers. In (1988) *Handatlas über alle Theile der Erde und über das Weltgebäude*, editor A. Stieler, Gotha, 95 charts.
617. Rouault, M., C.J.C. Reason, J.R.E. Lutjeharms and A. Belaars (2003). Underestimation of latent and sensible heat fluxes above the Agulhas Current in NCEP and

ECMWF analyses. *Journal of Climate*, **16**(2): 776–782.
618. Reason, C.J.C. (2001). Evidence for the influence of the Agulhas Current on regional atmospheric circulation patterns. *Journal of Climate*, **14**(6): 2769–2778.
619. Rouault, M., S.A. White, C.J.C. Reason and J.R.E. Lutjeharms and I. Jobard (2002). Ocean–atmosphere interaction in the Agulhas Current and a South African extreme weather event. *Weather and Forecasting*, **17**(4): 655–669.
620. Josey, S.A., E.C. Kent and P.K. Taylor (1999). New insights into the ocean heat budget closure problem from analysis of the SOC air–sea flux climatology. *Journal of Climate*, **12**( ): 2856–2880.
621. Xue, H., J.M. Bane and L.M. Goodman (1995). Modification of the Gulf Stream through strong air–sea interaction in winter: Observations and numerical simulations. *Journal of Physical Oceanography*, **25**( ): 533–557.
622. Kondo, J. (1976). Heat balance of the East China Sea during the Air Mass Transformation Experiment. *Journal of the Meteorological Society of Japan*, **54**( ): 382–398.
623. Bosart, L.F. and S.C. Lin (1984). A diagnostic analysis of the Presidents' Day Storm of February 1979. *Monthly Weather Review*, **112**( ): 2148–2177.
624. Holt, T. and S. Raman (1990). Marine boundary layer structure and circulation in the region of offshore redevelopment of a cyclone during GALE. *Monthly Weather Review*, **118**( ): 392–410.
625. Rouault, M., I. Jobard, S.A. White and J.R.E. Lutjeharms (2001). Studying rainfall events over South Africa and adjacent oceans using the TRMM satellite. *South African Journal of Science*, **97**(11/12): 455–460.
626. Morrow, R., F. Birol, D. Griffen and J. Sudre 2004. Divergent pathways of cyclonic and anti-cyclonic eddies. *Geophysical Research Letters*, **31**(24): L24311.
627. Richardson, P.L., J.R.E. Lutjeharms and O. Boebel (2003). Introduction to the "Inter-ocean exchange around southern Africa". *Deep-Sea Research II*, **50**(1): 1–12.
628. Boebel, O., J. Lutjeharms, C. Schmid, W. Zenk, T. Rossby and C. Barron (2003). The Cape Cauldron: a regime of turbulent inter-ocean exchange. *Deep-Sea Research II*, **50**(1): 57–86.
629. Lutjeharms, J.R.E., O. Boebel and T. Rossby (2003). Agulhas cyclones. *Deep-Sea Research II*, **50**(1): 13–34.
630. Penven, P., J.R.E. Lutjeharms, P. Marchesiello, S.J. Weeks and C. Roy (2001). Generation of cyclonic eddies by the Agulhas Current in the lee of the Agulhas Bank. *Geophysical Research Letters*, **28**(6): 1055–1058.
631. Lutjeharms, J.R.E., P. Penven, C. Roy (2003). Modelling the shear-edge eddies of the southern Agulhas Current. *Continental Shelf Research*, 23(11–13): 1099–1115.
632. Lutjeharms, J.R.E., O. Boebel, P.C.F. van der Vaart, W.P.M. de Ruijter, T.H. Rossby and H.L. Bryden (2001). Evidence that the Natal Pulse controls the Agulhas Current over its full depth. *Geophysical Research Letters*, **28**(18): 3449–3452.
633. Penven, P., C. Roy, G.B. Brundrit, A. Colin de Verdière, P. Fréon, A.S. Johnson, J.R.E. Lutjeharms and F.A. Shillington (2001). A regional hydrodynamic model of the Southern Benguela. *South African Journal of Science*, **97** (11/12): 472–475.
634. Ravenstein, E.G. (1900). The voyages of Diogo Cão and Bartholomeu Dias, 1482–1488. *Geographical Journal*, **712**( ): 625–655.
635. Axelson, E. (1973). *Congo to Cape, Early Portuguese Explorers*. Harper and Row, New York, 224 pp.
636. Thibault-Botha, D., J.R.E. Lutjeharms and M.J. Gibbons (2004). Siphonophore assemblages along the east coast of South Africa; mesoscale distribution and seasonal variations. *Journal of Plankton Research*, **26**(9): 1–14.
637. Liltved, W.R., E. Harley and J.R.E. Lutjeharms (2000). A new subspecies of *Cypraeovula* occurring off the northern part of the Eastern Cape, South Africa, with notes on possible genetic and oceanographic influences on its distribution. Supplement to *Cowries and Their Relatives of Southern Africa*, Seacomber Publications, Cape Town, pp. 225–232.
638. Gibbons, M.J. and D. Thibault-Botha (2002). The match between ocean circulation and zoogeography of epipelagic siphonophores around southern Africa. *Journal of the Marine Biological Association of the United Kingdom*, **82**(15): 801–810.
639. Gibbons, M.J., M. Barange and L. Hutchings (1995). Zoogeography and diversity of euphausiids around southern Africa. *Marine Biology*, **123**( ): 257–268.
640. Underhill, L.G., R.J.M. Crawford and C.J. Camphuysen (2002). Leach's storm petrels *Oceanodroma leucorhoa* off southern Africa: breeding and migratory status, and measurements and mass of the breeding population. *Transactions of the Royal Society of South Africa*, **57**(1&2): 43–46.
641. Nel, D.C., J.R.E. Lutjeharms, E.A. Pakhomov, I.J. Ansorge, P.G. Ryan and N.T.W. Klages (2001). Exploitation of mesoscale oceanographic features by grey-headed albatross *Thalassarche chrystomota* in the southern Indian Ocean. *Marine Ecology Progress Series*, **217**(6): 15–26.
642. Weimerskirch, H., M. le Corre, S. Jaquemet, M. Potier and F. Marsac (2004). Foraging strategy of a top predator in tropical waters: great frigatebirds in the Mozambique Channel. *Marine Ecology Progress Series*, **275**( ): 297–308.
643. Kostianoy, A.G., A.I. Ginzburg, S.A Lebedev, M. Frankignoulle and B. Delille (2003). Fronts and mesoscale variability in the southern Indian Ocean as inferred from the *TOPEX/POSEIDON* and *ERS-2* altimetry data. *Marine Physics*, **43**(5): 671–682.
644. Kostianoy, A.G., A.I. Ginzburg, M. Frankignoulle and B. Delille (2004). Fronts in the southern Indian Ocean

as inferred from satellite sea surface temperature data. *Journal of Marine Systems*, **45**(1): 55–73.

645. Anonymous (1983). Cruise report R/V "Dr. Fridtjof Nansen", Fisheries Resources Survey, Madagascar, 16–28 June 1983. *Reports on Surveys with the R/V Dr Fridtjof Nansen*, Institute of Marine Research, Bergen, 9 pp.
646. DiMarco, S.F., P. Chapman and W.D. Nowlin (2000). Satellite observations of upwelling on the continental shelf south of Madagascar. *Geophysical Research Letters*, **27**(24): 3965–3968.
647. Machu, E., J.R.E. Lutjeharms, A.M. Webb and H.M. van Aken (2003). First hydrographic evidence of the southeast Madagascar upwelling cell. *Geophysical Research Letters*, **29**(21), 2009, doi: 10.1029/2002GL015381.
648. Lutjeharms, J.R.E. (2005). The ocean environment off south-eastern Africa. *South African Journal of Science*, submitted.
649. Quartly, G.D. and M.A. Srokosz (2004). Eddies in the southern Mozambique Channel. *Deep-Sea Research II*, **51**(1): 69–83.
650. Lutjeharms, J.R.E., W.P.M. de Ruijter, H. Ridderinkhof, H. van Aken, C. Veth, P.J. van Leeuwen, S.S. Drijfhout, J.H.F. Jansen and G.-J.A. Brummer (2000). MARE and ACSEX: new research programmes on the Agulhas Current system. *South African Journal of Science*, **96**(3): 105–110.
651. Ho, C.-R., Q. Zheng and N.-J. Kuo (2004). SeaWiFS observations of upwelling south of Madagascar: long-term variability and interaction with East Madagascar Current. *Deep-Sea Rsearch II*, **51**(1): 59–67.
652. De Ruijter, W.P.M., H.M. van Aken, E.J. Beier, J.R.E. Lutjeharms, R.P. Matano and M.W. Schouten (2003). Eddies and dipoles around South Madagascar : formation, pathways and large-scale impact. *Deep-Sea Research I*, **51**(3): 383–400.
653. Schouten, M.W., W.P.M. de Ruijter and P.J. van Leeuwen (2002). Upstream control of Agulhas ring shedding. *Journal of Geophysical Research*, **107**(C8): 10.1029/2001JC000804.
654. Kizner, Z. (1997). Solitary Rossby waves with baroclinic modes. *Journal of Marine Research*, **55**(4): 671–685.
655. Kizner, Z., D. Berson and R. Khvoles (2002). Baroclinic modon equilibria: stability and transitions. *Journal of Fluid Mechanics*, **486**: 239–270.
656. Lutjeharms, J.R.E., S. Jamaloodien and I.J. Ansorge (2002). The temporal displacement of ocean fronts south-east of Africa. *South African Journal of Science*, **98**(5/6): 304–306.
657. Lutjeharms, J.R.E. and C.S. Fillis (2003). Intrusion of sub-Antarctic water across the Subtropical Convergence south of Africa. *South African Journal of Science*, **99**(3/4): 173–176.
658. Ridderinkhof, H., J.R.E. Lutjeharms and W.P.M. de Ruijter (2001). A research cruise to investigate the Mozambique Current. *South African Journal of Science*, **97**(11/12): 461–464.
659. Hughes, G.R., P. Luschi, R. Mencacci and F. Papi (1998). The 7000-km oceanic journey of a leatherback turtle tracked by satellite. *Journal of Experimental Marine Biology and Ecology*, **229**(2): 209–217.
660. Luschi, P., A. Sale, R. Mencacci, G.R. Hughes, J.R.E. Lutjeharms and F. Papi (2003). Current transport of leatherback sea turtles (*Dermochelys coriacea*) in the ocean. *Proceedings of the Royal Society of London* (Supplement), **270**: S129–S132: doi: 10.1098/rsbl.203.0036.
661. Uenzelmann-Neben, G. (2002). Contourites on the Agulhas Plateau, SW Indian Ocean: indications for the evolution of currents since Palaeogene times. In *Deep-Water Contourite Systems: Modern Drifts and Ancient Series, Seismic and Sedimentary Characteristics*, editors D.A.V. Stow, C.J. Pudsey, J.A. Howe, J.-C. Faugères and A.R. Viana, The Geological Society of London, *Memoirs*, **22**: 271–288.
662. Tucholke, B., and R.E. Embley (1984). Cenozoic regional erosion of the abyssal sea floor off South Africa. In *Interregional unconformities and hydrocarbon accumulation*, editor J. S. Schlee, AAPG, Tulsa, *Memoir*, **36**: 145–164.
663. Cortese, G., A. Abelmann and R. Gersonde (2004). A glacial warm water anomaly in the subantarctic Atlantic Ocean, near the Agulhas Retroflection. *Earth and Planetary Science Letters*, **222**(3–4): 767–778.
664. Knorr, G. and G. Lohmann (2003). Southern Ocean origin for the resumption of Atlantic thermohaline circulation during deglaciation. *Nature*, **424**(6948): 532–536.
665. Flores, J.-A., R. Gersonde and F.J. Sierro (1999). Pleistocene fluctuations in the Agulhas Current Retroflection based on the calcareous plankton record. *Marine Micropa-leontology*, **37**(1): 1–22.
666. Rau, A.J., J. Rogers, J.R.E. Lutjeharms, J. Giraureau, J.A. Lee-Thorp, M.-T. Chen and C. Waelbroeck (2002). A 450-kyr record of hydrographical conditions on the western Agulhas Bank slope, south of Africa. *Marine Geology*, **180**(1–4): 183–201.
667. Peeters, F.J.C., R. Acheson, G.-J.A. Brummer, W.P.M. de Ruijter, G.M. Ganssen, R.R. Schneider, E. Ufkes and D. Kroon (2004). Vigorous exchange between Indian and Atlantic Ocean at the end of the last five glacial periods. *Nature*, 430(7000): 661–665.
668. Weijer, W., W.P.M. de Ruijter and H.A. Dijkstra (2001). Stability of the Atlantic overturning circulation: competition between Bering Strait freshwater flux and Agulhas heat and salt sources. *Journal of Physical Oceanography*, **31**(8): 2385–2402.
669. Weijer, W., W.P.M. de Ruijter, A. Sterl and S.S. Drijfhout (2002). Response of the Atlantic overturning circulation to South Atlantic sources of buoyancy. *Global and Planetary Change*, **34**(3–4): 293–311.
670. Quartly, G.D. and M.A. Srokosz (2002). SST observa-

tions of the Agulhas and East Madagascar retroflections by the TRMM Microwave imager. *Journal of Physical Oceanography*, **32**(5): 1585–1592.

671. Rouault, M. and J.R.E. Lutjeharms (2003). Estimation of sea-surface temperature around southern Africa from satellite-derived microwave observations. *South African Journal of Science*, **99**(9/10): 489–494.
672. Lutjeharms, J.R.E., M. Rouault, W.P.M. de Ruijter and H.M. van Aken (2005). Temporary cessation of inter-ocean exchange south of Africa. *Geophysical Research Letters*, in preparation.
673. Quartly, G.D. and M.A. Srokosz (2002). Satellite observations of the Agulhas Current system. *Philosophical Transactions of the Royal Society of London A*, **361**(1802): 51–56.
674. Longhurst, A. (2001). A major seasonal plankton bloom in the Madagascar Basin. *Deep-Sea Research I*, **48**(11): 2413–2422.
675. Srokosz, M.A., G.D. Quartly and J.J.H. Buck (2004). A possible plankton wave in the Indian Ocean. *Geophysical Research Letters*, **31**, L13301, doi: 10.1029/2004GLO19738.
676. Kara, A.B., P.A. Rochford, and H.E. Hurlburt (2003). Mixed layer depth variability over the global ocean. *Journal of Geophysical Research*, **108**(C3): 3079, doi: 10.1029/2000JC000736.
677. Boebel, O., S. Anderson-Fontana, C. Schmid, I. Ansorge. P. Lazarevich, J. Lutjeharms, M. Prater. T. Rossby, and W. Zenk (2000). KAPEX RAFOS Float Data Report 1997–1999. Part A: The Agulhas- and South Atlantic Current Components. *GSO Technical Report* 2000–2; *UCT Oceanography Report* 2000–1; *Berichte aus dem Institut für Meereskunde an der Christian-Albrechts-Universität – Kiel*, IFM 318, 194 pp.
678. Fqtterer, D. and cruise participants (1998). The expedition ANTARKTIS–XIV/4 of RV "Polarstern" in 1997: Physical Oceanography. Alfred-Wegener Institut für Polar- und Meeresforschung, Bremerhaven, *Berichte zur Polarforschung*, **259**.
679. Quartly, G.D., J.J.H. Buck, M.A. Srokosz and A.C. Coward (2004). Eddies around Madagascar – An end to the retroflection? *Journal of Physical Oceanography*, submitted.
680. De Cuevas, B.A., D.J. Webb, A.C. Coward, C. S. Richmond and E. Bourke (1999). The UK Ocean Circulation and Advanced Modelling Project (OCCAM). In *High Performance Computing*, Proceeding of HPCI Conference 1998, Manchester, 12 –14 January 1998, editors R.J. Allan, M.F. Guest, A.D. Simpson, D.S. Henty and D.A. Nicole, Plenum Press, pp. 325–335.
681. Machu, E. and V. Garçon (2001). Phytoplankton seasonal distribution from SeaWiFs data in the Agulhas Current system. *Journal of Marine Research*, **59**(5): 795–812.
682. Machu, E., A. Biastoch, A. Oschlies, M. Kawamiya, J.R.E. Lutjeharms and V. Garçon (2005). Phytoplankton distribution in the Agulhas system from a coupled physical–biological model. *Deep-Sea Research I*, in press.
683. Machu, E., B. Ferret and V. Garçon (1999). Phytoplankton pigment distribution from SeaWiFS data in the subtropical convergence zone south of Africa: a wavelet analysis. *Geophysical Research Letters*, **26**(10): 1469–1472.
684. Lentini, C.A.D., D.B. Olson and G.P. Podestá (2002). Statistics of Brazil Current rings observed from AVHRR: 1993 to 1998. *Geophysical Research Letters*, **29**(16): doi: 10.1029/2002GL015221.
685. Jochum, M., and P. Malanotte-Rizzoli (2003). On the generation of North Brazil Current rings. *Journal of Marine Research*, **61**(2): 147–173.
686. Goni, G.J. and W.E. Johns (2003). Synoptic study of warm rings in the North Brazil Current retroflection region using satellite altimetry. In *Interhemispheric Water Exchange in the Atlantic Ocean*, editors G. J. Goni and P. Malanotte-Rizzoli, Elsevier, pp. 335–356.
687. Garzoli, S.L., A. Ffield, W.E. Johns and Q. Yao (2004). North Brazil Current retroflection and transports. *Journal of Geophysical Research*, **109**(C01013): doi: 10.1029/2003JC001775.
688. Garzoli, S.L., A. Ffield and Q. Yao (2003). North Brazil Current rings and the variability in the latitude of retroflection. In *Interhemispheric Water Exchange in the Atlantic Ocean*, editors G. J. Goni and P. Malanotte-Rizzoli, Elsevier, pp. 357–373.
689. Llido, J., J. Sudre, I. Dadou and V. Garçon (2004). Variability of the biological front south of Africa from SeaWiFS and a coupled physical–biological model. *Journal of Marine Research*, **62**( ): 595–609.
690. Oschlies, A. and V. Garçon (1998). Eddy-enhancement of primary production in a model of the North Atlantic Ocean. *Nature*, **394**( ): 266–269.
691. Reason, C.J.C. (1998). Warm and cold events in the south-east Atlantic/southwest Indian Ocean region and potential impacts on circulation and rainfall over southern Africa. *Meteorology and Atmospheric Physics*, **69**(1–2): 49–65.
692. Reason, C.J.C. and J.R.E. Lutjeharms (2000). Modelling multidecadal variability in the South Indian Ocean region: local forcing or a near-global mode? *South African Journal of Science*, **96**(3): 127–135.
693. Reason, C.J.C. (2001). Subtropical Indian Ocean SST dipole events and southern African rainfall. *Geophysical Research Letters*, **28**(11): 2225–2227.
694. Lutjeharms, J.R.E., P.M.S. Monteiro, P.D. Tyson and D. Obura (2001). The oceans around southern Africa and regional effects of global change. *South African Journal of Science*, **97** (3/4): 119–130.
695. Majodina, M. and M.R. Jury (1996). Composite winter cyclones south of Africa: evolution during eastward transit over the Agulhas warm pool. *South African Journal of Marine Science*, **17**:241–252.
696. Reason, C.J.C. (2000). Multidecadal climate variabil-

ity in the subtropics/mid-latitudes of the Southern Hemisphere oceans. *Tellus*, **52A**(2): 203–223.

697. Reason, C.J.C. and H. Mulenga (1999). Relationship between South African rainfall and SST anomalies in the Southwest Indian Ocean. *International Journal of Climatology*, **19**(15): 1651–1673.
698. Reason, C.J.C. and C.R. Godfred-Spenning (1998). SST variability in the South Indian Ocean and associated circulation and rainfall patterns over southern Africa. *Meteorology and Atmospheric Physics*, **66**( ): 243–258.
699. Reason, C.J.C. (2002). The wet winter of 2001 over the southwestern Cape, South Africa: potential large-scale influences. *South African Journal of Science*, **98**(5/6): 307–310.
700. Rouault, M. and A.M. Lee-Thorp (1996). Fine-time resolution measurements of atmospheric boundary layer properties between Cape Town and Marion Island. *South African Journal of Marine Science*, **17**: 281–296.
701. Jury, M., M. Rouault, S. Weeks and M. Schormann (1997). Atmospheric boundary-layer fluxes and structure across a land–sea transition zone in south-eastern Africa. *Boundary-Layer Meteorology*, **83**(2): 311–330.
702. Shannon, L.V. (2001). The Benguela Current. In *Encyclopedia of Ocean Science*, editors: J. Steele, S. Thorpe and K. Turekian, Academic Press, London, volume 1, pp. 255–267.
703. Howard, W.R. and W.L. Prell (1992). Late quaternary surface circulation of the southern Indian Ocean and its relationship to orbital variation. *Paleoceanography*, **7**(1); 79–117.
704. New, A.L., K. Stansfield, D. Smyth-Wright, D.A. Smeed, A.L. Evans and S.G. Alderson (2005). Physical and biochemical aspects of the flow across the Mascarene Plateau in the Indian Ocean. *Philosophical Transactions the Royal Society of London*, **A363** (1826): 151–168.
705. Fieux, M. and J.C. Swallow (1988). Flow of deep water into the Somali Basin. *Deep-Sea Research*, **35**( ): 303–309.
706. Barton, E.D. and A.E. Hill (1989). Abyssal flow through the Amirante Trench (western Indian Ocean). *Deep-Sea Research*, **36**(7): 1121–1126.
707. Arhan, M., H. Mercier and Y.-H. Park (2003). On the deep water circulation of the eastern South Atlantic Ocean. *Deep-Sea Research I*, **50**(7): 889–916.
708. Donohue, K., and J. Toole (2003). A near-synoptic survey of the southwest Indian Ocean. *Deep-Sea Research II*, **50**(12–13): 1893–1931.
709. Beal, L.M., A. Ffield and A.L. Gordon (2000). Spreading of Red Sea overflow waters in the Indian Ocean. *Journal of Geophysical Research*, **105**(C4): 8549–8564.
710. Reid, J.R. (2003). On the total geostrophic circulation of the Indian Ocean: flow patterns, tracers and transports. *Progress in Oceanography*, **56**(1): 137–186.
711. Reid, J.R. (1996). On the circulation of the South Atlantic Ocean. In *The South Atlantic: Present and Past Circulation*, editors: G. Wefer, W.H. Berger, G. Siedler and D. Webb, Springer-Verlag, Berlin, pp. 13–44.
712. Witter, D.L., and A.L. Gordon (1999). Interannual variability of South Atlantic circulation from 4 years of TOPEX/POSEIDON satellite altimeter observations. *Journal of Geophysical Research*, **104**(C9): 20,927–20,948.
713. Wunsch, C. and D. Stammer (1995). The global frequency-wavenumber spectrum of oceanic variability estimated from TOPEX/POSEIDON altimetric measurements. *Journal of Geophysical Research*, **100**(C12): 24,895–24,910.
714. Johnson, G.C., D.L. Musgrave, B.A. Warren, A. Ffield and D.B. Olson (1998). Flow of bottom and deep water in the Amirante Passage and Mascarene Basin. *Journal of Geophysical Research*, **103**(C13): 30,973–30,984.
715. Drijfhout, S.S., E. Maier-Reimer and U. Mikolajewicz (1996). Tracing the conveyer belt in the Hamburg large-scale geostrophic ocean general circulation model. *Journal of Geophysical Research*, **101**(C10): 22,563–22,575.
716. Van Aken, H.M., H. Ridderinkhof and W.P.M. de Ruijter (2004). North Atlantic Deep Water in the southwestern Indian Ocean. *Deep-Sea Research I*, **51**(6): 755–776.
717. Warren, B.A., T. Whitworth and J.H. LaCasce (2002). Forced resonant undulations in the deep Mascarene Basin. *Deep-Sea Research II*, **49**(7–8): 1513–1526.
718. Park, Y.-H., E. Charriaud and M. Fieux (1998). Thermohaline structure of the Antarctic Surface Water/Winter Water in the Indian sector of the Southern Ocean. *Journal of Marine Systems*, **17**(1–4): 5–23.
719. You, Y. (1997). Seasonal variations of thermocline circulation and ventilation in the Indian Ocean. *Journal of Geophysical Research*, **102**(C5): 10,391–10,422.
720. You, Y. (1998). Intermediate water circulation and ventilation of the Indian Ocean derived from water-mass contributions. *Journal of Marine Research*, **56**(5): 1029–1067.
721. You, Y. (1999). Dianeutral mixing, transformation and transport of the deep water of the Indian Ocean. *Deep-Sea Research I*, **46**(1): 109–148.
722. You, Y. (2000). Implications of the deep circulation and ventilation of the Indian Ocean on the renewal mechanism of North Atlantic Deep Water. *Journal of Geophysical Research*, **105**(C10): 23,895–23,926.
723. Matano, R.P., E.J. Beier, P.T. Strub and R. Tokmakian (2002). Large-scale forcing of the Agulhas variability: the seasonal cycle. *Journal of Physical Oceanography*, **32**(4): 1228–1241.
724. Fu, L.-L. and R.D. Smith (1996). Global ocean circulation from satellite altimetry and high-resolution computer simulation. *Bulletin of the American Meteorological Society*, **77**(11): 2625–2636.
725. Ganachaud, A., C. Wunsch, J. Marotzke and J. Toole

(2000). Meridional overturning and large-scale circulation of the Indian Ocean. *Journal of Geophysical Research*, **105**(C11): 26,117–26,134.

726. McCarthy, M.S. and L.D. Talley (1999). Three-dimensional isoneutral potential vorticity structure in the Indian Ocean. *Journal of Geophysical Research*, **104** (C6): 13,251–13,267.

727. Robbins, P.E. and J.M. Toole (1997). The dissolved silica budget as a constraint on the meridional overturning circulation of the Indian Ocean. *Deep-Sea Research I*, **44**( ): 879–906.

728. De Ruijter, W.P.M., H. Ridderinkhof, J.R.E. Lutjeharms, M.W. Schouten and C. Veth (2002). Observations of the flow in the Mozambique Channel. *Geophysical Research Letters*, **29**(10): 1401–1403.

729. Warren, B.A., H. Stommel and J.C. Swallow (1966). Water masses and patterns of flow in the Somali Basin during the southwest monsoon of 1964. *Deep-Sea Research*, **13**( ): 825–860.

730. DiMarco, S.F., P. Chapman, W.D. Nowlin, P. Hacker, K.A. Donohue, M. Luther, G.C. Johnson and J.M. Toole (2002). Volume transport and property distributions of the Mozambique Channel. *Deep-Sea Research I*, **49**(7–8): 1481–1511.

731. Nauw, J.J., H.M. van Aken, J.R.E. Lutjeharms and W.P.M. de Ruijter (2006). Intra-thermocline eddies in the southern Indian Ocean. *Journal of Geophysical Research*, in press.

732. Chapman, P., S.F. DiMarco, R.E. Davis and A.C. Coward (2003). Flow at intermediate depths round Madagascar based on ALACE float trajectories. *Deep-Sea Research II*, **50**(12–13): 1957–1986.

733. Ridderinkhof, H. and W.P.M. de Ruijter (2003). Moored current observations in the Mozambique Channel. *Deep-Sea Research II*, 50(12–13): 1933–1955.

734. Heywood, K.J. and Y.K. Somayajulu (1997). Eddy activity in the South Indian Ocean from ERS-1 altimetry. *Proceedings of the 3rd ERS Symposium on "Space at the Service of our Environment"*, Florence, March 1997, *ESA SP-414*, pp. 1479–1483.

735. Schouten, M.W., W.P.M. de Ruijter, P.J. van Leeuwen and H. Ridderinkhof (2003). Eddies and variability in the Mozambique Channel. *Deep-Sea Research II*, **50**(12–13): 1987–2003.

736. Schouten, M.W., W.P.M. de Ruijter, P.J. van Leeuwen and H. Dijkstra (2002). A teleconnection between the equatorial and southern Indian Ocean. *Geophysical Research Letters*, **29**(16), 1812, doi: 10.1029/2001GL014542.

737. Quadfasel, D.R. and J. C. Swallow (1986). Evidence for 50-day period planetary waves in the South Equatorial current of the Indian Ocean. *Deep-Sea Research*, **33**( ): 1307–1312.

738. Donohue, K.A., E. Firing and L. Beal (2000). Comparison of three velocity sections of the Agulhas Current and Agulhas Undercurrent. *Journal of Geophysical Research*, **105**(C12): 28,585–28,593.

739. Waseda, T., H. Mitsudera, B. Taguchi and Y. Yoshikawa (2003). On the eddy-Kuroshio interaction: Meander formation process. *Journal of Geophysical Research*, **108**(C7): 13-1 – 13-18, 3220, doi:10.1029/2002JC001583.

740. Bryden, H.L., L.M. Beal and L.M. Duncan (2005). Structure and transport of the Agulhas Current and its temporal variability. *Journal of Oceanography*, 61(3): 479–492.

741. Dijkstra, H.A. and W.P.M. de Ruijter (2001). Barotropic instabilities of the Agulhas Current system and their relation to ring formation. *Journal of Marine Research*, **59**(4): 517–533.

742. Bryden, H.L. and L.M. Beal (2001). Role of the Agulhas Current in Indian Ocean circulation and associated heat and freshwater fluxes. *Deep-Sea Research I*, **48**(8): 1821–1845.

743. Webb, D.J. (1999). An analytical model of the Agulhas Current as a western boundary current with linearly varying viscosity. *Journal of Physical Oceanography*, **29**(7): 1517–1527.

744. Van der Vaart, P.C.F and W.P.M. de Ruijter (2001). Stability of western boundary currents with an application to pulslike behaviour of the Agulhas Current. *Journal of Physical Oceanography*, **31**(9): 2625–2644.

745. Demarcq, H., R.G. Barlow and F.A. Shillington (2003). Climatology and variability of sea surface temperature and surface chlorophyll in the Benguela and Agulhas ecosystems as observed by satellite imagery. *African Journal of Marine Science*, **25**: 363–372.

746. Blanke, B., C. Roy, P. Penven. S. Speich, J. McWilliams and G. Nelson (2002). Linking wind and interannual upwelling variability in a regional model of the southern Benguela. *Geophysical Research Letters*, **29**(24): article number 2188.

747. Boebel, O., and C. Barron (2003). A comparison of in-situ float velocities with altimeter derived geostrophic velocities. *Deep-Sea Research II*, **50**(1): 119–139.

748. Okkonen, S.R., T.J. Weingartner, S.L. Danielson and D.L. Musgrave (2003). Satellite and hydrographic observations of eddy-induced shelf-slope exchange in the northwestern Gulf of Alaska. *Journal of Geophysical Research*, **108**(C2): 3033, doi:10.1029/2002JC001342.

749. Snaith, H.M. and I.S. Robinson (1996). A study of currents south of Africa using Geosat satellite altimetry. *Journal of Geophysical Research*, **101**(C8): 18,141–18,154.

750. Boebel, O., T. Rossby, J. Lutjeharms, W. Zenk and C. Barron (2003). Path and variability of the Agulhas Return Current. *Deep-Sea Research II*, **50**(1): 35–56.

751. Read, J.F., M.I. Lucas, S.E. Holley and R.T. Pollard (2000). Phytoplankton, nutrients and hydrography in the frontal zone between the Southwest Indian Subtropical gyre and the Southern Ocean. *Deep-Sea Research I*, **47**(12): 2341–2368.

752. Fu, L.-L. (1996). The circulation and its variability of the South Atlantic Ocean: first results from the TOPEX/

POSEIDON mission. In *The South Atlantic: Present and Past Circulation*, editors: G. Wefer, W.H. Berger, G. Siedler and D. Webb, Springer-Verlag, Berlin, pp. 63–82.
753. Holfort, J. and G. Siedler (2001). The meridional oceanic transports of heat and nutrients in the South Atlantic. *Journal of Physical Oceanography*, **31**(1): 5–29.
754. Garnier, E., J. Verron and B. Barnier (2003). Variability of the South Atlantic upper ocean circulation: a data assimilation experiment with 5 years of TOPEX/POSEIDON altimeter observations. *International Journal of Remote Sensing*, **24**(5): 911–934.
755. Feron, R.C.V. (1995). The Southern Ocean western boundary currents: comparison of Fine Resolution Antarctic Model results with GEOSAT altimeter data. *Journal of Geophysical Research*, **100**(C10): 4959–4975.
756. Donners, J. and S. Drijfhout (2004). The Lagrangian view of South Atlantic interocean exchange in a global ocean model compared with inverse model results. *Journal of Physical Oceanography*, **34**( ): 1019–1035.
757. Donners, J., S.S. Drijfhout and A.C. Coward (2004). Impact of cooling on the water mass exchange of Agulhas rings in a high-resolution ocean model. *Geophysical Research Letters*, **31**: L16312, doi: 10.1029/2004GL020644.
758. Marchesiello, P., B. Barnier and A.P. de Miranda (1998). A sigma-coordinate primitive equation model for studying the circulation in the South Atlantic. Part II: Meridional transport and seasonal variability. *Deep-Sea Research I*, **45**(4–5): 573–608.
759. Mercier, H., M. Arhan and J.R.E. Lutjeharms (2003). Upper-layer circulation in the eastern Equatorial and South Atlantic Oceans in January–March 1995. *Deep-Sea Research I*, **50**(7): 863–887.
760. Treguier, A.M., O. Boebel, B. Barnier and G. Madec (2003). Agulhas eddy fluxes in a 1/6° Atlantic model. *Deep-Sea Research II*, **50**(1): 251–280.
761. Treguier, A.M., B. Barnier, A.P. de Miranda, J.M. Molines, N. Grima, M. Imbard, G. Madec, C. Messager, Y. Reynaud and S. Michel (2001). An eddy-permitting model of the Atlantic circulation: Evaluating open boundary conditions. *Journal of Geophysical Research*, **106**(C10): 22,115–22,129.
762. Penduff, T., P. Brasseur, C.-E. Testut, B. Barnier and J. Verron (2002). A four-year eddy-permitting assimilation of sea-surface temperature and altimetric data in the South Atlantic Ocean. *Journal of Marine Research*, **60**(6): 805–833.
763. Penduff, T., B. Barnier, K. Beranger and T. Verron (2001). Comparison of near-surface mean and eddy flows from two numerical models of the South Atlantic Ocean. *Journal of Geophysical Research*, **106**(C8): 16,857–16,867.
764. Barnier, B., P. Marchesielo, A. Pimenta de Miranda, J.-M. Molines and M. Coulibaly (1998). A sigma-coordinate primitive equaton model for studying the circulation in the South Atlantic. Part I: Model configuration with error estimates. *Deep-Sea Research I*, **45**(4–5): 573–608.
765. Skogen, M.D. (1999). A biophysical model applied to the Benguela upwelling system. *South African Journal of Marine Science*, **21**: 235–249.
766. Drijfhout, S.S., C.A. Katsman, L. de Steur, P.C.F. van der Vaart, P.J. van Leeuwen and C. Veth (2003). Modeling the initial, fast sea surface height decay of Agulhas ring "Astrid". *Deep-Sea Research II*, **50**(1): 299–319.
767. Katsman, C.A., P.C.F. van der Vaart, H.A. Dijkstra and W.P.M. de Ruijter (2003). Stability of multilayer ocean vortices: a parameter study including realistic Gulf Stream and Agulhas Rings. *Journal of Physical Oceanography*, **33**(6): 1197–1218.
768. Drijfhout, S.S. (2003). Why anticyclones can split. *Journal of Physical Oceanography*, **33**(8): 1579–1591.
769. Herbette, S., Y. Morel and M. Arhan (2003). Erosion of a surface vortex by a seamount. *Journal of Physical Oceanography*, **33**(8): 1664–1679.
770. De Steur, L., P.J. van Leeuwen and S.S. Drijfhout (2004). Tracer leakage from modeled Agulhas rings. *Journal of Physical Oceanography*, **34**(6): 1387–1399.
771. Stern, M.E. (2000). Scattering of an eddy advected by a current towards a topographic obstacle. *Journal of Fluid Mechanics*, **402**: 211–223.
772. Strub, P.T., F.A. Shillington, C. James and S.J. Weeks (1998). Satellite comparison of the seasonal circulation in the Benguela and California Current systems. In *Benguela Dynamics*, editors S.C. Pillar, C.L. Moloney, A.I.L. Payne and F.A. Shillington, *South African Journal of Marine Science*, **19**: 99–112.
773. Ganachaud, A. and C. Wunsch (2003). Large-scale ocean heat and freshwater transports during the World Ocean Circulation Experiment. *Journal of Climate*, **16**(2): 696–705.
774. You, Y.Z. (2002). Quantitative estimate of Antarctic Intermediate Water contributions from the Drake Passage and the southwest Indian Ocean to the South Atlantic. *Journal of Geophysical Research*, **107**(C4): 10.1029/2001JC000880.
775. You, Y., J.R.E. Lutjeharms, O. Boebel and W.P.M. de Ruijter (2003). Quantification of interocean exchange of intermediate water masses around southern Africa. *Deep-Sea Research II*, **50**(1): 197–228.
776. Matano, R.P. and E.J. Beier (2003). A kinematic analysis of the Indian/Atlantic interocean exchange. *Deep-Sea Research II*, **50**(1): 229–249.
777. Reason, C.J.C., J.R.E. Lutjeharms, J. Hermes, A. Biastoch and R.E. Roman (2003). Inter-ocean fluxes south of Africa in an eddy-permitting model. *Deep-Sea Research II*, **50**(1): 281–298.
778. Dijkstra, H.A. and W.P.M. de Ruijter (2001). On the physics of the Agulhas: steady retroflection regimes. *Journal of Physical Oceanography*, **31**(10): 2971–2985.

779. Stramma, L. (2001). Current systems of the Atlantic Ocean. In *Encyclopedia of Ocean Science*, editors: J. Steele, S. Thorpe and K. Turekian, Academic Press, London, pp. 589–598.

780. Anonymous (1861). De Agulhas-stroom, afgeleid uit de temperatuur van het zeewater aan de oppervlakte en de invloed, dien hij op de atmosphere uitoefent. In *Onderzoekingen met den Zeethermometer, als uitkomsten van wetenschap en ervaring, aangaande de winden en zeestromingen in sommige gedeelten van den ocean*, Koninklijk Nederlandsch Meteorologisch Insituut te Utrecht, Kemink & Zoon, Utrecht, pp. 51–117.

781. Richardson, P.L. and S.L. Garzoli (2003). Characteristics of intermediate water flows in the Benguela Current as measured with RAFOS floats. *Deep-Sea Research II*, **50**(1): 87–118.

782. Garzoli, S.L. and J. Goni [sic] (2000). Combining altimeter observations and oceanographic data for ocean circulation and climate studies. In *Satellites, Oceanography and Society*, editor D. Halpern, Elsevier Oceanography Series 63, Elsevier, Amsterdam, pp. 79–97.

783. Garzoli, S.L., P.L. Richardson, C.M. Duncombe Rae, D.M. Fratantoni, G.J. Goñi and A.J. Roubicek (1999). Three Agulhas rings observed during the Benguela Current Experiment. *Journal of Geophysical Research*, **104**(C9): 20,971–20,985.

784 Schmid, C., O. Boebel, W. Zenk, J.R.E. Lutjeharms, S. Garzoli, P.L. Richardson and C. Barron (2003). Early evolution of an Agulhas ring. *Deep-Sea Research II*, **50**(1): 141–166.

785. Van Aken, H., A.K. van Veldhoven, C. Veth, W.P.M. de Ruijter, P.J. van Leeuwen, S.S. Drijfhout, C.P. Whittle and M. Rouault (2003). Observations of a young Agulhas ring, Astrid, during MARE, the Mixing of Agulhas Rings Experiment, in March 2000. *Deep-Sea Research II*, **50**(1): 167–195.

786. Nof, D. (1990). The role of angular momentum in the splitting of isolated eddies. *Tellus*, **42A**(4): 469–481.

787. Lutjeharms, J.R.E. (1987). Die Subtropiese Konvergensie en Agulhasretrofleksievaart (SCARC). *South African Journal of Science*, **83**(8): 454–456.

788. Bryden, H.L. *et al.* (1995). Agulhas Current Experiment. RRS Discovery Cruise 214, 26 Feb–09 Mar 1995. Institute of Oceanographic Sciences, Deacon Laboratory, Wormley, *Cruise Report* **249**, 85 pp.

789. Angot, M. and M. Ménaché (1963). Premiéres données hydrologiques sur la région voisine de Nosy Bé (nord-ouest de Madagascar). *Cahiers O.R.S.T.O.M., série Océanographie*, **3**: 7–15.

790. Angot, M. and R. Gerard (1966). Charactères hydrologiques de l'eau de surface au Centre ORSTOM de Nosy-Bé de 1962 à 1965). *Cahiers O.R.S.T.O.M.*, série *Océanographie*, **4**(3): 37–54.

791. Beckley, L.E. (1998). The Agulhas Current ecosystem with particular reference to dispersal of fish larvae. In *Large Marine Ecosystems of the Indian Ocean*, editors K. Sherman, E. N. Okemwa and M. J. Ntiba, Blackwell Science, Oxford, pp. 255–276.

792. Carter, R.A., H.F. McMurray and J.L. Largier (1987). Thermocline characteristics and phytoplankton dynamics in Agulhas Bank waters. *South African Journal of Marine Science*, **5**: 327–336.

793. De Villiers, S. (1998). Seasonal and interannual variability in phytoplankton biomass on the southern African continental shelf: evidence from satellite-derived pigment concentrations. *South African Journal of Marine Science*, **19**: 169–179.

794. Gerard, R. (1964). Étude de l'eau de mer de surface dans une baie de Nosy-Bé. *Cahiers O.R.S.T.O.M., série Océanographie*, **2**(2): 5–26.

795. Fowler, J.L. and A.J. Boyd (1998). Transport of anchovy and sardine eggs and larvae from the western Agulhas Bank to the west coast during the 1993/94 and 1994/95 spawning seasons. *South African Journal of Marine Science*, **19**: 181–195.

796. Roy, C., S. Weeks, M. Rouault, G. Nelson, R. Barlow and C. van der Lingen (2001). Extreme oceanographic events recorded in the Southern Benguela during the 1999–2000 summer season. *South African Journal of Science*, **97**(11/12): 465–471.

797. Hutchings, L., 1994. The Agulhas Bank: a synthesis of available information and a brief comparison with other east-coast shelf regions. *South African Journal of Science*, **90**(3): 179–185.

798. Boyd, A.J., J. Tauton-Clark and G.P.J. Oberholster, 1992. Spatial features of the near-surface and midwater circulation patterns off western and southern South Africa and their role in the life histories of various commercially fished species. *South African Journal of Marine Science*, **12**: 189–206.

799. Roel, B.A., J. Hewitson, S. Kerstan and I. Hampton (1994). The role of the Agulhas Bank in the life cycle of pelagic fish. *South African Journal of Science*, **90**(3): 185–196.

800. Olivar, M. P. and L. E. Beckley (1994). Influence of the Agulhas Current on the distribution of Laternfish larvae off the south-east coast of Africa. *Journal of Plankton Research*, **16**(12): 1759–1780.

801. Oliff, W.D. (1973). Chemistry and productivity at Richard's Bay. *NPRL Oceanography Division, Contract Report*, **CFIS 37B**, Durban, South Africa.

802. Lutjeharms, J.R.E. (2001). The Agulhas Current. In *Encyclopedia of Ocean Science*, editors: J. Steele, S. Thorpe and K. Turekian, Academic Press, London, volume 1, pp. 104–113.

803. Lutjeharms, J.R.E. (2006). The coastal oceans of south-eastern Africa. In *The Sea*, Volume 14, editors: A.R. Robinson, K. Brink, J. Hall, G. Ingram, D. Mackas and D. Townsend, pp. 781–832.

804. Verheye, H.M., L. Hutchings, J.A. Huggett, R.A. Carter, W.T. Peterson and S.J. Painting (1994). Community structure, distribution and tropic ecology of zooplankton on the Agulhas Bank with special reference to copepods. *South African Journal of Science*,

90(3): 154–165.

805. Hunter, I. T. (1988). Climate and weather off Natal. In *Coastal Ocean Studies off Natal, South Africa,* editor E. H. Schumann, Lecture Notes on Coastal and Estuarine Studies 26, Springer-Verlag, Berlin, pp. 81–100.

806. Tyson, P.D. (1986). *Climate Change and Variability in Southern Africa.* Oxford University Press, Cape Town, 220 pp.

807. Stramma, L. and M. England (1999). On the water masses and mean circulation of the South Atlantic Ocean. *Journal of Geophysical Research,* **104**(C9): 20,863–20,883.

808. Li, X., W. Zheng, W.G. Pichel, C.-Z Zou, P. Clemente-Colón and K.S. Friedman (2004). A cloud line over the Gulf Stream. *Geophysical Research Letters,* **31**( ): doi:10.1029/2004GL019892.

809. Nehring, D., E. Hagen, A. Jorge da Silva, R. Schemainda, G. Wolf, N. Michelsen, W. Kaiser, L. Postel, F. Gosselck, U. Brenning, E. Kuhner, G. Arlt, H. Siegel, L. Gohs and G. Bublitz (1987). Results of oceanological studies in the Mozambique Channel in February–March 1980. *Beiträge zur Meereskunde,* **56**: 51–63.

810. Anonymous (1984). The oceanological conditions in the western part of the Mocambique [sic] Channel in February–March 1980. *Geodätische und geophysikalische Veröffentlichungen,* **IV**, 163 pp.811. Tripp, R.T. (~1967). *An Atlas of Coastal Surface Drifts; Cape Town to Durban.* South African Oceanographic Data Centre, Department of Oceanography, University of Cape Town, Rondebosch, South Africa, 12 pp.

812. Harris, T.F.W. (1978). Review of coastal currents in Southern African waters. *South African National Science Programmes Report, CSIR Report,* **30**, vii+103 pp.

813. Gordon, A.L. (2003). The brawniest retroflection. *Nature,* **421**: 904–905.

814. Gründlingh, M.L., J.R.E. Lutjeharms, G. Brundrit, J.J. Agenbag, A. Boyd, G. Nelson, A Johnson, M. Roberts, L. Shannon and E. H. Schumann (1999). Physical oceanographic activities in South Africa: 1995–1998. *South African Journal of Science,* **95**(6/7): 281–288.

815. Schumann, E.H., W.K. Illenberger and W.S. Goschen (1991). Surface winds over Algoa Bay. *South African Journal of Science,* **87**(5): 202–207.

816. Lutjeharms, J.R.E. and H.R. Valentine (1983). Die fisiese oseanologie van die Agulhasbank. Deel II: Vaart 185 van die N. S. Thomas B. Davie. *CSIR Research Report,* **557**, 15 pp.

817. Landman, W.A. and S.J. Mason (1999). Change in the association between Indian Ocean sea-surface temperatures and summer rainfall over South Africa and Namibia. *International Journal of Climatology,* **19**(13): 1477–1492.

818. Mason, S. J., J.A. Lindesay and P.D. Tyson (1994). Simulating drought in southern Africa using sea-surface temperature-variations. *Water SA,* **20**(1): 15–22.

819. Park, Y.-H., E. Charriaud, P. Craneguy and A. Kartavtseff (2001). Fronts, transport, and Weddell Gyre at 30°E between Africa and Antarctica. *Journal of Geophysical Research,* **106**(C2): 2857–2879.

820. Martin, J., P. Guibout, M. Crepon and J.-C. Lizaray (1965). Circulation superficielle dans l'ocean Indien. Résultats de mesures faites à électrodes remorguées G.E.K. entre 1955–1963. *Cahiers Oceanographie,* **17**, suppl. 3, 221–241.

821. Donguy, J.-R. and B. Piton (1991). The Mozambique Channel revisited. *Oceanologica Acta,* **14**(6): 549–558.

822. Jorge da Silva, A. (1984). Hydrology and fish distribution at the Sofala Bank (Mozambique). *Revista de Investigação Pesquira,* **12**, Instituto de Investigação Pesqueiro, Moçambique, pp. 5–36.

823. Sætre, R. and R. de Paula e Silva (1979). The marine fish resources of Mozambique. *Reports on surveys with the R/V Dr Fridtjof Nansen,* Serviço de Investigações Pesqueiras, Maputo, Institute of Marine Research, Bergen, 179 pp.

824. Brinca, L., A. Jorge da Silva, L. Sousa, I.M. Sousa and R. Sætre (1983). A survey of the fish resources at Sofala Bank, Mozambique. *Reports on Surveys with the R/V Dr Fridtjof Nansen,* Serviço de Investigações Pesqueiras, Maputo, Institute of Marine Research, Bergen, 70+c15 pp.

825. Gould, W.J. (2005). From Swallow floats to Argo – the development of neutrally buoyant floats. *Deep-Sea Research II,* **52**(3–4): 529–543.

826. Moore, J.K., M.R. Abbot and J.G. Richman (1999). Location and dynamics of the Antarctic Circumpolar Front from satellite sea surface temperature data. *Journal of Geophysical Research,* **104**(C): 3059–3073.

Bibliography

This Bibliography is not merely an alphabetical representation of the References section. Instead, its purpose is to present a comprehensive list of all significant publications on aspects of the greater Agulhas Current system to date. This includes publications on the biology, chemistry, geology, palaeoceanography and meteorology of the region where it is deemed that such publications make an important contribution to a better understanding of the system. It also includes cruise, data and contract reports which contain valuable information that can not be found elsewhere. Articles of a historical interest are incorporated, but papers on more wide-ranging matters, such as global circulation patterns, that make no explicit reference to the Agulhas Current system, are usually not. Selection has thus consistently been based on the contribution made to a better understanding of specifically the Agulhas Current system.

Abrams, R.W. and J.R.E. Lutjeharms (1986). Relationship between seabirds and meso-scale hydrographic features in the Agulhas Current retroflection region. In *Acta XIX Congressus Internationalis Ornithologici* Vol. I, Ottawa, Canada, pp. 991–996.

Allanson, B.R., R.C. Hart and J.R.E. Lutjeharms (1981). Observations on the nutrients, chlorophyll and primary production of the Southern Ocean south of Africa. *South African Journal of Antarctic Research*, **10/11**: 3–14.

Anderson, F.P. (1967). Time-variations in the Agulhas Current near Durban. National Physical Research Laboratory, Council for Scientific and Industrial Research, *internal publication* **IG 67/8**, 13 pp.

Anderson, F.P. (1972). Circulation patterns along the east coast (coastal zone and open ocean) in relation to pollution. *South African Journal of Science*, **68**(5): 121–123.

Anderson, F.P., M.L. Gründlingh and C.C. Stavropoulos (1988). Kinematics of the southern Natal coastal circulation: some historic measurements 1962–63. *South African Journal of Science*, **84**(10): 857–860.

Andrau, K.F.R. (1859). De stormen op het Kaapse rif. In *Maandelikse Zeilaanwijzingen van Java naar het Kanaal*, Koninklijk Nederlandsch Meteorologisch Instituut, Utrecht, pp. 25–56.

Angot, M. and M. Ménaché (1963). Premiéres données hydrologiques sur la région voisine de Nosy Bé (nord-ouest de Madagascar). *Cahiers O.R.S.T.O.M., série Océanographie*, **3**: 7–15.

Angot, M. and R. Gerard (1966). Charactères hydrologiques de l'eau de surface au Centre ORSTOM de Nosy-Bé de 1962 à 1965). *Cahiers O.R.S.T.O.M.*, série *Océanographie*, **4**(3): 37–54.

Anonymous (1965). Development of the International Indian Ocean Expedition. *International Indian Ocean Expedition, Collected Reprints*, **I**: ix–xi.

Anonymous (1970). Agulhas. In *Standard Encyclopaedia of Southern Africa*, editors D.J. Potgieter and P.C. du Plessis, Vol.1, Nasou, Cape Town, p. 254.

Anonymous (1983). Cruise report R/V "Dr. Fridtjof Nansen", Fisheries Resources Survey, Madagascar, 16–28 June 1983. *Reports on Surveys with the R/V Dr Fridtjof Nansen*, Institute of Marine Research, Bergen, 9 pp.

Anonymous (1984). The oceanological conditions in the western part of the Mocambique [sic] Channel in February–March 1980. *Geodätische und geophysikalische Veröffentlichungen*, **IV**, 163 pp.

Arhan, M., H. Mercier and J.R.E. Lutjeharms (1999). The disparate evolution of three Agulhas rings in the South Atlantic Ocean. *Journal of Geophysical Research*, **104** (C9): 20,987–21,005.

Augustyn, C.J., M.R. Lipinski, W.H.H. Sauer, M.J. Roberts and B.A. Mitchell-Innes (1994). Chokka squid on the Agulhas Bank: life history and ecology. *South African Journal of Science*, **90**(3): 143–154.

Bang, N.D. (1970). Dynamic interpretations of a detailed surface temperature chart of the Agulhas Current retroflexion [sic] and fragmentation area. *South African Geographical Journal*, **52**(12): 67–76.

Bang, N.D. (1970). Major eddies and frontal structures in the Agulhas Current retroflexion [sic] area in March, 1969. In *Symposium "Oceanography South Africa 1970"*, South African National Committee for Oceanographic Research, 16 pp.

Bang, N.D. (1972). Logic intensive versus data intensive methods in oceanographic planning, analysis and prediction. In *Proceedings of 'Conference 72'*, South African Geophysical Society, Johannesburg.

Bang, N.D. (1973). Oceanography: Oceanographic Environment of Southern Africa. In *Standard Encyclopaedia of Southern Africa*, editors J.J. Spies and P.C. du Plessis, Vol. 8, Nasou, Cape Town, pp. 282–286.

Bang, N.D. and W.R.H. Andrews (1974). Direct current measurements of a shelf-edge frontal jet in the southern Benguela system. *Journal of Marine Research*, **32**(3):

405–417.

Barange, M. (1994). Acoustic identification, classification and structure of biological patchiness on the edge of the Agulhas Bank and its relation to frontal features. *South African Journal of Marine Science*, **14**: 333–347.

Barange, M., E.A. Pakhomov, R. Perissinotto, P.W. Froneman, H.M. Verheye, J. Taunton-Clark and M.I. Lucas (1998). Pelagic community structure of the subtropical convergence region south of Africa and in the mid-Atlantic Ocean. *Deep-Sea Research I*, **45**(10): 1663–1687.

Barlow, E.W. (1931). Currents in the western portion of the Indian Ocean. Introduction. *Marine Observer*, **8**(90): 130–132.

Barlow, E.W. (1931). Currents in the western portion of the Indian Ocean. II – Currents during the S.W. Monsoon period. *Marine Observer*, **8**(93): 193–194.

Barlow, E.W. (1931). Currents in the western portion of the Indian Ocean. III – Currents during the N.E. Monsoon period (northern winter and southern summer) and general summary. *Marine Observer*, **8**(96): 254–259.

B[arlow], E.W. (1932). Major James Rennell. *Marine Observer*, **9**(101): 94–98.

Barlow, E.W. (1933). Currents in the southern Indian Ocean, winter season. *Marine Observer*, **10**(111): 99.

Barlow, E.W. (1933). Currents in the southern Indian Ocean, summer season and general summary. *Marine Observer*, **10**(112): 132–135.

Barlow, E.W. (1935). The 1910 to 1935 survey of the currents of the Indian Ocean and China Seas. *Marine Observer*, **12**(120): 153–163.

Barnier, B., P. Marchesielo, A. Pimenta de Miranda, J.-M. Molines and M. Coulibaly (1998). A sigma-coordinate primitive equaton model for studying the circulation in the South Atlantic. Part I: Model configuration with error estimates. *Deep-Sea Research I*, **45**(4–5): 573–608.

Barton, E.D. and A.E. Hill (1989). Abyssal flow through the Amirante Trench (western Indian Ocean). *Deep-Sea Research*, **36**(7): 1121–1126.

Basto, A. de M. (1945). *Diario da viagem de Vasco da Gama*. Vol. 1. Editora, Porto, 150 pp.

Bé, A.W.H. and J.C. Duplessy (1976). Subtropical convergence fluctuations and quaternary climates in the middle latitudes of the Indian ocean. *Science*, **194**(4263): 419–422.

Beal, L.M. (1997). Observations of the velocity structure of the Agulhas Current. Ph. D. thesis, University of Southampton, 158 pp.

Beal, L.M. and H.L. Bryden (1997). Observations of an Agulhas Undercurrent. *Deep-Sea Research I*, **44**(9–10): 1715–1724.

Beal, L.M. and H.L. Bryden (1999). The velocity and vorticity structure of the Agulhas Current at 32° S. *Journal of Geophysical Research*, **104**(C3): 5151–5176.

Beal, L.M., A. Ffield and A.L. Gordon (2000). Spreading of Red Sea overflow waters in the Indian Ocean. *Journal of Geophysical Research*, **105**(C4): 8549–8564.

Becker, R. (1938). Über den jährlichen Temperaturgang auf dem Indischen und Stillen Ozean. *Annalen der Hydrographie, Berlin*, **66**(7): 338–340.

Beckley, L.E. (1983). Sea-surface temperature variability around Cape Recife, South Africa. *South African Journal of Science*, **79**(11): 436–438.

Beckley, L.E. (1988). Spatial and temporal variability in sea temperature in Algoa Bay, South Africa. *South African Journal of Science*, **84**(1): 67–69.

Beckley, L.E. and R.C. van Ballegooyen (1992). Oceanographic conditions during three ichthyoplankton surveys of the Agulhas Current in 1990/91. *South African Journal of Marine Science*, **12**: 3–17.

Beckley, L.E. and J.D. Hewitson (1994). Distribution and abundance of clupeoid larvae along the east coast of South Africa in 1990/91. *South African Journal of Marine Science*, **14**: 205–212.

Beckley, L.E. (1998). The Agulhas Current ecosystem with particular reference to dispersal of fish larvae. In *Large Marine Ecosystems of the Indian Ocean*, editors K. Sherman, E. N. Okemwa and M. J. Ntiba, Blackwell Science, Oxford, pp. 255–276.

Beismann, J.-O., R.H. Käse and J.R.E. Lutjeharms (1999). On the influence of submarine ridges on translation and stability of Agulhas rings. *Journal of Geophysical Research*, **104**(C4): 7897–7906.

Belkin, I.M. and A.L. Gordon (1996). Southern Ocean fronts from the Greenwich meridian to Tasmania. *Journal of Geophysical Research*, **101**(C2): 3675–3696.

Bennett, S.L. (1988). *Where Three Oceans Meet: the Agulhas retroflection region*. Doctoral dissertation, Woods Hole Oceanographic Institution and Massachusetts Institute of Technology, WHOI-88-51, 367 pp.

Berghaus, H. (1867). Welt-karte zur Übersicht der Meeres-Strömungen und das Schnellverkehrs. In (1988) *Handatlas über alle Theile der Erde und über das Weltgebäude*, editor A. Stieler, Gotha, 95 charts.

Berghaus, H. (1845). *Physikalischer Atlas*. Erster Band. Verlag J. Perthes, Gotha, 234 pp.

Biastoch, A. (1998). Zirkulation und Dynamik in der Agulhasregion anhand eines numerischen Modells. *Berichte aus dem Institut für Meereskunde an der Christian-Albrechts-Universität, Kiel*, **301**, 118 pp.

Biastoch, A. and W. Krauss (1999). The role of mesoscale eddies in the source regions of the Agulhas Current. *Journal of Physical Oceanography*, **29**(9): 2303–2317.

Biastoch, A., C.J.C. Reason, J.R.E. Lutjeharms and O. Boebel (1999). The importance of flow in the Mozambique Channel to seasonality in the greater Agulhas Current system. *Geophysical Research Letters*, **26**(21): 3321–3324.

Blanke, B., C. Roy, P. Penven, S. Speich, J. McWilliams and G. Nelson (2002). Linking wind and interannual upwelling variability in a regional model of the southern Benguela. *Geophysical Research Letters*, **29**(24): article number 2188.

Boddem, J. and R. Schlitzer (1995). Interocean exchange and

meridional mass and heat fluxes in the South Atlantic. *Journal of Geophysical Research*, **100**(C8): 15,821–15,834.

Boden, B.P., C.M. Duncombe Rae and J.R.E. Lutjeharms (1988). The distribution of the diatoms of the S. W. Indian Ocean waters between Cape Town and the Prince Edward Island archipelago. *South African Journal of Science*, **84**(10): 811–818.

Boebel, O., C. Duncombe Rae, S. Garzoli, J. Lutjeharms, P. Richardson, T. Rossby, C. Schmid and W. Zenk (1997). Float experiment studies interocean exchanges at the tip of Africa. *Eos, Transactions of the American Geophysical Union*, **79**(1): 1, 7–8.

Boebel, O., C. Schmid and M. Jochum (1998). Physical Oceanography. In *Die Expedition ANTARKTIS-XIV/4 mit FS "Polarstern" 1997*, editor K. Fütterer, *Berichte zur Polarforschung* **259**, pp. 6–20.

Boebel, O., S. Anderson-Fontana, C. Schmid, I. Ansorge. P. Lazarevich, J. Lutjeharms, M. Prater. T. Rossby, and W. Zenk (2000). KAPEX RAFOS Float Data Report 1997–1999. Part A: The Agulhas- and South Atlantic Current Components. *GSO Technical Report* 2000–2; *UCT Oceanography Report* 2000–1; *Berichte aus dem Institut für Meereskunde an der Christian-Albrechts-Universität – Kiel*, IFM 318, 194 pp.

Boebel, O. and C. Barron (2003). A comparison of in-situ float velocities with altimeter derived geostrophic velocities. *Deep-Sea Research II*, **50**(1): 119–139.

Boebel, O., J. Lutjeharms, C. Schmid, W. Zenk, T. Rossby and C. Barron (2003). The Cape Cauldron: a regime of turbulent inter-ocean exchange. *Deep-Sea Research II*, **50**(1): 57–86.

Boebel, O., T. Rossby, J. Lutjeharms, W. Zenk and C. Barron (2003). Path and variability of the Agulhas Return Current. *Deep-Sea Research II*, **50**(1): 35–56.

Boudra, D.B. and W.P.M. de Ruijter (1986). The wind-driven circulation in the South Atlantic-Indian Ocean – II. Experiments using a multi-layer numerical model. *Deep-Sea Research*, **33**(4): 447–482.

Boudra, D.B. and E.P. Chassignet (1988). Dynamics of Agulhas retroflection and ring formation in a numerical model. I. The vorticity balance. *Journal of Physical Oceanography*, **18**(2): 280–303.

Boudra, D.B., K.A. Maillet and E.P. Chassignet (1989). Numerical modeling of Agulhas retroflection and ring formation with isopycnal outcropping. In *Meso-scale/Synoptic Coherent Structures in Geophysical Turbulence*, editors J.C.J. Nihoul and B.M. Jamart, Elsevier, Amsterdam, pp. 315–335.

Boyd, A.J., B.B.S. Tromp and D.A. Horstman (1985). The hydrology of the South African south-western coast between Cape Point and Danger Point in 1975. *South African Journal of Marine Science*, **3**: 145–168.

Boyd, A.J., J. Taunton-Clark and G.P.J. Oberholster (1992). Spatial features of the near-surface and midwater circulation patterns off western and southern South Africa and their role in the life histories of various commercially fished species. *South African Journal of Marine Science*, **12**: 189–206.

Boyd, A.J. and F.A. Shillington (1994). Physical forcing and circulation patterns on the Agulhas Bank. *South African Journal of Science*, **90**(3): 114–122.

Bravo, M. (1993). James Rennell: antiquarian of ocean currents. *Ocean Challenge*, **4**(1/2): 41–50.

Brennecke, W. (1915). Ozeanographische Arbeiten S.M.S. Möwe im westlichen Indischen Ozean 1913. *Annalen der Hydrographie und Maritimen Meteorologie*, **43**(8): 337–343.

Brinca, L., F. Rey, C. Silva and R. Sætre (1981). A survey on the marine fish resources of Mozambique. *Reports on Surveys with the R/V Dr Fridtjof Nansen*, Instituto de Desenvolvimento Pesqueiro, Maputo, Institute of Marine Research, Bergen, 58 pp.

Brinca, L., A. Jorge da Silva, L. Sousa, I.M. Sousa and R. Sætre (1983). A survey of the fish resources at Sofala Bank, Mozambique. *Reports on Surveys with the R/V Dr Fridtjof Nansen*, Serviço de Investigações Pesqueiras, Maputo, Institute of Marine Research, Bergen, 70 + 15 pp.

Brown, A.C. (1997). John D. F. Gilchrist and the early years of marine science in South Africa. In *A Century of Marine Science in South Africa*, editors A.I.L. Payne and J.R.E. Lutjeharms, Sea Fisheries Research Institute, Royal Society of South Africa, pp. 2–16 and *Transactions of the Royal Society of South Africa*, **52**(1): 2–16.

Brundrit, G.B. and L.V. Shannon (1989). Cape storms and the Agulhas Current: a glimpse of the future? *South African Journal of Science*, **85**(10): 619–620.

Brundrit, G.B. and L. Krige (1994). Heard Island signals through the Agulhas retroflection region. *Journal of the Acoustical Society of America*, **96**(4): 2464–2468.

Bryantsev, V.A., S.V. Pavlukhin and A.S. Pelevin (1979). Characteristics of geostrophic currents and the Subtropical Convergence zone in the Southwestern Indian Ocean. *Oceanology*, **19**(4): 395–399.

Bryden, H.L. and L.M. Beal (2001). Role of the Agulhas Current in Indian Ocean circulation and associated heat and freshwater fluxes. *Deep-Sea Research I*, **48**(8): 1821–1845.

Bryden, H.L., L. M. Beal and L.M. Duncan (2004). Structure and transport of the Agulhas Current and its temporal variability. *Journal of Oceanography*, submitted.

Buchan, A. (1895). Report on oceanic circulations based on observations made on board H.M.S. Challenger and other observations. *Challenger Reports*, II (Physics and Chemistry). Part VIII, appendix to Summary, 38 pp.

Bunker, A.F. (1980). Trends of variables and energy fluxes over the Atlantic Ocean from 1948 to 1972. *Monthly Weather Review*, **108**(6): 720–732.

Bunker, A.F. (1988). Surface energy fluxes of the South Atlantic Ocean. *Monthly Weather Review*, **116**(4): 809–823.

Byrne, D.A., A.L. Gordon and W.F. Haxby (1995). Agulhas Eddies: a synoptic view using Geosat ERM data. *Journal of Physical Oceanography*, **25**(5): 902–917.

Camden-Smith, F., L.-A. Perrins, R.V. Dingle and G.B.

Brundrit (1981). A preliminary report on long-term bottom-current measurements and sediment transport/erosion in the Agulhas Passage, Southwest Indian Ocean. *Marine Geology*, **39**: M81–M88.

Carter, R.A., H.F. McMurray and J.L. Largier (1987). Thermocline characteristics and phytoplankton dynamics in Agulhas Bank waters. *South African Journal of Marine Science*, **5**: 327–336.

Carter, R. and J. d'Aubrey (1988). Inorganic nutrients in Natal continental shelf waters. In *Coastal Ocean Studies off Natal, South Africa*, editor E.H. Schumann, Lecture Notes on Coastal and Estuarine Studies 26, Springer-Verlag, Berlin, pp. 131–151.

Carter, R.A. and M.H. Schleyer (1988). Plankton distributions in Natal coastal waters. In *Coastal Ocean Studies off Natal, South Africa*, editor E.H. Schumann, Lecture Notes on Coastal and Estuarine Studies 26, Springer-Verlag, Berlin, pp. 152–177.

Catzel, R. and J.R.E. Lutjeharms (1987). Agulhas Current border phenomena along the Agulhas Bank south of Africa. *CSIR Research Report*, **635**, 22 pp. + 30 plates + addenda.

Chapman, P., C.M. Duncombe Rae and B.R. Allanson (1987). Nutrients, chlorophyll and oxygen relationships in the surface layers at the Agulhas retroflection. *Deep-Sea Research*, **34**(8): 1399–1416.

Chapman, P. (1988). On the occurrence of oxygen-depleted water south of Africa and its implications for Agulhas–Atlantic mixing. *South African Journal of Marine Science*, **7**: 267–294.

Chapman, P. and J.L. Largier (1989). On the origin of Agulhas Bank bottom water. *South African Journal of Science*, **85**(8): 515–519.

Chapman, P., S.F. DiMarco, R.E. Davis and A.C. Coward (2003). Flow at intermediate depths round Madagascar based on ALACE float trajectories. *Deep-Sea Research II*, **50**(12–13): 1957–1986.

Chassignet, E.P. and D.B. Boudra (1987). Dynamics of Agulhas retroflection and ring formation in a quasi-isopycnic coordinate numerical model. In *Three Dimensional Models of Marine and Estuarine Dynamics*, editors J.C.J. Nihoul and B.M. Jamart, Elsevier, Amsterdam, pp. 169–194.

Chassignet, E.P. and D.B. Boudra (1988). Dynamics of Agulhas retroflection and ring formation in a numerical model. II. Energetics and ring formation. *Journal of Physical Oceanography*, **18**(2): 304–319.

Chassignet, E.P., D.B. Olson and D.B. Boudra (1989). Evolution of rings in numerical models and observations. In *Mesoscale/Synoptic Coherent Structures in Geophysical Turbulence*, editors J.C.J. Nihoul and B.M. Jamart, Elsevier, Amsterdam, pp. 337–356.

Cheney, R.A., J.G. Marsh and B.D. Beckley (1983). Global mesoscale variability from collinear tracks of SEASAT altimeter data. *Journal of Geophysical Research*, **88**(C7): 4343–4354.

Christensen, M.S. (1980). Sea-surface temperature charts for Southern Africa, south of 26°S. *South African Journal of Science*, **76**(2): 541–546.

Chun, C. (1900). *Aus den Tiefen des Weltmeeres. Schilderungen von der Deutschen Tiefsee – Expedition*. Gustav Fisher, Jena, 549 pp.

Chun, C. (*et al.*) (1902–1940). *Wissenschaftliche Ergebnisse der Deutschen Tiefsee – Expedition auf dem Dampfer Valdivia, 1898–1899*. Gustaf Fischer, Jena, 24 volumes.

Clement, A.C. and A.L. Gordon (1995). The absolute velocity field of Agulhas eddies and the Benguela Current. *Journal of Geophysical Research*, **100**(C11): 22,591–22,601.

Clowes, A J. and G.E.R. Deacon (1935). The deep-water circulation of the Indian Ocean. *Nature*, **136**(3450): 936–938.

Clowes, A.J. (1950). An introduction to the hydrology of South African waters. Fisheries and Marine, Biological Survey Division, Union of South Africa, *Investigational Reports*, **12**: 42 pp.

Cockroft, V.G., V.M. Peddemors, P.G. Ryan and J.R.E. Lutjeharms (1990). Cetacean sightings in the Agulhas Retroflection, Agulhas Rings and Subtropical Convergence. *South African Journal of Antarctic Research*, **20**(2): 64–67.

Colborn, J.G. (1975). *The thermal structure of the Indian Ocean*. International Indian Ocean Expedition Oceanographic Monographs, Number 2, An East-west Center Book, The University Press of Hawaii, Honolulu, 173 pp.

Cooke, A., J.R.E. Lutjeharms and P. Vasseur (2004). Marine and coastal ecosystems. In *The Natural History of Madagascar*, editors: S. Goodman and J. P. Benstead, The University of Chicago Press, Chicago, pp. 179–209.

Crimp, S.J., J.R.E. Lutjeharms and S.J. Mason (1998). Sensitivity of a tropical-temperate trough to sea-surface temperature anomalies in the Agulhas retroflection region. *Water SA*, **24**(2): 93–101.

CSIR (1985). *A decade in oceanology. Activities of the National Research Institute for Oceanology, 1974–1984*. Council for Scientific and Industrial Research, Stellenbosch, CSIR Report, **T/SEA 8514**, 28 pp.

CSIR (1988). *CSIR. Review of activities in oceanology, 1985–1987*. Council for Scientific and Industrial Research, Stellenbosch, 40 pp.

Daniault, N. and Y. Menard (1985). Eddy kinetic energy distribution in the Southern Ocean from altimetry and FGGE drifting buoys. *Journal of Geophyscial Research*, **90**(C6): 18,877–11,889.

Darbyshire, J. (1964). A hydrological investigation of the Agulhas Current area. *Deep-Sea Research*, **11**(5): 781–815.

Darbyshire, J. (1972). The effect of bottom topography on the Agulhas Current. *Reviews of Pure and Applied Geophysics*, **101**(9): 208–220.

Darbyshire, M. (1966). Agulhas Current (and Mozambique Current). In *The Encyclopaedia of Oceanography,* editor R. W. Fairbridge, Reinhold Publishing Corporation, New York, pp. 23–28,

Darbyshire, M. (1966). The surface waters near the coast of Southern Africa. *Deep-Sea Research*, **13**(1): 57–81.

Deacon, G.E.R. (1933). A general account of the hydrology of the South Atlantic Ocean. *Discovery Report*, **7**: 171–238.

Deacon, G.E.R. (1937). The hydrology of the Southern Ocean. *Discovery Report*, **15**: 124 pp.

Deacon, G.E.R. (1960). The Indian Ocean Expedition. *Nature*, **187**(4737): 561–562.

Deacon, G.E.R. (1982). Physical and biological zonation in the Southern Ocean. *Deep-Sea Research*, **29**(1):1–15.

De Cuevas, B.A., G.B. Brundrit and A.M. Shipley (1986). Low-frequency sea-level fluctuations along the coasts of Namibia and South Africa. *Geophysical Journal of the Royal Astronomical Society*, **87**(1): 1–15.

De Decker, A.H.B. (1973). Agulhas Bank plankton. In *The Biology of the Indian Ocean*, editor B. Zeitzschel, Springer-Verlag, Berlin, pp. 189–219.

De Decker, A.H.B. (1984). Near-surface copepod distribution in the south-western Indian and south-eastern Atlantic Ocean. *Annals of the South African Museum*, **93**(5): 303–370.

Defant, A. (1932–1960). *Wissenschaftlichen Ergebnisse der Deutschen Atlantischen Expedition auf dem Forschung- und Vermessungschiff "Meteor" 1925–1927*. Bände I–XVI, Berlin.

Defant, A. (1936). Ausbreitung- und Vermischungsvorgänge im antarktischen Bodenstrom und im subantarktischen Zwischenwasser. In *Quantitative Untersuchungen zur Statik und Dynamik des Atlantischen Ozeans*, in *Wissenschaftlichen Ergebnisse der Deutschen Atlantischen Expedition auf dem Forschung- und Vermessungschiff "Meteor" 1925–1927*, **6**(2): 53–96.

Defant, A. (1941). Die absolute Topografie des physikalischen Meeresniveaus und der Druckflächen sowie die Wasserbewegung im Atlantischen Ozean. In *Quantitative Untersuchungen zur Statik und Dynamik des Atlantischen Ozeans*, in *Wissenschaftlichen Ergebnisse der Deutschen Atlantischen Expedition auf dem Forschung- und Vermessungschiff "Meteor" 1925–1927*, **6**(2): 191–260.

Defant, A. (1941). Die relative Topografie eizelner Druckflächen im Atlantischen Ozean. *'Meteor' Forschungsergebnisse*, **6**(2/4): 183–190.

Demarcq, H., R.G. Barlow and F.A. Shillington (2003). Climatology and variability of sea surface temperature and surface chlorophyll in the Benguela and Agulhas ecosystems as observed by satellite imagery. *African Journal of Marine Science*, **25**: 363–372.

De Ruijter, W. (1982). Asymptotic analysis of the Agulhas and Brazil Current systems. *Journal of Physical Oceanography*, **12**(4): 361–373.

De Ruijter, W.P.M. and D.B. Boudra (1985). The wind-driven circulation in the South Atlantic–Indian Ocean – I. Numerical experiments in a one-layer model. *Deep-Sea Research*, **32**(5): 557–574.

De Ruijter, W.P.M., A. Biastoch, S.S. Drijfhout, J.R.E. Lutjeharms, R.P. Matano, T. Pichevin, P.J. van Leeuwen and W. Weijer (1999). Indian-Atlantic inter-ocean exchange: dynamics, estimation and impact. *Journal of Geophysical Research*, **104**(C9): 20,885–20,911.

De Ruijter, W.P.M., P.J. van Leeuwen and J.R.E. Lutjeharms (1999). Generation and evolution of Natal Pulses: solitary meanders in the Agulhas Current. *Journal of Physical Oceanography*, **29**(12): 3043–3055.

De Ruijter, W.P.M., H. Ridderinkhof, J.R.E. Lutjeharms, M.W. Schouten and C. Veth (2002). Observations of the flow in the Mozambique Channel. *Geophysical Research Letters*, **29**(10): 1401–1403.

De Ruijter, W.P.M., H.M. van Aken, E.J. Beier, J.R.E. Lutjeharms, R.P. Matano and M.W. Schouten (2003). Eddies and dipoles around South Madagascar : formation, pathways and large-scale impact. *Deep-Sea Research I*, **51**(3): 383–400.

De Steur, L., P.J. van Leeuwen and S.S. Drijfhout (2004). Tracer leakage from modeled Agulhas rings. *Journal of Physical Oceanography*, **34**(6): 1387–1399.

Deutsches Hydrographisches Institut (1960). *Monatskarten für den Indischen Ozean*. Deutsches Hydrographisches Institut, Hamburg, Public. 2422.

De Villiers, S. (1998). Seasonal and interannual variability in phytoplankton biomass on the southern African continental shelf: evidence from satellite-derived pigment concentrations. In *Benguela Dynamics*, editors S.C. Pillar, C.L. Moloney, A.I.L. Payne and F.A. Shillington, *South African Journal of Marine Science*, **19**: 169–179.

Dietrich, G. (1935). Aufbau und Dynamik des südlichen Agulhasstromgebietes. *Veröffentlichungen des Institut für Meereskunde an der Universität Berlin*, n.f. **A**(27): 79 pp.

Dietrich, G. (1935). Zur Dynamik des Atlantischen Zweiges des Agulhasstromes. *Annalen der Hydrographie und Maritimen Meteorologie*, **63**: 383–387.

Dietrich, G. (1936). Aufbau und Bewegung von Golfstrom und Agulhasstrom, eine vergleichende Betrachtung. *Naturwissenschaften*, **24**(15): 225–230.

Dijkstra, H.A. and W.P.M. de Ruijter (2001). Barotropic instabilities of the Agulhas Current system and their relation to ring formation. *Journal of Marine Research*, **59**(4): 517–533.

Dijkstra, H.A. and W.P.M. de Ruijter (2001). On the physics of the Agulhas: steady retroflection regimes. *Journal of Physical Oceanography*, **31**(10): 2971–2985.

DiMarco, S.F., P. Chapman and W.D. Nowlin (2000). Satellite observations of upwelling on the continental shelf south of Madagascar. *Geophysical Research Letters*, **27**(24): 3965–3968.

DiMarco, S.F., P. Chapman, W.D. Nowlin, P. Hacker, K.A. Donohue, M. Luther, G.C. Johnson and J.M. Toole (2002). Volume transport and property distributions of the Mozambique Channel. *Deep-Sea Research I*, **49**(7–8): 1481–1511.

Dingle, R.V., G.V. Birch, J.M. Bremner, R.H. de Decker, A. du Plessis, J.A. Engelbrecht, M.J. Fincham, T. Fitton, B.W. Flemming, R.I. Gentle, S.H. Goodlad, A.K. Martin, E.G. Mills, G.J. Moir, R.J. Parker, S.H. Robson,

J. Rogers, D.A. Salmon, W.G. Sieser, E.S.W. Simpson, C.P. Summerhayes, F. Westall, A. Winter and M.W. Woodborne (1987). Deep-sea sedimentary environments around southern Africa (South-East Atlantic and South-West Indian Oceans). *Annals of the South African Museum*, **98**(1): 1–27.

Discovery Committee (1929). Discovery Investigations Station List, 1925–1927. *Discovery Reports*, **1**: 3–138.

Donguy, J.R. and B. Piton (1969). A perçu des conditions hydrologiques de la partie nord du Canal de Mozambique. *Cahiers O.R.S. T. O. M., série Océanographie*, **7**(2): 3–26.

Donguy, J.R. (1975). Les eaux superficielles tropicales de la partie accidentale de l'Ocean Indien en 1966–1967. *Cahiers O.R.S.T.O.M.*, **13**(1): 31–47.

Donguy, J.-R. and B. Piton (1991). The Mozambique Channel revisited. *Oceanologica Acta*, **14**(6): 549–558.

Donohue, K.A., E. Firing and L. Beal (2000). Comparison of three velocity sections of the Agulhas Current and Agulhas Undercurrent. *Journal of Geophysical Research*, **105**(C12): 28,585–28,593.

Donohue, K. and J. Toole (2003). A near-synoptic survey of the southwest Indian Ocean. *Deep-Sea Research II*, **50**(12–13): 1893–1931.

Döös, K. (1995). Interocean exchange of water masses. *Journal of Geophysical Research*, **100**(C7): 13,499–13,514.

Drijfhout, S. S. (2003). Why anticyclones can split. *Journal of Physical Oceanography*, **33**(8): 1579–1591.

Drijfhout, S.S., C.A. Katsman, L. de Steur, P.C.F. van der Vaart, P.J. van Leeuwen and C. Veth (2003). Modeling the initial, fast sea surface height decay of Agulhas ring "Astrid". *Deep-Sea Research II*, **50**(1): 299–319.

Duncan, C.P. (1968). An eddy in the Subtropical Convergence southwest of South Africa. *Journal of Geophysical Research*, **73**(2): 531–534.

Duncan, C.P. (1970). *The Agulhas Current*. Ph. D. dissertation, University of Hawaii, 76 pp.

Duncan, C.P. and S.G. Schladow (1981). World surface currents from ship's drift observations. *International Hydrographic Review*, **58**(2): 101–112.

Duncombe Rae, C.M., L.V. Shannon and F.A. Shillington (1989). An Agulhas ring in the South Atlantic Ocean. *South African Journal of Science*, **85**(11): 747–748.

Duncombe Rae, C.M. (1991). Agulhas retroflection rings in the South Atlantic Ocean; an overview. *South African Journal of Marine Science*, **11**: 327–344.

Duncombe Rae, C.M., A.J. Boyd and R.J.M. Crawford (1992). "Predation" of anchovy by an Agulhas ring: a possible contributory cause of the very poor year class of 1989. *South African Journal of Marine Science*, **12**: 167–173.

Duncombe Rae, C.M., F.A. Shillington, J.J. Agenbag, J. Taunton-Clark and M.L. Gründlingh (1992). An Agulhas Ring in the South East Atlantic Ocean and its interaction with the Benguela upwelling frontal system. *Deep-Sea Research*, **39**(11/12): 2009–2027.

Duncombe Rae, C.M., S.L. Garzoli and A.L. Gordon (1996). The eddy field of the southeast Atlantic Ocean: a statistical census from the Benguela Sources and Transports Project. *Journal of Geophysical Research*, **101**(C5): 11,949–11,964.

Eagle, G.A. and M.J. Orren (1985). A seasonal investigation of the nutrients and dissolved oxygen in the water column along two lines of stations south and west of South Africa. *National Research Institute for Oceanology, CSIR Research Report*, **567**: 52 pp.

Engelmann, G. (1969). Christian Gottfried Ehrenberg, ein Wegbereiter der deutschen Tiefseeforschung. *Deutsche Hydrographische Zeitschrift*, **22**(2): 145–157.

Evenson, G. and P.J. van Leeuwen (1996). Assimilation of Geosat altimeter data for the Agulhas Current using the ensemble Kalman filter with a quasigeostrophic model. *Monthly Weather Review*, **124**(1): 85–96.

Feron, R.C.V., W.P.M. de Ruijter and D. Oskam (1992). Ring-shedding process in the Agulhas Current system. *Journal of Geophysical Research*, **97**(C6): 9467–9477.

Feron, R.C.V. (1995). The Southern Ocean western boundary currents: comparison of Fine Resolution Antarctic Model results with GEOSAT altimeter data. *Journal of Geophysical Research*, **100**(C10): 4959–4975.

Feron, R.C.V., W.P.M. de Ruijter and P.J. van Leeuwen (1998). A new method to determine the mean sea surface dynamic topography from satellite altimeter observations. *Journal of Geophysical Research*, **103**(C1): 1343–1362.

Ffield, A., J.M. Toole and W.D. Wilson (1997). Seasonal circulation in the South Indian Ocean. *Geophysical Research Letters*, **24**(22): 2773–2776.

Fieux, M., F. Schott and J.C. Swallow (1986). Deep boundary currents in the western Indian Ocean revisited. *Deep-Sea Research*, **33**(4): 415–526.

Findlay, A.G. (1866). *A directory for the navigation of the Indian Ocean*. Richard Holmes Laurie, London, 113 pp.

Findlay, K.P., P.B. Best, G.J.B. Ross and V.G. Cockcroft (1992). The distribution of small odontocete cetaceans off the coasts of South Africa and Namibia. *South African Journal of Marine Science*, **12**: 237–270.

Fine, R.A., M.J. Warner and R.F. Weiss (1988). Water mass modification of the Agulhas retroflection: chlorofluoromethane studies. *Deep-Sea Research*, **35**(3): 311–332.

Fine, R.A. (1993). Circulation of Antarctic Intermediate Water in the South Indian Ocean. *Deep-Sea Research I*, **40**(10): 2021–2042.

Flemming, B.W. (1978). Underwater sand dunes along the southeast African continental margin – observations and implications. *Marine Geology*, **26**(1/2): 177–198.

Flemming, B.W. (1980). Sand transport and bedform patterns on the continental shelf between Durban and Port Elizabeth (Southeast African Continental Margin). *Sedimentary Geology*, **26**(1/3): 179–205.

Flemming, B. and R. Hay (1988). Sediment distribution and dynamics of the Natal continental shelf. In *Coastal Ocean Studies off Natal, South Africa*, editor E.H. Schumann, Lecture Notes on Coastal and Estuarine Studies 26, Springer-Verlag, Berlin, pp. 47–80.

Florenchie, P. and J. Veron (1998). South Atlantic Ocean circulation: Simulation experiments with a quasi-geostrophic model and assimilation of TOPEX/POSEIDON and ERS 1 altimeter data. *Journal of Geophysical Research*, **103**(C11): 24,737–24,758.

Flores, J.-A., R. Gersonde and F.J. Sierro (1999). Pleistocene fluctuations in the Agulhas Current Retroflection based on the calcareous plankton record. *Marine Micropaleontology*, **37**(1): 1–22.

Fowler, J.L. and A.J. Boyd (1998). Transport of anchovy and sardine eggs and larvae from the western Agulhas Bank to the west coast during the 1993/94 and 1994/95 spawning seasons. *South African Journal of Marine Science*, **19**: 181–195.

Fu, L.-L. (1986). Mass, heat and freshwater fluxes in the South Indian Ocean. *Journal of Physical Oceanography*, **16**(10): 1683–1693.

Fu, L.-L. (1996). The circulation and its variability of the South Atlantic Ocean: first results from the TOPEX/POSEIDON mission. In *The South Atlantic: Present and Past Circulation*, editors: G. Wefer, W.H. Berger, G. Siedler and D. Webb, Springer-Verlag, Berlin, pp. 63–82.

Ganachaud, A., C. Wunsch, J. Marotzke and J. Toole (2000). Meridional overturning and large-scale circulation of the Indian Ocean. *Journal of Geophysical Research*, **105** (C11): 26,117–26,134.

Ganachaud, A. (2003). Large-scale mass transports, water mass formation and diffusivities estimated from World Ocean Circulation Experiment (WOCE) hydrographic data. *Journal of Geophysical Research*, **108**(C7): article number 3213.

Garnier, E., J. Verron and B. Barnier (2003). Variability of the South Atlantic upper ocean circulation: a data assimilation experiment with 5 years of TOPEX/POSEIDON altimeter observations. *International Journal of Remote Sensing*, **24**(5): 911–934.

Garraffo, Z., S.L. Garzoli, W. Haxby and D. Olson (1992). Analysis of a general circulation model. 2. Distribution of kinetic energy in the South Atlantic and Kuroshio/Oyashio systems. *Journal of Geophysical Research*, **97**(C12): 20,139–20,153.

Garrett, J. (1981). Oceanographic features revealed by the FGGE drifting buoy array. In *Oceanography from Space*, editor J. F. R. Gower, Plenum Press, New York, pp. 61–69.

Garrett, J. (1981). The performance of the FGGE drifting buoy system. *Advances in Space Research*, **1**(4): 87–94.

Garzoli, S.L. and A.L. Gordon (1995). Origins and variability of the Benguela Current. *Journal of Geophysical Research*, **101**(C11): 22,591–22,601.

Garzoli, S.L., A.L. Gordon, V. Kamenkovich, D. Pillsbury and C. Duncombe-Rae [sic] (1996). Variability and sources of the southeastern Atlantic circulation. *Journal of Marine Research*, **54**(6): 1039–1071.

Garzoli, S.L., G.J. Goñi, A.J. Mariano and D.B. Olson (1997). Monitoring the upper southeastern Atlantic transports using altimeter data. *Journal of Marine Research*, **55**(3): 453–481.

Garzoli, S.L., P.L. Richardson, C.M. Duncombe Rae, D.M. Fratantoni, G.J. Goñi and A.J. Roubicek (1999). Three Agulhas rings observed during the Benguela Current Experiment. *Journal of Geophysical Research*, **104**(C9): 20,971–20,985.

Garzoli, S.L. and J. Goni [sic] (2000). Combining altimeter observations and oceanographic data for ocean circulation and climate studies. In *Satellites, Oceanography and Society*, editor D. Halpern, Elsevier Oceanography Series 63, Elsevier, Amsterdam, pp. 79–97.

Gerard, R. (1964). Étude de l'eau de mer de surface dans une baie de Nosy-Bé. *Cahiers O.R.S.T.O.M., série Océanographie*, **2**(2): 5–26.

Gerber, M. (1993). The interaction of deep-water gravity waves and an annular current: Linear theory. *Journal of Fluid Mechanics*, **248**: 153–172.

Gilchrist, J.D.F. (1902). Observations on the temperature and salinity of the sea around the Cape Peninsula. *Marine Investigations in South Africa*, **1**: 181–216 + 10 plates + p. 238.

Gilchrist, J.D.F. (1904). Currents on the South African coast, as indicated by the course of drift bottles. *Marine Investigations in South Africa*, **2**: 101–113.

Gilchrist, J.D.F. (1905). The South African marine fauna and its environment. In *Science in South Africa*, editors W. Flint and J.D.F. Gilchrist, Maskew Miller, Cape Town, pp. 182–197.

Gilchrist, J.D.F. (1923). The South African seas. *South African Journal of Science*, **20**(1): 1–27.

Gilchrist, J.D.F. (1925). II. Report of Director of Survey. In *Report of the Fisheries and Marine Biological Survey, Union of South Africa, June, 1923 – June, 1925*. **4**. Cape Town; Government Printers: xix–lx.

Gill, A.E. (1977). Coastally trapped waves in the atmosphere. *Quarterly Journal of the Royal Meteorological Society*, **103**(437): 431–440.

Gill, A.E. and E.H. Schumann (1979). Topographically induced changes in the structure of an inertial coastal jet: application to the Agulhas Current. *Journal of Physical Oceanography*, **9**(5): 975–991.

Goni [sic], G.J., S.L. Garzoli, A.J. Roubicek, D.B. Olson and O.B. Brown (1997). Agulhas ring dynamics from TOPEX/POSEIDON satellite altimeter data. *Journal of Marine Research*, **55**(5): 861–883.

Gordon, A.L., K.-I. Horai and M. Donn (1983). Southern hemisphere western boundary current variability revealed by GEOS 3 altimeter. *Journal of Geophysical Research*, **88**(1): 755–762.

Gordon, A.L. (1985). Indian-Atlantic transfer of thermocline water at the Agulhas retroflection. *Science*, **227**(4690): 1030–1033.

Gordon, A.L. (1986). Inter-ocean exchange of thermocline water. *Journal of Geophysical Research*, **91**(C4): 5037–5046.

Gordon, A.L., J.R.E. Lutjeharms and M.L. Gründlingh (1987). Select hydrographic sections from the Agulhas research cruises of the research vessels *Knorr* and *Meiring*

Naude 0 1983. Lamont-Doherty Geological Observatory of Columbia University, *Technical Report* **LDGO-87 - 1**.
Gordon, A.L., J.R.E. Lutjeharms and M.L. Gründlingh (1987). Stratification and circulation at the Agulhas Retroflection. *Deep-Sea Research*, **34**(4): 565–599.
Gordon, A.L. and W.F. Haxby (1990). Agulhas eddies invade the South Atlantic – evidence from GEOSAT altimeter and shipboard conductivity–temperature–depth survey. *Journal of Geophysical Research*, **95**(C3): 3117–3125.
Gordon, A.L., R.F. Weiss, W.M. Smethie and M.J. Warner (1992). Thermocline and intermediate water communication between the South Atlantic and Indian Oceans. *Journal of Geophysical Research*, **97**(C5): 7223–7240.
Gordon, A.L., K.T. Bosley and F. Aikman (1995). Tropical Atlantic water within the Benguela upwelling system at 27°S. *Deep-Sea Research*, **24**(1): 1–12.
Gordon, A.L. (2003). The brawniest retroflection. *Nature*, **421**: 904–905.
Goschen, W.S. and E.H. Schumann (1988). Ocean current and temperature structures in Algoa Bay and beyond in November 1986. *South African Journal of Marine Science*, **7**: 101–116.
Goschen, W.S. and E.H. Schumann (1990). Agulhas Current variability and inshore structures off the Cape Province, South Africa. *Journal of Geophysical Research*, **95**(1): 667–678.
Goschen, W.S. and E.H. Schumann (1994). An Agulhas Current intrusion into Algoa Bay during August 1988. *South African Journal of Marine Science*, **14**: 47–57.
Goschen, W.S. and E.H. Schumann (1995). Upwelling and the occurrence of cold water around Cape Recife, Algoa Bay, South Africa. *South African Journal of Marine Science*, **16**: 57–67.
Gould, J. (1993). James Rennell's view of the Atlantic circulation: a comparison with our present knowledge. *Ocean Challenge*, **4**(1/2): 26–32.
Gründlingh, M.L. (1974). A description of inshore current reversals off Richards Bay based on airborne radiation thermometry. *Deep-Sea Research*, **21**(1): 47–55.
Gründlingh, M.L. (1977). A Nimbus VII satellite-tracked buoy moored on the Mozambique Ridge. *South African Journal of Science*, **73**(12): 384–385.
Gründlingh, M.L. (1977). Drift observations from Nimbus VI satellite-tracked buoys in the southwestern Indian Ocean. *Deep-Sea Research*, **24**(10): 903–913.
Gründlingh, M.L. (1978). Drift of a satellite-tracked buoy in the southern Agulhas Current and Agulhas Return Current. *Deep-Sea Research*, **25**(12): 1209–1224.
Gründlingh, M.L. (1979). Observation of a large meander in the Agulhas Current. *Journal of Geophysical Research*, **84**(C7): 3776–3778.
Gründlingh, M.L. and J.R.E. Lutjeharms (1979). Large scale flow patterns of the Agulhas Current system. *South African Journal of Science*, **75**(6): 269–270.
Gründlingh, M.L. (1980). On the volume transport of the Agulhas Current. *Deep-Sea Research*, **27**(7): 557–563.
Gründlingh, M.L. (1983). Eddies in the southern Indian Ocean and Agulhas Current. In *Eddies in Marine Science*, editor A.R. Robinson, Springer-Verlag, Berlin, pp. 245–264.
Gründlingh, M.L. (1983). On the course of the Agulhas Current. *South African Geographical Journal*, **65**(1): 49–57.
Gründlingh, M.L. (1984). An eddy over the northern Mozambique Ridge. *South African Journal of Science*, **80**(7): 324–329.
Gründlingh, M.L. and A.F. Pearce (1984). Large vortices in the northern Agulhas Current. *Deep-Sea Research*, **31**(9): 1149–1156.
Gründlingh, M.L. (1985). An intense cyclonic eddy east of the Mozambique Ridge. *Journal of Geophysical Research*, **90**(C4): 7163–7167.
Gründlingh, M.L. (1985). Features of the circulation in the Mozambique Basin in 1981. *Journal of Marine Research*, **43**(4): 779–792.
Gründlingh, M.L. (1985). Occurrence of Red Sea Water in the Southwestern Indian Ocean, 1981. *Journal of Physical Oceanography*, **15**(2): 207–212.
Gründlingh, M.L. (1986). Features of the northern Agulhas Current in spring, 1983. *South African Journal of Science*, **82**(1): 18–20.
Gründlingh, M.L. (1987). Anatomy of a cyclonic eddy of the Mozambique Ridge Current. *Deep-Sea Research*, **34**(2): 237–251.
Gründlingh, M.L. (1987). Cyclogenesis in the Mozambique Ridge Current. *Deep-Sea Research*, **34**(1): 89–103.
Gründlingh, M.L. (1987). On the seasonal temperature variation in the southwestern Indian Ocean. *South African Geographical Journal*, **69**(2): 129–138.
Gründlingh, M.L. (1988). Altimetry in the southwest Indian Ocean. *South African Journal of Science*, **84**(7): 568–573.
Gründlingh, M.L. (1988). Review of cyclonic eddies in the Mozambique Ridge Current. *South African Journal of Marine Science*, **6**: 193–206.
Gründlingh, M.L., C.C. Stavropoulos and L.J. Watt (1988). The R. V. Meiring Naudé: Two decades of support to physical oceanography. *South African Journal of Science*, **84**(9): 746–748.
Gründlingh, M.L. (1989). Drift of the SAA *Helderberg* wreckage in the Indian Ocean. *South African Journal of Science*, **85**(4): 209–211.
Gründlingh, M.L. (1989). Two contra-rotating eddies of the Mozambique Ridge Current. *Deep-Sea Research*, **36**(1): 149–153.
Gründlingh, M.L. and A.F. Pearce (1990). Frontal features of the Agulhas Current in the Natal Bight. *South African Geographical Journal*, **72**(1): 11–14.
Gründlingh, M.L. (1991). Draaikolke suid van die Mosambiekkanaal. *Suid-Afrikaanse Tydskrif vir Natuurwetenskap en Tegnologie*, **10**(3): 137–141.
Gründlingh, M.L., R.A. Carter and R.C. Stanton (1991). Circulation and water properties of the Southwest Indian Ocean, Spring 1987. *Progress in Oceanography*, **28**(4): 305–342.

Gründlingh, M.L. (1992). Agulhas Current meanders: review and a case study. *South African Geographical Journal*, **74**(1): 19–28.

Gründlingh, M.L. (1993). On the satellite-derived wave climate off southern Africa. *South African Journal of Marine Science*, **13**: 223–235.

Gründlingh, M.L. (1993). On the winter flow in the southern Mozambique Channel. *Deep-Sea Research I*, **40**(2): 409–418.

Gründlingh, M.L. (1994). Evidence of surface wave enhancement in the southwest Indian Ocean. *Journal of Geophysical Research*, **99**(C4): 7917–7927.

Gründlingh, M.L. (1994). Some characteristics of satellite-derived wave heights in the South Atlantic Ocean. *Deep-Sea Research I*, **41**(2): 413–428.

Gründlingh, M.L. (1995). Tracking eddies in the southeast Atlantic and southwest Indian oceans with TOPEX/POSEIDON. *Journal of Geophysical Research*, **100** (C12): 24,977–24,986.

Gründlingh, M.L., G.W. Bailey, F.A. Shillington, E.H. Schumann and J.J. Agenbag (1995). Physical oceanographic activities in South Africa. 1991–1994. *South African Journal of Science*, **91**(5): 247–254.

Gründlingh, M.(L.) and M. Rossouw (1995). Wave attenuation in the Agulhas Current. *South African Journal of Science*, **91**(7): 357–359.

Gründlingh, M.L., J.R.E. Lutjeharms, G. Brundrit, J.J. Agenbag, A. Boyd, G. Nelson, A. Johnson, M. Roberts, L. Shannon and E.H. Schumann (1999). Physical oceanographic activities in South Africa: 1995–1998. *South African Journal of Science*, **95**(6/7): 281–288.

Gutshabash, Ye. Sh. and I.V. Lavrenov (1986). Swell transformation in the Cape [sic] Agulhas Current. *Izvestiya, Atmospheric and Oceanic Physics*, **22**(6): 494–497.

Haines, M., M. Luther, Z. Ji and R. Fine (1995). Particle trajectories in an Indian Ocean Model. In *1995 U. S. WOCE Report*, U. S. WOCE Implementation Report Number 10, pp. 34–35.

Harries, H.D. (1932). Über die Veränderlichkeit von Monatswerten meteorologischer und hydrologischer Elementen der Äquatorialsee. *Annalen der Hydrographie, Berlin*, **60**(12): 496–499.

Harris, T.F.W. (1964). Notes on Natal coastal waters. *South African Journal of Science*, **60**(8): 237–241.

Harris, T.F.W. and C.C. Stavropoulos (1967). Some experience with a radiation thermometer over the Agulhas Current. *South African Journal of Science*, **63**(4): 132–136.

Harris, T.F.W. (1970). Planetary-type waves in the South West Indian Ocean. *Nature*, **227**(5262): 1043–1044.

Harris, T.F.W. (1972). Sources of the Agulhas Current in the spring of 1964. *Deep-Sea Research*, **19**(9): 633–650.

Harris, T.F.W. and N.D. Bang (1974). Topographic Rossby waves in the Agulhas Current. *South African Journal of Science*, **70**(7): 212–214.

Harris, T.F.W. (1978). Review of coastal currents in Southern African waters. *South African National Science Programmes Report, CSIR Report*, **30**, vii + 103 pp.

Harris, T.F.W. and C.C. Stavropoulos (1978). Satellite tracked drifters between Africa and Antarctica. *Bulletin of the American Meterological Society*, **59**(1): 51–59.

Harris, T.F.W. and D. van Foreest (1978). The Agulhas Current in March 1969. *Deep-Sea Research*, **25**(6): 549–561.

Harris, T.F.W., R. Legeckis and D. van Forest [sic] (1978). Satellite infra-red images in the Agulhas Current system. *Deep-Sea Research*, **25**(6): 543–548.

Hastenrath, S. and P.J. Lamb (1979). *Climate Atlas of the Indian Ocean. Part I: Surface climate and atmospheric circulation*. University of Wisconsin Press, Madison, xix pp + 97 charts.

Hastenrath, S. and P.J. Lamb (1979). *Climatic Atlas of the Indian Ocean. Part II: The ocean heat budget*. University of Wisconsin Press, Madison, xv pp + 84 charts.

Herbette, S., Y. Morel and M. Arhan (2003). Erosion of a surface vortex by a seamount. *Journal of Physical Oceanography*, **33**(8): 1664–1679.

Heydorn, A.E.F., N.D. Bang, A.F. Pearce, B.W. Flemming, R.A. Carter, M.H. Schleyer, P.F. Berry, G.R. Hughes, A.J. Bass, J.H. Wallace, R.P. van der Elst, R.J.M. Crawford and P.A. Shelton (1978). Ecology of the Agulhas Current region: an assessment of biological responses to environmental parameters in the South-West Indian Ocean. *Transactions of the Royal Society of South Africa*, **43**(2): 151–190.

Heywood, K.J. and Y.K. Somayajulu (1997). Eddy activity in the South Indian Ocean from ERS-1 altimetry. *Proceedings of the 3rd ERS Symposium on "Space at the Service of our Environment"*, Florence, March 1997, *ESA SP-414*, 1479–1483.

Ho, C.-R., Q. Zheng and N.-J. Kuo (2004). SeaWiFS observations of upwelling south of Madagascar: long-term variability and interaction with East Madagascar Current. *Deep-Sea Rsearch II*, **51**(1): 59–67.

Hofmann, E.C. (1985). The large-scale horizontal structure of the Antarctic Circumpolar Current from FGGE drifters. *Journal of Geophysical Research*, **90**(C7): 7087–7097.

Holfort, J. and G. Siedler (2001). The meridional oceanic transports of heat and nutrients in the South Atlantic. *Journal of Physical Oceanography*, **31**(1): 5–29.

Holland, W.R., V. Zlotnicki and L.-L. Fu (1991). Modelled time-dependent flow in the Agulhas retroflection region as deduced from altimeter data assimilation. *South African Journal of Marine Science*, **10**: 407–427.

Holliday, N.P. and J.F. Read (1998). Surface oceanic fronts between Africa and Antarctica. *Deep-Sea Research I*, **45**(2–3): 217–238.

Howard, W.R. and W.L. Prell (1992). Late quaternary surface circulation of the southern Indian Ocean and its relationship to orbital variation. *Paleoceanography*, **7**(1); 79–117.

Hughes, G.R., P. Luschi, R. Mencacci and F. Papi (1998). The 7000-km oceanic journey of a leatherback turtle tracked by satellite. *Journal of Experimental Marine Biology and Ecology*, **229**(2): 209–217.

Hulley, P.A. and J.R.E. Lutjeharms (1995). The southwestern limit for the warm water, mesopelagic ichthyofauna of the Indo-West Pacific: Lanternfish (*Myctophidae*), a case study. *South African Journal of Marine Science*, **15**: 185–205.

Hunter, I.[T.] (1981). On the land breeze circulation off the Natal coast. *South African Journal of Science*, **77**(8): 376–378.

Hunter, I.T. (1987). The weather of the Agulhas Bank and Cape south coast. (South African) Council for Scientific and Industrial Research. *CSIR Research Report*, **634**, 184 pp.

Hunter, I.T. (1988). Climate and weather off Natal. In *Coastal Ocean Studies off Natal, South Africa,* editor E.H. Schumann, Lecture Notes on Coastal and Estuarine Studies 26, Springer-Verlag, Berlin, pp. 81–100.

Hutson, W.H. (1980). The Agulhas Current during the late Pleistocene: analysis of modern faunal analogs. *Science*, **207**(4426): 64–66.

Hydrographic Department (1896). *List of oceanic depths and serial temperature observations during the years 1895.* Great Britain, Hydrographic Department, London.

Irvine, D.E. and D.G. Tilley (1988). Ocean wave directional spectra and wave-current interaction in the Agulhas from the Shuttle Imaging Radar-B synthetic aperture radar. *Journal of Geophysical Research*, **93**(C12): 15,389–15,401.

Jacobs, S.S. and D.T. Georgi (1977). Observations on the Southwest Indian/Antarctic Ocean. *Deep-Sea Research*, **24** (supplement): 43–84.

Johnson, G.C., D.L. Musgrave, B.A. Warren, A. Ffield and D.B. Olson (1998). Flow of bottom and deep water in the Amirante Passage and Mascarene Basin. *Journal of Geophysical Research*, **103**(C13): 30,973–30,984.

Jorge da Silva, A., A. Mubango and R. Sætre (1981). Information on oceanographic cruises in the Mozambique Channel. *Revista de Investigação Pesquira*, **2**, Instituto de Desenvolvimento Pesqueiro, Maputo, República Popular de Moçambique, 89 pp.

Jorge da Silva, A. (1984). Circulation system and areas of potentially successful tuna fishing with surface methods off Mozambique. *Revista de Investigação Pesquira*, **11**, Instituto de Investigação Pesqueiro, Moçambique, pp. 5–40.

Jorge da Silva, A. (1984). Hydrology and fish distribution at the Sofala Bank (Mozambique). *Revista de Investigação Pesquira*, **12**, Instituto de Investigação Pesqueiro, Moçambique, pp. 5-36.

Jorge da Silva, A. (1984). Report on the oceanographic investigations carried out at the Sofala Bank by the Soviet trawler "Sevastopolsky Rybak" in September–December 1982. *Revista de Investigação Pesquira*, **10**, Instituto de Investigação Pesqueiro, Maputo, Moçambique, pp. 5–35.

Jury, M.(R.) and N. Walker (1988). Marine boundary layer modification across the edge of the Agulhas Current. *Journal of Geophysical Research*, **93**(C1): 647–654.

Jury, M.R. and S. Courtney (1991). A transition in weather over the Agulhas Current. *South African Journal of Marine Science*, **19**: 159–171.

Jury, M.R. and J.R.E. Lutjeharms (1993). Die struktuur en die aandrywingskragte van die 1991–1992 droogte in suidelike Afrika. *Suid-Afrikaanse Tydskrif vir Natuurwetenskap en Tegnologie*, **12**(1): 8–16.

Jury, M.R., H.R. Valentine and J.R.E. Lutjeharms (1993). Influence of the Agulhas Current on summer rainfall on the southeast coast of South Africa. *Journal of Applied Meteorology*, **32**(7): 1282–1287.

Jury, M.R. (1994). A review of the meteorology of the eastern Agulhas Bank. *South African Journal of Science*, **90**(3): 109–113.

Jury, M.R. (1994). A thermal front within the marine atmospheric boundary layer over the Agulhas Current south of Africa: Composite aircraft observations. *Journal of Geophysical Research*, **99**(C2): 3297–3304.

Jury, M.R. (1995). A review of research on ocean-atmosphere interactions and South African climate variability. *South African Journal of Science*, **91**(6): 289–294.

Jury, M.R., B. Pathack, C.J. de W [sic] Rautenbach and J. vanHeerden [sic] (1996). Drought over South Africa and Indian Ocean SST: statistical and GCM results. *The Global Atmosphere and Ocean System*, **4**(1):47–63.

Jury, M.R. and M. Majodina (1997). Preliminary climatology of southern Africa extreme weather 1973–1992. *Theoretical and Applied Climatology*, **56**(1–2): 103–112.

Jury, M., M. Rouault, S. Weeks and M. Schormann (1997). Atmospheric boundary-layer fluxes and structure across a land–sea transition zone in south-eastern Africa. *Boundary-Layer Meteorology*, **83**(2): 311–330.

Kamenkovich, V.M., Y.P. Leonov, D.A. Nechaev, D.A. Byrne and A.L. Gordon (1996). On the influence of bottom topography on the Agulhas eddy. *Journal of Physical Oceanography*, **26**(6): 892–192.

Katsman, C.A., P.C.F. van der Vaart, H.A. Dijkstra and W.P.M. de Ruijter (2003). Stability of multilayer ocean vortices: a parameter study including realistic Gulf Stream and Agulhas Rings. *Journal of Physical Oceanography*, **33**(6): 1197–1218.

Kerhallet, M.C.P. de (1851). Considérations générales sur l'océan Indien. *Annales Hydrographiques*, **2e Sem**, pp. 219–264.

Kerhallet, M.C.P. de (1852). Considérations générales sur l'océan Atlantique. *Annales Hydrographiques*, **1er Sem**. pp. 33–143.

Knauss, J.A. (1961). The International Indian Ocean Expedition. *Science*, **134**(3491): 1674–1676.

Kolla, V., L. Sullivan, S.S. Streeter and M.G. Langseth (1976). Spreading of Antarctic Bottom Water and its effects on the floor of the Indian Ocean inferred from bottom-water potential temperature, turbidity, and seafloor photography. *Marine Geology*, **21**: 171–189.

Koninklijk Nederlandsch Meteorologisch Instituut (1856). *Maury's passatkaart van den Atlantischen Oceaan, vermeerderd met Hollandsche gegevens door den lt. t. z. K. F. R. Andrau.* Koninklijk Nederlandsch Meteorologisch

Instituut.

Koninklijk Nederlandsch Meteorologisch Instituut (1857). De Agulhas stroom, afgeleid uit de temperatuur van het zeewater aan de oppervlakte; en de invloed dien deze op de atmosteer uitoefent. In *Uitkomsten van Wetenschap en Ervaring, aangaande winden en zeestromingen in sommige gedeelten van den oceaan,* Koninklijk Nederlandsch Meteorologisch Instituut, L. E. Bosch en Zoon, Utrecht, pp. 40–50.

Kortum, G. (1990). An unpublished manuscript of Alexander von Humboldt on the Gulf Stream. In *Ocean Sciences. Their history and relation to man,* editors W. Lenz and M. Deacon. *Deutsche Hydrographische Zeitschrift,* **B**(22): 122–129.

Kostianoy, A.G., A.I. Ginzburg, S.A Lebedev, M. Frankignoulle and B. Delille (2003). Fronts and mesoscale variability in the southern Indian Ocean as inferred from the *TOPEX/POSEIDON* and *ERS-2* altimetry data. *Marine Physics,* **43**(5): 671–682.

Kostianoy, A.G., A.I. Ginzburg, M. Frankignoulle and B. Delille (2004). Fronts in the southern Indian Ocean as inferred from satellite sea surface temperature data. *Journal of Marine Systems,* **45**(1): 55–73.

Krümmel, O. (1882). Bemerkungen über die Meereströmungen und Temperaturen in der Falklandsee. *Aus den Archiv der Deutschen Seewarte,* **V**, 2, 25 pp. (as shown in Peterson *et al.*, 1993).

Krümmel, O. (1911). *Handbuch der Ozeanographie. Band II. Die Bewegungsformen des Meeres.* von J. Engelhorns Nachf., Stuttgart, 766 pp.

Landman, W.A. and S.J. Mason (1999). Change in the association between Indian Ocean sea-surface temperatures and summer rainfall over South Africa and Namibia. *International Journal of Climatology,* **19**(13): 1477–1492.

Largier, J.L. and V.P. Swart (1987). East–west variation in thermocline breakdown on the Agulhas Bank. *South African Journal of Marine Science,* **5**: 263–272.

Largier, J.L.,P. Chapman, W.T. Peterson and V.P. Swart (1992). The western Agulhas Bank: circulation, stratification and ecology. *South African Journal of Marine Science,* **12**: 319–339.

Lazarus, B.I. and D. Dowler (1979). Pelagic tunicata off the west and south-west coasts of South Africa, 1964–1965. *Fisheries Bulletin,* **12**: 93–119.

Lee-Thorp, A.M., M. Rouault and J.R.E. Lutjeharms (1998). Cumulus cloud formation above the Agulhas Current. *South African Journal of Science,* **94**(7): 351–354.

Lee-Thorp, A.M., M. Rouault and J.R.E. Lutjeharms (1999). Moisture uptake in the boundary layer above the Agulhas Current: a case study. *Journal of Geophysical Research,* **104**(C1): 1423–1430.

Liltved, W.R., E. Harley and J.R.E. Lutjeharms (2000). A new subspecies of *Cypraeovula* occurring off the northern part of the Eastern Cape, South Africa, with notes on possible genetic and oceanographic influences on its distribution. Supplement to *Cowries and Their Relatives of Southern Africa,* Seacomber Publications, Cape Town, pp. 225–232.

Llido, J., J. Sudre, I. Dadou and V. Garçon (2004). Variability of the biological front south of Africa from SeaWiFS and a coupled physical–biological model. *Journal of Marine Systems,* in press.

Luschi, P., A. Sale, R. Mencacci, G.R. Hughes, J.R.E. Lutjeharms and F. Papi (2003). Current transport of leatherback sea turtles (*Dermochelys coriacea*) in the ocean. *Proceedings of the Royal Society of London* (Supplement), **270**: S129–S132; DOI 10.1098/rsbl.203.0036.

Lütgens, R. (1905). Oberflächentemperaturen im südlichen Indischen Ozean 1901 bis 1903. *Annalen der Hydrographie Berlin,* **33**(11): 498–513.

Luther, M.E. and J.J. O'Brien (1989). Modelling the variability in the Somali Current. In *mesoscale/Synoptic Coherent Structures in Geophysical Turbulence,* editors J. C. J. Nihoul and B. M. Jamart, Elsevier, Amsterdam, pp. 373–386.

Lutjeharms, J.R.E. (1972). *A Guide to Research Done Concerning Ocean Currents and Water Masses in the South West Indian Ocean.* University of Cape Town, Cape Town, 577 pp.

Lutjeharms, J.R.E. (1972). A quantitative assessment of year-to-year variability in water movement in the Southwest Indian Ocean. *Nature,* **239**(91): 59–69.

Lutjeharms, J.R.E. (1976). The Agulhas Current system during the Northeast Monsoon Season. *Journal of Physical Oceanography,* **6**(5): 665–670.

Lutjeharms, J.R.E. (1977). The need for oceanological research in the South-west Indian Ocean. *South African Journal of Science,* **73**(2): 40–43.

Lutjeharms, J.R.E. (1980). Fisiese Oseanologie van die Suidwes Indiese Oseaan: 'n bronnelys en oorsig van 1970 tot 1980. *WNNR Verslag,* **T/SEA 8016**, 82 pp.

Lutjeharms, J.R.E. (1980). The influence of the Agulhas Current. *CSIR Research Report,* **376**, 6 pp.

Lutjeharms, J.R.E. and D.J. Baker (1980). A statistical analysis of the meso-scale dynamics of the Southern Ocean. *Deep-Sea Research,* **27**(2): 145–159.

Lutjeharms, J.R.E., G.A.W. Fromme and C.P. Duncan (1980). Large scale motion in the SW Indian Ocean. *CSIR Research Report,* **375**, 9 pp.

Lutjeharms, J.R.E. (1981). Features of the southern Agulhas Current circulation from satellite remote sensing. *South African Journal of Science,* **77**(5): 231–236.

Lutjeharms, J.R.E. (1981). Satellite studies of the South Atlantic upwelling system. In *Oceanography from Space,* editor J. F. R. Gower, Plenum Press, New York, pp. 195–199.

Lutjeharms, J.R.E. (1981). Spatial scales and intensities of circulation in the ocean areas adjacent to South Africa. *Deep-Sea Research,* **28**(11): 1289–1302.

Lutjeharms, J.R.E. (1981). Interaction between the Agulhas Current and the Subtropical Convergence. *CSIR Research Report,* **384**, 39 pp.

Lutjeharms, J.R.E. and H.R. Valentine (1981). Ocean circulation studies in the vicinity of southern Africa: preliminary results using FGGE drifters and remote sensing. *Advances in Space Research*, **1**(4): 211–223.

Lutjeharms, J.R.E., N.D. Bang and C.P. Duncan (1981). Characteristics of the currents east and south of Madagascar. *Deep-Sea Research*, **28**(9): 879–899.

Lutjeharms, J.R.E., N.D. Bang and H.R. Valentine (1981). Die fisiese oseanologie van die Agulhasbank. 1. Vaart 170 van die N. S. *Thomas B. Davie*. South African Council for Scientific and Industrial Research *CSIR Research Report*, **386**, 38 pp.

Lutjeharms, J.R.E. (1982). Die geskiedkundige ontwikkeling van fisiese navorsing oor die oseane grensend aan Suider-Afrika. *Suid-Afrikaanse Tydskrif vir Natuurwetenskap en Tegnologie*, **1**(1): 42–52.

Lutjeharms, J.R.E. (1983). An Agulhas Current source book. *CSIR Report*, **T/SEA 8306**, 24 pp.

Lutjeharms, J.R.E. and H. R. Valentine (1983). Die fisiese oseanologie van die Agulhasbank. Deel II: Vaart 185 van die N. S. Thomas B. Davie. *CSIR Research Report*, **557**, 15 pp.

Lutjeharms, J.R.E. and H.R. Valentine (1984). Southern Ocean thermal fronts south of Africa. *Deep-Sea Research*, **31**(12): 1461–1476.

Lutjeharms, J.R.E. and R.C. van Ballegooyen (1984). Topographic control in the Agulhas Current system. *Deep-Sea Research*, **31**(11): 1321–1337.

Lutjeharms, J.R.E. (1985). Location of frontal systems between Africa and Antarctica: some preliminary results. *Deep-Sea Research*, **32**(12): 1499–1509.

Lutjeharms, J.R.E. and N.M. Walters (1985). Ocean colour and thermal fronts south of Africa. In *South African Ocean Colour and Upwelling Experiment*, editor L.V. Shannon, Sea Fisheries Research Institute, Cape Town, pp. 227–237.

Lutjeharms, J.R.E. and L.H. McQuaid (1986). Changes in the structure of thermal ocean fronts south of Africa over a three-month period. *South African Journal of Science*, **82**(9): 470–476.

Lutjeharms, J.R.E., B.R. Allanson and L. Parker (1986). Frontal zones, chlorophyll and primary production patterns in the surface waters of the Southern Ocean south of Africa. In *Marine Interfaces Ecohydrodynamics*, editor J.C.J. Nihoul, Elsevier, Amsterdam, pp. 105–117.

Lutjeharms, J.R.E., D. Baird and I.T. Hunter (1986). Seeoppervlak dryfgedrag aan die Suid-Afrikaanse suidkus in 1979. *South African Journal of Science*, **82**(6): 324–326.

Lutjeharms, J.R.E., R.D. Mey and I.T. Hunter (1986). Cloud lines over the Agulhas Current. *South African Journal of Science*, **82**(11/12): 635–640.

Lutjeharms, J.R.E. (1987). Die Subtropiese Konvergensie en Agulhasretrofleksievaart (SCARC). *South African Journal of Science*, **83**(8): 454–456.

Lutjeharms, J.R.E. (1987). Meridional heat transport across the Sub-Tropical Convergence by a warm eddy. *Nature*, **331**(6153): 251–253.

Lutjeharms, J.R.E. (1987). The Physical Oceanology of the South West Indian Ocean; a bibliography for 1980 to 1985. *WNNR-verslag*, **T/SEA 8712**, 131 pp.

Lutjeharms, J.R.E. and F.P. Anderson (1987). The World Ocean Circulation Experiment (WOCE): a South African perspective. *South African Journal of Science*, **83**(1): 10–14.

Lutjeharms, J.R.E. and A.L. Gordon (1987). Shedding of an Agulhas Ring observed at sea. *Nature*, **325**(7000): 138–140.

Lutjeharms, J.R.E. and J.M. Meeuwis (1987). The extent and variability of the South East Atlantic upwelling. *South African Journal of Marine Science*, **5**: 51–62.

Lutjeharms, J.R.E. and P.L. Stockton (1987). Kinematics of the upwelling front off Southern Africa. *South African Journal of Marine Science*, **5**: 35–49.

Lutjeharms, J.R.E. (1988). Examples of extreme circulation events of the Agulhas Retroflection. *South African Journal of Science*, **84**(7): 584–586.

Lutjeharms, J.R.E. (1988). On the role of the East Madagascar Current as a source of the Agulhas Current. *South African Journal of Science*, **84**(4): 236–238.

Lutjeharms, J.R.E. (1988). Remote sensing corroboration of retroflection of the East Madagascar Current. *Deep-Sea Research*, **35**(12): 2045–2050.

Lutjeharms, J.R.E. and A. Jorge da Silva (1988). The Delagoa Bight eddy. *Deep-Sea Research*, **35**(4): 619–634.

Lutjeharms, J.R.E. and H.R. Roberts (1988). The Natal Pulse; an extreme transient on the Agulhas Current. *Journal of Geophysical Research*, **93**(C1): 631–645.

Lutjeharms, J.R.E. and H.R. Valentine (1988). Eddies at the Sub-Tropical Convergence south of Africa. *Journal of Physical Oceanography*, **18**(5): 761–774.

Lutjeharms, J.R.E. and H.R. Valentine (1988). Evidence for persistent Agulhas rings southwest of Cape Town. *South African Journal of Science*, **84**(9): 781–783.

Lutjeharms, J.R.E. and H.R. Valentine (1988). On mesoscale ocean eddies at the Agulhas Plateau. *South African Journal of Science*, **84**(3): 194–200.

Lutjeharms, J.R.E. and R.C. van Ballegooyen (1988). Anomalous upstream retroflection in the Agulhas Current. *Science*, **240**(4860): 1770–1772.

Lutjeharms, J.R.E. and R.C. van Ballegooyen (1988). The retroflection of the Agulhas Current. *Journal of Physical Oceanography*, **18**(11): 1570–1583.

Lutjeharms, J.R.E. (1989). The role of mesoscale turbulence in the Agulhas Current system. In *Mesoscale/synoptic Coherent Structures in Geophysical Turbulence,* editors J. C. J. Nihoul and B. M. Jamart, Elsevier, Amsterdam, pp. 357–372.

Lutjeharms, J.R.E. and A.D. Connell (1989). The Natal Pulse and inshore counter currents off the South African east coast. *South African Journal of Science*, **85**(8): 533–535.

Lutjeharms, J.R.E., R. Catzel and H.R. Valentine (1989). Eddies and other border phenomena of the Agulhas Current. *Continental Shelf Research*, **9**(7): 597–616.

Lutjeharms, J.R.E., M.L. Gründlingh and R.A. Carter (1989). Topographically induced upwelling in the Natal Bight.

South African Journal of Science, **85**(5): 310–316.

Lutjeharms, J.R.E. (1990). Temperatuurstruktuur van die oseaanbolaag tussen Kaapstad en Marion-eiland. *South African Journal of Antarctic Research*, **20**(1): 21–32.

Lutjeharms, J.R.E. (1991). The temperature/salinity relationships of the South West Indian Ocean. *South African Geographer*, **18**(1/2): 15–31.

Lutjeharms, J.R.E. (1991). Twintig jaar fisiese oseanologie in Suid-Afrika: 'n persoonlike beskouing. *South African Journal for Marine Research*, **10**: 305–320.

Lutjeharms, J.R.E. and P.L. Stockton (1991). Aspects of the upwelling regime between Cape Point and Cape Agulhas. *South African Journal of Marine Science*, **10**: 91–102.

Lutjeharms, J.R.E., J. Olivier and E. Lourens (1991). Surface fronts of False Bay and vicinity. *Transactions of the Royal Society of South Africa*, **47**(415): 433–445.

Lutjeharms, J.R.E., F.A. Shillington and C.M. Duncombe Rae (1991). Observations of extreme upwelling filaments in the South East Atlantic Ocean. *Science*, **253**(5021): 774–776.

Lutjeharms, J.R.E., D.J. Webb and B.A. de Cuevas (1991). Applying the Fine Resolution Antarctic Model (FRAM) to the ocean circulation around southern Africa. *South African Journal of Science*, **87**(8): 346–349.

Lutjeharms, J.R.E. (1992). Hoe gaan dit met die fisiese oseanologie in Suid-Afrika? *South African Journal of Science*, **88**(2): 73.

Lutjeharms, J.R.E. (1992). The Physical Oceanology of the South West Indian Ocean; a bibliography from 1986 to 1990. *CSIR Research Report*, **707**, 95 pp.

Lutjeharms, J.R.E., W.P.M. de Ruijter and R.G. Peterson (1992). Interbasin exchange and the Agulhas retroflection; the development of some oceanographic concepts. *Deep-Sea Research*, **39**(10): 1791–1807.

Lutjeharms, J.R.E., S.J. Weeks, R.D. [sic] van Ballegooyen and F.A. Shillington (1992). Shedding of an eddy from the seaward front of the Agulhas Current. *South African Journal of Science*, **88**(8): 430–433.

Lutjeharms, J.R.E., C.-T. Liu, W.-S. Chuan and C.-Z. Shyu (1993). On some similarities between the oceanic circulations off southern Africa and off Taiwan. *South African Journal of Science*, **89**(8): 367–371.

Lutjeharms, J.R.E., M.I. Lucas, R. Perissinotto, R.C. van Ballegooyen and M. Rouault (1994). Oceanic processes at the Subtropical Convergence: report of research cruise SAAMES III. *South African Journal of Science*, **90**(7): 367–370.

Lutjeharms, J.R.E. and C.P. Matthysen (1995). A recurrent eddy in the upwelling front off Cape Town. *South African Journal of Science*, **91**(7): 355–357.

Lutjeharms, J.R.E. and D.J. Webb (1995). Modelling the Agulhas Current system with FRAM (*Fine Resolution Antarctic Model*). *Deep-Sea Research I*, **42**(4): 523–551.

Lutjeharms, J.R.E. (1996). The exchange of water between the South Indian and the South Atlantic. In *The South Atlantic: Present and Past Circulation*, editors G. Wefer, W.H. Berger, G. Siedler and D. Webb, Springer-Verlag, Berlin, pp. 125–162.

Lutjeharms, J.R.E. and J. Cooper (1996). Interbasin leakage through Agulhas Current filaments. *Deep-Sea Research I*, **43**(2): 213–238.

Lutjeharms, J.R.E. and W.P.M. de Ruijter (1996). The influence of the Agulhas Current on the adjacent coastal zone: possible impacts of climate change. *Journal of Marine Systems*, **7**(2–4): 321–336.

Lutjeharms, J.R.E., D.J. Webb, B.A. de Cuevas and S.R. Thompson (1996). Large-scale modelling of the South-East Atlantic upwelling system. *South African Journal of Marine Science*, **16**: 205–225.

Lutjeharms, J.R.E. and L.V. Shannon (1997). A century of physical oceanography in South Africa; in search of the legacy of John D. Gilchrist. In *A Century of Marine Science in South Africa*, editors A.I.L. Payne and J.R.E. Lutjeharms, Sea Fisheries Research Institute, Royal Society of South Africa, pp. 17–30 and *Transactions of the Royal Society of South Africa*, **52**(1): 17–30.

Lutjeharms, J.R.E., A.A. Meyer, I.J. Ansorge, G.A. Eagle and M.J. Orren (1996). The nutrient characteristics of the Agulhas Bank. *South African Journal of Marine Science*, **17**: 253–274.

Lutjeharms, J.R.E., O. Boebel and T. Rossby (1997). KAPEX: an international experiment to study deep water movement around southern Africa. *South African Journal of Science*, **93**(9): 377–381.

Lutjeharms, J.R.E. (1998). Coastal hydrography. In *A Field Guide to the Eastern and Southern Cape Coast*, editors R. Lubke and I. de Moor, The Wildlife and Environment Society of Southern Africa, Grahamstown, University of Cape Town Press, Rondebosch, pp. 50–61.

Lutjeharms, J.R.E. (1999). Frontal structures and eddy formation in the greater Agulhas Current system. In *Oceanic Fronts and Related* Phenomena, Konstantin Federov Memorial Symposium, Proceedings, I O C Workshop Report No. 159, UNESCO 2000, pp. 320–325.

Lutjeharms, J.R.E. and E. Machu (2000). An upwelling cell inshore of the East Madagascar Current. *Deep-Sea Research I*, **47**(12): 2405–2411.

Lutjeharms, J.R.E. and M. Rouault (2000). Observations of cloud formation above Agulhas Current intrusions in the South-east Atlantic. *South African Journal of Science*, **96**(11/12): 577–580.

Lutjeharms, J.R.E., J. Cooper and M. Roberts (2000). Upwelling at the inshore edge of the Agulhas Current. *Continental Shelf Research*, **20**(7): 737–761.

Lutjeharms, J.R.E., H.R. Valentine and R.C. van Ballegooyen (2000). The hydrography and water masses of the Natal Bight, South Africa. *Continental Shelf Research*, **20**(14): 1907–1939.

Lutjeharms, J.R.E., P.M. Wedepohl and J.M. Meeuwis (2000). On the surface drift of the East Madagascar and the Mozambique Currents. *South African Journal of Science*, **96**(3): 141–147.

Lutjeharms, J.R.E., W.P.M. de Ruijter, H. Ridderinkhof, H. van Aken, C. Veth, P.J. van Leeuwen, S.S. Drijfhout,

J.H.F. Jansen and G.-J.A. Brummer (2000). MARE and ACSEX: new research programmes on the Agulhas Current system. *South African Journal of Science*, **96**(3): 105–110.

Lutjeharms, J.R.E. (2001). The Agulhas Current. In *Encyclopedia of Ocean Science*, editors: J. Steele, S. Thorpe and K. Turekian, Academic Press, London, volume 1, pp. 104–113.

Lutjeharms, J.R.E. and I. Ansorge (2001). The Agulhas Return Current. *Journal of Marine Systems*, **30**(1/2): 115–138.

Lutjeharms, J.R.E., P.M.S. Monteiro, P.D. Tyson and D. Obura (2001). The oceans around southern Africa and regional effects of global change. *South African Journal of Science*, **97**(3/4): 119–130.

Lutjeharms, J.R.E., O. Boebel, P.C.F. van der Vaart, W.P.M. de Ruijter, T.H. Rossby and H.L. Bryden (2001). Evidence that the Natal Pulse controls the Agulhas Current over its full depth. *Geophysical Research Letters*, **28**(18): 3449–3452.

Lutjeharms, J.R.E., S. Jamaloodien and I.J. Ansorge (2002). The temporal displacement of ocean fronts south-east of Africa. *South African Journal of Science*, **98**(5/6): 304–306.

Lutjeharms, J.R.E. and C.S. Fillis (2003). Intrusion of sub-Antarctic water across the Subtropical Convergence south of Africa. *South African Journal of Science*, 99(3/4): 173–176.

Lutjeharms, J.R.E. and G. Kortum (2003). On the influence of German marine research on South African physical oceanography since 1890. *Historisch-meereskundliches Jahrbuch*, **9**: 81–100.

Lutjeharms, J.R.E., O. Boebel and T. Rossby (2003). Agulhas cyclones. *Deep-Sea Research II*, **50**(1): 13–34.

Lutjeharms, J.R.E., P. Penven, C. Roy (2003). Modelling the shear-edge eddies of the southern Agulhas Current. *Continental Shelf Research*, 23(11–13): 1099–1115.

Lutjeharms, J.R.E. (2004). The coastal oceans of south-eastern Africa. In *The Sea*, Volume 14, editors: A.R. Robinson, K. Brink, J. Hall, G. Ingram, D. Mackas and D. Townsend, accepted for publication.

Lutjeharms, J.R.E. and G. Kortum (2004). On the German influence in South African physical oceanography since 1890. In *Ocean Sciences Bridging the Millenia – A spectrum of Historical Accounts*, IOC, in press.

Lutjeharms, J.R.E. and G. Kortum (2004). German research on the Agulhas Current system between the World Wars; a lost scientific achievement. *Historisch-meereskundliches Jahrbuch*, in preparation.

Lutjeharms, J.R.E. and A.A. Meyer (2004). On the origin and circulation of bottom water on the Agulhas Bank, South Africa. *Continental Shelf Research*, in preparation.

Luyten, J.R. and P.F. Smith (1985). Agulhas Current trajectory from new Argos drifter compared with simultaneous shipboard measurements. In *Oceans '85 Ocean Engineering and the Environment*, San Diego, 12–24 November 1985, 2, pp. 1165–1167.

Luyten, J., A. Spencer, S. Tarbell, K. Luetkemeyer, P. Flament, J. Toole, M. Francis and S. Bennett (1990). Moored current meter, AVHRR, CTD, and drifter data from the Agulhas Current and Retroflexion [sic] region (1985–1987), Vol, XLII. Woods Hole Oceanographic Institution, Technical Report, WHOI-90-30, 100 pp.

Machu, E., B. Ferret and V. Garçon (1999). Phytoplankton pigment distribution from SeaWiFS data in the subtropical convergence zone south of Africa: a wavelet analysis. *Geophysical Research Letters*, **26**(10): 1469–1472.

Machu, E. and V. Garçon (2001). Phytoplankton seasonal distribution from SeaWiFs data in the Agulhas Current system. *Journal of Marine Research*, **59**(5): 795–812.

Machu, E., J.R.E. Lutjeharms, A.M. Webb and H.M. van Aken (2003). First hydrographic evidence of the southeast Madagascar upwelling cell. *Geophysical Research Letters*, **29**(21), 2009, doi:10.1029/2002GL015381.

Machu, E., A. Biastoch, A. Oschlies, M. Kawamiya, J.R.E. Lutjeharms and V. Garçon (2004). Phytoplankton distribution in the Agulhas system from a coupled physical–biological model. *Deep-Sea Research I*, in revision.

Magnier, Y. and B. Piton (1973). Les masses d'eau de l'Ocean Indien a l'ouest et au nord de Madagascar au début de l'été austral (Novembre–Décembre). *Cahiers ORSTOM.*, **11**: 97–113.

Majodina, M. and M.R. Jury (1996). Composite winter cyclones south of Africa: evolution during eastward transit over the Agulhas warm pool. *South African Journal of Marine Science*, **17**: 241–252.

Malan, O.G. and E.H. Schumann (1979). Natal shelf circulation revealed by Landsat imagery. *South African Journal of Science*, **75**(3): 136–137.

Mallory, J.K. (1961). Bathymetric and hydrographic aspects of marine studies off the Natal Coast. In *Marine Studies off the Natal Coast*, Natal Regional Research Committee, *CSIR Symposium*, **S2**, pp. 31–39.

Mallory, J.K. (1984). Abnormal waves off the south-east coast of South Africa. *Marine Observer*, **54**(283): 29–38.

Maltrud, M.E., R.D. Smith, A. J. Semtner and R.C. Malone (1998). Global eddy-resolving ocean simulations driven by 1985–1995 atmospheric winds. *Journal of Geophysical Research*, **103**(C13): 30,825–30,853.

Mantyla, A.W. and J.L. Reid (1995). On the origins of deep and bottom waters of the Indian Ocean. *Journal of Geophysical Research*, **100**(C2): 2417–2439.

Marchesiello, P., B. Barnier and A.P. de Miranda (1998). A sigma-coordinate primitive equation model for studying the circulation in the South Atlantic. Part II: Meridional transport and seasonal variability. *Deep-Sea Research I*, **45**(4–5): 573–608.

Markham, C.R. (1901). *Major James Rennell and the rise of modern English geography.* Cassell and Company, London, 232 pp.

Martin, A.K. (1981). Evolution of the Agulhas Current and its palaeo–ecological implications. *South African Journal of Science*, **77**(12): 547–554.

Martin, A.K. (1981). The influence of the Agulhas Current

on the physiographic development of the northernmost Natal Valley (S.W. Indian Ocean). *Marine Geology*, **39**: 259–276.

Martin, A.K. and B.W. Flemming (1988). Physiography, structure and geological evolution of the Natal continental shelf. In *Coastal Ocean Studies off Natal, South Africa*, editor E.H. Schumann, Lecture Notes on Coastal and Estuarine Studies 26, Springer-Verlag, Berlin, pp. 11–46.

Martin, J., P. Guibout, M. Crepon and J.-C. Lizaray (1965). Circulation superficielle dans l'ocean Indien. Résultats de mesures faites à électrodes remorguées G.E.K. entre 1955–1963. *Cahiers Oceanographie*, **17**, suppl. 3, 221–241.

Mason, S.J. (1990). Temporal variability of sea surface temperatures around Southern Africa: a possible forcing mechanism for the 18-year rainfall oscillation. *South African Journal of Science*, **86**(5/6): 243–252.

Mason, S.J. and P.D. Tyson (1992). The modulation of sea surface temperature and rainfall associations over southern Africa with solar activity and the quasi-biennial oscillation. *Journal of Geophysical Research*, **97**(D5): 5847–5856.

Mason, S.J., J.A. Lindesay and P.D. Tyson (1994). Simulating drought in southern Africa using sea-surface temperature-variations. *Water SA*, **20**(1): 15–22.

Mason, S.J. (1995). Sea-surface temperature – South African rainfall associations, 1910–1989. *International Journal of Climatology*, **15**(2): 119–135.

Matano, R.P. and S.G.H. Philander (1993). Heat and mass balances of the South Atlantic Ocean calculated from a numerical model. *Journal of Geophysical Research*, **98**(C1): 977–984.

Matano, R.P. (1996). A numerical study of the Agulhas retroflection: the role of bottom topography. *Journal of Physical Oceanography*, **26**(10): 2267–2279.

Matano, R.P., C.G. Simionato, W.P.(M.) de Ruijter, P.J. van Leeuween [sic], P.T. Strub, D.B. Chelton and M.G. Schlax (1998). Seasonal variability in the Agulhas Retroflection region. *Geophysical Research Letters*, **25**(23): 4361–4364.

Matano, R.P., C.G. Simionato and P.T. Strub (1999). Modelling the wind driven variability of the South Indian Ocean. *Journal of Physical Oceanography*, **29**(2): 217–230.

Matano, R.P., E.J. Beier, P.T. Strub and R. Tokmakian (2002). Large-scale forcing of the Agulhas variability: the seasonal cycle. *Journal of Physical Oceanography*, **32**(4): 1228–1241.

Matano, R.P. and E.J. Beier (2003). A kinematic analysis of the Indian/Atlantic interocean exchange. *Deep-Sea Research II*, **50**(1): 229–249.

Maury, M.F. (1855). *The Physical Geography of the Sea*. Harper & Brothers, New York, 274 pp., 12 plates.

Maury, M.F. (edited by J. Leighly, 1963). *The physical geography of the sea; and its meteorology*. The Belknap Press of Harvard University Press, Cambridge, MA, 432 pp., 10 plates.

McCartney, M.S. and M.E. Woodgate-Jones (1991). A deep-reaching anticyclonic eddy in the subtropical gyre of the eastern South Atlantic. *Deep-Sea Research*, **38** (Supplement 1): s411–s443.

McCarthy, M.S. and L.D. Talley (1999). Three-dimensional isoneutral potential vorticity structure in the Indian Ocean. *Journal of Geophysical Research*, **104**(C6): 13,251–13,267.

McDonagh, E.L. and K.J. Heywood (1999). The origin of an anomalous ring in the Southeast Atlantic. *Journal of Physical Oceanography*, **29**(8): 2025–2064.

McDonagh, E.L., K.J. Heywood and M.P. Meredith (1999). On the structure, paths and fluxes associated with Agulhas rings. *Journal of Geophysical Research*, **104**(C9): 21,007–21,020.

Menaché, M. (1963). Premiere campagne océanographique du "Commandant Robert Giraud" dans le Canal de Mozambique 11 Octobre – 28 Novembre 1957. *Cahiers Océanographiques*, **15**(4): 224–235.

Mercier, H., M. Arhan and J.R.E. Lutjeharms (2003). Upper-layer circulation in the eastern Equatorial and South Atlantic Oceans in January–March 1995. *Deep-Sea Research I*, **50**(7): 863–887.

Merz, A. (1925). Die Deutsche Atlantische Expedition auf dem Vermessungs- und Forschungsschiff "Meteor". *Sitzungsberichten der Physikalish-Matematischen Klasse der Preussischen Akademie der Wissenschaften*, **31**, pp. 562–586.

Metzl, N., B. Moore and A. Poisson (1990). Resolving the intermediate and deep convective flows in the Indian Ocean by using temperature, salinity, oxygen and phosphate data: the interplay of biogeochemical and geophysical tracers. *Palaeogeography, Palaeoclimatology, Palaeoecology*, **89**(1–2): 81–111.

Mey, R.D., N.D. Walker and M.R. Jury (1990). Surface heat fluxes and marine boundary layer modification in the Agulhas retroflection region. *Journal of Geophysical Research*, **95**(C9): 15,997–16,015.

Meyer, A.A., J.R.E. Lutjeharms and S. de Villiers (2002). The nutrient characteristics of the Natal Bight, South Africa. *Journal of Marine Systems*, **35**(1–2): 11–37.

Michaelis, G. (1923). Die Wasserbewegung an der Oberfläche des Indischen Ozeans im Januar und Juli. *Veröffentlichungen des Instituts für Meereskunde an der Universität Berlin*, n.f. **A8**(16): 32 pp.

Minster, J.-F. and M.C. Gennero (1995). High-frequency variability of western boundary currents using ERS 1 three-day repeat altimeter data. *Journal of Geophysical Research*, **100**(C11): 22,603–22,612.

Möller, L. (1929). Die Zirkulation des Indischen Ozeans; auf Grund von Temperatur- und Salzgehaltstiefenmessungen und Oberflächenstrombeobachtungen. *Veröffentlichungen des Instituts für Meereskunde an der Universität Berlin*, n.f. **A21**: 48 pp.

Möller, L. (1933). Zur Frage der Tiefenzirkulation im Indischen Ozean. *Annalen des Hydrographie Berlin*, **61**(8,9): 233–236.

Morrow, R., F. Birol, D. Griffen and J. Sudre (2004). Divergent pathways of cyclonic and anti-cyclonic eddies. *Geophysical Research Letters*, to be submitted.

Mühry, A. (1864). Die Meeresströmungen an der Südspitze Afrikas. *Petermanns Geographische Mitteilungen*, **10**: 34–35.

Naeije, M.C., K.F. Wakker, R. Scharoo and B.C.C. Ambrosius (1992). Observation of mesoscale ocean currents from GEOSAT altimeter data. *ISPRS Journal of Photogrammetry and Remote Sensing*, **47**: 347–368.

Naeije, M.C., E. Wisse, R. Scharoo and K.F. Wakker (1994). Ocean dynamics from the ERS-1 35-day repeat mission. *Proceedings Second ERS-1 Symposium – Space at the Service of our Environment*, **ESA SP–361**, pp. 501–506.

Nafpaktitis, B.G. (1978). Systematics and distribution of laternfishes of the genera Lobianchia – and Diaphus (*Myctophidae*) in the Indian Ocean, *Scientific Bulletin, Natural History Museum, Los Angeles*, **30**: 1–91.

Natal Regional Research Committee (1961). *Marine studies off the Natal coast.* Council for Scientific and Industrial Research, CSIR Symposium **S2**, 134 pp.

Naval Oceanographic Office (1977). *Surface Currents. Southeast South Atlantic Ocean.* U.S. Naval Oceanographic Office, Special Publication 1401–SA4.

Naval Oceanographic Office (1977). *Surface Currents. Southwest Indian Ocean.* U.S. Naval Oceanographic Office, Special Publication. 1404–IN6.

Naval Oceanographic Office (1977). *Surface Currents. West Central Indian Ocean Including Mozambique Channel.* U.S. Naval Oceanographic Office, Special Publication 1404–IN3.

Nehring, D. (editor) (1984). The oceanological conditions in the western part of the Mozambique Channel in February–March 1980. *Geodätische und geophysikalische Veröffentlichungen*, **4**(9): 163 pp.

Nehring, D., E. Hagen, A. Jorge da Silva, R. Schemainda, G. Wolf, N. Michelsen, W. Kaiser, L. Postel, F. Gosselck, U. Brenning, E. Kuhner, G. Arlt, H. Siegel, L. Gohs, G. Bublitz (1987). Results of oceanological studies in the Mozambique Channel in February–March 1980. *Beiträge zur Meereskunde*, **56**: 51–63.

New, A.L., K. Stansfield, D. Smyth-Wright, D.A. Smeed, A.L. Evans and S.G. Alderson (2004). Physical and biochemical aspects of the flow across the Mascarene Plateau in the Indian Ocean. *Philosophical Transactions*, submitted.

Nof, D. (2000). Does the wind control the import and export of the South Atlantic? *Journal of Physical Oceanography*, **30**(11): 2650–2667.

Olbers, D. (editor) (1992). *Hydrographic Atlas of the Southern Ocean.* Alfred-Wegener-Institut für Polar- und Meeresforschung, Bremerhaven, xvii + 82 pp. + plates.

Oliff, W.D. (1973). Chemistry and productivity at Richard's Bay. *NPRL Oceanography Division, Contract Report*, **CFIS 37B**, Durban, South Africa.

Olivar, M.P. and L.E. Beckley (1994). Influence of the Agulhas Current on the distribution of Laternfish larvae off the south-east coast of Africa. *Journal of Plankton Research*, **16**(12): 1759–1780.

Olson, D.B. and R.H. Evans (1986). Rings of the Agulhas Current. *Deep-Sea Research*, **33**(1): 27–42.

Olson, D.B., R.A. Fine and A.L. Gordon (1992). Convective modifications of water masses in the Agulhas. *Deep-Sea Research*, **39** (Supplement 1): s163–s181.

Orren, M.J. (1963). Hydrological observations in the South West Indian Ocean. *Investigational Reports of the Division of Sea Fisheries* (of South Africa), **45**: 61 pp.

Ou, H.W. and W.P.M. de Ruijter (1986). Separation of an inertial boundary current from a curved coastline. *Journal of Physical Oceanography*, **16**(2): 280–289.

Paech, H. (1926). Die Oberflächenströmungen um Madagascar in ihrem Jährlichen Gang. *Veröffentlichungen des Instituts für Meereskunde an der Universität Berlin*, n.f. **A**(16): 39 pp.

Park, Y.-H., L. Gambéroni and E. Charriaud (1991). Frontal structure and transport of the Antarctic Circumpolar Current in the South Indian Ocean sector, 40–80° E. *Marine Chemistry*, **35**(1): 45–62.

Park, Y.-H., L. Gambéroni and E. Charriaud (1993). Frontal structure, water masses, and circulation in the Crozet Basin. *Journal of Geophysical Research*, **98**(C7): 12,361–12,385.

Park, Y.-H. and L. Gambéroni (1995). Large-scale circulation and its variability in the South Indian Ocean from TOPEX/POSEIDON altimetry. *Journal of Geophysical Research*, **100**(C12): 24,911–24,929.

Park, Y.-H. and L. Gambéroni (1997). Cross-frontal exchanges of Antarctic Intermediate Water and Antarctic Bottom Water in the Crozet Basin. *Deep-Sea Research I*, **44**(5): 963–986.

Park, Y.-H., E. Charriaud and M. Fieux (1998). Thermohaline structure of the Antarctic Surface Water/Winter Water in the Indian sector of the Southern Ocean. *Journal of Marine Systems*, **17**(1–4): 5–23.

Park, Y.-H., E. Charriaud, P. Craneguy and A. Kartavtseff (2001). Fronts, transport, and Weddell Gyre at 30° E between Africa and Antarctica. *Journal of Geophysical Research*, **106**(C2): 2857–2879.

Patterson, S.L. (1985). Surface circulation and kinetic energy distribution in the southern hemisphere oceans from FGGE drifting buoys. *Journal of Physical Oceanography*, **15**(7): 865–884.

Pearce, A.F. (1975). Ode to an eddy. *Nature*, **258**(5535): 486.

Pearce, A.F. (1977). The shelf circulation off the east coast of South Africa. *Council for Scientific and Industrial Research, CSIR Research Report*, **361**, 220 pp.

Pearce, A.F. (1977). Some features of the upper 500m of the Agulhas Current. *Journal of Marine Research*, **35**(4): 731–753.

Pearce, A.F. (1978). Seasonal variations of temperature and salinity on the northern Natal continental shelf. *South African Geographical Journal*, **60**(2): 135–143.

Pearce, A.F., E.H. Schumann and G.S.H. Lundie (1978). Features of the shelf circulation off the Natal coast. *South*

African Journal of Science, **74**(9): 328–331.
Pearce, A.F. (1980). Early observations and historical notes on the Agulhas Current circulation. *Transactions of the Royal Society of South Africa*, **44**(2): 205–212.
Pearce, A.F. and M.L. Gründlingh (1982). Is there a seasonal variation in the Agulhas Current? *Journal of Marine Research*, **40**(1): 177–184.
Pearce, A.F. (1983). Adrift in the Agulhas Return Current. *Deep-Sea Research*, **30**(3): 343–347.
Peeters, F.J.C., R. Acheson, G.-J.A. Brummer, W.P.M. de Ruijter, G.M. Ganssen, R.R. Schneider, E. Ufkes and D. Kroon (2004). Vigorous exchange between Indian and Atlantic Ocean at the end of the last five glacial periods. *Nature*, accepted.
Penduff, T., B. Barnier, K. Beranger and T. Verron (2001). Comparison of near-surface mean and eddy flows from two numerical models of the South Atlantic Ocean. *Journal of Geophysical Research*, **106**(C8): 16,857–16,867.
Penduff, T., P. Brasseur, C.-E. Testut, B. Barnier and J. Verron (2002). A four-year eddy-permitting assimilation of sea-surface temperature and altimetric data in the South Atlantic Ocean. *Journal of Marine Research*, **60**(6): 805–833.
Penven, P., J.R.E. Lutjeharms, P. Marchesiello, S.J. Weeks and C. Roy (2001). Generation of cyclonic eddies by the Agulhas Current in the lee of the Agulhas Bank. *Geophysical Research Letters*, **27**(6): 1055–1058.
Penven, P., C. Roy, G.B. Brundrit, A. Colin de Verdière, P. Fréon, A.S. Johnson, J.R.E. Lutjeharms and F.A. Shillington (2001). A regional hydrodynamic model of the Southern Benguela. *South African Journal of Science*, **97** (11/12): 472–475.
Petermann, A. (1850). *Atlas of Physical Geography*. Wm. S. Orr and Co., London, 124 pp., 15 plates.
Petermann, A. (1865). Karte der Arktischen und Antarktischen Regimen zur Übersicht des geographischen Standpundtes im Jahre 1865. *Petermanns Mitteilungen*, Tafel 5.
Peterson, R.G. and L. Stramma (1991). Upper-level circulation in the South Atlantic Ocean. *Progress in Oceanography*, **26**: 1–75.
Peterson, R.G., L. Stramma and G. Kortum (1996). Early concepts and charts of ocean circulation. *Progress in Oceanography*, **37**(1): 1–115.
Pichevin, T., D. Nof and J. Lutjeharms (1999). Why are there Agulhas rings? *Journal of Physical Oceanography*, **29**(4): 693–707.
Pillsbury, R.D., J.M. Bottero, G. Pittock, D.C. Root, J. Simpkins and R.E. Still (1994). Benguela Source and Transport Project (BEST): Current measurements off the coast of South Africa. WOCE ACM-4, June 1992–October 1993. *OSU Data Report* 157, Reference 94-3, Corvallis, 69 pp.
Piola, A.R., H.A. Figueroa and A.A. Bianchi (1987). Some aspects of the surface circulation south of 20°S revealed by First GARP Global Experiment drifters. *Journal of Geophysical Research*, **92**(C5): 5101–5114.
Piton, B. and J.F. Poulain (1974). Resultat des mesures de courant superficiels au GEK effectuees avec le N.O. "Vauban" dans le sud-ouest de l'océan Indien, 1973–74. *ORSTOM Centre de Nosy Bé, Document Scientifique*, **47**, 77 pp.
Piton, B., J.-H. Pointeau and J.-S. Nigoumbi (1981). Atlas hydrologique du Canal de Mozambique (Océan Indien). *Travaux et Documents de l'ORSTOM*, **132**: 41 pp.
Piton, B. (1989). Quelques aspects nouveaux sur la circulation superficielle dans le canal de Mozambique (Ocean Indien). *Document Scientifique, Centre Brest ORSTOM*, **54**: 31 pp.
Pollak, M.J. (1958). Frequency distribution of potential temperatures and salinities in the Indian Ocean. *Deep-Sea Research*, **5**(2): 128–133.
Pollard, R. and G. Griffiths (1993). James Rennell, the father of oceanography. *Ocean Challenge*, **4**(1/2): 24–25.
Prell, W.L. and W.H. Hutson (1979). Zonal temperature-anomaly maps of Indian Ocean surface waters – modern and ice-age patterns. *Science*, **206**(4417): 454–456.
Prell, W.L., W.H. Hutson and D.F. Williams (1979). The subtropical convergence and Late Quaternary circulation in the southern Indian Ocean. *Marine Micropaleontology*, **4**: 225–234.
Prell, W.L., W.H. Hutson, D.F. Williams, A.W.H. Bé, K. Geitzenauer and B. Molfino (1980). Surface circulation of the Indian Ocean during the last glacial maximum, approximately 18,000 yr. B.P. *Quarternary Research*, **14**(3): 309–336.
Preston-Whyte, R.A. (1974). Land breezes and mountain–plain winds over the Natal coast. *South African Geographical Journal*, **56**(1): 27–35.
Preston-Whyte, R.A. and P.D. Tyson (1988). *The Atmosphere and Weather of Southern Africa*. Oxford University Press, Cape Town, pp. 374.
Probyn, T.A., B.A. Mitchell-Innes, P.C. Brown, L. Hutchings and R.A. Carter (1994). A review of primary productivity and related processes on the Agulhas Bank. *South African Journal of Science*, **90**(3): 166–173.
Quartly, G.D. and M.A. Srokosz (1993). Seasonal variations in the region of the Agulhas retroflection: studies with Geosat and FRAM. *Journal of Physical Oceanography*, **23**(9): 2107–2124.
Quartly, G.D. and M.A. Srokosz (2002). Satellite observations of the Agulhas Current system. *Philosophical Transactions of the Royal Society of London A*, **361**(1802): 51–56.
Quartly, G.D. and M.A. Srokosz (2002). SST observations of the Agulhas and East Madagascar retroflections by the TRMM Microwave imager. *Journal of Physical Oceanography*, **32**(5): 1585–1592.
Quartly, G.D. and M.A. Srokosz (2004). Eddies in the southern Mozambique Channel. *Deep-Sea Research II*, **51**(1): 69–83.
Quartly, G.D., J.J.H. Buck, M.A. Srokosz and A.C. Coward (2004). Eddies around Madagascar – An end to the retroflection? *Journal of Physical Oceanography*, submitted.
Ramage, C.S. (1984). Climate of the Indian Ocean north of

35°S. In *Climates of the Oceans,* editor H. van Loon, World Survey of Climatology, Volume 15, Elsevier, Amsterdam, pp. 693–697.

Randall, R.M., B.M. Randall and D. Baird (1981). Speed of movement of Jackass penguins over long distances and their possible use of ocean currents. *South African Journal of Science*, **77**(9): 420–421.

Rau, A.J., J. Rogers, J.R.E. Lutjeharms, J. Giraureau, J.A. Lee-Thorp, M.-T. Chen and C. Waelbroeck (2002). A 450-kyr record of hydrographical conditions on the western Agulhas Bank slope, south of Africa. *Marine Geology*, **180**(1–4): 183–201.

Read, J.F. and R.T. Pollard (1993). Structure and transport of the Antarctic Circumpolar Current and Agulhas Return Current at 40°E. *Journal of Geophysical Research*, **98** (C7): 12,281–12,295.

Read, J.F. and R.T. Pollard (1999). Deep inflow into the Mozambique Basin. *Journal of Geophysical Research*, **104**(C2): 3075–3090.

Read, J.F., M.I. Lucas, S.E. Holley and R.T. Pollard (2000). Phytoplankton, nutrients and hydrography in the frontal zone between the Southwest Indian Subtropical gyre and the Southern Ocean. *Deep-Sea Research I*, **47**(12): 2341–2368.

Reason, C.J.C., R.J. Allan and J.A. Lindesay (1996). Dynamical response of the oceanic circulation and temperature to interdecadal variability in the surface winds over the Indian Ocean. *Journal of Climate*, **9**(1): 97–114.

Reason, C.J.C., R.J. Allan and J.A. Lindesay (1996). Evidence for the influence of remote forcing on interdecadal variability in the southern Indian Ocean. *Journal of Geophysical Research*, **101**(C5): 11,867–11,882.

Reason, C.J.C. (1998). Warm and cold events in the southeast Atlantic/southwest Indian Ocean region and potential impacts on circulation and rainfall over southern Africa. *Meteorology and Atmospheric Physics*, **69**(1–2): 49–65.

Reason, C.J.C. and C.R. Godfred-Spenning (1998). SST variability in the South Indian Ocean and associated circulation and rainfall patterns over southern Africa. *Meteorology and Atmospheric Physics*, **66**(3–4): 243–258.

Reason, C.J.C. and J.R.E. Lutjeharms (1998). Variability of the South Indian Ocean and implications for southern African rainfall. *South African Journal of Science*, **94**(3): 115–123.

Reason, C.J.C., C. R. Godfred-Spenning, R.J. Allan and J.A. Lindesay (1998). Air–sea interaction mechanisms and low frequency climate variability in the South Indian Ocean region. *International Journal of Climatology*, **18**(4): 391–405.

Reason, C.J.C. and H. Mulenga (1999). Relationships between South African rainfall and SST anomalies in the Southwest Indian Ocean. *International Journal of Climatology*, **19**(15): 1651–1673.

Reason, C.J.C. (2000). Multidecadal climate variability in the subtropics/midlatitudes of the Southern Hemisphere oceans. *Tellus*, **52**(A): 203–223.

Reason, C.J.C. and J.R.E. Lutjeharms (2000). Modelling multidecadal variability in the South Indian Ocean region: local forcing or a near-global mode? *South African Journal of Science*, **96**(3): 127–135.

Reason, C.J.C. (2001). Evidence for the influence of the Agulhas Current on regional atmospheric circulation patterns. *Journal of Climate*, **14**(6): 2769–2778.

Reason, C.J.C. (2001). Subtropical Indian Ocean SST dipole events and southern African rainfall. *Geophysical Research Letters*, **28**(11): 2225–2227.

Reason, C.J.C. (2002). The wet winter of 2001 over the south-western Cape, South Africa: potential large-scale influences. *South African Journal of Science*, **98**(5/6): 307–310.

Reason, C.J.C., J.R.E. Lutjeharms, J. Hermes, A. Biastoch and R.E. Roman (2003). Inter-ocean fluxes south of Africa in an eddy-permitting model. *Deep-Sea Research II*, **50**(1): 281–298.

Reichs-marine Ampt (1909). *Forschungsreise S.M.S. "Planet" 1906–1907.* Reichs-marine Ampt, Berlin, Bände I–V.

Reid, J.L. (1989). On the total geostrophic circulation of the South Atlantic Ocean: flow patterns, tracers and transports. *Progress in Oceanography*, **23**(3): 149–244.

Reid, J.R. (1996). On the circulation of the South Atlantic Ocean. In *The South Atlantic: Present and Past Circulation*, editors: G. Wefer, W. H. Berger, G. Siedler and D. Webb, Springer-Verlag, Berlin, pp. 13–44.

Reid, J.R. (2003). On the total geostrophic circulation of the Indian Ocean: flow patterns, tracers and transports. *Progress in Oceanography*, **56**(1): 137–186.

Rennell, J. (1778). *Chart of the Bank of Lagullus and southern Coast of Africa,* London.

Rennell, J. (1832). *An investigation of the currents of the Atlantic Ocean, and of those which prevail between the Indian Ocean and the Atlantic.* J.G. and F. Rivington, London, editor J. Purdy, viii + 359 pp. + 2 folding maps.

Richardson, P.L., M.E. Pacheco and C.M. Wooding (2002). KAPEX RAFOS float data report 1997–1997. Part B. Float trajectories at 750 m in the Benguela Current. Woods Hole Oceanographic Institution Technical Report, in preparation.

Richardson, P.L. and S.L. Garzoli (2003). Characteristics of intermediate water flows in the Benguela Current as measured with RAFOS floats. *Deep-Sea Research II*, **50**(1): 87–118.

Richardson, P.L., J.R.E. Lutjeharms and O. Boebel (2003). Introduction to the "Inter-ocean exchange around southern Africa". *Deep-Sea Research II*, **50**(1): 1–12.

Ridderinkhof, H. (2000). RV Pelagia cruise report: Cruise 64PE156, Project ACSEX-1. NIOZ, Texel, pp. 25 + appendix.

Ridderinkhof, H., J.R.E. Lutjeharms and W.P.M. de Ruijter (2001). A research cruise to investigate the Mozambique Current. *South African Journal of Science*, **97**(11/12): 461–464.

Ridderinkhof, H. and W.P.M. de Ruijter (2003). Moored current observations in the Mozambique Channel. *Deep-*

Sea Research II, 50(12–13): 1933–1955.

Rintoul, S. R. (1991). South Atlantic interbasin exchange. *Journal of Geophysical Research*, **96**(C2): 2675–2692.

Rochford, D. (1964). Salinity maximum in the upper 1000 waters of the north Indian Ocean. *Australian Journal of Marine and Freshwater Research*, **15**(1): 1–24.

Roel, B. A., J. Hewitson, S. Kerstan and I. Hampton (1994). The role of the Agulhas Bank in the life cycle of pelagic fish. *South African Journal of Science*, **90**(3): 185–196.

Römer, E. (1939). The counter current off the south and southeast African coast. *Hydrographic Review*, **16**: 88–91.

Rossouw, J. (1984). Review of existing wave data, wave climate and design waves for South African and South West African (Namibian) coastal waters. *National Research Institute for Oceanology, CSIR Report*, **T/SEA 8401**, 66 pp. + tables + figures.

Rouault, M. and J.R.E. Lutjeharms (1994). Air-sea interaction in the marine atmosphere boundary layer: a new South African research venture. *South African Journal of Science*, **90**(1): 11–12.

Rouault, M., A. Lee-Thorp, I. Ansorge and J. Lutjeharms (1995). Agulhas Current Air-Sea Interaction Experiment. *South African Journal of Science*, **91**(10): 493–496.

Rouault, M. and A.M. Lee-Thorp (1996). Fine-time resolution measurements of atmospheric boundary layer properties between Cape Town and Marion Island. *South African Journal of Marine Science*, **17**: 281–296.

Rouault, M., A.M. Lee-Thorp, I.J. Ansorge and J.R.E. Lutjeharms (1997). Data report on the Agulhas Current Air–Sea Exchange Experiment, April–May 1995. University of Cape Town, Department of Oceanography, Report **DO-97-1**, 160 pp.

Rouault, M., J.R.E. Lutjeharms, A.M. Lee-Thorp, M.R. Jury and M. Majodina (1999). Air–sea Interaction over the Agulhas Current and Implication for South African Weather. Report to the Water Research Commission, Department of Oceanography, University of Cape Town, *WRC Report* **No 374/1/99**, 104 pp.

Rouault, M. and J.R.E. Lutjeharms (2000). Air–sea exchange over an Agulhas eddy at the Subtropical Convergence. *The Global Atmosphere and Ocean System*, **7**(2): 125–150.

Rouault, M., A.M. Lee-Thorp and J.R.E. Lutjeharms (2000). The atmospheric boundary layer above the Agulhas Current during alongcurrent winds. *Journal of Physical Oceanography*, **30**(1): 40–50.

Rouault, M., I. Jobard, S.A. White and J.R.E. Lutjeharms (2001). Studying rainfall events over South Africa and adjacent oceans using the TRMM satellite. *South African Journal of Science*, **97**(11/12): 455–460.

Rouault, M., S.A. White, C.J.C. Reason and J.R.E. Lutjeharms and I. Jobard (2002). Ocean–atmosphere interaction in the Agulhas Current and a South African extreme weather event. *Weather and Forecasting*, **17**(4): 655–669.

Rouault, M., C.J.C. Reason, J.R.E. Lutjeharms and A. Belaars (2003). Underestimation of latent and sensible heat fluxes above the Agulhas Current in NCEP and ECMWF analyses. *Journal of Climate*, **16**(2): 776–782.

Rouault, M., C. J. Reason, J.R.E. Lutjeharms, H. Mulenga, Y. Richard, N. Fauchereau and S. Trzaksa (2003). Role of the oceans in South Africa's rainfall. Water Research Commission, *WRC Report* No. **953/1/03**, 55 pp.

Rouault, M. and J.R.E. Lutjeharms (2003). Estimation of sea-surface temperature around southern Africa from satellite-derived microwave observations. *South African Journal of Science*, **99**(9/10): 489–494.

Roubicek, A.J., S.L. Garzoli, P.L. Richardson, C.M. Duncombe Rae and D.M. Fratantoni (1998). Benguela Current Experiment: RV *Seward Johnson* SJ 9705, Cape Town September 4, 1997–Recife September 30, 1997. *NOAA Data Report* ERL AOML-33.

Ryan, P.G. (1988). The characteristics and distribution of plastic particles at the sea-surface off the southwestern Cape Province, South Africa. *Marine Environmental Research*, **25**(4): 249–273.

Sætre, R. and R. de Paula e Silva (1979). The marine fish resources of Mozambique. *Reports on surveys with the R/V Dr Fridtjof Nansen*, Serviço de Investigações Pesqueiras, Maputo, Institute of Marine Research, Bergen, 179 pp.

Sætre, R. and A. Jorge da Silva (1982). Water masses and circulation of the Mozambique Channel. *Revista de Investigação Pesquira*, **3**, Instituto de Desenvolvimento Pesqueiro, Maputo, República Popular de Moçambique, 83 pp.

Sætre, R. and A. Jorge da Silva (1984). The circulation of the Mozambique Channel. *Deep-Sea Research*, **31**(5): 485–508.

Sætre, R. (1985). Surface currents in the Mozambique Channel. *Deep-Sea Research*, **32**(12): 1457–1467.

Schell, I.I. (1968). On the relation between the winds off southwest Africa and the Benguela Current and the Agulhas Current penetration in the South Atlantic. *Deutsche Hydrographische Zeitschrift*, **21**(3): 109–117.

Schemainda, R. and E. Hagen (1983). On steady state intermediate vertical currents induced by the Mozambique Current. *Océanographie Tropicale*, **18**(1): 81–88.

Schleyer, M.H. (1985). Chaetognaths as indicators of water masses in the Agulhas Current system. *Investigations Reports, Oceanographic Research Institute, Durban*, **61**: 1–20.

Schlitzer, R. (1996). Mass and heat transport in the South Atlantic derived from historical hydrographic data. In *The South Atlantic: present and past circulation*, editors G. Wefer, W.H. Berger, G. Siedler and D.J. Webb, Springer-Verlag, Berlin, pp. 289–304.

Schmid, C., O. Boebel, W. Zenk, J.R.E. Lutjeharms, S. Garzoli, P.L. Richardson and C. Barron (2003). Early evolution of an Agulhas ring. *Deep-Sea Research II*, **50**(1): 141–166.

Schmitz, W.J. and J.R. Luyten (1991). Spectral time scales for mid-latitude eddies. *Journal of Marine Research*, **49**(1): 75–107.

Schmitz, W.J. (1996). On the eddy field in the Agulhas Retroflection, with some global considerations. *Journal of Geophysical Research*, **101**(C7): 16,259–16,271.

Schmitz, W.J. (1996). On the World Ocean Circulation: Volume II. The Pacific and Indian Oceans/A Global Update. Woods Hole Oceanographic Institution, Technical Report WHOI-96-08, 237 pp.

Schott, F., M. Fieux, J. Kindle, J. Swallow and R. Zantopp (1988). The boundary currents east and north of Madagascar. Part II. Direct measurements and model comparisons. *Journal of Geophysical Research*, **93**(C5): 4963–4974.

Schott, G. (1926). Die Tiefenwasserbewegung des Indischen Ozeans. *Annalen der Hydrographie, Berlin*, **54**(2): 417–431.

Schott, G. (1928). Die Verteilung des Salzgehaltes im Oberflachenwasser der Ozeane. *Annalen der Hydrographie, Berlin*, **5**: 145–166.

Schott, G. (1935). *Geographie des Indischen und Stillen Ozeans*. C. Boysen, 413 pp., 37 plates.

Schott, G. (1944). *Geographie des Atlantischen Ozeans*. C. Boysen, 438 pp., 38 plates.

Schouten, M.W., W.P.M. de Ruijter, P.J. van Leeuwen and J.R.E. Lutjeharms (2000). Translation, decay and splitting of Agulhas rings in the south-eastern Atlantic ocean. *Journal of Geophysical Research*, **105**(C9): 21,913–21,925.

Schouten, M.W., W.P.M. de Ruijter and P.J. van Leeuwen (2002). Upstream control of Agulhas ring shedding. *Journal of Geophysical Research*, **107**(C8): 10.1029/2001JC000804.

Schouten, M.W., W.P.M. de Ruijter, P.J. van Leeuwen and H. Dijkstra (2002). A teleconnection between the equatorial and southern Indian Ocean. *Geophysical Research Letters*, **29**(16), 1812, doi: 10.1029/2001GL014542.

Schouten, M.W., W.P.M. de Ruijter, P.J. van Leeuwen and H. Ridderinkhof (2003). Eddies and variability in the Mozambique Channel. *Deep-Sea Research II*, **50**(12–13): 1987–2003.

Schouten, M.W., W.P.M. de Ruijter, P.J. van Leeuwen and H.A. Dijkstra (2003). An oceanic connection between Indian Ocean equatorial winds and eddy variability around Southern Africa. *Deep-Sea Research I*, resubmitted.

Schumann, E.H. (1976). Changes in energy of surface gravity waves in the Agulhas Current. *Deep-Sea Research*, **23**(6): 509–518.

Schumann, E.H. (1981). Low frequency fluctuations off the Natal coast. *Journal of Geophysical Research*, **86**(C7): 6499–6508.

Schumann, E.H. (1982). Inshore circulation of the Agulhas Current off Natal. *Journal of Marine Research*, **40**(1): 43–55.

Schumann, E.H. and L.-A. Perrins (1982). Tidal and inertial currents around South Africa. In *Proceedings of the Eighteenth International Coastal Engineering Conference*, American Society of Civil Engineers, Cape Town, South Africa, Nov. 14–19, 1982, pp. 2562–2580.

Schumann, E.H., L.-A. Perrins and I.T. Hunter (1982). Upwelling along the south coast of the Cape Province, South Africa. *South African Journal of Science*, **78**(6): 238–242.

Schumann, E.H. (1983). Long period coastal trapped waves off the southeast coast of Southern Africa. *Continental Shelf Research*, **2**(2/3): 97–107.

Schumann, E.H. and L.J. Beekman (1984). Ocean temperature structures on the Agulhas Bank. *Transactions of the Royal Society of South Africa*, **34**(2): 191–203.

Schumann, E.H. (1986). The bottom boundary layer inshore of the Agulhas Current off Natal in August 1975. *South African Journal of Marine Science*, **4**: 93–102.

Schumann, E.H. (1987). The coastal ocean off the east coast of South Africa. *Transactions of the Royal Society of South Africa*, **46**(3): 215–229.

Schumann, E.H. (editor) (1988). *Coastal Ocean Studies off Natal, South Africa*, Lecture Notes on Coastal and Estuarine Studies 26, Springer-Verlag, Berlin, 271 pp.

Schumann, E.H. (1988). Physical oceanography off Natal. In *Coastal Ocean Studies off Natal, South Africa*, editor E.H. Schumann, Lecture Notes on Coastal and Estuarine Studies 26, Springer-Verlag, Berlin, pp. 101–130.

Schumann, E.H. and I. Li [sic] van Heerden (1988). Observations of Agulhas Current frontal features south of Africa, October 1983. *Deep-Sea Research*, **35**(8): 1355–1362.

Schumann, E.H., G.J.B. Ross and W.S. Goschen (1988). Cold water events in Algoa Bay and along the Cape south coast, South Africa, in March/April 1987. *South African Journal of Science*, **84**(7): 579–584.

Schumann, E.H. (1989). The propagation of air pressure and wind systems along the South African coast. *South African Journal of Science*, **85**(6): 382–385.

Schumann, E.H. and K.H. Brink (1990). Coastal-trapped waves off the coast of South Africa: generation, propagation and current structures. *Journal of Physical Oceanography*, **20**(8): 1206–1218.

Schumann, E.H. and J.A. Martin (1991). Climatological aspects of the coastal wind field at Cape Town, Port Elizabeth and Durban. *South African Geographical Journal*, **73**(1): 48–51.

Schumann, E.H., W.K. Illenberger and W.S. Goschen (1991). Surface winds over Algoa Bay. *South African Journal of Science*, **87**(5): 202–207.

Schumann, E.H., J.R.E. Lutjeharms, A.J. Boyd, M.L. Gründlingh and G.B. Brundrit (1991). Physical oceanography in South Africa: 1987 to 1990. *South African Journal of Science*, **87**(10): 486–492.

Schumann, E.H. (1992). Interannual wind variability on the south and east coasts of South Africa. *Journal of Geophysical Research*, **97**(D18): 20,397–20,403.

Schumann, E.H., A.L. Cohen and M.R. Jury (1995). Coastal sea surface temperature variability along the south coast of South Africa and the relationship to regional and global climate. *Journal of Marine Research*, **53**(2): 231–248.

Schumann, E.H. (1998). The coastal ocean off southeast Africa, including Madagascar. In *The Sea*, Volume 11,

Chapter 19, editors A.R. Robinson and K.H. Brink, John Wiley & Sons, pp. 557–581.

Schumann, E.H. (1999). Wind-driven mixed layer and coastal upwelling processes off the south coast of South Africa. *Journal of Marine Research*, **57**(4): 671–691.

Semtner, A.J. and R.M. Chervin (1988). A simulation of the global ocean circulation with resolved eddies. *Journal of Geophysical Research*, **93**(C12): 15,502–15,522.

Semtner, A.J. and R.M. Chervin (1992). Ocean general circulation from a global eddy-resolving model. *Journal of Geophysical Research*, **97**(C4): 5493–5550.

Shannon, L.V. (1966). Hydrology of the South and West coasts of South Africa. *Investigational Report, Division of Sea Fisheries*, **58**: 1–52.

Shannon, L.V. (1979). The Cape Nimbus-7 CZCS programme: An overview and preliminary results. *South African Journal of Science*, **75**(12): 564.

Shannon, L.V. and P. Chapman (1983). Suggested mechanisms for the chronic pollution by oil of beaches east of Cape Agulhas, South Africa. *South African Journal of Marine Science*, **1**: 231–244.

Shannon, L.V., P. Chapman, G.A. Eagle and T.P. McClurg (1983). A comparative study of tar ball distribution and movement in two boundary current regimes: implication for oiling of beaches. *Oil & Petrochemical Pollution*, **1**(4): 243–259.

Shannon, L.V., S.A. Mostert, N.M. Walters and F.P. Anderson (1983). Chlorophyll concentrations in the southern Benguela Current region as determined by satellite (Nimbus-7 Coastal Zone Colour Scanner). *Journal of Plankton Research*, **5**(4): 565–583.

Shannon, L.V., P. Schlittenhardt and S.A. Mostert (1984). The NIMBUS 7 CZCS experiment in the Benguela Current region off southern Africa, February 1980. 2. Interpretation of imagery and oceanographic implications. *Journal of Geophysical Research*, **89**(D4): 4968–4976.

Shannon, L.V. (editor) (1985). *South African Ocean Colour and Upwelling Experiment*. Sea Fisheries Research Institute, Cape Town, 270 pp.

Shannon, L.V. (1985). The Benguela Ecosystem. 1. Evolution of the Benguela, physical features and processes. *Oceanography and Marine Biology, An Annual Review*, **23**: 105–182.

Shannon, L.V. and D. Hunter (1988). Notes on Antarctic Intermediate Water around Southern Africa. *South African Journal of Marine Science*, **6**: 107–117.

Shannon, L.V., J.R.E. Lutjeharms and J.J. Agenbag (1989). Episodic input of Subantarctic water into the Benguela region. *South African Journal of Science*, **85**(5): 317–332.

Shannon, L.V., J.R.E. Lutjeharms and G. Nelson (1990). Causative mechanisms for intra-annual and inter-annual variability in the marine environment around Southern Africa. *South African Journal of Science*, **86**(7/8/9/10): 356–373.

Shannon, L.V., J.J. Agenbag, N.D. Walker and J.R.E. Lutjeharms (1990). A major perturbation in the Agulhas retroflection area in 1986. *Deep-Sea Research*, **37**(3): 493–512.

Shannon, L.V. and G. Nelson (1996). The Benguela: large scale features and processes and system variability. In *The South Atlantic: Present and Past Circulation*, editors G. Wefer, W.H. Berger, G. Siedler and D. Webb, Springer-Verlag, Berlin, pp. 163–210.

Shannon, L.V. (2001). The Benguela Current. In *Encyclopedia of Ocean Science*, editors: J. Steele, S. Thorpe and K. Turekian, Academic Press, London, volume 1, pp. 255–267.

Shillington, F.A. (1984). Long period edge waves off southern Africa. *Continental Shelf Research*, **3**(4): 343–357.

Shillington, F.A., W.T. Peterson, L. Hutchings, T.A. Probyn, H.N. Waldron and J.J. Agenbag (1990). A cool upwelling filament off Namibia, southwest Africa: preliminary measurements of physical and biological features. *Deep-Sea Research*, **37**(11): 1753–1772.

Shillington, F.A. (1998). The Benguela upwelling system off southwestern Africa. In *The Sea*, Volume 11, Chapter 20, editors A.R. Robinson and K.H. Brink, John Wiley & Sons, pp. 583–604.

Skogen, M.D. (1999). A biophysical model applied to the Benguela upwelling system. *South African Journal of Marine Science*, **21**: 235–249.

Smythe-Wright, D., A.L. Gordon, P. Chapman and M.S. Jones (1996). CFC – 113 shows Brazil eddy crossing the South Atlantic to the Agulhas retroflection region. *Journal of Geophysical Research*, **101**(C1): 885–895.

Smythe-Wright, D., P. Chapman, C. Duncombe Rae, L.V. Shannon and S.M. Boswell (1998). Characteristics of the South Atlantic subtropical frontal zone between 15°W and 5°E. *Deep-Sea Research I*, **45**(1): 167–192.

Snaith, H.M. and I.S. Robinson (1996). A study of currents south of Africa using Geosat satellite altimetry. *Journal of Geophysical Research*, **101**(C8): 18,141–18,154.

Sparrow, M.D., K.J. Heywood, J. Brown and D.P. Stevens (1996). Current structure of the south Indian Ocean. *Journal of Geophysical Research*, **101**(C3): 6377–6391.

Stavropoulos, C.C. and C.P. Duncan (1974). A satellite-tracked buoy in the Agulhas Current. *Journal of Geophysical Research*, **79**(C18): 2744–2746.

Stevens, D.P. and S.R. Thompson (1994). The South Atlantic in the Fine-Resolution Antarctic Model. *Annales Geophysicae*, **12**: 826–839.

Stramma, L. (1992). The South Indian Ocean Current. *Journal of Physical Oceanography*, **22**(4): 421–430.

Stramma, L. and R.G. Peterson (1990). The South Atlantic Current. *Journal of Physical Oceanography*, **20**(6): 846–859.

Stramma, L. and J.R.E. Lutjeharms (1997). The flow field of the subtropical gyre of the South Indian Ocean. *Journal of Geophysical Research*, **102**(C3): 5513–5530.

Stramma, L. and M. England (1999). On the water masses and mean circulation of the South Atlantic Ocean. *Journal of Geophysical Research*, **104**(C9): 20,863–20,883.

Stramma, L. (2001). Current systems of the Atlantic Ocean. In *Encyclopedia of Ocean Science*, editors: J. Steele, S.

Thorpe and K. Turekian, Academic Press, London, pp. 589–598.

Srokosz, M.A., G.D. Quartly and J.J.H. Buck (2004). A possible plankton wave in the Indian Ocean. *Geophysical Research Letters*, **31**, L13301, doi: 10.1029/2004GLO19738.

Strub, P.T., F.A. Shillington, C. James and S.J. Weeks (1998). Satellite comparison of the seasonal circulation in the Benguela and California Current systems. In *Benguela Dynamics*, editors S.C. Pillar, C.L. Moloney, A.I.L. Payne and F.A. Shillington, *South African Journal of Marine Science*, **19**: 99–112.

Stutzer, S. (1997). Modellierung der mittleren Zirkulation im Südatlantik. *Berichte aus dem Institut für Meereskunde an der Christian-Albrecht-Universität, Kiel*, **287**, 130 pp.

Suga, T. and L.D. Talley (1995). Antarctic Intermediate Water circulation in the tropical and subtropical South Atlantic. *Journal of Geophysical Research*, **100**(C7): 13,441–13,453.

Swallow, J.C. and R.T. Pollard (1988). Flow of bottom water through the Madagascar Basin. *Deep-Sea Research*, **35**(8): 1437–1440.

Swallow, J.C., M. Fieux and F. Schott (1988). The boundary currents east and north of Madagascar. Part I. Geostrophic currents and transports. *Journal of Geophysical Research*, **93**(C5): 4951–4962.

Swallow, J.C., F. Schott and M. Fieux (1991). Structure and transport of the East African Coastal Current. *Journal of Geophysical Research*, **96**(C12): 22,245–22,257.

Swart, V.P. and J.L. Largier (1987). Thermal structure of Agulhas Bank water. *South African Journal of Marine Science*, **5**: 243–253.

Taljaard, J.J. (1977). South Africa's contribution to the FGGE drifting buoy programme for the southern hemisphere. *News Letter, South African Weather Bureau*, **339**: 198–207.

Taljaard, J.J., J. van Heerden, R.D. Sewell and W. Jordaan (1983). South African drifting buoy programmes. *South African Weather Bureau, Technical Paper*, **12**: 1–57.

Taljaard, J.J. and H. van Loon (1984). Climate of the Indian Ocean south of 35°S. In *Climates of the Oceans*, editor H. van Loon, World Survey of Climatology Volume 15, Elsevier, Amsterdam, pp. 505–601.

Taunton-Clark, J. and L.V. Shannon (1988). Annual and interannual variability in the South-East Atlantic during the 20th century. *South African Journal of Marine Science*, **6**: 97–196.

Tennant, W.J. (1996). Influence of Indian Ocean sea-surface temperature anomalies on the general circulation of southern Africa. *South African Journal of Science*, **92**(6): 289–295.

Thibault-Botha, D., J.R.E. Lutjeharms and M.J. Gibbons (2004). Siphonophore assemblages along the East Coast of South Africa; mesoscale distribution and seasonal variations. *Journal of Plankton Research*, 26(9): 1–14.

Thomsen, H. (1933). The circulation in the depths of the Indian Ocean. *Journal Conseil*, **8**(1): 73–79.

Thomsen, H. (1935). Entstehung und Verbreitung einiger charakteristischer Wassermassen in dem Indischen und südlichen Pazifischen Ozean. *Annalen der Hydrographie und Maritimen Meteorologie*, **63**(8): 293–305.

Thomson, C.W. and J. Murray (editors) (1881–1895). *Report on the scientific results of the voyage of H.M.S. Challenger during the years 1873–1876*. His Majesty's Stationery Office, London, 50 volumes.

Thompson, S.R., D.P. Stevens and K. Döös (1997). The importance of interocean exchange south of Africa in a numerical model. *Journal of Geophysical Research*, **102**(C2): 3303–3315.

Tomczak, M. (1984). Ausbreitung und Vermischung der Zentralwassermassen in den Tropengebieten der Ozeane. 2. Indischer und Pazifischer Ozean. *Oceanologica Acta*, **7**(3): 271–288.

Toole, J.M. and B.A. Warren (1993). A hydrographic section across the subtropical South Indian Ocean. *Deep-Sea Research I*, **40**(10): 1973–2019.

Toole, J.M. and M.E. Raymer (1985). Meridional heat and fresh water budgets of the South Indian Ocean – revisited. *Deep-Sea Research*, **32**(8): 917–928.

Treguier, A.M., B. Barnier, A.P. de Miranda, J.M. Molines, N. Grima, M. Imbard, G. Madec, C. Messager, Y. Reynaud and S. Michel (2001). An eddy-permitting model of the Atlantic circulation: Evaluating open boundary conditions. *Journal of Geophysical Research*, **106** (C10): 22,115–22,129.

Treguier, A.M., O. Boebel, B. Barnier and G. Madec (2003). Agulhas eddy fluxes in a 1/6° Atlantic model. *Deep-Sea Research II*, **50**(1): 251–280.

Tripp, R.T. (~1967). *An Atlas of Coastal Surface Drifts; Cape Town to Durban*. South African Oceanographic Data Centre, Department of Oceanography, University of Cape Town, Rondebosch, South Africa, 12 pp.

Tyson, P.D. (1986). *Climate Change and Variability in Southern Africa*. Oxford University Press, Cape Town, 220 pp.

Tyson, P.D., T.G.J. Dyer and M.N. Manetse (1975). Secular changes in South African rainfall: 1880 to 1972. *Quarterly Journal of the Royal Meteorological Society*, **101** (430): 817–833.

United Kingdom Hydrographic Office (1970). *Routing Charts – Indian Ocean*. British Admiralty, Hydrographic Office, London, No. 5126.

Valentine, H.R. and J.R.E. Lutjeharms (1986). Eddy generation at the Sub-Tropical Convergence south of South Africa. *CSIR Research Report*, **631**, 23 pp. + 27 figures.

Valentine, H.R., C.M. Duncombe Rae, R.C. van Ballegooyen and J.R.E. Lutjeharms (1988). The Subtropical Convergence and Agulhas Retroflection Cruise (SCARC); Data Report. *CSIR Report*, **T/SEA 8804**, 10 pp. + 56 tables.

Valentine, H.R., R.C. van Ballegooyen and J.R.E. Lutjeharms (1991). Data report on the Natal Bight Cruise, July 1989. *CSIR Report*, **EMA-D 9107**, ii + 15 pp. + figures + tables.

Valentine, H.R., J.R.E. Lutjeharms and G.B. Brundrit (1993). The water masses and volumetry of the southern Agulhas

Current region. *Deep-Sea Research I*, **40**(6): 1285–1305.

Van Aken, H.M. (2001). RV Pelagia shipboard report: Cruise 64PE176, Agulhas Current Sources Experiment (ACSEX). NIOZ, Texel, pp. 22 + appendix.

Van Aken, H., A.K. van Veldhoven, C. Veth, W.P.M. de Ruijter, P.J. van Leeuwen, S.S. Drijfhout, C.P. Whittle and M. Rouault (2003). Observations of a young Agulhas ring, Astrid, during MARE, the Mixing of Agulhas Rings Experiment, in March 2000. *Deep–Sea Research II*, **50**(1): 167–195.

Van Aken, H.M., H. Ridderinkhof and W.P.M. de Ruijter (2004). North Atlantic Deep Water in the south-western Indian Ocean. *Deep-Sea Research I*, **51**(6): 755–776.

Van Ballegooyen, R.C., M.L. Gründlingh and J.R.E. Lutjeharms (1994). Eddy fluxes of heat and salt from the southwest Indian Ocean into the southeast Atlantic Ocean: a case study. *Journal of Geophysical Research*, **99**(C7): 14,053–14,070.

Van der Elst, R.P. (1988). Shelf ichthyofauna off Natal. In *Coastal Ocean Studies off Natal, South Africa*, editor E.H. Schumann, Lecture Notes of Coastal and Estuarine Studies, 26, Springer-Verlag, Berlin, pp. 209–225.

Van der Vaart, P.C.F and W.P.M. de Ruijter (2001). Stability of western boundary currents with an application to pulslike behaviour of the Agulhas Current. *Journal of Physical Oceanography*, **31**(9): 2625–2644.

Van Gogh, J. (1857). *Uitkomsten van Wetenskap en Ervaring aangaande winden en zeestromingen in sommige gedeelten van den oseaan*. Koninklijk Nederlandsch Meteorologisch Instituut, Bosch en Zoon, 74 pp.

Van Gogh, J. (1858). De stormen nabij de Kaap de Goede Hoop in verband beschouwd met de temperatuur der zee. *Verhandelingen en Mededelingen der Koninklijke Akademie, afdeling Natuurkunde*, **8**, 23 pp.

Van Heerden, J. and J.J. Taljaard (1998). Africa and surrounding waters. In *Meteorology of the Southern Hemisphere*, editors D. J. Karoly and D. G. Vincent, *Meteorological Monographs*, **27**(49): 141–174.

Van Leeuwen, P.J. (1997). A monitoring system for the Indian-Atlantic connection. In *Operational Oceanography, The Challenge for European Co-operation*, editors J.H. Stel, H.W.A. Behrens, J.C. Borst, L.J. Droppert and J. van der Meulen, Elsevier, Amsterdam, pp. 596–602.

Van Leeuwen, P.J. (1999). The time-mean circulation in the Agulhas region determined with the ensemble smoother. *Journal of Geophysical Research*, **104**(C1): 1393–1404.

Van Leeuwen, P.J., W.P.M. de Ruijter and J.R.E. Lutjeharms (2000). Natal Pulses and the formation of Agulhas rings. *Journal of Geophysical Research*, **105**(C3): 6425–6436.

Van Leeuwen, P.J. (2001). An ensemble smoother with error estimates. *Monthly Weather Review*, **129**(4): 709–728.

Van Riel, P.M. (1932). Einige ozeanographische Beobachtungen im Roten Meer, Golf van Aden und Indischen Ozean. *Annalen der Hydrographie, Berlin*, **60**(10): 401–407.

Van Sittert, L. (1995). 'The Handmaiden of Industry': marine science and fisheries development in South Africa 1895–1939. *Studies of History and Philosophy of Science*, **26**(4): 531–558.

Verheye, H.M., L. Hutchings. J.A. Huggett, R.A. Carter, W.T. Peterson and S.J. Painting (1994). Community structure, distribution and tropic ecology of zooplankton on the Agulhas Bank with special reference to copepods. *South African Journal of Science*, **90**(3): 154–165.

Villacastin-Herrero, C.A., L G. Underhill, R.J.M. Crawford and L.V. Shannon (1996). Sea surface temperature of oceans surrounding subequatorial Africa: seasonal patterns, spatial coherence and long-term trends. *South African Journal of Science*, **92**(4): 189–197.

Vincent, E. and N.J. Shackleton (1980). Agulhas Current temperature distribution delineated by oxygen isotope analysis of foraminifer in surface sediments. *Cushman Foundation Special Publication*, **19**: 84–95.

Von Drygalski, E. (editor) (1908–1952). *Ergebnisse der Deutschen Südpolar-Expedition 1901–1903*. Reichsamtes des Innern, Berlin, Bände I – XX.

Von Schleinitz, B. (1889–1890). *Die Forschungsreise S.M.S. "Gazelle" in den Jahren 1874 bis 1876 unter Kommando des Kapitän zur Zee Freiherrn von Schleinitz*. Hydrographische Amt, Reichs-Marine-Ampts, Berlin, Bände I–V.

Wakker, K.F., R.C.A. Zandbergen, M.C. Naeije and B.A.C. Ambrosius (1990). GEOSAT altimeter data analysis for the oceans around South Africa. *Journal of Geophysical Research*, **95**(C3): 2991–3006.

Wakker, K.F., M.C. Naeije, R. Scharroo and B.A.C. Ambrosius (1991). Extraction of mesoscale ocean currents information from Geosat altimeter data. *Proceedings of the Space & Sea Colloquium*, Paris, ESA SP-312, pp. 221–226.

Walker, N.D. (1986). Satellite observations of the Agulhas Current and episodic upwelling south of Africa. *Deep-Sea Research*, **33**(8): 1083–1106.

Walker, N.D. and R.D. Mey (1988). Ocean/atmosphere heat fluxes within the Agulhas Retroflection region. *Journal of Geophysical Research*, **93**(C12): 15473–15483.

Walker N.D. and J.A. Lindesay (1989). Preliminary observations of oceanic influences on the February–March 1988 floods in central South Africa. *South African Journal of Science*, **85**(3): 164–169.

Walker, N.D. (1990). Links between South African summer rainfall and temperature variability of the Agulhas and Benguela Current Systems. *Journal of Geophysical Research*, **95**(C3): 3297–3319.

Walker, N.D. and F.A. Shillington (1990). The effect of oceanographic variability on South African weather and climate. *South African Journal of Science*, **86**(7/8/9/10): 382–386.

Wang, L. and C.J. Koblinsky (1996). Low-frequency variability in the region of the Agulhas Retroflection. *Journal of Geophysical Research*, **101**(C2): 3597–3614.

Warren, B.A. (1974). Deep flow in the Madagascar and Mascarene basins. *Deep-Sea Research*, **21**(1): 1–21.

Warren, B.A. (1978). Bottom water transport through the

Southwest Indian Ridge. *Deep-Sea Research*, **25**(3): 315–321.

Warren, B.A. (1981). The shallow oxygen minimum of the South Indian Ocean. *Deep-Sea Research*, **28**(8): 859–864.

Warren, B.A. (1981). Transindian hydrographic section at Lat. 18° S: Property distributions and circulation in the South Indian Ocean. *Deep-Sea Research*, **28**(8): 759–788.

Warren, B.A., T. Whitworth and J.H. LaCasce (2002). Forced resonant undulations in the deep Mascarene Basin. *Deep-Sea Research II*, **49**(7–8): 1513–1526.

Webb, D.J. (1999). An analytical model of the Agulhas Current as a western boundary current with linearly varying viscosity. *Journal of Physical Oceanography*, **29**(7): 1517–1527.

Wedepohl, P.M., J.R.E. Lutjeharms and J.M. Meeuwis (2000). Surface drift in the South East Atlantic ocean. *South African Journal of Marine Science*, **22**: 71–79.

Weeks, S.J. and F.A. Shillington (1994). Interannual scales of variation of pigment concentrations from coastal zone colour scanner data in the Benguela upwelling system and the Subtropical Convergence zone south of Africa. *Journal of Geophysical Research*, **99**(C4): 7385–7399.

Weeks, S.J. and F.A. Shillington (1996). Phytoplankton pigment distribution and frontal structure in the subtropical convergence region south of Africa. *Deep-Sea Research I*, **43**(5): 739–768.

Weeks, S.J., F.A. Shillington and G. B. Brundrit (1998). Seasonal and spatial SST variability in the Agulhas retroflection and Agulhas return current. *Deep-Sea Research I*, **45**(10): 1611–1625.

Weijer, W., W.P.M. de Ruijter, H.A. Dijkstra and P.J. van Leeuwen (1999). Impact of interbasin exchange on the Atlantic overturning circulation. *Journal of Physical Oceanography*, **29**(9): 2266–2284.

Weijer, W., W.P.M. de Ruijter and H.A. Dijkstra (2001). Stability of the Atlantic overturning circulation: competition between Bering Strait freshwater flux and Agulhas heat and salt sources. *Journal of Physical Oceanography*, **31**(8): 2385–2402.

Weijer, W., W.P.M. de Ruijter, A. Sterl and S.S. Drijfhout (2002). Response of the Atlantic overturning circulation to South Atlantic sources of buoyancy. *Global and Planetary Change*, 34(3–4): 293–311.

Weimerskirch, H., M. le Corre, S. Jaquemet, M. Potier and F. Marsac (2004). Foraging strategy of a top predator in tropical waters: great frigatebirds in the Mozambique Channel. *Marine Ecology Progress Series*, **275**( ): 297–308.

Wells, N.C., V. Ivchenko and S.E. Best (2000). Instabilities in the Agulhas retroflection current system – A comparative model study. *Journal of Geophysical Research*, **105**(C2): 3233–3241.

Whitworth, T. and W.D. Nowlin (1987). Water masses and currents of the Southern Ocean at the Greenwich meridian. *Journal of Geophysical Research*, **92**(C6): 6462–6476.

Wilkin, J.L. and R.A. Morrow (1994). Eddy kinetic energy and momentum flux in the Southern Ocean: comparison of a global eddy-resolving model with altimeter, drifter, and current-meter data. *Journal of Geophysical Research*, **99**(C4): 7903–7916.

Willimzik, M. (1929). Die Strömungen im Subtropischen Konvergenzgebiet des Indischen Ozeans. *Veröffentlichungen des Instituts für Meereskunde an der Universität Berlin, n.f.* **A**(14): 1–27.

Winter, A. and K. Martin (1990). Late quaternary history of the Agulhas Current. *Paleoceanography*, **5**(4): 479–486.

Witter, D. L. and A.L. Gordon (1999). Interannual variability of South Atlantic circulation from 4 years of TOPEX/POSEIDON satellite altimeter observations. *Journal of Geophysical Research*, **104**(C9): 20,927–20,948.

Wüst, G. (1934). Anzeichen von Beziehungen zwischen Bodenstrom und Relief in der Tiefsee des Indischen Ozeans. *Naturwissenschaften*, **22**(16): 241–244.

Wüst, G. (1935). Die Ausbreitung des antarktischen Bodenwassers im Atlantischen und Indischen Ozean. *Zeitschrift zur Geophysik*, **11**: 40–49.

Wüst, G. (1935). Zur Frage des Indischen Tiefenstroms. *Naturwissenschaften*, **23**(9): 137–139.

Wüst, G. (1960). Proposed International Indian Ocean Expedition, 1962–1863. *Deep-Sea Research*, **6**(3): 245–249.

Wyrtki, K. (1971). *Oceanographic Atlas of the International Indian Ocean Expedition*. National Science Foundation, Washington, D.C. 531 pp.

Wyrtki, K. (1973). Physical oceanography of the Indian Ocean. In *Biology of the Indian Ocean*, editor B. Zeitzschel, Springer-Verlag, Berlin, pp. 18–36.

Wyrtki, K., L. Magaard and J. Hagar (1976). Eddy energy in the oceans. *Journal of Geophysical Research*, **81**(15): 2641–2646.

You, Y. and M. Tomczak (1993). Thermocline circulation and ventilation in the Indian Ocean derived from water mass analysis. *Deep-Sea Research I*, **40**(1): 13–56.

You, Y. (1996). Dianeutral mixing in the thermocline of the Indian Ocean. *Deep-Sea Research I*, **43**(3): 291–320.

You, Y. (1997). Seasonal variations of thermocline circulation and ventilation in the Indian Ocean. *Journal of Geophysical Research*, **102**(C5): 10,391–10,422.

You, Y. (1998). Dianeutral mixing and transformation of Antarctic Intermediate Water in the Indian Ocean. *Journal of Geophysical Research*, **103**(C13): 30,941–30,971.

You, Y. (1998). Intermediate water circulation and ventilation of the Indian Ocean derived from water-mass contributions. *Journal of Marine Research*, **56**(5): 1029–1067.

You, Y. (1999). Dianeutral mixing, transformation and transport of the deep water of the Indian Ocean. *Deep-Sea Research I*, **46**(1): 109–148.

You, Y. (2000). Implications of the deep circulation and ventilation of the Indian Ocean on the renewal mechanism of North Atlantic Deep Water. *Journal of Geophysical Research*, **105**(C10): 23,895–23,926.

You, Y.Z. (2002). Quantitative estimate of Antarctic Intermediate Water contributions from the Drake Passage and the southwest Indian Ocean to the South Atlantic. *Jour-*

nal of Geophysical Research, **107**(C4): 10.1029/2001JC000880.

You, Y., J.R.E. Lutjeharms, O. Boebel and W.P.M. de Ruijter (2003). Quantification of interocean exchange of intermediate water masses around southern Africa. *Deep-Sea Research II*, **50**(1): 197–228.

Zahn, W. (1984). Eine Abschätzung des Volumentransportes im Kanal von Moçambique während des Zeitraumes Oktober–November 1957. *Beiträge zur Meereskunde*, **51**: 67–74.

Zahn, W. (1984). Influence of bottom topography on currents in the Mozambique Channel. *Tropical Ocean–Atmosphere Newsletter*, **26**: 22–23.

Zimmermann, W.F.A. (1865). *Der Erdball und Seine Naturwunder*, Gustav Hempel, Berlin, 544 pp. (as shown in Peterson *et al.*, 1993).

Name Index

In order to facilitate ease of reading of the text, references have as a rule been indicated by numerical superscripts. However, some key contributors to current knowledge on the greater Agulhas Current system are mentioned by name where their most important work is discussed. This index therefore lists the names of this subjectively selected collection of individuals only; a full list of all significant contributors as well as their works is to be found in the Bibliography. Wherever a picture of a person has been used in the text, this is indicated by a page number given in bold.

Andrau 155, 225
Aken, van 174
Arhan 180

Ballegooyen, van 174, 187, 193, 194
Bang 132, 155, **155**
Barlow 54, 92
Beal 38, 95
Belin 2
Bennett 170, 192
Berghaus 53
Biastoch 47
Boddem 197
Boebel 178, 211
Boudra 11, 199
Boyd 140
Bryden 95
Byrne 180, 193, 194

Cantino 2
Chapman 140, 167
Chassignet 11
Cheney 154
Chervin 47, 66, 197, 204
Chun 5, 19
Clement 193, 194
Clowes 19
Cooper 193, 194
Covens 2

Da Gama 1
d'Anville 2
Darbyshire 155
Defant 2
De Ruijter 75, 77, 81, 118, 157, **157**, 199
Dias 1, 2
Dietrich v, 2, 19, 155, 156
Donohue 24, 81
Duncan 57, 80, 84, 214
Duncombe Rae 171, 174, 193, 194

Emery 40
Evans 174, 180, 193, 194

Feron 85, 194
Findlay 2
Flemming 96
Foreest, van 192
Foulis 103
Fu 80, 197

Garzoli 162, 174, 192, 194
Georgi 170
Gibbons 15
Gilchrist 4
Gill 110, 138, 148
Godfred-Spenning 11
Gogh, van 225
Goñi 192–194
Gordon 85, 122, 157, **157**, 166, 168, 170, 191–194
Goschen 129
Grant 225
Gründlingh 44, 85, 88, 91, 93, 100, 101, 105

Harris 65, 68, 74, 80, 81, 85, 125, 133, 192, 220, **220**
Hasio 2
Haxby 193, 194
Holfort 197
Homan 2
Hondio 2
Humboldt, von 1

Jacobs 170
Jansen 225
Jorge da Silva 71, 73, 82, 83
Jury 12

Kerhallet 2, 53
Kostianoy 217
Krümmel 2, 53

Largier 140, 145
Leeuwen, van 203
Levitus 205
Linschoten, van 2
Luyten 152

MacDonald 197
Machu 18
Mallory 16
Martin 130
Mason 223
Matano 47, 197
Maury 2, 225
McCartney 103
McDonagh 174
Meincke 40
Menache 80, 81
Mercier 44
Michaelis 2, 54
Möller 2, 19
Mortier 2
Murray 19

Olson 174, 180, 193, 194

Paech 2, 54
Park 218
Pearce 91, 96, 98, 109, 113
Petermann 53
Peterson 192
Philander 197
Pichevin 200
Prell 15

Quartly 65

Raymer 170
Reason 11, 12
Rennell 1, 2, 3, **3**, 155
Richardson 187
Ridderinkhof 77
Rintoul 197
Rouault 135

Sætre 71, 73, 82, 83
Schlitzer 197
Schmitz 152
Schott 3, 19
Schouten 89, 187
Schumann 105, 108, 110, 129, 138, 149
Semtner 47, 66, 197, 204
Shannon 133, 162
Shillington 190
Simpson 64
Snyman 133
Srokosz 65
Stavropoulos 133, 152
Stramma 192, 215
Swallow 60
Swart 145

Tchernia ix
Thompson 197
Thomsen 2
Thomson 19
Toole 24, 32, 40, 81, 170
Tyson 223

Valentine 164
Valk 2
Van Aken 34, 174
Van Ballegooyen 174, 187, 193, 194
Van Foreest 192
Van Gogh 225
Van Leeuwen 203
Van Linschoten 2
Von Humboldt 1

Walker 12, 223
Warren 32, 40, 167
Weeks 44
Weijer 197
Winter 130
Woodgate-Jones 193
Wüst 19
Wyrtki 56

You 40, 191

Zahn 75, 80
Zimmerman 2, 53

Subject Index

This Subject Index is considerably more comprehensive than is normal for a textbook of this kind. This calls for a brief justification. The text attempts, first, to give a pre-digested summation of all that is currently known about the greater Agulhas Current system. In addition, it also aims at facilitating easy access to the extensive, but scattered, literature. If a particular subject is therefore mentioned only in passing in this text, but original publications on the subject exist and are referenced here, this subject has been included in the index. Copious cross-referencing is also offered. It is hoped that interested individuals will, by making use of both index and text in this way, be able easily to find out much more about many subjects than is directly available in the text of this monograph itself. Normal numbers indicate page numbers where a subject is mentioned in the text; bold numbers where a subject appears in a figure.

ACE (*see* Agulhas Current Experiment)
ACSEX (*see* Agulhas Current Sources Experiment)
ARC (*see* Agulhas Retroflection Cruise)
ASTTEX (*see* Agulhas-South Atlantic Thermohaline Transport Experiment)
abyssal boundary currents
– east of Madagascar 69
acoustic Doppler current profiler (*see* current measurements)
Africana II (vessel) 21
Agulhas Bank
– African penguins swimming across 14
– Agulhas Current's effect on 143, 145, 146
– Agulhas Bank Autumn Bottom Water on **146**
– Agulhas Bank Autumn Surface Water on **146**
– Agulhas Bank Spring Mixed Water on **146**
– Agulhas filaments advected along 139, 192
– Agulhas rings lodged against 139
– atmospheric conditions over **142**,143, 146, 148
– bathymetry of 91, **139**, **140**
– biology of 142, 143, 146
– bottom movement over 141
– bottom water on 129, 135, 1**129**, **140**, 141, 142, **145**
– bottom ridge, cold, on **140**, 141, 142, **143**, **145**
– chlorophyll enhancement at shelf edge of 139
– Central Water moving onto 140, 142, 143
– coastal upwelling over 139, 140, 141, **142**, 190
– cold ridge on **140**, 141, **143**, **144**, **145**
– currents observed on 145, 146, **148**
– current jet off western 146
– eastern 140
– ecosystem influence of Agulhas Current on 15, 146
– filament movement past 131
– fisheries, effect on 146
– fishing grounds on 4
– flow across 137
– hydrographic observations on 121, 122, **129**, **140**, 141, **144**, **145**, **147**
– hydrographic provinces of 139
– hydrographic sections across **129**, **140**, 141, **144**, **145**, **147**
– hydrography of 138
– ichthyo-ecology, effect on 146
– Indian Ocean Central Water on **146**
– inertial fluctuations on 145, **148**
– internal tides over 145
– lee eddy, modelling of 127
– lee eddy off 127, 128, **128**, 146, 176, 177 (*see also* lee eddy)
– location of **6**, **139**
– mixed layer over 140, **141**, 143
– model of circulation on 142, 144
– Natal Pulse, motion along **161**
– nutrients over 168
– oxygen along shelf edge of **167**, 167
– plumes, warm water, at **116**, 125, **126**, 127, 128, **129**, 129, 132, 139
– satellite studies of 133, 135
– seasonality of waters on 110, 129, 140, **141**, 144, **147**
– sea surface temperatures over **142**, **147**
– shear edge eddies off 127, **127**, **128**, **129**, 176
– shelf edge current of 139, 143, 176
– shelf edge front of **126**, 140
– shelf edge upwelling at 139
– South Indian Central Water moving onto 142, 143
– stratification over 140, **141**, 142, **147**
– Subtropical Surface Water on **146**
– temperature–salinity relations for water over 137, 143, **146**
– thermocline over 129, **141**, 143, **144**, 144, **147**
– tides on **148**
– Tropical Surface Water over **146**
– upwelling at capes and headlands of 140, 141
– upwelling on 133, 141, **145**, **159**
– volume flux at southern tip of 121, **122**
– western 140
– winds at 141, **142**, 143, 144, 145, 148, 149

Agulhas Basin
– Antarctic Bottom Water, movement in 32
– sea height variability in 69
Agulhas Bight
– location of **139**
– current location at **109**, 109, **112**
– hydrographic section at 123
– shear edge eddies in 127, **127**, 131
Agulhas, Cape (*see* Cape Agulhas)
Agulhas Current (*see also* Agulhas Undercurrent; Northern Agulhas Current; Southern Agulhas Current)
– age of 130
– Agulhas Bank, influence on 143, 145
– Agulhas Plateau, at 120
– Antarctic Circumpolar Current, merging with 218
– Antarctic Intermediate Water in 44
– "Atlantic branch" of 2
– atmosphere, influence on overlying 9, 137, 235
– atmospheric circulation over 149
– baroclinic instability of 118
– Benguela Current, volume flux contribution to 190
– bifurcation of 155
– biological influence of 14, 15, 16
– boundary, offshore, of 97, **97**, 98
– chaetognath distribution, influence on (*see also* chaetognaths) 15
– characteristics, unique, of 7
– cloud formation over 12
– coastal rainfall, influence on **13**
– compared to Gulf Stream 2
– current edge, inshore, of 97, **97**, 98
– current velocities for **93**, **94**, 94, **95**, **97**, **98**, 98
– depth of **43**
– drifters, surface, in 98, 104, 112, 120, 152, 158, 211, **212**
– early measurements of 53, 54, **55**
– eddies of (*see also* Agulhas (Current) eddies) 144, 153, 170, 198, **222**, **226**, **227**
– eddies, adsorption of, onto **116**, 118
– eddy kinetic energy for 91, **92**, 151
– Ekman veering under 110, 131, 132
– extent, latitudinal, of 5
– filaments of (*see* Agulhas filaments)
– first written account of 1
– first truly scientific investigation of 2
– fish larvae, influence on distribution of 14
– fully constituted at 91
– German, early research on 2, 3, 19
– giant waves, influence on 16, 105
– global thermohaline overturning, role in 130, 151, 191
– growth in knowledge of ix, x, 3
– heat fluxes to atmosphere from (*see also* heat fluxes) 9, 12, 13, 223
– historical concepts on 1
– hydrographic section across 122, **123**, **134**, **213**, **224**
– IIOE, impact of, on research on 21, 157
– indicator species for 15
– inflow into 104
– influence on Agulhas Bank of 143, 145, 146
– inhibiting propagation of shelf waves 149
– inshore counter current to 116
– kinetic energy for 91
– knowledge on ix
– latitudinal extent of 5
– leakage from (*see* inter-basin exchange)
– location of current core of **93**, 93, 97, **98**, **126**
– meandering of 7, 125, **126**, 127
– modelled as inertial jet 94, 120, 199, **200**
– modelled 47, **49**, **66**, 151, 154, 199, **201**, **202**, **204**, **205**
– moisture fluxes to the atmosphere from 12
– myctophids, influence on distribution of (*see* myctophids)
– name, origin of its 2
– Natal Bight, off the **93**, **99**, 109, 111, **112**, 113
– northern (*see* Northern Agulhas Current)
– nutrients at rims of 168
– oxygen, subsurface minimum layer in 191, **213**, 215
– path of **93**, 93, 94, **100**, **126**, 126, 127, **213**, 213
– pollution, plastic, by (*see also* pollution) 14
– publication rate on ix, x, 4
– rainfall, influence on **13**, 13, 137, 223
– recirculation, contribution of, to 85
– Red Sea Water in 38, **38**, 40,101, **102**, 106, **107**, 107, 123, **124**
– Red Sea Water, flux of, in 40
– retroflection of (*see* Agulhas (Current) retroflection)
– retroflection, upstream, on 6, 120, 196
– return flow of (*see* Agulhas Return Current)
– rings of (*see* Agulhas rings)
– salinity of **97**, **99**, 106, **106**, **107**, **114**, **124**, **125**, **146**
– salinities, surface, of 5
– salinities, surface, off Durban **99**
– satellite remote sensing, studied by **100**, **116**, 125, **126**, 133
– seasonal variability of 91, 92
– sections, vertical, across **86**, 93, 122, 123, **123**, **134**, **213**, 213, **224**
– sediments below 96, 130
– sheer edge features of 7, 100, **100**, 127, **127**, 131, 176, **177**, 177
– shelf circulation, effect on, knowledge gained on 235
– shelf waves, inhibiting progression of 149
– ships' drift measurements in sources of 53, 54
– siphonophores in 137
– sources of 5, 53, 85
– sources, lack of knowledge on 236
– sources, conceptual image for **67**
– speeds decreased, simulated for 203
– speeds simulated for 95, 203
– speeds in 4, **43**, 92, **95**, 97, 98, **99**, **108**, 144
– speeds, surface, of **93**, **94**
– stabilisation, by continental slope, of 94
– storms, intensity of, influence on 13, 14
– structure function analysis of 152
– subgyre, fed by, knowledge gained on 235
– Subtropical Convergence, effect on (*see also* Subtropical

Convergence) 218, **219**, **221**, 221, 222
– seasonality of surface drift of 55
– surface drift of 55, **55**, 91, 92, **93**
– temperature of **134**
– temperatures, surface, of 5
– temperatures, surface, off Durban **97**, 97, **99**
– temperature–salinity relationships of **39**, **102**, 101, **102**, **106**, 106, **107**, **124**, **179**
– temperature sections across 93, 122, **123**, 123, **125**, **134**, **224**
– textbook treatment of ix
– trajectory of **93**, 93, 94, **100**, **126**, 126, 127, **213**, 213
– Tropical Surface Water in 101, **102**, **106**, 106, 107, 122
– tunicates, influence on distribution of pelagic, in (*see* tunicates)
– upstream retroflection of 6, 120, 196
– upwelling inshore of (*see also* Port Alfred upwelling cell) 132, **134**, **136**, **138**, 138
– variability, surface, off Durban **97**, 98
– velocities, surface, off Durban **94**, **97**, **99**, 99
– velocity of 4, **43**, 92, **93**, **94**, 95, **95**, 97, 98, **99**, **108**, 144, 203
– vertical extent of 5, **94**, **95**, **123**
– vertical profiles for **43**, 93, 122, **123**, 123, **134**, **213**, 213, **224**
– vertical velocity structure of **43**, **94**, **95**, **99**
– volume flux of 6, 11, 42, **87**, 100, **101**, 144, 168, **170**, 214, 215, **215**
– volume fluxes simulated for 202, **203**
– volume transport of 100, 121, 156, 168, **170**
– volume transport of, downstream increase in 122, 168, **170**
– volume transport of, modelled **48**, 203
– water of (*see* Agulhas (Current) water)
– waves, huge, on 16, 105
– wave amplification on 16, 105
– western boundary current nature of 7, 8, 9
– western limb of anti-cyclonic gyre, as 5, 17, **24**
– width of 4, 94, **97**, 98
– winds, influence on 1, 22, 23, 164
Agulhas (Current) eddies (*see also* eddy)
– Agulhas retroflection, at 198, **226**, **227**
– current speeds of 227
– dimensions of 227
– drifter in 227
– heat flux, meridional, by 198
– heat loss to atmosphere over 12, 13, 198, 226
– hydrographic section across **163**, 198, **222**
– nomenclature of 170
– Subtropical Convergence, presence at 153, 170, **222**, **227**
– volume transport of 227
Agulhas Current Experiment (ACE)
– current measurements in Agulhas Current by 157
– Agulhas Undercurrent discovered during 157
Agulhas Current filaments (*see* Agulhas filaments)
Agulhas (Current) retroflection (*see also* Upstream retroflection)
– altimetric observations of 25, 154, **154**, 204
– Antarctic Bottom Water in **166**
– Antarctic Intermediate Water in 164, 165, **166**
– atmospheric boundary layer over 185
– beta (ß) effect, role of, on 199
– borders, thermal, of **153**
– Circumpolar Deep Water in **166**
– concept, historical development of, on 155
– configuration of **158**, 158
– continental rainfall patterns, influence on 151
– CTD observations in **165**
– current, simulated, in 199, **200**, **202**, **204**
– current measurements in 152, 154, 204
– drifters in 152, 157, 204, **212**
– dynamic height of **211**
– dynamic topography of **171**, **211**
– dynamics of 199 ff
– eddy kinetic energy for 152
– eddy kinetic energy modelled at 204
– eddy shedding at **226**, **227**, 227
– eddy variability modelled at 204
– heat fluxes to atmosphere at 151, 168, 185, 223
– hydrographic section across **163**, 167, **169**
– inertial effect in 199
– inter-basin exchange at 151
– kinetic energy of 152, **156**, 156
– kinetic variability for 152
– limit, easternmost, of 158
– limit, westernmost, of 158, **160**
– location, extreme, of 158, **160**
– loop, diameter of 151, 154, 158
– loop, locations of 151, 158, **160**
– loop, progradation of 7, 158, **160**
– Lower Circumpolar Deep Water in **166**
– mesoscale turbulence simulated for 154
– model of 47, 151, 199, **202**, **204**
– modelled as inertial jet 199, **200**
– moisture fluxes to atmosphere at 151
– momentum flux in model of 200
– Natal Pulse, influence of, on 154, 159, 160
– North Atlantic Deep Water in 164, **166**
– nutrients in 168
– origin of term 155
– oxygen minimum, shallow, in **167**, 215
– portrayal of **6**, **158**, **159**, **163**, **177**
– progradation events of 159, **160**
– rainfall, influence on 11, 151
– Red Sea Water in 30, **166**, 166
– ring shedding at 159, **162**
– ring shedding frequency at 162
– ring shedding, model of, at **201**, **202**, **204**, **205**
– satellite study of 133, 158, **159**, **160**
– sea surface temperatures of 153, **153**, **159**
– seasonality in 154, 195
– speeds in 158, 218, **219**
– stability, locational, of 156, 158, 159, **163**, 164
– (South East Atlantic) Central Water in 164, 165
– South Indian Central Water in **166**

– (South West Indian) Central Water in 164, 165
– Subantarctic Mode Water in **166**, 166, 167, 168
– Subantarctic Surface Water, equatorward penetration at **159**, **162**, 162, **163**, 164
– Subtropical Convergence, effect on 218, **219**, **221**, 221, 222
– Subtropical Surface Water in **166**, 166, 167
– temperatures, sea surface, of 153, **153**, **159**
– temperature section across **167, 169**
– temperature–salinity relationships for 164, 165, **165**, **166**
– Tropical Thermocline Water in **166**, 166, 167
– thermostads, source of, found at 168
– Upper Circumpolar Deep Water in **166**
– upstream (*see* upstream retroflection)
– variability at 25, 151, 152, 153, **154**, 194, **196**
– viscous stress in modelling of 200
– volume transport of 151, **170**, 170, 218
– volumetry of 164
– water masses of 164, 165, **165**, **166**
– water mass modifications at 168
– wind influence on Agulhas water penetration from 164
Agulhas Current return flow (*see* Agulhas Return Current)
Agulhas Current rings (*see* Agulhas rings)
Agulhas Current Sources Experiment (ACSEX)
– investigating sources of the Agulhas Current 157
Agulhas Current system
– biology modelled by NORWECOM for 201
– climate variations, role in 11, 223
– conduit for inter-ocean exchanges 233
– copepoda, influence on distribution of 15
– currents of, according to ships' drift 53, 54, **54**
– currents, simulated for **49**
– early portrayal of 53, 54, **54**
– eddy kinetice energy for 91, **92**
– kinetic energy for 91, **92**
– kinetic energy, simulated for 67
– knowledge on, growth in ix, **ix**, x
– marine turtles, influence on 14
– mesoscale features, lack of information on 237
– model simulation of 101, 104
– models of **65**, 101
– papers on ix, **ix**, x
– portrayal of **6**
– rainfall, effects on 11, 223, 234
– South African rainfall, effects on 11, 223
– subtropical gyre, part of 17
– temporal changes of, lack of knowledge on 237
– treatment of in text books ix
– whale distribution, influence on 14, 209
Agulhas (Current) water 10
Agulhas eddies (*see* Agulhas (Current) eddy)
Agulhas filaments
– advection of 7, 131, 139, 186, 192
– advection of, studied by satellite remote sensing **126**, 133
– Agulhas Bank, western edge, at 192
– biological impact of 132
– cloud formation over 12, 150, 192
– dimensions of 132
– encircling Agulhas rings 132, 171, 192
– fluxes, inter- ocean, of heat and salt by 131, 132
– frequency of 194
– heat flux, inter-ocean, due to 194
– heat fluxes to atmosphere from 192
– hydrographic stations across **132**, 132
– image, thermal infrared, of **126**, **132**
– inter-ocean exchanges by 192, 193
– salt flux, inter-ocean, due to 194
– salt flux by 192
– satellite studies of 133
– South Atlantic Ocean, in 132
– trajectories of 131
– volume fluxes from 192
Agulhas Front
– associated with Agulhas Return Current 217, 221
– characteristics of 210, **218**, 220
– characteristics, thermal, of 217, **218**, **222**, 222, 223
– eddy shedding at 221
– extent, zonal, of 222
– hydrographic section across 220, **224**
– location of 223
– satellite studies of 217
– simulation by model of 218
– surface expression of 217, 223
– termination, eastern, of 217, **218**, 222
– volume transport of 215
– width of 223
– zonal jet at 220
Agulhas Plateau
– Agulhas Current at 120
– Agulhas Return Current at 120, 210
– Agulhas Return Current, being crossed by 7, 85, 213
– Agulhas Return Current, effect in model of, on 228, **229**
– Antarctic Bottom Water at 32
– drifters at **212**
– eddies formed at 211, **212**, **226**, **227**, 227, **230**
– flow excursions, northward caused by 210
– hydrographic section near 213, **213**
– inflow to Agulhas Current north of 122
– meander dimensions at 225
– meander downstream of 225, 226
– meander at, simulated in model 228
– meander at, stability of 211, 226
– name, origin of 2
– Rossby waves west of 213, 226
– satellite investigations at **230**
– upstream retroflection of Agulhas Current, causing 196
Agulhas retroflection (*see* Agulhas (Current) retroflection)
Agulhas Retroflection Cruise (ARC)
– investigating the Agulhas Current 157
Agulhas Return Current
– Agulhas Plateau, at 120, 210
– Agulhas Plateau, crossing of, by 85, 213
– altimetric study of 204, 217, 220
– amalgamation with Agulhas Current 7, 156
– Antarctic Intermediate Water in 44

– baroclinic instability of 226
– chlorophyll, enhanced levels at 15
– crossing Agulhas Plateau 85
– currents simulated for 228, 229
– current observations in 204, 213, **214**
– current variability of **68**
– deflection of water from 7, 86
– depth of **43**, 210
– dissolved oxygen in 213
– drifters in 204, 209, 211, **212**, 213, 220, **229**
– dynamic topography of **210**, **211**
– dynamics of 228
– eastward extent of 209, 216, 218, 220
– eastward flow of 209
– eddies, Agulhas, at (*see also* Agulhas (Current) eddies) 7, **226**, **227**
– eddies associated with 211, **212**, **226**, **227**, **228**
– eddy, hydrographic section across **228**
– eddy kinetic energy modelled at 204, 230
– eddy shedding from 226, 227
– floats in 211, **212,** 213, **214**
– geostrophic velocities in **43, 212**, 213, 214, **214**, 218, **219**
– heat loss from 7, 12
– historical mention, first, of 225
– horizontal gradients across **222**
– hydrographic section across 24, **24**, **163**, 209, 213, **213**, **216**, 218, **222**, **224**, **228**
– inertial jet model of 228, **229**
– kinetic variability for **68**, **92**, 153, **154**
– knowledge, lack of, on 236
– leakage from 24, 216, 217, **217**
– low pressure systems, agreement with 11
– meander, stability of 211
– meanders in 160, 210, **210**, 210, 211, **211**, 225, **229**
– meanders modelled for 228
– meanders, wavelengths of 211, 213
– mixing in 216
– model simulation of 46, **202**, **204**, 218, 228, 230
– Mozambique Plateau, crossing of 86
– myctophids carried by 15
– nomenclature for 220
– North Atlantic Deep Water movement in 32, **33**
– nutrients at rims of 168
– oxygen in 213, **213**, 215
– portrayal of **6**, 7, **158**, **159**, **210**
– path stability of 24, **24**, 210, **210**
– planetary wave on 220
– RAFOS floats in 211, **212,** 213, **214**
– Rossby waves in 213, 226
– satellite studies of **159**, 210, 226, **227**
– sea surface temperatures at 153, **153**, 225
– ship's drift for 220
– speed decline eastward of 213
– speeds in **43, 212**, 213, 214, **214**, 218, **219**
– south of Madagascar 214, 215
– Subtropical Convergence, juxtapositioned to 7, 24, 209, 210, 218
– surface expression of **210**, **211**, 223
– termination of 209, 216, 218, 220
– thermal borders of 223
– trajectory of 24, **24**, 210, **210**
– treatment of, by Maury 225
– turbulence at 25
– turbulence, mesoscale, simulated for **320**
– variability of 25, 217, 227
– velocities, zonal 7, 214, **214**, 218, **219**
– vertical profiles for **43**, 122, 213, **213**, **214**
– volume transport of 11, 24, **24**, 42, 121, 170, 214, 215, **215**, 216, **217**, 218, **219**
– volume transport, eastward diminution of **215**, 216, **217,** **219**
– volume transport, variability of 215
– wave attenuation on 105
– width of 214
Agulhas return flow (*see* Agulhas Return Current)
Agulhas Return Front (*see* Agulhas Front)
Agulhas rings
– advection rate of (*see* drift rate for)
– ageing of **183**
– Agulhas Bank, lodged against 139, **181**, 186
– Agulhas filaments, encircled by 132, 192
– altimetric studies of 133, 173, 180, **183, 186**, 187, 189, **189**, **190**, 193
– atmospheric boundary layer over 185
– Angola Basin, in 180, **182**
– at shelf edge of Agulhas Bank **181**
– available potential energy of 194
– azimuthal speeds in 162, 171, **172**, 174, 179
– baroclinicity of 180
– barotropicity of 180
– Benguela Current, in 189
– birds feeding at 15
– bottom ridges interaction with 187
– Brazil, reaching 187, 233
– Cape Basin, in **173**, 173, **183**, 186
– characteristics of 162, 194
– coastal upwelling, interacting with 190, **191**
– convective mixing in 170, 180, 185
– corridor of 194
– current measurements in 180, 186
– cyclonic eddy at 173, 174, 176 (*see also* Cape Basin cyclones)
– decay of 180, 182, **183**, **185**, 185, 189
– depth of 171
– diameter of 158, 170, 171, **173**, 177, 180
– dipoles associated with 174, 202
– dissipation of 180, 182, **183**, **185**, 185, 189
– dissipation in Cape Basin of 180, 185
– dissipation of, modelled 182, 184, **184**, 185, **185**
– dissolved oxygen in 179, **181**, **182**
– drift direction of 186, **188**
– drift, studied by altimetry **188**
– drift rate for 186, 187
– drift trajectories of 186, **188**

- durability of 7, 180
- dynamic height of 171, **211**
- encircled by Agulhas filaments 132, 171
- ENSO influence on shedding of 160
- erosion of 180, 182, **183**, **185**, 185, 189
- filaments associated with 184, **184**, 189
- filament, upwelling, wrapped around 190, **191**
- fish recruitment, hypothesised role on, of 190
- floats, studies using, on 173, 185, 186, 187, 234
- fluxes, inter-ocean, caused by (*see also* inter-basin exchange) 7, 10, 193
- formation of 7
- frequency, of shedding of 10, 160, 162, 193, 194
- global thermohaline circulation cell, role in 191
- heat content of 174, 193
- heat flux, inter-ocean, caused by 193, 194
- heat losses to atmosphere from 162, 180, 185, 186
- hydrographic sections across **163**, **172**, **181**, **182**, 186
- Indian Ocean dipole, influence on shedding of 160
- influence on fish recruitment 190
- interacting in Cape Basin 233
- interaction with seamounts 188, **188**
- inter-ocean exchanges due to 7, 192, 193
- inverted echo sounders in 162, 186
- KAPEX, studies by, of 173
- kinetic energy of 171, 174, 180, 194
- kinetic energy flux by 193, 194
- kinematics of 179
- knowledge, lack of, on 236
- lifetimes of 7, 180
- METEOSAT images for **153**, 171
- migration of 186, **188**
- mixed layer in 185
- modelling of 44, 173, 174, **175**, 182, 184, **184**, **185**, 185, 186, 187, 188, 189
- nitrite in 180, **181**
- nomenclature for 170
- nutrients in 180, **181**
- oxygen in 179, **181**, **182**
- pathways of 186, 187, **188**, **206**
- phosphate in 180, **181**
- potential energy of 174, 180, **183**, 194
- potential energy fluxes due to 193, 194
- portrayal of **158**, **159**, **163**, **175**
- presence in Cape Basin of **173**, 173, **175**
- propagation rates for 186, 187
- radial speeds in 162, 171, **172**, 174, 179
- radius of 171, 174, 180
- RAFOS floats in 173, 185, 186, 187, 234
- Red Sea Water in 40
- Rossby waves radiating from 182
- salinity modification in 185
- salt content of 174, 193
- salt flux, inter-ocean, caused by 193, 194
- satellite investigations of 234
- Scanfish undulator, observations by 185, **186**
- sea surface height at 171
- seamounts, interaction with 180, **189**, 189, 204
- self-steering of 187
- shedding at Agulhas retroflection 10, 11, 128, 159, 162, **162**, **177**, **205**
- shedding, discovery of 159
- shedding, frequency of 10, 160, 162, 193, 194
- shedding, modelled 48, **201**, **204**, 204, 205, **205**
- shedding, precipitated by Natal Pulse 11, 160
- shedding, seen in satellite imagery **205**
- shedding, triggering processes for 11, 128, 206
- shelf edge currents, influence on, by 146
- south-west of South Africa 170, **173**
- spawning frequency of 10, 11, 128, 159, 162, **162**, **177**, **205**
- spawning process of 10, 11, 128, 159, 162, **162**, **177**, **205**
- splitting of 180, 182, **189**
- splitting of, simulated 204
- stability of 174
- Stokes' drift at 182
- Subantarctic Surface Water wedge at spawning of **159**, **162**, 162
- tangential speeds of 162, 171, **172**, 174, 179
- thermal structure of 162, **163**, 185
- thermostads in 179, 180, 185
- tracer in 184, **184**
- trajectory of 186, 187, **188**, **206**
- trajectories simulated for 204, 205, **206**
- translation of 186, 187, **188**
- upwelling, interaction with 190, 205
- variables of 174
- velocity structure of **172**, **175**
- vertical penetration of **163**, 171, **172**, 173
- volume of 171, 174
- volume transport in 156, 170, 193
- west of Cape Town **181**, 186
- Walvis Ridge, beyond 180, **182**, 182
- Walvis Ridge, effect of crossing, on 171, 187, 188, 233

Agulhas Undercurrent
- depth of **95**, 95
- discovered during ACE 157, 235
- first observations of **95**
- modelling of 95
- Red Sea Water in 95
- simulated velocities for 95
- speeds of 95
- volume flux of 95

Agulhas–South Atlantic Thermohaline Transport Experiment (ASTTEX)
- studying the flux of Agulhas water in South Atlantic Ocean 157

airborne radiation thermometry
- Agulhas Current investigations, first use for 133
- Natal Bight, over **109**, **110**

Algoa Bay
- Agulhas Current at 121
- birds feeding in 137
- cold water advection into 137

– current edge features at **134**
– current speed at 121
– sand movement at 121
– temperature, decrease southward of 124, **134**
– temperature section at **134**
– upwelling at 141
– upwelling upstream of (*see* Port Alfred upwelling cell)
– warm plume penetration into 129
Almirante Lacerda (vessel)
– in Mozambique Channel 81
Alphard Banks
– division between water characteristics on Agulhas Bank, role as **140**
– Agulhas Bank bathymetry, shallowest part of **139**
altimetry (*see also* satellite investigations)
– Agulhas Basin, over 68, **68**
– Agulhas retroflection, of 154, **154**, **161**, 204
– Agulhas Return Current, of 204, 217, 220
– Agulhas rings, of 133, 180, **183**, 186, 187, 193
– assimilation in model of 199
– Brazil rings, of 178
– history of use of 4, 133
– kinetic variability, according to 91
– Madagascar, south of 68, **68**
– Mozambique Channel, over 88, **88**
– Natal Pulse, of 114
– Natal Valley, over 68, **68**
– South Atlantic circulation variability, of 194, 195
– South Indian Ocean, of **25**
– South West Indian Ocean, of 88
– southern Agulhas Current, of 204
– surface waves, of 105
Amazon River
– volume flux of 151
Amirante Passage
– Antarctic Bottom Water moving through 30
– Indian Deep Water moving through 34
anchovy
– recruitment affected by Agulhas ring 190
Andriamanao
– upwelling cell at 63
Angola Basin
– Agulhas rings in 180, **182**
Angola–Benguela Front
– northern border to Benguela upwelling system, as 190
Angoche
– lee eddy at 73, 78, **79**
– chlorophyll in eddy at 73
– nutrients in eddy at 78
Antarctic Bottom Water
– Agulhas Basin, movement of, in 32
– Agulhas Current, northern, in 106
– Agulhas retroflection, in **166**
– Crozet Basin, movement in 32
– depth of 30
– Enderby Basin, from 32
– fluxes across South Atlantic Ocean 30
– heat and salt fluxes in South Atlantic modelled for 197
– northern Agulhas Current, in 106
– Mascarene Basin, movement in 32
– Mozambique Basin, in 32
– Mozambique Plateau, movement past 32
– percentage by volume in Indian Ocean **27**, **28**
– percentage by volume in South West Indian Ocean **27**, **28**
– Prince Edward Fracture Zone, movement through 32
– South West Indian Ocean, movement in 30, **31** 32
– temperature–oxygen relationship in western Indian Ocean, of **29**, 30
– temperature–salinity relationship in western Indian Ocean, of **28**, 30
– volume flux in South Atlantic of 197
Antarctic Circumpolar Current
– location of, according to altimetry 24, **25**
– Agulhas Return Current, merging with 218
– global thermohaline circulation, component of 10
– mass transport function of South Indian, component of **122**
– speeds in 231
Antarctic Intermediate Water
– Agulhas Current, in 44
– Agulhas Current, in northern **107**, 107
– Antarctic Current, in southern 123
– Agulhas retroflection, in 164, 165, **166**
– Agulhas Return Current, in 44
– Agulhas rings, in 197
– characteristics in South West Indian Ocean 19, **28**, **29**, 30, **35**, **36**, 37
– characteristics in tropical South Atlantic Ocean 9
– circulation in South West Indian Ocean 19, **37**, **84**
– circulation constrained by ridges 84
– depth of, in South West Indian Ocean 37
– distribution of, in South West Indian Ocean **37**, **83**
– Drake Passage, from 192
– East Madagascar Current, in 44, 61
– erosion of **36**, 37
– fluxes across South Atlantic Ocean 197, 198
– heat and volume fluxes across South Atlantic, modelled 197
– Indian Ocean, in 192
– Indonesian Seas, from 192
– inter-ocean leakage of 30, 197
– modelled simulation of 48
– Mozambique Channel, in 37, **37**, **82**, 82, **83**, 83
– Mozambique eddies, in 77
– Mozambique Undercurrent, in 37, **38**, 78
– northern Agulhas Current, in 101, **102**
– South Atlantic Ocean, in 192
– South Indian Ocean, circulation in 40, 192
– South Pacific Ocean, source of, from 192
– South West Indian Ocean, erosion in **36**, 37, 83
– South West Indian Ocean, salinity of, in **28**, **37**
– southern Agulhas Current, in 123, **124**
– subduction at Antarctic Polar Front 30
– subduction below Subtropical Convergence 30

– temperature–salinity relationships in western Indian Ocean of **28**
– temperature–oxygen relationships in western Indian Ocean of **29**
– upward mixing of 21
– volume transport in Agulhas Current of 101, **102**
– volume flux in South Atlantic of 197
– volume transport south of Africa of 192
Antarctic Polar Front
– Antarctic Intermediate Water, production of, at 30
– location of 223
– characteristics, surface, of **218**, 223
– characteristics, thermal, of 223
– volume transport at 122
– width of 223
Antarctic Surface Water
– temperature–salinity relationships, in western Indian Ocean, of **28**
– temperature–oxygen relationships, in western Indian Ocean, of **29**
anti-cyclonic eddy (*see* eddy)
Antilles Current
– as precursor to Gulf Stream (*see also* Gulf Stream) 8
annual variation (*see* seasonality)
Arabian Sea
– as source of high salinity water 17
Arabian Sea Water (*see* Red Sea Water; North Indian Intermediate Water)
Argo
– profiling floats of 152
Atlantis II (vessel)
– in Mozambique Channel 80
Atlantic Ocean
– heat fluxes of, at equator 196
– overturning circulation in 197, **198**, 233
– surface energy fluxes of 9
atmosphere
– Agulhas Current, affected by 149
– coastally trapped waves in 148
atmospheric boundary layer
– Agulhas Current, over 149, 150
– Agulhas rings, over 185
– southern Agulhas Current, over 121, 149, 150
atmospheric pressure system
– South African coastline, movement along 148
– South West Indian Ocean, over 148
– south-west of Africa 105
– Southern Ocean, over 105

BEST (*see* Benguela Sources and Transport)
baroclinic instability
– Agulhas Return Current, in 226
– meanders on southern Agulhas Current, of 125
– model of Agulhas Current, in 66, 67
– northern Agulhas Current, of 118
barotropic instability
– northern Agulhas Current, of 116
– model of Agulhas Current, in 67
bathymetry (*see also* bottom topography)
– Agulhas Bank, of **139**
– South West Indian Ocean, of **211**
Benguela Current
– Agulhas Current, volume flux contribution to 190
– Agulhas rings, influence of, on 189
– Agulhas rings, steering of, by 189
– floats in 187
– jet in 190
– leakage into 176, 191, 193, 194, 195, 197, 198, 199, 202
– model of 202
– origin in Cape Basin of 189
– rainfall, South African summer, influence on 223
– seasonal behaviour of 190
– sources of 190
– South Atlantic Current feeding of 221
– South Atlantic gyre, part of 189
– South African rainfall, influence on 223
– Walvis Ridge, flow over 188
Benguela Sources and Transport (BEST)
– Agulhas rings identified during **175**
– Cape Basin cyclones identified during **175**
– programme to investigate Agulhas ring fluxes 157
– NOAA participating in 157
– Sea Fisheries Research Institute participating in 157
– Woods Hole Oceanographic Institution participating in 157
Benguela upwelling system
– Agulhas rings, interaction with 234
– Lüderitz, central point of 190
– model of 190
– northern border of 190
– South African rainfall related to 223
– southernmost upwelling cell of 140, 190
– upwelling eddies in 190
– upwelling filaments in 190
– vortex dipoles in 190
berg winds (*see also* winds)
– upwelling filaments, influence on 190
Bering Sea
– inter-ocean exchanges modelled for 197
beta effect
– role in Agulhas retroflection 199
biological impact
– Agulhas filaments, of 132
– Port Alfred upwelling cell, of 137
– cold ridge on Agulhas Bank, of 141, 142
birds
– Subtropical Convergence, along 209
– Agulhas rings, feeding at 15
bottom topography (*see also* bathymetry)
– Agulhas Bank, of **139**
– South West Indian Ocean, of **6**, **25**, **31**, **34**
bottom shelf water
– Natal Bight, in 110, **111**
– Agulhas Bank, on 135, 141, 142

boundary current (*see* abyssal boundary current, deep boundary current)
Brazil Current (southern)
- Agulhas rings reaching 187, 233
- confluence with Falkland Current 8, 178
- continental constraints on 151
- publications on x
- rings shed from 8, 162
- sources of 8, 53
- temperature–salinity characteristics of **179**
- unique characteristics of 8
- volume flux of 6, 8
- western boundary current, as 8
Brazil Current (northern) (*see also* Brazil (Current) ring)
- retroflection of 8, 155
- rings shed from 8
- vertical penetration of 8
- volume transport of 8
Brazil (Current) ring
- advection in South Atlantic gyre 178
- age of 178
- Cape Basin, found in **175**
- dimensions of 8
- temperature–salinity relations of 178, **179**
- life-time, estimated, of 8
- origin at Brazil-Falkland Current confluence 178
- shedding frequency of 8
- volume transport of 8
- shedding trigger for 9
- Subantarctic Mode Water in 178, **179**
British Hydrographic Office
- study of surface currents at 54

CFC
- in presumed Brazil rings 178
CSIR (*see* Council for Scientific and Industrial Research)
CTD (*see* Conductivity–Temperature–Depth instrument)
CZCS (*see* Coastal Zone Color Scanner)
Cape Agulhas
- coastal upwelling west of 139, 140
- name, origin of 2
- temperature section from 122, **123**
- volume transport south of **215**
- separating coast in upwelling and non-upwelling regimes 139
Cape Amber (Cap d'Ambre)
- surface speeds past 57
- volume transport past 60, 72
- East Madagascar Current moving past 69
Cape Basin
- Agulhas retroflection loop extending into 158
- Agulhas rings, interaction amongst, in 233
- Agulhas rings dissipating in **183**
- Agulhas rings present in **173**, 173, **175**, 186
- Agulhas rings splitting in **188**
- Benguela Current originating in 189
- cyclones in (*see also* Cape Basin cyclones) 173, 174, **175**
- intermediate waters, sources of, in 190
- model of water movement through 192
- North Atlantic Deep Water movement through 32, **33**
- Red Sea Water in 190
- variability in **196**
- volume transport across 192
Cape Basin cyclones
- advection rates of 174, 177
- Agulhas ring, attendant on 173, 174
- azimuthal speeds of 177
- comparable to cyclones at other locations 187
- ß-plane influence on 174
- continental shelf edge, generated at 175, 176
- diameter of 177
- floats in 187
- kinetic energy of 177
- lee eddies, from 174, 176
- lifetimes of 177
- mixing in 177
- modelled 176
- Natal Pulses, from 174
- portrayal of **175**
- tracks of 178
- Walvis Ridge, crossing 177, 187
Cape Hatteras
- Gulf Stream fully constituted at 8
Cape of Good Hope Experiment (KAPEX)
- floats placed during 151, 173
- RAFOS floats placed as part of 157
- studying the Agulhas Current system 157, 173
- studying Agulhas rings 173
Cape Point
- coastal upwelling east of 140
- coastal upwelling north of (*see also* Benguela upwelling system) 140
Cape St Lucia (*see also* St Lucia upwelling cell)
- Agulhas Current characteristics at 96, 97
- bathymetric similarity to south-eastern Madagascar, of 62
- chlorophyll at 109
- hydrographic sections south of 93
- large Natal Pulse off 114
- nutrient rich water off 109
- plankton production at 109
- siphonophores at 109
- temperature-salinity relations near 109
- upwelling at 109
- volume transport off 100
Cape St Marie (Cap Ste Marie)
- currents south of 64
Cape Town
- Agulhas filaments south of 131, **132**
- Agulhas Bank lee eddy south-west of **176**
- Agulhas ring off **181**, 186
- observational stations close to 2
- upwelling at **159**
Caretta caretta (*see also* Loggerhead turtle)
- interaction with Agulhas Current 14

Cargados Carojos shoals
– current shadow downstream of 47
Central Water (*see also* South Indian Central Water)
– Agulhas Bank, on 136
– Agulhas retroflection, in 164, 165
– formation of, in Indian Ocean 40
– Mozambique Channel, in 40
– northern Agulhas Current, in **106**, 106
– volume in South West Indian Ocean 27, **27**
cetaceans (*see also* whales)
– Agulhas Current system, distribution influenced by 14
– Subtropical Convergence, at 209
Chaetognaths (*see also Sagitta enflata*)
– Agulhas Current water, indicator species for 15
Challenger (vessel)
– making pioneering observations close to South Africa 2, 19
Chlorophyll-*a*
– Angoche lee eddy, in 73
– Agulhas Current, on the inshore edge of the 139
– filaments of **60**, 65
– Madagascar, off south coast of **60**, **62**, 63, 65
– Mozambique Channel, in lee eddy in 73
– Natal Bight upwelling cell, in 109, **110**, 110
– Port Alfred upwelling cell, in 137
– seasonality, south-eastern Madagascar upwelling cell, in 63
– subsurface maximum **62**, 63
– Subtropical Convergence, at the 209, 224
circulation (*see also* currents)
– continental shelf off south-eastern Africa, on 98, 101, 107, 108, 109, **112**, 145
– deep 34
– Indian Ocean, of **18**
– Natal Bight, over 109, 111, **112**
– South Indian Ocean, of **20**, **24**, **25**
– South West Indian Ocean, of 6, **34**
Circumpolar Deep Water (*see also* Upper Circumpolar Deep Water; Lower Circumpolar Deep Water)
– Agulhas retroflection, in **166**
– circulation in Madagascar basin 34
– circulation in Mascarene basin 34
– South West Indian Ocean, in 32
climate
– fluctuations in global thermohaline cell, role of in 130
– upstream retroflection, importance of, on 196
– southern Africa, of 11, 12, **13**, 14, 223
climate change
– role of ocean 4
clouds (*see also* cumulus clouds)
– formation over Agulhas Current 149, **150**, 150
– formation over Agulhas filaments 150, 192
– formation over Gulf Stream 149
coastal counter current
– observations for 94, **94**, 98, **99**, 108, 112, 116
coastal lee eddies (*see* lee eddies)
coastal lows (*see* winds; waves)
coastal upwelling (*see also* upwelling; upwelling cell)
– Agulhas Bank, along 139, 140, **142**
– Benguela regime, of 140
– Cape Agulhas, near 139, 140
– Cape Point, north of 140
– Cape Point, south of 140
– headlands of South Coast, at 140
– Natal Bight, in (*see* St Lucia upwelling cell)
– south-eastern Madagascar, off **61**, 62, **62**, 63
– west coast, along **6**, 154
coastal winds (*see* winds)
Coastal Zone Color Scanner (CZCS) (*see also* NIMBUS 7)
– use made of by South Africans 133
cold ridge
– Agulhas Bank, on **140**, 141, **143**
Commandant Robert Giraud (vessel)
– research cruise by, in Mozambique Channel 71, **72**
Comores
– East Madagascar moving past 69
Comores Basin
– circulation in 71, **71**, 72
– modelled circulation for 70
– origin of Mozambique eddies in 77
conductivity–temperature–depth instrument (CTD)
– Agulhas retroflection, at **165**
continental shelf
– Agulhas Current stability, influence on 6
– bottom water of 129
– circulation on 96, 142
– East Madagascar Current stability, influence on 60
– northern Agulhas Current, flow along 105, 106, 108
– sediments on 96, 121
– south-eastern Africa, off **93**, **96**, 107 ff., 138 ff.
continental shelf currents (*see also* currents)
– Agulhas Bank, at edge of 96
– Agulhas Bank, over 142
– Durban, at 108
– Durban and Port Elizabeth, between 107, 108, 109, 111, **112**
– Natal Bight, in 108, **112**
– Port Edward, at 108
continental slope
– Agulhas Current path, stabilising influence on 6
– south-eastern Africa, off 94, 95, 96, 121, 122, 130
convective overturning
– Agulhas rings, in 180
conveyor belt (*see* thermohaline circulation cell)
copepoda
– distribution influenced by Agulhas Current system of 15
Council for Scientific and Industrial Research (CSIR)
– airborne radiation thermometry, use of by 133
– commercialisation of 64, 103
– contributor to research on Agulhas Current, as 64
– drifters, deployment by 151
– IIOE, co-ordinating South African participation in 21
– marine pollution studies by 64
– National Aeronautical and Space Administration, collabo-

rating with 152
– National Centre for Atmospheric Research, collaborating with 152
– National Physical Research Laboratory of 64, 133
– National Research Institute for Oceanology, establishing 64
– R. V. *Meiring Naudé*, owner of 64, 103, 157
– Satellite Remote Sensing Centre of 133
counter current (*see also* onshore counter current)
– at Durban 98
Crozet Basin
– bottom water, origin of, at 30
– Antarctic Bottom Water movement in 32
– circulation in **71**
Crozet Island
– Antarctic Bottom Water flow of, past 32
– volume transport between Durban and 215
cumulus clouds (*see also* clouds)
– Agulhas Current, formed over 149, **150**, 150
– Agulhas filaments, formed over 150
currents (*see also* current measurements; counter current; Agulhas Undercurrent; circulation; surface currents; continental shelf currents)
– Agulhas Bank, on 145
– Agulhas Current, speeds simulated for 203
– Agulhas retroflection, in **158**
– Agulhas Return Current, speeds of 213, 214, **214**, 218, **219**
– Cape St Marie, south of 64
– continental shelf, on 97
– Durban, at **97**, **98**, **108**
– Durban and Port Elizabeth, velocities, between 93, 108
– greater Agulhas Current system, simulated for 203
– Madagascar, south of **63**, 64
– Natal, off 97, 97, **98**
– Natal Bight, over the 111, **112**
– northern Agulhas Current, for **94**, **97**, **98**, **101**, **108**
– Port Edward, at **108**
– Richards Bay, at **113**
– South West Indian Ocean, seasonal variability in, for 102
– variability of, in South West Indian Ocean **68**, 68
current measurements
– Agulhas Bank, on 145
– Agulhas Current, in northern 92, **94**, 94, **95**, **97**, **98**, **108**, 157
– Agulhas Current, in southern **153**, 204
– Agulhas retroflection, in 152, 204
– Agulhas Return Current, in 204, 213, 214, **214**
– Agulhas rings, of 180, 186
– Cape Basin, in **153**
– Durban, at 108
– East Madagascar Current, in 92
– Mozambique Channel, in 157
– Natal, off 149
– Natal Pulse, in 114, 176
– Port Alfred, at 108
– Port Edward, at **95**, **97**, **98**, 108, 157
– Richards Bay, at 108
– shear edge eddies, in 131
– South East Atlantic, in 157
cyclonic eddy (*see* eddy)

deep basins
– South West Indian Ocean, of **31**
deep boundary current
– along west coast of southern Africa 164
– North Atlantic Deep Water in 164
deep circulation
– South West Indian Ocean, of 30, **31**, 32, **33**, **34**, 34
deep water
– South West Indian Ocean, volume of, in 27, **27**
deep-sea eddy (*see* eddy)
Deep-Sea Research
– special edition of, on KAPEX 157
Delagoa Bight
– lee eddy in 73, 79, **80**, 81
– lee eddy, movement of, in 81
– Delagoa Pulses, source of, in 116, 118
– Mozambique eddies, influence of, on 79
– upwelling cell in 79
Delagoa Pulses
– seen in satellite images 81, 116, 118
– off the Natal Bight **119**
– vertical sections across **119**
Dermochelys coriacea (*see* leatherback turtles)
diatoms
– Agulhas Current, in southern 209
– Subtropical Convergence, at 209
dipoles (*see* vortex dipoles)
Discovery (vessel)
– making pioneering observations near South Africa 2
dissolved oxygen content (*see* oxygen content)
dissolved oxygen minimum layer (*see* oxygen minimum layer)
dissolved oxygen section (*see* oxygen section)
Division for Earth, Marine and Atmospheric Science and Technology (EMATEK)
– establishment of 64
Doholum denticulatum (pelagic tunicate)
– Agulhas Current water, indicator species for, as 15
Drake Passage
– Antarctic Intermediate Water from 192
– heat fluxes through 196
– inter-ocean exchanges modelled for 197
– water from 198
drift
– Agulhas rings, of 186, 187, **188**, **206**
drift bottles, use of 4
drifters
– Agulhas Current at Durban, offshore of 98, 120, **211**
– Agulhas Current system, use in studying 104, 152
– Agulhas Current, southern, in 158, **212**
– Agulhas eddy, in 227
– Agulhas Plateau, over 152, 211, **212**
– Agulhas retroflection region, in the 158, 204

- Agulhas Return Current, in 204, 209, 211, **212**, **213**, 220, **229**
- Durban eddy, in **112**
- East Madagascar Current, in 67, 152
- FGGE, placed during 152
- GARP, placed during 152
- Indian Mid-Ocean Ridge, over **212**
- Madagascar, east of, from 68
- Madagascar Ridge, over **212**
- Mozambique Channel, in 81
- Mozambique eddy, in 119
- Mozambique Plateau, moored on 152
- Mozambique Plateau, moving over **212**
- northern Agulhas Current, in 98, 104
- pioneering studies using 152
- South West Indian Ocean subgyre, in 84, 86
- southern Agulhas Current, in 204
- Southern Ocean, in 152
- Subtropical Convergence, at 224
- upstream retroflection, at 196

drifting buoys (*see* drifters)

dunes
- shelf sediments, in 96, **96**

Durban
- CSIR research establishment, location of, at 64, 103, 133, 151, 157
- current measurements at 96, **97**
- current speeds at 96, **97**, **99**
- drifters at 112, 120
- hydrographic surveys south of 122, **123**
- hydrographic section across Agulhas Current at 122, **123**
- inshore counter current off 98, **99**
- lee eddy off 98, **99**, 108, 109, **111**, 111, 137
- location of **6**
- nitrate values off **111**
- recurrent eddy off 98, **99**, **111**
- temperature section at **86**, **99**
- temperature–salinity relationship of water at **99**, **106**
- upwelling filament off 110
- volume transport off 100

ECMWF (*see* European Centre for Medium-Range Prediction)

EMATEK (*see* Division for Earth, Marine and Atmospheric Science and Technology)

ENSO (*see* El Niño- Southern Oscillation)

ERS 1 (satellite)
- model assimilation of altimeter data from 203

early portrayals
- sources to the Agulhas Current, of 53, 54, **54**, 55, **55**
- South Equatorial Current, of **54**, **55**, 55

East African Coastal Current (*see also* East African Current)
- model simulation of 47
- seasonal variability of 91
- surface drift of 54

East African Current (*see also* East African Coastal Current)
- depth of 42, **43**
- formed by East Madagascar Current 69
- seasonal variability of 55
- speed of **43**, 69
- surface speeds of **43**
- vertical profiles for **43**
- volume transport of 69
- width of 69

East Australian Current
- meridional extent of 8
- publications on x
- ring formation by 8, 162
- speeds of 8
- Tasman Front, flow along (*see also* Tasman Front) 8

East London
- shelf/current arrangement at **134**
- temperature sections at **134**
- upwelling cell downstream of (*see* Port Alfred upwelling cell)

East Madagascar Current
- characteristics of 57
- Antarctic Intermediate Water in 44, 61
- bifurcation of 57
- bifurcation of northern branch of 69
- chlorophyll at **61**
- Comores, flow past, of 69, 72
- conceptual portrayal of **6**
- current measurements in 92
- deep sea eddies, impact of, on 60
- depth of **43**, 60, 61
- drifters in 67, 151
- early nomenclature for 53
- eddies, impact on 60
- eddies generated by 65, 66, **66**, 88, 133
- eddies generated by, in model 60
- eddy kinetic energy for **92**
- filaments from 65
- float tracks in 64
- kinetic energy for 91, **92**
- knowledge, lack of, on 235
- model simulation of 47, 67
- model simulation of retroflection of 66, **66**
- modelled as inertial jet 67
- need for more information on 233
- northern branch of 60, 72
- northern branch, modelled, of **65**
- nutrients, enhanced in associated upwelling cell of 63
- point of divergence between branches of 57, 58
- retroflection of 5, **6**, **60**, 65, 66
- retroflection of, modelled 67
- retroflection discovered for 133
- rings of **6**
- satellite studies of 133
- seasonal variations in 58, 92
- source for Agulhas Current, as 5
- sources of 57
- southern branch of **6**, 60, 64
- speeds of 8, **43**, 57, 60, **60**, 61

– Subtropical Surface Water in 61
– surface speeds of 55, **55**, 57
– temperature–salinity relationships for 61
– Tropical Surface Water of 44, 61
– upwelling associated with **60**, 62, 63, 132
– velocity, downstream increase of 58
– vertical penetration of 8, **43**
– vertical profiles for **43**
– volume flux of 8, 57, 58, 72
– volume transport modelled for **48**
– volume transport variations of 60
– vortex dipoles of 81, **89**, 89
– western boundary current, as 5
– width of 8, 60, **60**, 61
East Madagascar Undercurrent
– depth of 69
– volume transports of 69
– width of 69
eastern boundary current
– Australia, off 17
eddy (*see also* lee eddy; shear edge eddy; eddy kinetic energy)
– Agulhas Current 198
– Agulhas Current, adsorption onto **116**, 118
– Agulhas Current, spun off seaward side of 196
– Agulhas Plateau, at the 211, **212**, **226**, **227**
– Agulhas retroflection, at the **226**, **227**
– Agulhas retroflection, cyclonic, at the (*see* Cape Basin cyclones)
– Agulhas Return Current, at the 211, **212**, 225, **226**, **227**
– Agulhas Return Current, modelled at 225
– Agulhas ring, attendant on (*see* Cape Basin cyclones)
– Benguela upwelling front, in 190
– Durban, off 98, **99**, 108, 109, **111**, 111, 137
– East Madagascar Current, generated by 65, 66, **66**, 88
– East Madagascar Current, impact on 60
– East Madagascar Current, simulated for 60
– heat loss from 198
– intra-thermocline 89, **90**
– kinetic energy of 91, **92**
– Madagascar, from east of 60, **88**
– movement studied by altimetry 88, **88**
– Mozambique Basin, in 88, 89, **89**, 118
– Mozambique Channel, from 76, 77, **77**, **78**, 88, **88**, 118
– Mozambique Channel, in **65**, 65, 81, 88, **88**
– Mozambique Plateau, at 81, 82, 86, **87**, 118
– Natal Basin, in 68, 82, **88**, 89, **89**, 118
– Natal Bight, over 111, **112**
– Natal Pulse, in 114, 116, 176
– shelf interaction, with 65
– South Atlantic Current, shed from 187
– South West Indian Ocean, in **86**
– Subtropical Convergence, at the 156, 221, **222**, 225, **226**, **227**, **228**
– Subtropical Convergence, formation at the 88, 118, 187, 198, 221, **222**, 225, **226**, **227**, **230**
– Subtropical Convergence, simulation of formation at the **230**
– upwelling in 190
eddy kinetic energy
– Agulhas Current, for 91, **92**, 152, **153**
– Agulhas Current, modelled in southern **153**
– Agulhas Current system, of, modelled for 67, **75**
– Agulhas retroflection, at 91, **92**, 151, 152
– Agulhas Return Current, at 91, **92**
– Cape Basin, for **153**
– East Madagascar Current, at **92**
– Gulf Stream, for 152
– Kuroshio, for 152
– Mozambique Channel, in **92**
– South West Indian Ocean, for **92**
Ekman transport
– Port Alfred upwelling cell, at 138
– South West Indian Ocean, over 23
Ekman veering
– Agulhas Current, under 110, 131, 132, 138
El Niño (see also El Niño–Southern Oscillation)
– influence on East Madagascar Current dipoles 89
El Niño–Southern Oscillation (ENSO)
– influence on Agulhas ring shedding 160
Enderby Basin
– Antarctic Bottom Water, as origin of 32
Environmentek
– establishment of 64
Eole (satellite)
– drifter tracking by 152
Equatorial Counter Current
– location of 17, **18**
– seasonality of 17
Equatorial Indian Ocean Water
– Mozambique Channel, in **82**, 82, 83
Erica Seamount
– Agulhas rings, splitting of, at 180, 189, **189**
Europa Island
– frigate birds from 78
European Centre for Medium-Range Prediction (ECMWF)
– model for heat transfer from Agulhas Current by 149

FGGE (*see* First Global GARP Experiment)
FRAM (*see* Fine Resolution Antarctic Model)
Falkland Current
– Brazil Current, confluence with 8
– rings shed from 178
False Bay
– thermal front at 140
– upwelling water moving past 140
– water temperature of 4, 5
filaments (*see also* upwelling filaments; Agulhas filament)
– Agulhas rings, associated with 184, **184**, 190, **191**
– off western Madagascar **65**
– of East Madagascar Current 65
Fine Resolution Antarctic Model (FRAM)
– Agulhas Return Current, modelling by, for 230
– Agulhas ring trajectory, modelled by, for 205, **206**
– Agulhas rings, simulation of, by 204, 205, **205**

– South West Indian Ocean circulation, modelled with 47, 48, 49, **50**
– Subtropical Convergence simulated by **205**
– Subtropical Convergence, simulation of eddy formation at, by **230**, 230
First Global GARP Experiment (FGGE) (*see also* drifters; GARP)
– buoys forming part of 152
– drift tracks of buoys forming part of **229**
fish stocks
– off eastern Madagascar 63
floats
– Agulhas Bank, off 127, 128
– Agulhas Current, off Durban, in 127, 151
– Agulhas Return Current, in 211, **212**
– Agulhas ring, in 185, 186, 234
– Benguela Current, in 187
– East Madagascar Current, in 64
– KAPEX, placed during
– lee eddy, in 128
– Natal Pulse, in 114, 115, 176
– profiling, Argo 152
– shear edge eddies, in 128
– Walvis Ridge, crossing 187, 192
Florida
– Gulf Stream, downstream of 53
Florida Current
– Gulf Stream, as precursor to (*see also* Gulf Stream) 8
Florida Straits 8
flotsam
– drift of 68
Fraay (vessel)
– current measurements from 94
FRAM (*see also* Fine Resolution Antarctic Model)
– modelling the South West Indian Ocean with 47, 48, **49**, **50**
– modelling the South West Indian Ocean subgyre using 48, **49**
"freak waves" (*see* waves)
fronts (*see* by name of individual fronts)

GARP (*see* Global Atmospheric Research Programme)
GCM (*see* global circulation model)
GEK (*see* geomagnetic electrokinetograph)
Gauss (vessel)
– making pioneering measurements near South Africa 2
Gazelle (vessel)
– making pioneering measurements near South Africa 2
geomagnetic electrokinetograph (GEK)
– measurements by 70, **71**
Geosat (satellite)
– Agulhas rings, used to study **161**, **188**
– altimeter data of, assimilated in model 203
– eddy kinetic energy according to 154
– observations south of Madagascar by 68
geostrophic velocity
– South West Indian Ocean circulation, components of, for 20, 43
German Hydrographic Office
– surface current atlas drawn up by 23
German research
– Second World War, before 19
– influence of 19
"giant waves" (*see* waves)
Global Atmospheric Research Programme (GARP)
– drifters placed during 152
global circulation model (GCM) (*see also* model)
– simulating Agulhas retroflection with 199, 204
– simulating variability along Agulhas Return Current using 217
global thermohaline circulation (cell) (*see also* thermohaline circulation cell)
– Agulhas Current's role in 191
– modelling of 9
Gouriqua (*see* Mossel Bay)
Grand Banks
– Gulf Stream leaves the shelf edge at 8
Gulf of Mexico 8, 53
Gulf Stream
– atmospheric circulation cells over 149
– cloud formation over 12
– cold-core rings, shedding of, by 8
– compared to Agulhas Current 2
– continental constraints on 151
– eddies, absorption of, by 118
– eddy kinetic energy for 152
– global thermohaline circulation, contribution to 10
– growth in knowledge on ix, x
– heat fluxes to atmosphere from 13
– kinetic energy of **156**, 156
– loop occlusion in 8
– meandering of 8, 94, 131
– origin of 8
– plumes associated with 131
– publication on x
– ring formation by 159
– shear edge disturbances on 100, 131
– sources of 8, 53
– speed of 8
– storms, intensity of, influence on 13
– text book treatment of ix
– trajectory of 8
– undercurrent of 95
– upwelling at 122, 144
– volume flux of 8
– warm-core rings, shedding of, by 8
– width of 8

H. M. S. Tiger (*see Tiger*)
Hartebeesthoek
– Satellite Remote Sensing Centre located at 133
heat fluxes (*see also* heat loss)
– Agulhas Current, to atmosphere from southern 9, 12, 13
– Agulhas filaments, by 192

– Agulhas retroflection, to atmosphere from 12, 13, 151, 185, 223
– Agulhas rings, by 193, 194
– Atlantic Ocean, across line of latitude in 196, 197
– Drake Passage, through the 196
– inter-ocean, by Agulhas rings 193
– model, simulated in 196
– Port Alfred upwelling cell, to atmosphere at 149
– South Atlantic subtropical gyre, modelled for 197
– Southern Ocean, from 196
– Southern Ocean, to, from Agulhas system 199
Heat loss to atmosphere
– Agulhas Current, over southern 149
– Agulhas eddies, by 12, 13, 198
– Agulhas filaments, from 192
– Agulhas retroflection, at 180, 185
– Agulhas rings, from 162, 180, 185
– Port Alfred upwelling cell, at 149
Heezen Ridge
– obstruction to Agulhas retroflection loop, as 158
hydrographic section
– Agulhas Bank, across **129**, **140**, 141, **144**, **145**, **147**
– Agulhas Bank, along edge of **167**
– Agulhas Bank lee eddy, across **176**
– Agulhas Bight, at 123, **123**
– Agulhas Current, across northern 93, 122, **123**
– Agulhas Current, across southern 122, 123, **123**, **134**, **213**, 213, **224**
– Agulhas Current at Cape Agulhas, across 122, **123**
– Agulhas Current at Durban, across 122, **123**
– Agulhas Current at East London, across **134**
– Agulhas Current at Port Alfred **134**
– Agulhas Current at Port Elizabeth, across 122, **123**, **134**
– Agulhas Current at Port St Johns, across 122, **123**
– Agulhas Current meander, across **115**, **138**
– Agulhas eddy, across **163**, **222**
– Agulhas filament, across **132**, 132
– Agulhas Front, across 220, **222**, **224**
– Agulhas Plateau, near 213, **213**
– Agulhas retroflection, across **163**, **169**
– Agulhas Return Current, across **163**, 209, 213, **213**, **216**, 218, **224**, **228**
– Agulhas ring, across **163**, **172**, **181**, **182**, **186**
– Algoa Bay, at **134**, **138**
– Cape St Lucia, south of 93
– continental shelf north-east of Port Elizabeth, across **134**
– East London, at **134**
– Natal Bight, along 110, **111**
– Natal Pulse, across **138**, 138
– Port Alfred, at 133, **134**, **135**, **138**
– Port Alfred upwelling cell, across 133, **134**, **135**, **138**, 138
– Port Elizabeth, north of 93, **134**
– shear edge eddy and plume, across 129, **129**
– South Indian Ocean, across **25**, **26**, 26
– southern Agulhas Current, across **228**
– Subantarctic Front, across **224**
– Subtropical Convergence, across **182**, **216**, 220, **222**, **224**
– Subtropical Convergence eddy, across **228**
IGY (*see* International Geophysical Year)
IIOE (*see* International Indian Ocean Expedition)
IOC (*see* Intergovernmental Oceanographic Commission)
ichthyo-ecology
– Agulhas Bank, influence of, on 121
Indian Central Water
– South West Indian Ocean, in 41
– temperature-salinity characteristics of 41
Indian Equatorial Water
– Indian Ocean, in 40
Indian Ocean
– Antarctic Intermediate Water in **35**, 35
– circulation of, according to ships' drift **18**, 23, 53
– Indian Equatorial Water in 40
– Indonesian Intermediate Water in 35, **35**
– mesoscale variability in 21
– mid-ocean front in 17
– Red Sea Water in **35**
– sea surface temperature anomalies in 11
– rainfall, influence on 11
– surface circulation of 17, **18**, 18, 19
Indian Deep Water
– Amirante Passage, movement through 34
– South West Indian Ocean, in the **28**, **29**
– Mozambique Channel, flow into 32
– temperature–oxygen relationships of in western Indian Ocean **29**
– temperature–salinity relationships of in western Indian Ocean **28**
Indian Equatorial Water
– location of 40
Indian Mid-ocean Ridge
– location of **6**, **25**, **31**
Indian Ocean Central Water
– Agulhas Bank, on **146**
Indian Ocean Dipole
– influence on East Madagascar Current dipole shedding 89
– influence on ring shedding 160
Indian Tropical Surface Water
– Agulhas Bank, on **146**
Indonesian Intermediate Water
– Indian Ocean, in 35
– South West Indian Ocean, in 35
Indonesian throughflow
– Antarctic Intermediate Water from 192
– influence on inter-ocean exchange south of Africa 233
inertial jet
– Agulhas Current, modelling, as 199, **200**
– Agulhas retroflection, modelling, as 199, **200**
– Agulhas Return Current, model of 228, 229, **229**
– East Madagascar Current, modelling, as 67
inertial motion
– KwaZulu-Natal, on continental shelf off 108
– Agulhas Bank, on 145, **148**
inshore counter current (*see also* counter current)
– Durban, off 98, **99**
– Agulhas Current, to 116

Institut für Meereskunde
– Berlin, in 19, 54
– Kiel, in 19
inter-decadal changes
– data sets on winds for 40
Intergovernmental Oceanographic Commission (IOC)
– establishment of 21
internal tides
– Agulhas Bank, at the 145
International Geophysical Year (IGY)
– first global geophysical programme, as 21
International Indian Ocean Expedition (IIOE)
– Agulhas Current research, impact on, of 21, 157
– atlas based on 56, **56**
– data, use of, from 57, 84, 156
inter-ocean exchange
– Agulhas Bank water, by 191
– Agulhas filaments, by 192, 193
– Agulhas retroflection, at 151
– Agulhas rings, by 191, 192, 193, 194
– Antarctic Intermediate Water, by 192
– biogeographical patterns, role in 234
– biota, of 234
– direct leakage, by 191, 192
– flux between Indian and Atlantic, estimates of direct, by 191, 192
– heat fluxes of 193, 194, 234
– models of **195**, 197
– role in climate in 151
– salt fluxes of 193, 194, 324
– volume fluxes of 193
inter-ocean leakage (*see* inter-ocean exchange)
intra-thermocline eddy
– depth of 89, **90**
– diameter of 89
– found east of Madagascar 89, **90**
– speeds of 89
– temperature–salinity characteristics of 89, **90**
inverted echo-sounders
– monitoring Agulhas ring movement by 162, 186, 192

KAPEX (*see* Cape of Good Hope Experiment)
KNMI (*see Koninklijk Nederlandsch Meteorologisch Instituut*)
Kalman filter
– used to assimilate data into model 203
Kelvin waves
– propagation of Agulhas Current anomalies into Atlantic Ocean by 198
Kerguelen Plateau
– Agulhas Return Current, as termination of 209
– Antarctic Bottom Water, flow of, past 32
kinetic energy
– Agulhas Current, for **75**, 91, **92**, **156**
– Agulhas retroflection, at **75**, 91, **92**, 152, **156**
– Agulhas Return Current, for **75**, 91, **92**
– Agulhas rings, of 171, 174, 194
– East Madagascar Current, for **75**, 91, **92**
– Gulf Stream, of **156**
– Kuroshio, of **156**
– Mozambique Current, for **75**, 91, **92**
– New England seamounts, off 156
– South West Indian Ocean, of **75**, **92**
Knorr (vessel)
– used for Agulhas Retroflection Cruise 157
Koninklijk Nederlandsch Meteorologisch Instituut (KNMI)
– early investigation on Agulhas Current at 1
– establishment of 225
Kuroshio
– characteristics, unique, of 8
– continental constraints of 151
– eddy absorption by 118
– eddy kinetic energy for 152
– heat fluxes from 13
– kinetic energy of **156**, 156
– knowledge, growth in, on x
– meanderings of 8, 94
– origin of 8
– shear edge disturbances on 100, 131
– speed of 8
– trajectory of 8
– undercurrent of 95
– volume transport of 8
– width of 8

LOCO (*see* Long Term Ocean Climate Observations)
Labrador Sea
– North Atlantic Deep Water, origin of, in 9
Lagrangian drifters (*see* drifters)
Lamont-Doherty Geological Observatory
– source of Agulhas Retroflection Cruise, as 157
land–sea breeze (*see also* winds)
– KwaZulu-Natal coast, off 108
leakage (*see also* inter-ocean exchange)
– Agulhas Bank water, by 191
– Agulhas Current water, of 191
– Agulhas Return Current leakage, from 216, 217, **217**
– Agulhas rings, by 191, 192, 193, 194
– Antarctic Intermediate Water, by 192
– Subantarctic Water, across Subtropical Convergence, of 157
leatherback turtles (*Dermochelys coriacea*) (*see also* marine turtles)
– Agulhas Current system, interacting with 14
lee eddy (*see also* eddy)
– Agulhas Bank, off 127, **128**, 146, 158, 174, **176**, 176, 177
– Angoche, at 73, **79**
– Cape Basin, forming cyclonic eddies in 174
– Delagoa Bight, in 73, 79, **80**, 81
– Durban, off 98, **99**, 108, 109, **111**, **112**, 138
– floats in 128
– nutrients, enhanced, in 78
– Maputo, at 73
– model of **176**, 176

– Mozambique, off 73
– Mozambique Channel, in 73, 78
– Mozambique Current, associated with 73, 78
– Natal Bight, over 79, 111, **112**
life-time
– assumed Brazil ring, estimated for 178
– Agulhas rings, of 180
loggerhead turtles (*Caretta caretta*) (*see also* marine turtles)
– Agulhas Current system, interacting with 14
Long Term Ocean Climate Observations (LOCO)
– in Mozambique Channel 157
Lower Circumpolar Deep Water
– temperature-oxygen relationships, in western Indian Ocean, of **29**
– Agulhas retroflection, in **166**
low-pressure cells (*see* winds)
Lüderitz
– Benguela upwelling system, as central point to 190
Luzon
– Kuroshio, as start of 8

MARE (*see* Mixing in Agulhas Rings Experiment)
Madagascar
– Agulhas Return Current south of 214, 215
– cruises south of, on research vessel *Pelagia* 157
– currents simulated around 58, **59**
– currents south of **63**, 64
– currents west of **72**, **71**, 73, 81
– deep western boundary current off 32
– eddies west of 81
– influence on Agulhas Current system by 9
– intra-thermocline eddies west of 89, **90**
– North Atlantic Deep Water, north of 32
– Subantarctic Front south of 222
– Subtropical Convergence south of 222
– upwelling cell off 62, **62**, 63
– volume flux south of 68
Madagascar Basin
– Circumpolar Deep Water, circulation of, in 34
Madagascar Ridge
– Antartic Intermediate Water, constrained by 84
– drifter tracks, influence on **212**
– model simulated variability, as boundary to 47
– northward currents at 220
– portrayal of **6**
– sub-gyre west of (*see* South West Indian Ocean sub-gyre)
Madagascar upwelling cell (*see* East Madagascar Current; upwelling; upwelling cell; upwelling filament)
Maputo
– Mozambique Current at 69
– lee eddy at 73
– poleward flow off 81
marine boundary layer
– southern Agulhas Current, over 149
marine pollution
– CSIR, investigations by 64
marine turtles (*see also* leatherback turtles; loggerhead turtles)
– Agulhas Current system, affected by 14
Mascarene Basin
– Antarctic Bottom Water movement through 30, 32
– Circumpolar Deep Water circulation in 34
Mascarene Plateau
– effect on South Equatorial Current, of 19
mass transport function (*see* volume transport)
Mauritius
– cruises by reearch vessel *Meiring Naudé* to 103
meanders
– Agulhas Current, of 92, 93, 125, **126**, 126
– Agulhas Current, on northern 92, 93
– Agulhas Current, on southern 125, **126**, 126
– Agulhas Return Current, of 160, 210, **210**, 210, 211, **211**, 225, **229**
– bottom topography, caused by 160, 210, **210**, 210, 211, **211**, 225, **229**
– Gulf Stream, on 8, 94, 131
– Kuroshio, on 8, 94
– solitary (*see* Natal Pulse)
Meiring Naudé (vessel)
– crew of 103
– cruises on 103, 157
– description of 103
– equipment on 103
– ownership of 64, 103
– portrayal of **103**
– sale of 64, 103
mesoscale variability (*see* variability)
Meteor (vessel)
– pioneering observations by 2, 19
METEOSAT (satellite)
– Agulhas Current, use in studying 133,
– Agulhas retroflection, images for, from **153**, **160**
– Agulhas rings, images for, from **153**, 171
– images from **132**
mixed layer
– Agulhas Bank, over 140, **141**, 143
– South West Indian Ocean, in 44, **45**
Mixing in Agulhas Rings Experiment (MARE)
– studying Agulhas rings as part of 157
model (*see also* inertial jet; global circulation model)
– Agulhas Bank, of circulation on 142
– Agulhas Current, of eddy kinetic energy in 75
– Agulhas Current, of northern 94, 95, 102, 115, 119, 120
– Agulhas Current, of southern 125, 127, 142, 144, 149
– Agulhas Current as inertial jet, of 94, 120
– Agulhas Current retroflection, of 47, 151, 154, 199
– Agulhas Current system, of greater **49**
– Agulhas Current volume flux, of 47, **48**
– Agulhas Bank lee eddy, of 176
– Agulhas Current system, biology of 201
– Agulhas Front, of 218
– Agulhas retroflection, of **201**, **202**, **204**, **205**

– Agulhas retroflection, of eddy kinetic energy at 154
– Agulhas Return Current, of 46, **202**, **204**, 218, 225, 228, 229, **229**, 230
– Agulhas Return Current, of eddy kinetic energy at 225, 230
– Agulhas rings, of 46, 182, 184, **184**, **202**, 205, **205**
– Agulhas rings, shedding of, of 48, **201**, **202**, **204**, 204, **205**
– Agulhas ring dissipation, of 182, 184, **184**, 185, **185**, 189, 205
– Agulhas ring drifts, of 187, 188, 189, **206**
– Agulhas ring interaction with Walvis Ridge, of 188
– Agulhas ring shedding, of 48, **201**, **202**, **204**, 204, **205**
– Agulhas Undercurrent, of 95
– altimetric data assimilated by 199
– Antarctic Intermediate Water movement, of 48
– Atlantic Ocean, of heat transport in 196, 205
– Benguela Current, of 202
– Benguela upwelling system, of 190
– Cape Basin, sources of water moving through 192
– Comores Basin, circulation in 70
– data assimilation with 199, 203, **204**
– East African Coastal Current, simulation of 47
– East Madagascar Current, of 47, 60, 65
– East Madagascar Current, of volume flux of **48**
– East Madagascar Current, of 66, **66**, 67
– East Madagascar Current retroflection, of 67
– eddy kinetic energy, of distribution of 67, **75**, **153**
– eddy variability, of 75
– ensemble Kalman filter, use of, in 203
– global circulation, of (*see* global circulation model)
– global thermohaline circulation, of 204
– heat fluxes, of 196
– inertial jet, of Agulhas retroflection as 199, **200**
– inter-ocean exchanges, of **195**, 197
– inverse 196
– lee eddy off Agulhas Bank, of **176**, 176
– meridional heat flux simulated by 196, 205
– momentum flux in 200
– Mozambique Channel throughflow, of 47
– Mozambique Channel, variability in, of 47, 66, **66**, 70, **75**
– Mozambique Current, volume flux of **48**
– Mozambique eddies, of 66, **66**, 75, 78
– Natal Pulse in 176
– North Atlantic Deep Water in 48, **198**, 198
– North Madagascar Current, volume flux of **48**
– OCCAM, by 58
– Princeton Ocean 200
– seasonal forcing of South Indian Ocean, of 47
– shear edge eddies at Agulhas Bank, of 128
– Somali Current, of seasonality of 70
– South Atlantic, of heat and salt fluxes of 197
– South Atlantic, of heat and volume fluxes of 197
– South Equatorial Current, simulation of 47
– South Indian Ocean Current, simulation of 46, 230
– South West Indian Ocean, of movements in 46, 47, 48, **49**
– South West Indian Ocean subgyre, of **49**
– Subtropical Convergence, of 47, **202**
– Subtropical Convergence, of eddy shedding at 48, **230**, 230
– Subtropical Convergence, of phytoplankton blooms at 224
– Subtropical Surface Water penetration at Agulhas retroflection, of 164
– using data assimilation in 192
– viscous stress curl, incorporating 200
– West Indian Ocean, of currents in the **66**, **75**
– wind driven 199
MODIS (satellite)
– image from **61**
moisture flux
– from Agulhas Current termination 151
momentum flux
– ring shedding, as a cause for 200
monsoon (*see also* winds)
– circulation of **18**
– ocean circulation due to **18**, 92
– winds of 8, 17
"monster waves" (*see* waves)
Mossel Bay
– surface temperatures at 141, **142**
– wind speeds at 141, **142**
Möwe (vessel)
– early investigations by 2
Mozambique Basin
– Antarctic Bottom Water in 32
– eddies in 89, **89**, 118
– North Atlantic Deep Water in 32
Mozambique Channel
– Antarctic Intermediate Water in 37, **37**, **82**, 82, **83**, 83
– birds feeding in 78
– circulation in 69, 71, **71**, **72**, 73, **73**, **74**
– circulation in, according to ships' drift 53
– cruises by research vessel *Meiring Naudé* to 103
– cruises by research vessel *Pelagia* in 157
– current dynamics, need for more information on 233
– current meter array in 157
– currents in 5, **71**, 72, 73, **73**, **74**, 74
– currents simulated for **66**
– current variability in **68**, 68
– drifters in 81
– eddies emerging from 88, **88**, 118
– eddies in 5, **65**, 81, 88, **88**
– eddies simulated in 66, **66**, 70
– Equatorial Indian Ocean Water in **82**, 82, 83
– flow into northern mouth of **70**, **71**, **72**, 72
– flow out of southern mouth of **70**, **72**, 80
– flow through 42, **42**, **70**, **71**, **72**
– hydrographic observations, lack of, for 235, 236
– Indian Deep Water in 32
– inflow, width of, of 72
– lee eddies in 73, 78, **79**, **80**
– kinetic energy values for **75**, 91, **92**
– knowledge, growth in, of 235
– mesoscale variability in **75**

– mixed layer depth in 44, **45**
– modelling of flow in 69, 70
– North Indian Central Water in 40
– North Indian Deep Water in **82**, **83**
– North Atlantic Deep Water in 32, **82**, 83
– Red Sea Water in 30, 37, 40, 107, 166
– shelf circulation of 73
– simulated volume flux for 69, 70
– seasonality in 92
– southern mouth, flow through **70**, **72**, 80
– southern mouth, volume flux through 80, 81
– Subtropical South Indian Ocean Water in **82**, 83
– surface drift of waters in 54
– temperature–oxygen relationships of waters in **82**
– temperature–salinity relationships of waters in **82**, 82
– throughflow modelled for 47
– Tropical Surface Water flowing through 103, 106, 107
– variability in, according to altimetry 25, 75, **75**, **78**, 78
– vertical temperature structure in 44, **45**
– volume flux into 72
– volume transport through 42, **42**
– water mass characteristics in (*see* temperature-salinity relationships of waters in)
– wind steadiness over 22
Mozambique Current (*see also* Mozambique Undercurrent)
– Agulhas Current, continuity with 5
– Agulhas Current, volume flux contribution to, by **24**, **85**
– depth of **43**
– lee eddies associated with 73
– location of 74
– model simulation of 47, 66, **66**, 88
– source for Agulhas Current, as 5
– speeds in **43**, 74
– surface drift of 54
– seasonality, surface drift of, of 55, 91
– seasonality, modelled for 91
– Tropical Surface Water, as source of 77, 102, 103
– vertical profiles for **43**
– variability at 75, **75**, **78**, 78
– volume transport of 75
– volume transport model simulation for **48**
Mozambique eddies
– Antarctic Intermediate Water in 77
– birds feeding from 78
– current measurements of **76**, 76
– Delagoa Bight eddy, influence on 79
– depth of 77
– diameters of 76
– drifter in 78, **78**, 119
– energy increase in **77**
– frequency of formation of 77
– Red Sea Water in 37, **38**, 40
– mixing in 77
– model simulations of 66, **66**, 75
– origin of 77
– speeds of 76
– Red Sea Water in 77
– Subtropical Surface Water in 77
– trajectories of **77**, 77, **78**, 88, **88**
– translation speed of 76, **88**
– Tropical Surface Water in 77
– volume transport by 76
Mozambique Plateau (previously Mozambique Ridge)
– Agulhas Return Current crossing 86
– Antarctic Water movement past 32
– buoy moored on 152
– deep-sea eddies at 65, 81, 82, 86, **87**, 87, 89, 116, 118
– deflection caused by 7, 87
– drifters at **212**
– Delagoa Pulses dissipating over 81
– drifter tracks, influence on 211, **212**, 228, **229**
– eddy movement, influence on 65, 87
– extension of 7
– hypothesised current along 88
– "Mozambique Ridge Current" at 88
– planetary type wave at 220
"Mozambique Ridge Current"
– concept of 88
Mozambique Undercurrent (*see also* Mozambique Current)
– Antarctic Intermediate Water, in 37, 78
– carrying water equatorward 78
– depth of 78
– North Atlantic Deep Water in 78
– speed of 78
myctophids
– Agulhas Current, distribution influenced by 15

NASA (*see* National Aeronautical and Space Administration)
NCAR (*see* National Centre for Atmospheric Research)
NCEP (*see* National Centers for Medium-Range Prediction)
NOAA (*see also* NOAA satellite) (*see* National Oceanic and Atmospheric Administration)
NPRL (*see* National Physical Research Laboratory)
NRIO (*see* National Research Institute for Oceanology)
NIMBUS VI (satellite)
– Coastal Zone Color Scanner of 133
– relayed moored buoy data by 152
NOAA (satellite)
– Agulhas Current, use of in studying **100**, 133, **161**
– Agulhas retroflection, image of, from 158, **159**, **162**
– Agulhas ring, image of, from **191**
– Subantarctic Surface Water intrusion **162**
– thermal infrared imagery from **100**, 133, **161**
NORWECOM
– biology of Agulhas Current system, modelling by 201
Natal (vessel)
– participating in the International Indian Ocean Expedition 21
Natal Basin (*see* Natal Valley)
Natal Bight
– Agulhas Current, ecosystem influenced by 15, 16
– airborne radiation thermometry over **109**, **111**
– bottom shelf water on 110, **111**
– chlorophyll-*a* distribution over **110**, 110

– circulation over 109, 111, **112**
– cyclonic eddies at **88**
– endemic species of 110
– flow distribution over 109, 111, **11**
– lee eddy in 79, 111, **112**
– morphology of 6, 108
– Natal Pulse, generated at **116**, 159, 234
– nutrient distribution over **110**, **111**
– sea surface salinities over 105, 106
– sea surface temperatures over **109**, 109, 110, **111**
– shear edge features at **100**, 109, 113
– solitary meander north of 81, 116, 118, **119**
– South Indian Central Water in 113
– Subtropical Surface Water over 110, 113
– temperature–salinity relations for 113, **114**
– Tropical Surface Water over 113
– upwelling cell in 109, **109**
– upwelling filament at 110
– vertical section along 110, **111**
Natal Pulse
– Agulhas Bank, along **116**
– Agulhas ring, shedding of, influence on 11, 159, **161**, 206
– amplitude of 114
– characteristics brought to light by satellite investigations 234
– current measurements in 114, 176
– cyclones shed into Cape Basin 174
– cyclonic motion in 116, **116**
– depth of 114
– diameter of 114
– discovery of 114
– downstream progress of 114, 115, **117**, 159, **161**
– floats in 114, 115, 176
– hydrographic section across **138**, 138
– inception of 113
– inshore current reversals due to 116
– lateral growth of 6, **117**
– Natal Bight, generated at **116**, 116
– occurrence frequency of 114
– origin of 6, 114
– phase velocity of 114, 115, **117**, 159, **161**
– Port Elizabeth, off 6, **116**, **158**
– portrayal of **158**
– RAFOS floats in 176
– ring shedding, influence on 11, 159, **161**, 206
– satellite thermal imagery, in **115**, **116**, 133, **161**, 176
– sea-surface temperatures of **115**
– solitary meander, as 114, 127
– thermohaline circulation, influence on 196
– trigger for **116**, 118, 119, 235
– trigger for Agulhas ring shedding, as **161**
– upstream retroflection, influence on 196
– vertical temperature profile across **115**
– vortex shedding as generator of 116
Natal Valley
– sea height variability in 68, **68**
– eddies propagating into 68, 82, 88, **89**, 118
– North Atlantic Deep Water in 32
National Aeronautical and Space Administration (NASA)
– CSIR, collaboration with 152
National Centre for Atmospheric Research (NCAR)
– CSIR, collaboration with 152
National Centers for Medium-Range Prediction (NCEP)
– model for heat transfer from Agulhas Current 149
National Oceanic and Atmospheric Administration (NOAA)
– CSIR, collaboration with 152
– participating in the Benguela Sources and Transports programme 157
– satellite images from **100**, **116**, **126**, **159**, **162**, **191**
National Physical Research Laboratory (NPRL)
– Physical Oceanography Division of 64
– use of airborne radiation thermometry by 133
National Research Institute for Oceanology (NRIO) 64
– establishment of 64
– demise of 64
– role in Agulhas Current research 64
– owner of research vessel *Meiring Naudé* 64
Nazareth Bank
– bifurcation of South Equatorial Current at 58, **58**
New England seamounts
– current kinetic energy at 156
Newfoundland
– Gulf Stream off 8
nitrate (*see also* nutrients)
– Agulhas retroflection, section across **169**
– Agulhas rings, in 180
– Natal Bight, distribution over **110**, 110
– Natal Bight, section across **110**, 110
– south-eastern Madagascar upwelling cell, in 63
– Durban, off **111**
North Atlantic Current
– Gulf Stream, as termination of the 8
North Atlantic Deep Water
– Agulhas Current, in northern 101, **102**, 106
– Agulhas Current, in southern 123, **124**
– Agulhas retroflection, in 164, **166**
– Agulhas Return Current, in 32, **33**
– Cape Basin, movement through 32, **33**
– deep boundary current off west coast, in 164
– depth of 30, 32
– flow path of 9, 30
– heat and volume fluxes in South Atlantic modelled for 197
– influx to South West Indian Ocean of 32, **33**
– Labrador Sea, origin in 9
– Madagascar, north of 32
– modelled simulation of 48
– Mozambique Basin, in 32
– Mozambique Channel, in 32, **82**, 83
– Mozambique Undercurrent, in 78
– Natal Basin, in 32
– production of, modelled **198**, 198
– salinity of 9
– Somali Basin, flow of, into 32
– South Atlantic Ocean, fluxes across 9

– Southern Ocean, in 9
– South Indian Ocean, erosion in 32
– South West Indian Ocean, in **28**, **29**, 34, **34**
– temperature of 9, **28, 29**
– temperature–salinity relationships of, in western Indian Ocean **28**, 30
– temperature–oxygen relationships of, in western Indian Ocean **29**, 30
– Tropical Surface Water in **102**
– volume flux in Agulhas Current of 101, **102**
– volume flux in South Atlantic of 197
– Walvis Ridge, moving through 32, **33**
North Atlantic Ocean
– currents of ix
– surface salinity of 9
– heat flux through 196, 197
North East Monsoon (*see also* winds; monsoon winds)
– Somali Current, as driving force for 8, 17
North East Pacific Ocean
– crossing of cyclones and anti-cyclones in 174
North Indian Central Water (*see also* Central Water)
– formation of 40
– Indian Ocean, volume percentage in 28
– Mozambique Channel, in 40
– temperature-salinity characteristics of 41
North Indian Deep Water
– South West Indian Ocean, circulation in 34, **34**, **83**
– Mozambique Channel, in **82**, **83**
North Indian Intermediate Water (*see* Red Sea Water)
North Madagascar Current
– volume transport modelled for **48**, 60
North Subtropical Convergence (*see also* Subtropical Convergence)
– South Atlantic Ocean, in 221
– surface characteristics of **218**
Northern Agulhas Current (*see also* Agulhas Current)
– adsorption of eddies onto 118
– Antarctic Bottom Water in 106
– Antarctic Intermediate Water in 101, **102**, **107**, 107
– boundaries of **97**, 126
– boundary, offshore, of **97**
– characteristics of, learnt to date 235
– core of **93**, 94, **97**, 98, **98**
– clouds over 149
– Durban, at 96, **97**
– counter current at 98
– current edge inshore of **97**, 126
– current measurements in 92**, 94**, 94, **95**, **97**, 103**, 108**, 108
– current structure of **94**, 94, **95**, 96, **97**
– current velocities of 91, 92, **93**, **94**, 94, **95**, 96, **97**, **98**, 98, **99**, 104, **108**, 108
– depth of 94, **94**, **95**, 95, 101
– downstream intensification of 105
– drifters in 9
– Ekman veering under 110
– floats in 127
– hydrographic sections across 93, **123**
– hydrography of 105
– inflow into 104
– kinetic energy for 91, **92**
– meander in 93, 114
– mesoscale variability in **68**, **75**, 91, **92**
– Natal Bight, off the **93**, **99, 104**, **109**, 111, **112**
– North Atlantic Deep Water in 101, **102**, 106
– path stability of 94, **100**
– plumes, warm water, of **100**, 100
– Port Edward, at 96
– rainfall, coastal, influence on 223
– Red Sea Water in 40, 101, **102**, **107**, 107
– Richards Bay, at 96
– river runoff along 106, 107, 111, 113
– satellite portrayal of **100**
– sea surface salinities of **97**
– seasonal variability of 91, 92, **100**, 102, 106
– shear edge features on **100**, 100
– speeds of (*see* current velocities of)
– South Indian Central Water in **102**, **106**, **107**, 107
– South Indian Intermediate Water in 106, 107
– storm intensification by 223
– Subtropical Surface Water in 101, **102**, **106**, 106, 107, **107**
– surface characteristics of 96, 97, **97**, 105
– surface drift speeds of 91, 94, 121
– surface drifters in 104
– surface variability of **97**
– temperature–salinity relationships of **97**, **99**, 101, **102**, 105, **106**, 107, **107**
– transport, downstream increase in, of **104**, 104, 105
– transport, variability of 103
– Tropical Surface Water in 101, **102**, **106**, 106
– Tropical Thermocline Water in 107
– vertical penetration of 94, **94**, **95**
– vertical sections across 93, 94, **94**, **95**, **99**
– vertical velocity structure of 93, **94**, 94, **95**
– volume transport of 42, 100, **101**, **102**, **104**
– volume transport of water masses of **102**
– water masses in 101, **102**
– waves on 105
– width of 94, **97**, 98
Northern Subtropical Front
– nomenclature for 222
numerical model (*see* model)
nutrients (*see also* nitrate; silicate; phosphate)
– Agulhas Bank, on 168
– Agulhas retroflection, in 168
– Agulhas ring, in 180, **181**
– Cape St Lucia, -rich water off 109, **110**
– Natal Bight, distribution over **110**, 111, **111**
– Natal Bight, section across **111**
– Port Alfred upwelling cell, enhancement in 130, 135, 150
– south-eastern Madagascar upwelling cell, in 63
– Subtropical Convergence, at 168, 220

OCCAM (*see* Ocean Circulation and Advanced Modelling Project)
Ocean Circulation and Advanced Modelling Project (OCCAM)
– simulation point of bifurcation of South Equatorial Current modelled by 58, **59**
– simulation of currents around Madagascar by **59**
– simulation of connection between East Madagascar and Agulhas Currents by 65
– simulation of eddy kinetic energy by 67, **75**
– simulation of flow in Mozambique Current by 70
oceanic conveyor belt (*see* thermohaline circulation cell)
oceanic eddies (*see also* eddy)
– Natal Basin, in 68, 82, 88, **89**, 118
oil slicks
– Durban, to trace water movement off 112
oxygen content (*see also* oxygen minimum layer)
– Agulhas Bank, along edge of 167, **167**, 168
– Agulhas Current retroflection, in 166, 167, **167**,
– Agulhas Return Current, in **213**, 215
– Agulhas ring, in 170, 179, 180, **181**, **182**
– Antarctic Intermediate Water, in **29**, **82**, 82, 83, 123, **166**
– Antarctic Bottom Water, in **29**, **166**
– Benguela upwelling regime, in 190, 191
– Indian Ocean, of western 17, 19, 28, **29**, 29, 30, 32, 38, 40
– North Atlantic Deep Water, in **29**, **82**, 83, **124**, **166**
– North Indian Deep Water, in **82**, 83
– Red Sea Water **82**, 83, **124**, **166**
– Subantarctic Mode Water, of 29, **29**, **123**, **124**, **166**
– Subtropical Surface Water, of 29, **29**, **82**, 83, 123
– Tropical Thermocline Water, of **124**, **166**
– Tropical Surface Water, of **29**, **82**
oxygen minimum layer
– Agulhas Current, in **167**, **213**
– Agulhas Current water, as tracer of 168, 191, 215
– Agulhas Return Current, in 215
– Agulhas retroflection, in **167**
– Agulhas rings, in 170, 179, 180, **181**, **182**
– shallow 29, **29**, 107, 166, 167, **167**, 170, 171, 191
oxygen section (*see* hydrographic section)

palaeoceanography
– Agulhas Current, studies of 15, 130
Pelagia (vessel)
– used in ACSEX programme 157
Persian Gulf
– source of high salinity water, as 17, 30
phosphate
– Agulhas retroflection, section across 168, **169**
– Agulhas ring, in 180
– south-eastern Madagascar upwelling cell, in 63
Pieter Faure (vessel)
– used for pioneering exploration by John Gilchrist 4
pigment concentrations (*see* chlorophyll-*a*)
Planet (vessel)
– used in pioneering observations 2, 19
planetary wave
– Agulhas Return Current, on 220
– shifting east of Agulhas Plateau 226
plankton bloom
– seasonal 44
– St Lucia upwelling cell, at 109
plastic pollution (*see also* pollution; marine pollution)
– Agulhas Current, transportation by 14
plumes
– Agulhas Bank, at 7, 125, 127, 128, **129**, 129, 139
– Agulhas Bank, dimensions at 127, 129, 132
– Agulhas Current, on northern 100, **100**
– Agulhas Current, on southern **116**, 125, **126**, 126, 127
– Gulf Stream, on edges of 131
– Natal Bight, at 100, **100**
– warm water, of 100, **100**, **116**, 125, **129**, 129
Polar Front (*see* Antarctic Polar Front)
pollution (*see also* plastic pollution; marine pollution)
– Agulhas Current, transportation by 14
Ponto do Ouro
– sediments downstream of 96
– temperature decrease downstream of 124
– Agulhas Current at 91
Port Alfred (*see also* Port Alfred upwelling cell)
– Agulhas Current, volume transport at 101
– air-sea interaction at 135, 150
– bathymetric similarity to south-eastern Madagascar 62
– cloud formation at **150**
– hydrographic sections at 133, **134**
– nutrients in waters at 135
– temperature–salinity relationships at **136**, 136
– temperature sections at 133, **134**
– upwelling at (*see* Port Alfred upwelling cell)
Port Alfred upwelling cell (*see also* Port Alfred)
– Agulhas Bank bottom water, as origin of 144, **145**
– air-sea interaction at 149, 150
– chlorophyll in 137
– cowries in 137
– driving forces for 138, 139
– Ekman drift at 138
– geographic extent of 132, **135**, **137**
– heat fluxes to atmosphere from 135, 149, 150
– hydrographic section across 133, **134**, **136**, **138**, 138
– influence of 137
– nutrient enhancement at 135
– satellite studies of **116**, 133, 135, **135**
– siphonophores at 137
– South Indian Central Water in 135, **136**, 136
– South Indian Subtropical Surface Water in 135, **136**
– surface temperatures of 132, 133, **134**, 135
– temperature-salinity relationships at 135, **136**
– Tropical Indian Surface Water in 135, **136**
– water upwelled in 131, **134**, 135
Port Edward
– Agulhas Current characteristics at 96, **99**, 108
– current measurements at **95**, **97**, **98**, 108, 157
– current velocities at **95**, **97**, **98**, **99**, **104**

– volume flux, growth in, downstream of 104
– volume transport at 100, **104**
Port Elizabeth
– Agulhas retroflection south of 196
– bottom water at 142
– current border at 122
– drifter south of 158
– hydrographic section at 122, **123**
– hydrographic sections north of 93, 122
– hydrographic surveys north of **123**
– location of **6**
– Natal Pulse, phase velocity change of, at 15, 117, 159, **161**
– Natal Pulse at 6, **116**
– shear instabilities downstream of 125, **126**
– shelf/current arrangement at 138
– shelf waves, termination of, at 149
– temperature–salinity relationships of water at 105
– temperature section at 122
– upstream retroflection at 6, 196
– upwelling close to **159**
– volume transport of Agulhas Current at 103
– winds at 149
Port St Johns
– temperature section at 122, **123**
potential energy
– Agulhas ring, of 174, 194
– Agulhas rings, fluxes due to 193
potential temperatures
– Antarctic Bottom Water, of **28**, 30, **31**, **39**, **166**
– Antarctic Intermediate Water, of **28**, 30, **36**, **82**, 82, **106**, 107, **166**
– bottom waters in South West Indian Ocean, of 30, **31**
– North Atlantic Deep Water, of **28**, 30, 35, 37, **39**, 40, **82**, 82, 123, **124**, **166**
– postulated Brazil Current ring, of 179
– Red Sea Water, of **28**, 30, 35, 37, **39**, 40, **82**, 82, **106**, 106, **107**, 123, **124**, **166**
– Subtropical Surface Water, of **28**, **39**, 41, **82**, 82, **106**, 106, 113, **114**, **146**, **166**
– western Indian Ocean, related to oxygen of **29**, **82**, **124**, **166**
– western Indian Ocean, related to salinities of **28**, **82**, **106**, **107**, **124**, **165**
potential temperature–salinity relationships (*see* temperature–salinity relationships)
primary productivity
– in South Indian Ocean 18
– seasonality of, in South Indian Ocean 18
– Subtropical Convergence, at 21, 223
Prince Edward Fracture Zone
– Antarctic Bottom Water, as conduit for 32
Princeton Ocean Model (*see* model)

quasi-biennial atmospheric oscillation
– impact on rainfall modulated by sea surface temperatures 223

RSA (vessel)
– drifters launched from 152
radiation thermometry (*see* airborne radiation thermometry)
RAFOS floats (*see* floats)
rainfall
– Agulhas Current, affected by 137, 223, 234
– coastline, at 223
– oscillations of 223
recirculation
– South West Indian Ocean subgyre, in 84, **84**, 104
recruitment
– pelagic fish in Benguela upwelling regime, of 190
– Agulhas rings, influence of, on 190
Red Sea
– source of high salinity water 17
– pre-war German research in 19
Red Sea-Persian Gulf Intermediate Water (*see* North Indian Intermediate Water)
Red Sea Intermediate Water (*see* Red Sea Water)
Red Sea Water (*see also* North Indian Intermediate Water)
– Agulhas Current, flux of, in 40
– Agulhas Current, in northern 40, 101, **102**, 106, **107**, 107
– Agulhas Current, in southern 38, **38**, 123, **124**
– Agulhas retroflection, in **166**, 166
– Agulhas rings, in 40
– Agulhas undercurrent, in 95
– Cape Basin, in 190
– depth of 37, 38
– Indian Ocean, in 35, **35**, **38**
– Mozambique Channel, in 30, **36**, 37, **38**, 38, 40, 83, 107, 166
– Mozambique Channel, depth in 37
– Mozambique Basin, in 37, **38**, 40, 77, **82**, 83
– Mozambique eddy, in 37, **38**, 40, 77
– oxygen content of **82**, 83
– South West Indian Ocean, in 17, **35**, **36**, **38**
– South West Indian Ocean, erosion of, in 17, **35**, **36**, **38**
– temperature–salinity characteristics of, in Mozambique Channel 30
– temperature–salinity characteristics of, in western Indian Ocean **28**, 30
– temperature–oxygen relationships of, in western Indian Ocean **29**, 29
– volume of 28
– volume transport in Agulhas Current of 101, **102**
Red Sea-Persian Gulf Intermediate Water (*see* Red Sea Water)
retroflection
– Agulhas Current, of (*see* Agulhas (Current) retroflection) 155
– East Madagascar Current, of **60**, 65, 66
– East Madagascar Current, modelling of 66, **66**, 67
– upstream 6, 120, 196
Réunion (island of)
– divergence of South Equatorial Current at 54
Richards Bay

– Agulhas Current characteristics at 96, **99**
– currents at 96, **99**, 99, **113**
– current measurements at 96, 99, 108
– nutrients at 109
– sea surface temperatures at **109**, 109
– upwelling at 109, **109**
– winds at **113**
rings (*see also* Agulhas rings)
– Agulhas Current, formed by (*see* Agulhas rings)
– East Australian Current, formed by 162
– Gulf Stream, formed by 159
– shedding process of (*see* Agulhas rings)
river runoff
– Agulhas Current, along northern 106, 107
– Natal Bight, at 111, 113
Rossby radius of deformation
– Agulhas rings, and 200
Rossby wave
– Agulhas Return Current, in 213
– Agulhas Plateau, propagating west of 213
– Agulhas rings, radiating from 182

SAF (*see* Synthetic Aperture Radar)
SANCOR (*see* South African National Committee for Oceanographic Research)
S.A.S. Natal (*see Natal*)
SAR (*see* Synthetic Aperture Radar)
SCARC (*see* Subtropical Convergence and Agulhas Retroflection Cruise)
SCOR (*see* Scientific Committee on Oceanic Research)
SeaWiFS (*see* Sea-viewing of Wide Field Sensor)
SA Agulhas (vessel)
– used for Subtropical Convergence and Agulhas Retroflection Cruise 157
Sagitta enflata (chaetognath) (*see also* chaetognaths)
– Agulhas Current water, indicator species for 15
Saldanha Bay
– cruises by research vessel *Meiring Naudé* to 103
salinity distribution (*see also* temperature-salinity relationships)
– Antarctic Bottom Water, in 30, **31**
– Antarctic Intermediate Water in South West Indian Ocean, in 37, **37**
– North Atlantic Deep Water in South West Indian Ocean, in 32, **33**, 34, **34**
– Red Sea Water in South West Indian Ocean, in 34, **34**
– South West Indian Ocean, at bottom of 30, **31**
salinity profiles
– Agulhas Return Current, in **213**, **216**
– South West Indian Ocean, of current elements of **43**
salinity section (*see* hydrographic section)
salt fluxes
– Agulhas filaments, by 132, 192
– inter-ocean, by Agulhas filaments 192, 194
– inter-ocean, by Agulhas rings 193
– Sofala Bank, on 74
– South Atlantic subtropical gyre, in, modelled for 197
salt marshes
– at Sofala Bank 74
sand dunes
– bedform, as part of 96, **96**
– bottom sediments, in 96, **96**
– current speeds, to indicate 96
satellite altimetry (*see* altimetry)
satellite investigations (*see also* altimetry; TRMM)
– Agulhas Bank, of upwelling on **116**, 133
– Agulhas Current, of southern **116**, 125, **126**, 133, **159**, 235
– Agulhas Current, of northern **100**, 133
– Agulhas Current, to study edge, of **116**, **126**
– Agulhas Current retroflection, of 151, 158, **159**, **160**, **162**, 234
– Agulhas filaments, of **126**, **132**, 133
– Agulhas Front, of 217
– Agulhas Plateau, at **230**
– Agulhas Return Current, of **159**, 210, 211, 226, **227**
– Agulhas retroflection, of 133, 158, **159**, **162**
– Agulhas ring shedding and movement of 133, 159, **161**, **162**, **188**, **191**, **205**, **206**, 234
– altimetry, using 133, 178
– Brazil rings, of 178
– Delagoa Pulse, of 118
– East Madagascar Current retroflection, of 61,133
– giant waves, to study 105, 133
– history of use of 133
– Natal Bight, using colour imagery in 110
– Natal Pulse, of 114, **117**, 133, **161**, 176, 234
– pioneering use of 133
– Port Alfred upwelling cell, of 133, 135, **135**
– ring shedding process, of 234
– sea level, of 133, 178
– shear edge features, of **116**, **126**, 133
– Subtropical Convergence, of **221**, **227**
– surface waves, of 105
– thermal infrared, usefulness of 3, 133
– upstream retroflection, of 196
– upwelling filament, of **191**
Satellite Remote Sensing Centre (*see also* Hartebeesthoek)
– CSIR, of 133
Scanfish undulator
– observations by, in Agulhas ring 185, **186**
Scientific Committee on Oceanic Research (SCOR)
– International Indian Ocean Expedition, role in 21
– South African presidency of 64
Sea Fisheries Research Institute
– participating in Benguela Sources and Transports programme 157
sea level observations
– shelf waves, to establish 149
sea level variability (*see* altimetry)
sea surface salinities (*see also* temperature–salinity relationships)
– Agulhas Current, of northern 105, **106**
– Agulhas Current, of northern shelf off 105
– Agulhas rings, modification in 180

– Subantarctic Surface Water, in wedge of 162, **163**, 168
sea surface temperatures (*see also* temperature–salinity relationships)
– Agulhas Bank, over 140, **142**, **147**
– Agulhas Current, of northern 105, **106**
– Agulhas Current, of southern 153, **153**, 223
– Agulhas Front, of 217, **218**
– Agulhas retroflection, of 153, **153**, **159**
– Agulhas Return Current, of 153, **153**
– Antarctic Polar Front, of 218
– cycles in 223
– Durban, off 105
– Gouriqua (Mossel Bay), at 141, **142**
– Natal Bight, over 109, **109**, 110, **111**
– Natal Pulse, across a **115**
– North Subtropical Convergence, of **218**
– Port Alfred upwelling cell, at 132, 133
– rainfall, affected by 223
– South East Atlantic Ocean, of 194
– South West Indian Ocean, of, and rainfall 223
– South Subtropical Convergence, of **218**
– Subantarctic Front, of **218**
– Subantarctic Surface Water, in wedge of **163**, 164
– Subtropical Convergence, of 153, **218**
– variability of 217, 230
sea surface temperature anomalies (*see also* sea surface temperatures)
– Agulhas retroflection, in 223
– cycles in 223
– Indian Ocean, in 223
– rainfall, effect on, of 223, 235
– South Indian Ocean, in 223
– Southern Oscillation, dampening effect of, on 223
SEASAT (satellite)
– Agulhas retroflection, altimeter observations of 154
– sea height variability south of Madagascar, observations of 68, **68**
seasonality
– Agulhas Bank, of waters on 110, 129, 140, **141**, 144, **147**
– Agulhas Current, of 91, 92
– Agulhas Current, of northern 91, 92
– Agulhas Current, of southern 124
– Agulhas Current, of volume transport of **100**, 102
– Agulhas Current retroflection, of 154, 195
– Agulhas Current system, of 102
– Agulhas retroflection variability, of 154, 195
– East African Coastal Current, of 55
– East Madagascar Current, of 58, 91, 92
– Mozambique Channel, in 92
– Mozambique Current, of 91
– Mozambique Current, modelled for 91
– Somali Current, modelled for 70
– South East Atlantic Ocean, of sea surface temperatures, in 194
– South Equatorial Current, of 91
– South Equatorial Current, modelled for 91
– South Indian Ocean gyre, of 102
– South West Indian Ocean, of temperature structure in the 44, **45**
– south-eastern Madagascar upwelling cell, of 63
sea turtles (*see* marine turtles)
Sea-viewing of Wide Field Sensor (SeaWiFS) (satellite)
– South Africans, use made of by 133
sediments
– bottom current directions, showing 96, **96**, 111, 112, 130
– continental shelf, on 96, 121
– Natal Bight, in water over 111
self-steering
– Agulhas rings, by 187
Seychelles–Mauritius Ridge
– flow obstruction, as 47
– current shadow behind 47
shear edge eddies
– Agulhas Bank, off 100, 127, **127**, 131, 176, **177**, 177
– Agulhas Bight, in 127, **127**, 131
– Agulhas Current, on northern **100**, 100
– Agulhas Current, on southern 125, 127
– floats in 176, **177**,
– Gulf Stream, of 100, 131
– modelled 127, **128**
– Natal Bight, at **100**, 109, 113
– Kuroshio, of 100
– South Indian Ocean Central Water in 131
– studies by satellite of **126**, 133
shelf sediment (*see* continental shelf)
shelf waves
– location of occurrence of 149
– propagation, termination to, of 149
shelf currents (*see* continental shelf currents)
shipping routes
– early 53
ships' drift observations
– Agulhas Current, in northern 91, 92, **93**
– Agulhas Current, in southern 1, 121, **228**
– Agulhas Current, in sources of 53, 54
– Agulhas Return Current, for 220, **228**
– Indian Ocean circulation, of **18**, 23
– kinetic energy of flow, used to establish 151
– South West Indian Ocean circulation, of 23, 53, **55**
– Subtropical Convergence eddy, in **228**
silicate (*see also* nutrients)
– Port Alfred upwelling cell, in 135
– Subantarctic Surface Water, in 168
siphonophore
– Agulhas Current, influence on distribution of 14, 137
– Cape St Lucia upwelling cell, at 109
– Port Alfred upwelling cell, at 137
Sofala Bank
– circulation on 73
– freshwater influence on 74
– salt marshes at 74
– Zambezi runoff onto 73
Somali Current
– characteristics, unique, of 8, 17, **18**

– depth of 8
– seasonality of 8, 17
– surface speeds of 8
– volume flux of 8
– western boundary current, as 8
Somali Basin
– Antarctic Bottom Water movement into 30
– North Atlantic Deep Water, flow of, into 32
sources
– Agulhas Current, of the 53 ff.
– East Madagascar Current, of the 57
South African National Committee for Oceanographic Research (SANCOR)
– formation of 21
South African National Oceanographic Symposium
– first held 21
South African Weather Bureau
– deploying drifters 152
– predicting hazardous waves on Agulhas Current 105
South Atlantic Current (*see* South Atlantic Ocean Current)
South Atlantic Ocean (*see also* South East Atlantic Ocean)
– Agulhas Current, tracing water from, in 167
– Agulhas retroflection loop protruding into **159**, **160**
– Antarctic Bottom Water fluxes in 197
– Antarctic Intermediate Water fluxes in 192, 197
– Benguela Current, forming part of gyre of, in 189
– circulation variability of 194, 195
– general mean motion of **188**
– geostrophic velocities in **188**
– heat fluxes across 30° S meridian of 197
– heat fluxes modelled for 197
– Indian Ocean water, direct flux of, into 191, 192
– Indian Ocean water, flux of, by rings into 192, 193, 194
– North Atlantic Deep Water fluxes in 197
– salt fluxes modelled for 197
– seasonality in 194, 195
– Subtropical Convergence in 220, **221**, 221
– Subtropical Mode Water in 166, **166**, 167, 168, 178
– tropical water from 190, 192
– volume transport across 30° S meridian of 197
– volume transport in 101, 102, 190, 192, 197, 198, **219**
South Atlantic (Ocean) Current
– Benguela Current, feeding 221
– depth of 44
– eddies shed from 187
– geostrophic velocities of **219**
– South Atlantic gyre, as part of 220, 221
– South Indian Ocean Current, analogy to 220
– volume transport of 44, 192, 218, **219**
South (East) Atlantic (Ocean) Central Water
– Agulhas retroflection, in 164, 165
– temperature and salinity range of 40
South Atlantic Surface Water
– heat fluxes for 197
– heat and volume fluxes modelled for 197
– volume flux of 197
South East Atlantic Ocean (*see also* South Atlantic Ocean)
– Agulhas water, influence of warm, on 157, 164, 189, 192, 199
– sea surface temperatures of 194
– cold pulse, influence of, on 164
– current observations in 157
South East Indian Ocean
– crossing of cyclones and anti-cyclones, in 174
South Equatorial Current
– Agulhas Current, volume flux contribution to 68
– bifurcation at Madagascar of (*see* splitting at Madagascar coast of)
– bifurcation of, modelled at Nazareth Bank 58, **58**
– depth of 42, **43**
– early portrayals of 54, **55**
– East Madagascar Current, as source of 19, 57
– Mascarene Plateau, effect on 19
– model simulation of 47
– Mozambique Channel, input to 70
– seasonality of 17, 91
– seasonality, modelled for 91
– speed of 43
– splitting at Madagascar coast of 19, 57, 58
– Subtropical Surface Water in 44
– Tropical Surface Water, as source of 44
– vertical profiles for 43
South Indian Central Water
– Agulhas Bank, on 142
– Agulhas Current, volume transport in 101, **102**
– Natal Bight, in 113
– northern Agulhas Current, in **102**, **107**, 107
– Port Alfred upwelling cell, in 135, **136**
– southern Agulhas Current, in 123
– temperature-salinity relationships of, in western Indian Ocean **28**, 29, 40
– temperature-oxygen relationships of, in western Indian Ocean **29**, 29
– volume transport in Agulhas Current, in 101, **102**
South Indian (Ocean) Intermediate Water
– Agulhas Current, in northern 106, 107
– Port Elizabeth, at 107
South Indian Ocean (*see also* South West Indian Ocean)
– Agulhas water, course of, across 209, 231
– Antarctic Intermediate Water, circulation of, in 40
– atmospheric pressure over **22**
– baroclinic volume flux of 24, 41, **41**
– circulatory cells in 47
– Current (*see* South Indian Ocean Current)
– flow patterns of 23
– flow perturbations moving across 233
– fronts of **218**
– gyre invigoration of 130
– hydrographic section across 26, **26**
– leakage from (*see* inter-ocean exchange)
– mean dynamic topography of **20**, **25**
– mixed layer depths of 44, **45**
– modes of variability, model simulation of, for 47
– North Atlantic Deep Water erosion in 32

– primary productivity in 18
– sea surface temperature anomalies on 50, 223
– seasonality of gyre in 102
– Subtropical Convergence in **221**
– subtropical gyre of 17
– Subtropical Mode Water in 19, 28, 29, **29**
– surface currents of **18**, **20**, **25**
– surface temperatures of 11, 12
– thermocline ventilation, in 44
– volume transport in gyre of 24
South Indian (Ocean) Central Water
– Agulhas Bank, on 142, 143
– Agulhas retroflection, in **166**
– depth of, in South West Indian Ocean 40
– Natal Bight upwelling cell, in 113
– northern Agulhas Current, in 101, **102**, 106, **106**, **107**, 107
– shear edge eddies, in 131
– southern Agulhas Current, in 131
– temperature–oxygen characteristics of **29**
– temperature–salinity characteristics of **28**
– volume of **27**
South Indian Ocean Current
– depth of 231
– eddies associated with 231
– fronts associated with 231
– geostrophic velocities of 230, 231
– Kerguelen Plateau, initiated at 7, 209
– location of 230, 231
– model simulation of 44, 230
– nomenclature on 220
– South Atlantic Current, analogy to 220
– subtropical gyre, as part of 19, 231
South Indian (Ocean) Subtropical Surface Water (*see* Subtropical Surface Water)
"South Madagascar Current"
– as origin of deep sea eddies 88
South Pacific Ocean
– Antarctic Intermediate Water, source of 192
South Subtropical Convergence (*see also* Subtropical Convergence)
– South Atlantic Ocean, in 221
– surface characteristics of **218**
South West Indian (Ocean) Central Water
– Agulhas retroflection, in 164, 165
South West Indian Ocean (*see also* South Indian Ocean)
– altimetric observations of 24, **25**, 88
– Antarctic Bottom Water, temperature–salinity characteristics of, in **28**
– Antarctic Bottom Water, temperature–salinity characteristics of, in **29**
– Antarctic Intermediate Water, erosion of, in **36**, 37
– Antarctic Intermediate Water in **28**, **29**, **33**, **35**, **36**, **83**
– Antarctic Intermediate Water, temperature–salinity relationships of, in **28**
– Antarctic Intermediate Water, temperature–oxygen relationships of, in **29**
– Antarctic Intermediate Water, circulation of, in 32, **33**, 37, **37**, **83**
– Antarctic Intermediate Water, depth of, in 37
– Antarctic Intermediate Water, distribution of, in 37
– bottom temperatures of 30, **32**
– bottom topography of **6**
– bottom salinities of 30, **32**
– circulation at 1000 m depth in **20**, 23, **84**
– circulation of, according to ship's drift 53
– circulation in, modelled 47, 48, **49**
– Circumpolar Deep Water in 32
– continental rainfall, and 11, 12, 16, 223
– current variability in **68**, 68
– deep-sea eddies in
– downwelling in 23
– drifters in 84, 86
– eddies in 86, **86**, **87**
– eddy kinetic energy for **92**
– Ekman transport over 23
– flow patterns of **18**, **20**, 23, **24**, **25**
– general circulation in **18**, **20**, **24**, **25**
– geostrophic velocity of current components of 56, **56**
– high pressure systems over 148
– Indian Deep Water, temperature-salinity characteristics of, in **28**
– Indian Deep Water, temperature-oxygen characteristics of, in **29**
– Indonesian Intermediate Water, in 35, **35**
– kinetic energy of **92**
– large-scale circulation of 17, **18**, **20**, **24**, **25**
– mass transport function of **41**
– mixed layer of 44, **45**
– modelling with FRAM 47
– models of movement in 46, 47
– North Atlantic Deep Water, temperature–salinity relationships of, in **28**
– North Atlantic Deep Water, temperature–oxygen relationships of, in **29**
– North Atlantic Deep Water influx into 32
– North Indian Deep Water, circulation in 34, **34**, **83**
– permanent thermocline of 40, **43**, 44
– publications, annual number of, on ix, **ix**, x
– rainfall, influence on 11
– recirculation in 23
– recirculation, modeled, of 46
– Red Sea Water, in **35**, **36**, **38**
– Red Sea Water, salinity of, in **28**, **36**, **38**
– Red Sea Water, temperature–salinity relationships of, in **28**, **36**
– Red Sea Water, temperature–oxygen relationships of, in **29**, 29
– salinities of **28**, 30, **32**
– salinity of Red Sea Water, in **28**, **36**, **38**
– salinity profiles in **43**
– sea surface temperatures of 19, **28**, 28, **29**, **45**, **46**
– seasonal thermocline of 44

– seasonal variability in 24
– seasonal variability in currents of 24
– ships' drift measurements in 53
– South Indian Central Water, temperature-salinity relationships of, in **28**
– South Indian Central Water, temperature-oxygen relationships of, in **29**, 29
– South Indian Central Water, depth of, in 40
– Subantarctic Mode Water, temperature-salinity relationships of, in **28**, 28, 29
– Subantarctic Mode Water, temperature-oxygen relationships of, in **29**, 29
– subgyre of (*see* South West Indian Ocean subgyre) **6**, 17, 23
– Subtropical Convergence in 19 ff., 24, 30, 40, 41, 44, 47, 48, 209, 210, 213, 214, 215, 216, 217, 218 ff.
– Subtropical Surface Water, temperature–salinity relationships of, in **28**
– Subtropical Surface Water, temperature–oxygen relationships of, in **29**, 29
– surface flow of 55, **55**
– temperature–salinity relationships of 28, **28**, **146**
– temperatures of 28, **28**, 30, **32**
– thermal structure of upper layers of **43**, 44, **45**, **46**
– thermocline of 40, **43**, 44
– thermocline ventilation of 168
– Tropical Surface Water, temperature–salinity relationships of, in 28, **28**
– Tropical Surface Water, temperature–oxygen relationships of, in 29, **29**
– Tropical Thermocline Water, temperature–salinity relationships of, in **28**, 28
– Tropical Thermocline Water, temperature–oxygen relationships of, in **29**, 29
– variability of 24
– ventilation of thermocline waters in 168
– vertical temperature structure in 44, **45**
– volume flux of **24**, **41**, 41, 42, **42**,
– volume transport of **24**, **41**, 41, 42, **42**
– volumetry of 26, 27, **27**
– water masses of 26, **27**
– wind regime of 22, **22**
– wind steadiness over 22
– wind stress over 23
South West Indian Ocean eddies
– depth of 87
– dimensions of 86, **87**, 87
– hydrographic section across **86**
– kinetic energy of 87
– life spans of 87
– Mozambique Plateau, at 65, 81, 82, 86, **87**, 87
– origin of 88
– potential energy of 87
– speeds in 86
– volume transport of 86, **87**
South West Indian Ocean Ridge
– border to model simulated variability, as 47
– drifters, influenced by **212**
– leakage through 69
– portrayal of **6**
South West Indian Ocean sub-gyre
– Agulhas Current, contribution to, by 6, 84, 85
– Agulhas retroflection, recirculating thermostads from 168
– altimetry, from 24, **25**
– depth of 23
– drifters in 84, 86
– geostrophic circulation of **84**
– hydrographic data, from historical 23, **24**, 24
– intra-thermocline eddies in 89
– model simulation of 48, **49**
– numerical model, from a 48, **49**
– portrayal of **6**, 6, **20**, **84**
– recirculation in 84, **84**, 104, 107, 209
– surface drift of 84
– volume flux of 85
South West Monsoon (*see also* winds; monsoon winds)
– Somali Current, as driving force for 8, 17
Southern Agulhas Current (*see also* Agulhas Current)
– Agulhas Bank, influence on 143, 145, 146
– altimetric observations of 204
– Antarctic Intermediate Water in 123
– atmosphere, effect on overlying, of 135
– atmospheric circulation over 138, 141, **142**, **143**, 148, 149, **150**, 164
– baroclinic instability, modelled for 125
– clouds over 149, **150**, 150
– defined, generically 121
– drifters in 158, 204, **212**
– dynamic topography of **210**, **211**
– eddy kinetic energy modelled at 204
– floats in 128
– heat fluxes from 135, 149
– hydrographic section across 122, **123**, **125**, **213**, **224**, **228**
– hydrography of 121 ff
– marine boundary layer over 149
– meanderings on 125, 126, **126**
– meander movement of 125, **126**
– meander wavelength of 125, 127
– mesoscale variability in **68**, **75**, **92**
– mixed layer of 143
– modelling of 125
– North Atlantic Deep Water in 123, **124**
– portrayal of **158**, **159**
– plumes on **116**, 125, **126**, 126, 127
– Red Sea Water in 123, **124**
– satellite imagery of **116**, 125, **126**
– seasonality of 124, 143
– shear edge eddies at 125, **126**, 127, **127**, 176
– shelf edge front of **126**
– silica in 135
– South Indian Central Water in 123
– South Indian Subtropical Surface Water in 123
– Subantarctic Mode Water in 123, **124**
– surface characteristics of 124

– surface speeds of 121, 122
– temperature–oxygen characteristics of 122, 123, **124**
– temperature–salinity characteristics of **39**, 122, **124**
– thermal borders of **126**
– thermocline of 143
– trajectory of **126**
– Tropical Indian Surface Water in 122
– Tropical Thermocline Water in 123, **124**
– upwelling inshore of **116**, 132, **135**
– variability of **68**, **75**, **92**
– vertical stratification of 143
– volume flux of 103, **122**, 122, 214
– winds over 149
Southern oscillation (*see* El Niño–Southern Oscillation (ENSO))
Southern Ocean
– drifters in 152
– frontal systems of 216, **218**
– heat fluxes from 196
– heat loss to, from Agulhas Current system 199
– variability in 152
– volumetry of 28
South Subtropical Front
– nomenclature for 222
space shuttle
– use of SAR on board of 133
splitting
– Agulhas rings, of 180, 182, **189**, 204
St Lucia (*see* Cape St Lucia)
St Lucia upwelling cell
– chlorophyll at 109
– hydrographic sections south of 93
– nutrient rich water of 109
– plankton production at 109
– siphonophores at 109
– South Indian Central Water in 113
– South Indian Subtropical Surface Water in 113
– temperature-salinity relations near 109
– upwelling in 109
Stokes' drift
– Agulhas rings, at 182
Subantarctic Front
– location of 223
– meridional range of 222
– characteristics, surface, of **218**, 223
– characteristics, thermal, of 223
– hydrographic section across **224**
– width of 223
Subantarctic Mode Water
– Agulhas Current, in southern 123, **124**
– Agulhas retroflection region, in **166**, 166, 167
– Agulhas rings, in 168
– Brazil rings, in 178, **179**
– oxygen content of 123, **124**, 166, **166**, 167
– South West Indian Ocean, in **28**, 29, **29**
– Subtropical Convergence, formation at 166
– temperature–oxygen characteristics of **29**, 29
– temperature–salinity characteristics of **29**, 29
– thermostad formed by 19, 29
Subantarctic Surface Water
– Agulhas ring shedding, northward penetration at 157, **159**, 162, **163**, 164, 168
– wedge at Agulhas retroflection, models of 164
– temperature-salinity characteristics at Agulhas retroflection 162
– Subtropical Convergence, subducting at 30, 218
– temperature-salinity relationships of, in western Indian Ocean **28**
– temperature-oxygen relationships of, in western Indian Ocean **29**
sub-gyre (*see* South West Indian Ocean subgyre)
submarine canyons
– channelling shelf sediments, role in 96
Subtropical Convergence (see also North Subtropical Convergence; South Subtropical Convergence)
– Agulhas Current, affected by 220, 221
– Agulhas Current eddies at 153, 170, **222**, **227**
– Agulhas Return Current, juxtapositioned to 7, 24, 209, 210
– Antarctic Intermediate Water, subduction of, at 30, 166
– biological habitat at 231
– birds at 209
– blooms, phytoplankton at 224
– characteristics of 221, **222**, 222, 223
– characteristics, surface, of 223
– characteristics, thermal, of 223
– chlorophyll levels, enhanced at 209, 213
– cetaceans at 209
– colour at 213
– convergence at 224
– diatoms at 209
– drifters at 224
– eddies shed from 88, 187, 198, **216**, 221, **226**, **227**, **228**, **230**
– eddies shed from, model simulation of 48
– eddy at 156, **222**, **226**, **227**, **228**
– flow along 19
– geostrophic speed at 218, **219**
– hydrographic section across 182, **222**, **216**, 220, **222**, **224**, **228**
– location of 19, 130, **221**, 223
– meanders of 225, 227, 229, **229**
– meanders modelled for **229**
– meridional range of **221**, 222, 223, **226**
– mesoscale variability at 21, **154**, 224
– mixed layer depths at 19, 44, **45**
– model simulation of 47, **202**, **204**, 224, 225, 229, **229**
– movement, meridional, of 220, 221
– nomenclature on 222
– nutrients at 168, 220
– palaeo movement of 130
– pelagic community at 221
– phytoplankton blooms at 224
– portrayal of **6**, **226**
– primary productivity, enhanced, at 21, 224

- research, in pre-war 19
- salinity gradients across 221, **222**
- satellite studies of **221**
- sea surface temperatures at 153, **153**, **221**, **222**
- South Atlantic Ocean, in 220
- South East Pacific Ocean, missing in 218
- Southern Ocean, boundary to 218
- speed along 218, **219**
- Subantarctic Mode Water, formation of, at 19, 178
- Subantarctic Water leakage across 157
- Subantarctic Water subduction at 30, 218
- Subtropical Surface Water at 218
- subsurface gradients of 221
- surface expressions of **218**, **221**
- surface gradients across **218**, 221, **222**
- temperature gradients, horizontal, of 209, 221, **222**
- temperatures of surface waters of 153, **218**
- temperature variability, high, in **153**
- thermal characteristics of **222**
- thermostads at 19
- vertical stability at 224
- vertical temperature structure at 44, **45**
- volume transport along 214, 215, 216, 218, **219**
- whales at 209
- width of 223, **226**
- wind steadiness at 23
- zonal jet at 220

Subtropical Convergence and Agulhas Retroflection Cruise (SCARC)
- investigating Agulhas retroflection 157
- investigating Subtropical Convergence 157

Subtropical front (*see* Subtropical Convergence)

Subtropical South Indian Ocean Water
- Mozambique Channel, in **82**, 83

Subtropical Mode Water
- Agulhas retroflection, in and formed at **166**, 166, 167, 168, 178
- South Atlantic Ocean, in 178
- South Indian Ocean, in 19, 28, 29, **29**

Subtropical South Indian Ocean Water (*see also* Subtropical Surface Water)
- Mozambique Channel, in 82, **82**

Subtropical Surface Water (*see also* Subtropical South Indian Ocean Water)
- Agulhas Bank, on **146**
- Agulhas Current, in northern 101, **103**, **106**, 106, 107, **107**
- Agulhas Current, in southern 123
- Agulhas Current, volume transport in 101, **102**
- Agulhas retroflection, in 166, **166**, 167
- East Madagascar Current, in 61
- formation of, at 19
- Mozambique eddies, in 77
- Natal Bight, on bottom over 110, 113
- Natal Bight upwelling cell, in 110
- oxygen content of 29, **29**, **82**, 83, 123
- Port Alfred upwelling cell, in 135, **136**
- south-eastern Madagascar upwelling cell, in 62, 63
- South Equatorial Current, in 44
- South West Indian Ocean, in **28**, 29, **29**
- Subtropical Convergence, at 218
- temperature–salinity relationships of **28**
- temperature–oxygen relationships of **29**, 29
- thermostad in 107

Subtropical Underwater
- formed in 89
- in intra-thermocline eddies 89

surface currents (*see also* currents)
- East Madagascar Current, of 57
- South Indian Ocean, of 55, **55**
- Agulhas Current, of northern 121
- Agulhas Current, of southern 121, 122

surface drifter (*see* drifters)

surface temperature (*see* sea surface temperature)

surface waves (*see* waves)

suspension
- particulate matter in Natal Bight, of 111, 113

Synthetic Aperture Radar (SAR)
- surface waves, use of, to study 133

TRMM (*see* Tropical Rainfall Measuring Mission)

Table Bay
- water temperature of 5

Taiwan
- Kuroshio, start of, at 8
- upwelling cell at 132

Tasman Front
- East Australian Current, relationship to (*see also* East Australian Current) 8

temperature section (*see* hydrographic section)

temperature–oxygen relationships
- Agulhas Current, for southern 122, 123, **124**
- Agulhas retroflection, in **166**
- Indian Ocean, for western 28, **29**, 29, 30
- Mozambique Channel, of waters in 82, **82**

temperature–salinity relationships
- Agulhas Bank waters, for 143, **146**
- Agulhas Current, across the **179**
- Agulhas Current, of the northern 101, **102**, **106**, 106, **107**
- Agulhas Current, for the southern **39**, **124**
- Agulhas retroflection, at the 164, 165, **165**
- Brazil Current, of the **179**
- Brazil ring, of postulated 178, **179**
- East Madagascar Current, for 61
- Indian Central Water, for 4
- Indian Ocean, of western 28, **28**, 29, 30, **36**
- Mozambique Channel, of waters in 82, **82**
- Natal Bight, for the 113, **114**
- North Indian Central Water, for 41
- Port Alfred (upwelling cell), at 135, **136**
- South Atlantic Central Water, of 40
- South Indian Central Water, of 40
- South West Indian Ocean, of **36**, **146**
- south-eastern Madagascar upwelling cell, of 62, 63

Thalia democrita (pelagic tunicate) (*see also* tunicates)
– Agulhas Current water, as indicator species for 15
thermal structure
– Agulhas ring, of new 162
– Natal Pulse, of a **115**, 116, 138, **138**
– South West Indian Ocean, of upper layers of 44, **45**, **46**
thermocline
– Agulhas Bank, seasonal, over 129, **141**, 143, **147**
– Agulhas Current, seasonal, over 143
– ventilation of in South Indian Ocean 44
– South West Indian Ocean, permanent in 40, **43**, 44
– South West Indian Ocean, seasonal in 44
thermohaline circulation cell
– Agulhas Current's role in 130, 191
– Agulhas rings, role in 191
– Atlantic Ocean, of 197, **198**, 198, 233
– climate fluctuations in 130
– inflow from Indian to Atlantic Ocean, relation to 9, 10
– Natal Pulse, role in 196
– meridional heat transport, relation to 10
– model of 204
– overturning component of 10
thermostad
– Agulhas retroflection, in the 168
– Agulhas rings, in 179, 180, 185
– Brazil rings, in 179
– Subantarctic Mode Water, of 19, 29
– Tropical Surface Water, in 107
Thomas B. Davie (vessel)
– Agulhas Return Current, making observations in the 228
tide
– Agulhas Bank, on **148**
– shelf current, influence on 145
– Sofala Bank, at 74
Tiger (vessel)
– making observations across the Agulhas Bank 140
Timor Sea
– Indonesian throughflow, as conduit for 10, 233
TOPEX/Poseidon (satellite)
– model incorporating altimeter data from 203
– observations south of Madagascar 68
– establishing current variability in South West Indian Ocean **68**, 68, **154**
trajectory
– East Madagascar Current, of 60, 67, 88
– Agulhas Current, of northern **93**, 93, 94
– Agulhas Current, of southern 125, **126**, 127
Transkei Basin
– eddies in **86**, 86, **87**
transport (*see* volume transport; heat fluxes)
trapped lee eddy (*see* lee eddy)
trapped shelf wave (*see* shelf waves)
Tropical (Indian) Surface Water (*see also* Indian Tropical Surface Water)
– Agulhas Bank, on **146**
– Agulhas Current, in northern 101, **102**, **106**, 106, 107
– Agulhas Current, in southern 122
– Agulhas Current, volume transport in, of 101, **102**
– Agulhas retroflection, in **166**
– East African Current, in 44
– East Madagascar Current, in 44, 61
– south-eastern Madagascar upwelling cell, in 62
– Mozambique Channel, movement through 103, 106
– Mozambique eddies, in 77
– Natal Bight, in 113
– Port Alfred upwelling cell, in 135, **136**
– South Equatorial Current, in 44
– South West Indian Ocean, in 28, **28**
– temperature-salinity characteristics of **28**
– temperature-oxygen characteristics of **29**, 29
– volume transport in Agulhas Current of 101, **102**
Tropical Thermocline Water
– Agulhas Current, in southern **124**
– Agulhas Current, in northern **107**
– Agulhas retroflection, in **166**, 166, 167
– oxygen content of **166**
– South West Indian Ocean, in **28**, 29, **29**
– temperature-salinity relationships of **28**, **124**
– temperature-oxygen relationships of **29**, **124**
Tropical Rainfall Measuring Mission (TRMM)
– use of 133
tunicates
– Agulhas Current water, as indicator species for 15

UNESCO (*see* United Nations Educational, Scientific and Cultural Organisation)
United Nations Educational, Scientific and Cultural Organisation (UNESCO)
United States Navy
– Lieutenant Maury of the 225
University of Cape Town (UCT)
– pioneering use of satellite remote sensing at 133
– naming of the Agulhas Return Current at 220
Upper Circumpolar Deep Water
– temperature-oxygen relationships of, in western Indian Ocean **29**
– Agulhas retroflection, in **166**
upstream retroflection
– Agulhas Plateau, intersecting 196
– depth of 196
– drifters, shown by 196
– duration of 196
– incomplete 196
– Natal Pulse, caused by 6, 120, 196
– Port Elizabeth, south of 196
upwelling (*see also* coastal upwelling; upwelling cell; upwelling filament)
– Agulhas Bank, over **159**
– Agulhas Current, along edge of 122, 132, **135**
– Agulhas filament, along edge of 131
– Benguela regime 140, 190
– Cape Town, off **159**
– eddies in 190
– filaments on 190, **191**

- Madagascar, south-east of **60**, 61, **62**
- Natal Bight, at 110
- Port Alfred, at 132, **135**
- south coast headlands, at 140, 141

upwelling cell
- Agulhas Current, inshore of southern **159**
- Benguela regime, part of 190
- Cape St Lucia, off 109
- Delagoa Bight, in 79
- East Madagascar Current, inshore of **60**, 61, 132
- Madagascar, off south-eastern **60**, 61, **62**, 132
- Natal Bight, in 109
- Port Alfred, at **116**, 131, 132, **135**, 135, **136**, 136, 137, 138, **138**
- Richards Bay, off 109
- Taiwan, off 132
- temperature–salinity relationships of, at Port Alfred 135, **136**
- subsurface chlorophyll maxima in **62**, 63

upwelling filament
- Agulhas ring, wrapping around 190
- Benguela upwelling system, in frontal zone of 190
- dimensions of 190
- modelling of 190
- Natal Bight, at 110
- Madagascar, south-east of **61**, 63, 65

US Naval Oceanographic Office 23

Utrecht University
- involvement in MARE 157

Valdivia (vessel)
- pioneering observations near South Africa 2, **5**, 19

variability
- Agulhas Current retroflection, at 25, **68**, 151, 152, **154**
- Agulhas retroflection, of **68**, 194, **196**
- Agulhas Return Current, of **68**, 217, **154**
- Cape Basin, in **68**, 196
- drifter movement, based on 152
- modelled **75**
- South Atlantic circulation, of 194
- South West Indian Ocean, for **68**, **154**
- South West Indian Ocean, seasonal in 44, **45**
- Southern Ocean, in 152
- Subtropical Convergence, along 21, **154**, 224, 225

Vema Seamount
- Agulhas ring, splitting of, at 180, **189**, 189
- Agulhas rings, splitting at, simulated 204
- models, in 202

vertical (hydrographic) sections (*see also* hydrographic sections)
- Agulhas, southern, across 122, **134**
- Natal Bight, along 110, **111**
- Natal Pulse, across a **115**
- meander in northern Agulhas Current, across 114, **115**

vertical temperature structure
- Agulhas Bank, over **141**, **144**
- Agulhas ring, of 162
- Mozambique Channel, in 44, **45**
- South West Indian Ocean, in 44, **45**
- Subtropical Convergence, at 44, **45**

viscous stress curl
- model, in 200

volume flux (*see* volume transport)

volume transport
- abyssal boundary current east of Madagascar, of 69
- Agulhas Current, contribution to, by sources **24**, **41**, 57, 58, 60, 72, 75, 85
- Agulhas Current, downstream increase in 104, 121, 168, **170**, 215, **215**
- Agulhas Current, modelled for 202, **203**
- Agulhas Current, of **87**, 100, **101**, 146, **170**, 214, 215, **215**
- Agulhas Current, of northern 42, **101**, **102**, 103, **104**, 104, **215**, 215
- Agulhas Current, of southern **122**, 151, 214
- Agulhas Current, seasonality for, of **101**, 102, 144
- Agulhas Current, simulated for **195**
- Agulhas Current penetration into South Atlantic, affecting 190, **198**, 198
- Agulhas eddy, around 227
- Agulhas Front, at 215
- Agulhas Retroflection, of **170**, 170, 218
- Agulhas Return Current, of 42, **122**, 214, 215, **215**, 216, **217**, 218, **219**
- Agulhas Current, eastward diminution in **215**, 216, **217**
- Agulhas Return Current, simulated for **195**
- Agulhas ring off Cape Town, of an 156, 170
- Agulhas Undercurrent, of 95
- Antarctic Bottom Water in South Atlantic, of 197
- Antarctic Intermediate Water south of Africa, of 192
- Antarctic Intermediate Water in South Atlantic, of 197
- Antarctic Polar Front, at **122**
- Benguela Current, Agulhas Current contribution to 190
- Cape Agulhas, from **215**
- Cape Basin, across 192
- Cape St Lucia, off 100
- Durban, off **87**, 100, **215**
- East African (Coastal) Current, of 69
- East Madagascar Current, of 57, 58, 60, 72
- eddies in Natal Basin, of **87**
- inter-ocean, by Agulhas filaments 193
- inter-ocean, by Agulhas rings 193
- inter-ocean, modelled **195**
- Madagascar, south of 68, 215
- modelled, south of Africa **195**
- Mozambique Channel, into the northern 70
- Mozambique Channel, southern mouth, through 80, 81
- Mozambique Channel, through 42, **42**
- Mozambique Current, of 75
- Mozambique Current, simulated for 69, 70
- Mozambique Undercurrent, of 69
- North Atlantic Deep Water in South Atlantic, of 101, **102**, 197
- Port Alfred, off 100, 101
- Red Sea Water in Agulhas Current, of 101, **102**

– South Atlantic Current, of 192, **219**
– South Atlantic Surface Water, of 197
– South Atlantic Ocean, in 197
– South Indian Central Water in Agulhas Current, of 101, **102**
– South Indian Ocean, of 23
– South West Indian eddies, of **87**
– South West Indian Ocean subgyre, of **41**, 85
– Subtropical Convergence, along 41, 214, 215, 216, **219**
– Subtropical Surface Water in Agulhas Current, of 101, **102**
– Tropical Surface Water in Agulhas Current, of 101, **102**
– water masses of northern Agulhas Current, of **102**
volumetry
– Agulhas retroflection, of 164
– Indian Ocean, of 27, **28**
– South West Indian Ocean, of 27, **27**
vortex (*see* eddy)
vortex dipole
– altimetric data, seen in 204
– Agulhas retroflection, modelled at 204
– Benguela upwelling front, in 190
– East Madagascar Current, from 81, **89**, 89
vortex shedding
– Natal Pulse, as mechanism for 116

WOCE (*see* World Ocean Circulation Experiment)
Waikato (vessel)
– in Agulhas Return Current 209
Walvis Ridge
– Agulhas rings beyond 180, **182**, 182
– Agulhas rings influenced by 187, 188, 233
– Benguela Current moving through 188
– Cape Basin cyclone crossing 177
– Cape of Good Hope Experiment investigating up to 157
– floats crossing 187, 192
– models, in 202
– North Atlantic Deep Water moving through 32, **33**
warm water plumes (*see* plumes)
water masses
– Agulhas retroflection, of 164, **165**, **166**, 166, 167, 168
– Agulhas retroflection, modifications at 168
– South West Indian Ocean, of 26, 28, **28**, 29, **29**, **30**, **31**
– South West Indian Ocean, volumes for 27, **27**
Waterwich (vessel)
– pioneer observations from 2
waves
– Agulhas Current, relationship to 105
– altimetry, studied using 133
– amplification of 105, 133
– attenuation of 105, 133
– climate of 105
– coastally trapped in atmosphere 148
– coastally trapped shelf 149
– energy of 105
– "monster", "freak" or "giant" 105
– shelf 149
– synthetic aperture radar, studied using 133
Weather Bureau (*see* South African Weather Bureau)
Weddell Sea
– Antarctic Bottom Water, formation region for 9
western Agulhas Bank (*see* Agulhas Bank)
western boundary currents
– Agulhas Current as (*see also* Agulhas Current) ix, 9
– Brazil Current as 8, 9
– deep 32
– East Australian Current as 8, 9
– Gulf Stream as ix, 8, 9
– Kuroshio as 8, 9
western Indian Ocean (*see also* Indian Ocean; South West Indian Ocean)
– current simulations for 57, **57**
– Indian Central Water in 41
– North Indian Central Water in 41
– temperature–oxygen relationships of water masses in **29**
– temperature–salinity relationships of water masses in **28**
whales
– Agulhas Current, distribution influenced by 14
– Agulhas retroflection, in 209
– Subtropical Convergence, along 209
wind (*see also* land–sea breezes; South West Monsoon; North East Monsoon; wind steadiness; wind stress)
– Agulhas Bank, over the 141, **142**, **143**, 148, 149
– Agulhas Bank, spreading Agulhas plumes over 129
– Agulhas water entering South Atlantic, influence on 164
– berg, and upwelling filaments 190
– circulation cells over southern Agulhas Current 149
– circulation cells over Gulf Stream 149
– coastal 23, 148
– Gouriqua, at 141, **142**
– jet over southern Agulhas Current 149
– Natal Bight, longshore, at 112, **113**
– Port Alfred, at 138, **150**
– Port Elizabeth, at 149
– Richards Bay, at 113
– South Atlantic, over 194
wind steadiness (*see also* winds)
– South West Indian Ocean, over 23
wind stress
– controlling Agulhas retroflection in model 199
– South West Indian Ocean, patterns over 23, 154, 168
Woods Hole Oceanographic Institution
– participating in Benguela Sources and Transports programme 157
World Ocean Circulation Experiment (WOCE)
– South African participation in 21
– measurements in Mozambique Channel during 81, 236

Zambezi
– runoff from 73
Zanzibar Current (*see* East African Current)